METHODS IN MOLECULAR BIOLOGY

Series Editor
John M. Walker
School of Life and Medical Sciences
University of Hertfordshire
Hatfield, Hertfordshire, AL10 9AB, UK

For further volumes:
http://www.springer.com/series/7651

Schwann Cells

Methods and Protocols

Edited by

Paula V. Monje

Department of Neurological Surgery, University of Miami, Miami, FL, USA

Haesun A. Kim

Department of Biological Sciences, Rutgers University, Newark, NJ, USA

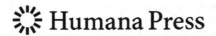 Humana Press

Editors
Paula V. Monje
Department of Neurological Surgery
University of Miami
Miami, FL, USA

Haesun A. Kim
Department of Biological Sciences
Rutgers University
Newark, NJ, USA

ISSN 1064-3745   ISSN 1940-6029   (electronic)
Methods in Molecular Biology
ISBN 978-1-4939-9254-6   ISBN 978-1-4939-7649-2   (eBook)
https://doi.org/10.1007/978-1-4939-7649-2

Cover image: Dorsal root ganglion explant grown in vitro displaying Schwann cells (*green*) and neurons (*red*). Courtesy of Dr. Felipe A. Court

Printed on acid-free paper

This Humana Press imprint is published by the registered company Springer Science+Business Media, LLC part of Springer Nature.
The registered company address is: 233 Spring Street, New York, NY 10013, U.S.A.

Preface

If we have normal cell types available in culture and if a nervous system disease stems from an absence or deficiency of one of these cell types, then it is an obvious step to transfer cultured cells into the diseased animal to attempt to correct the deficiency. Whereas at present this would not seem practical for many types of neurons, it may be practical for supporting cells (e.g. Schwann or glial).

–Richard P. Bunge (The changing uses of nerve tissue culture 1950–1975, in The Nervous System, 1975)

The Schwann cell has traditionally been viewed as the ensheathing cell type of the peripheral nervous system associated with a multilamellar myelin sheath or a Remak bundle. This somewhat static view of the Schwann cell has changed in recent years upon the continued realization that besides their role in ensheathment and myelination, Schwann cells contribute to a variety of functions during nerve development, regeneration, and neurodegenerative disease. Schwann cells have many more facets than initially expected. These fascinating glial cells have an active role in immunomodulation, axon maintenance and regeneration, synaptic transmission, cancer, and disease progression in many inherited and acquired neuropathies. As wisely anticipated by Dr. Richard Bunge over 40 years ago, cultured Schwann cells can serve as a therapeutic product for cell replacement strategies aimed at treating central and peripheral nervous system injuries. Indeed, the safety of autologous human Schwann cell transplantation is currently being tested as a strategy for spinal cord repair through clinical trials in the United States. This is an unprecedented milestone for the clinical testing of a glial cell product.

A new era in the study of Schwann cells has already emerged with the glia-neuron field cross-talking with areas of more recent development such as bioengineering, materials science, stem cell biology, and regenerative medicine. An exciting wave of knowledge on Schwann cell biology in health and disease has resulted from the development of state-of-the-art technologies in cell culture, genetic manipulation of cells in vitro and in vivo, and high-resolution imaging techniques for live and fixed cells, tissues, and whole organisms. The development of mammalian and nonmammalian animal models targeting Schwann cells in peripheral nerves has greatly improved our understanding of myelination, Schwann cell-axon communication, and mechanisms of disease at the cellular and molecular levels. The availability of the Schwann cells themselves along with the products they secrete or generate has motivated novel applications with potential for translation. The current feasibility to generate Schwann cells "in a dish" from virtually any given type of cell has revolutionized our concepts and greatly expanded the possibilities for basic and applied research.

The book *Schwann Cells: Methods and Protocols* intends to present an assortment of traditional and emerging experimental procedures relevant to Schwann cell research. The chapter collection consists of 31 chapters divided into four discrete parts. Part I contains protocols for in vitro culture, purification, and characterization of primary Schwann cells from diverse species and stages of nerve development. It also contains protocols to derive cancer cell lines and create engineered Schwann cells from unconventional sources via chemical conversion or induced differentiation. Parts II and III outline a wide range of methodologies used to study Schwann cells within in vitro and in vivo systems, respectively, relevant to the analysis of peripheral nerve development, cancer, axon degeneration/regeneration, and myelination. Part II encompasses protocols for assessing the myelinating Schwann cell phenotype and the interactions of Schwann cells with neurons, nonneuronal cell types, and the physical environment using organotypic and neuron-free cultures. Part III focuses on the analysis and visualization of Schwann cells in dissected peripheral nerve and human skin biopsies, as well as the whole organism in adult and larvae zebra fish, using standard immunochemical approaches, high-resolution image analysis, and time-lapse microscopy. Lastly, Part IV outlines protocols for Schwann cell production, collection, labeling, and transplantation in the injured peripheral nerve and spinal cord of experimental animals and human subjects.

We hope this selection provides a practical framework to aid both experienced and new investigators make progress in their research endeavors involving Schwann cells. We are immensely indebted to all of the scientists who contributed to this book by preparing the chapters and sharing their unique experience in the development and use of these methodologies. We greatly appreciate their valuable time, insights, and dedication to the joint effort of materializing this project. Our sincere thanks to John Walker first and foremost for his recognition that the fascinating cell of Schwann merited an independent book in the prestigious series *Methods in Molecular Biology* and for bringing this initiative to our attention, facilitating the preparation of the book, and mentoring us on editing the chapters. Many thanks to all members of the Springer team who contributed to this project. A whole-hearted special recognition to Patrick Marton for the expeditious assistance and support in helping us navigate the miscellaneous aspects of editing and publishing.

To conclude, our understanding of Schwann cells, their function, and potential uses relies primarily on the methods we use to approach their study. The scientific literature on the subject of Schwann cells is currently advancing at an incredible pace. If this trend continues, a wave of groundbreaking research is likely to reshape our concepts sooner than expected. We hope that readers find this comprehensive set of step-by-step protocols both timely and useful in their experimental approaches while making a contribution to the research and innovation that lies ahead in the Schwann cell field.

Miami, FL, USA   *Paula V. Monje*
Newark, NJ, USA   *Haesun A. Kim*

Contents

Contributors

VANIA W. ALMEIDA • *The Miami Project to Cure Paralysis, University of Miami Miller School of Medicine, Miami, FL, USA*

INGE M. AMBROS • *Children's Cancer Research Institute, Vienna, Austria*

PETER F. AMBROS • *Children's Cancer Research Institute, Vienna, Austria*

NATALIA D. ANDERSEN • *The Miami Project to Cure Paralysis, Department of Neurological Surgery, University of Miami Miller School of Medicine, Miami, FL, USA*

GAGANI ATHAUDA • *Department of Cellular Biology and Pharmacology, Herbert Wertheim College of Medicine, Florida International University, Miami, FL, USA*

MARGARET L. BATES • *The Miami Project to Cure Paralysis, University of Miami Miller School of Medicine, Miami, FL, USA*

FRANCISCO D. BENAVIDES • *The Miami Project to Cure Paralysis, University of Miami Miller School of Medicine, Miami, FL, USA*

PETER BOND • *Electron Microscopy Centre, Plymouth University, Plymouth, Devon, UK*

ADRIANA E. BROOKS • *The Miami Project to Cure Paralysis and Department of Neurological Surgery, University of Miami Miller School of Medicine, Miami, FL, USA; Interdisciplinary Stem Cell Institute, University of Miami Miller School of Medicine, Miami, FL, USA*

MARY BARTLETT BUNGE • *The Miami Project to Cure Paralysis, University of Miami Miller School of Medicine, Miami, FL, USA; Department of Cell Biology, University of Miami Miller School of Medicine, Miami, FL, USA; Department of Neurological Surgery, University of Miami Miller School of Medicine, Miami, FL, USA; Lois Pope LIFE Center, Miami, FL, USA*

SA CAI • *School of Biomedical Sciences, Li Ka Shing Faculty of Medicine, The University of Hong Kong, Hong Kong, China*

ALEJANDRA CATENACCIO • *Center for Integrative Biology, Faculty of Sciences, Universidad Mayor, Santiago, Chile, FONDAP Center for Geroscience, Brain Health and Metabolism, Santiago, Chile*

SUSANA R. CERQUEIRA • *The Miami Project to Cure Paralysis, Lois Pope LIFE Center, University of Miami Miller School of Medicine, Miami, FL, USA*

GÜRALP O. CEYHAN • *Department of Surgery, Klinikum rechts der Isar, Technische Universität München, Munich, Germany*

YING-SHING CHAN • *School of Biomedical Sciences, Li Ka Shing Faculty of Medicine, The University of Hong Kong, Hong Kong, China*

FELIPE A. COURT • *Center for Integrative Biology, Faculty of Sciences, Universidad Mayor, Santiago, Chile; FONDAP Center for Geroscience, Brain Health and Metabolism, Santiago, Chile*

REBECCA L. CUNNINGHAM • *Department of Developmental Biology, Washington University in St. Louis, St. Louis, MO, USA*

YING DAI • *Burke Medical Research Institute, White Plains, NY, USA; Weill Cornell Medicine, Feil Family Brain and Mind Research Institute, New York, NY, USA*

IHSAN EKIN DEMIR • *Department of Surgery, Klinikum rechts der Isar, Technische Universität München, Munich, Germany*

PAULA DIAZ • *Center for Integrative Biology, Faculty of Sciences, Universidad Mayor, Santiago, Chile*

XIN-PENG DUN • *Plymouth University Peninsula Schools of Medicine and Dentistry, Plymouth, Devon, UK*

LAURA FANGMANN • *Department of Surgery, Klinikum rechts der Isar, Technische Universität München, Munich, Germany*

M. LAURA FELTRI • *Department of Biochemistry and Neurology, Hunter James Kelly Research Institute, Jacobs School of Medicine and Biomedical Sciences, University at Buffalo Buffalo, NY, USA*

CRISTINA FERNÁNDEZ-VALLE • *Neuroscience Division, Burnett School of Biomedical Science, College of Medicine, University of Central Florida, Orlando, FL, USA*

HELMUT FRIESS • *Department of Surgery, Klinikum rechts der Isar, Technische Universität München, Munich, Germany*

CRISTIAN DE GREGORIO • *Center for Integrative Biology, Faculty of Sciences, Universidad Mayor, Santiago, Chile; FONDAP Center for Geroscience, Brain Health and Metabolism, Santiago, Chile*

JAMES D. GUEST • *The Miami Project to Cure Paralysis, University of Miami Miller School of Medicine, Miami, FL, USA; Neurological Surgery, University of Miami Miller School of Medicine, Miami, FL, USA*

COREY HEFFERNAN • *Department of Biological Sciences, Rutgers University, Newark, NJ, USA*

CAITLIN E. HILL • *Burke Medical Research Institute, White Plains, NY, USA; Weill Cornell Medicine, Feil Family Brain and Mind Research Institute, New York, NY, USA*

KRISTJAN R. JESSEN • *Department of Cell and Developmental Biology, University College London, London, UK*

AISHA KHAN • *The Miami Project to Cure Paralysis and Department of Neurological Surgery, University of Miami Miller School of Medicine, Miami, FL, USA; Interdisciplinary Stem Cell Institute, University of Miami Miller School of Medicine, Miami, FL, USA*

HAESUN A. KIM • *Department of Biological Sciences, Rutgers University, Newark, NJ, USA*

JIHYUN KIM • *Department of Biological Sciences, Rutgers University, Newark, NJ, USA*

YEE-SHUAN LEE • *The Miami Project to Cure Paralysis, Lois Pope LIFE Center, University of Miami Miller School of Medicine, Miami, FL, USA*

RODRIGO LÓPEZ-LEAL • *Center for Integrative Biology, Faculty of Sciences, Universidad Mayor, Santiago, Chile; FONDAP Center for Geroscience, Brain Health and Metabolism, Santiago, Chile*

PATRICIO MANQUE • *Center for Integrative Biology, Faculty of Sciences, Universidad Mayor, Santiago, Chile*

PATRICE MAUREL • *Department of Biological Sciences, Rutgers University, Newark, NJ, USA*

CARMEN V. MELENDEZ-VASQUEZ • *Department of Biological Sciences, Hunter College, New York, NY, USA; The Graduate Center, City University of New York, New York, NY, USA*

RHONA MIRSKY • *Department of Cell and Developmental Biology, University College London, London, UK*

PAULA V. MONJE • *The Miami Project to Cure Paralysis, Department of Neurological Surgery, University of Miami Miller School of Medicine, Miami, FL, USA*

KELLY R. MONK • *Vollum Institute, Oregon Health & Science University, Portland, OR, USA; Department of Developmental Biology, Washington University in St. Louis, St. Louis, MO, USA*

RENATA DE MORAES MACIEL • *Department of Neurology, University of Miami Miller School of Medicine, Miami, FL, USA; Department of Human Genetics, University of Miami Miller School of Medicine, Miami, FL, USA*

DAVID B. PARKINSON • *Plymouth University Peninsula Schools of Medicine and Dentistry, Plymouth, Devon, UK*

ALEJANDRA M. PETRILLI • *Neuroscience Division, Burnett School of Biomedical Science, College of Medicine, University of Central Florida, Orlando, FL, USA*

CHRISTINE D. PLANT • *Department of Neurosurgery, School of Medicine, Stanford University, Stanford, CA, USA*

GILES W. PLANT • *Department of Neurosurgery, School of Medicine, Stanford University, Stanford, CA, USA*

YANNICK POITELON • *Department of Neuroscience and Experimental Therapeutics, Albany Medical College, Albany, NY, USA*

KRISTINE M. RAVELO • *The Miami Project to Cure Paralysis, Department of Neurological Surgery, University of Miami Miller School of Medicine, Miami, FL, USA*

ANDREA J. SANTAMARÍA • *The Miami Project to Cure Paralysis, University of Miami Miller School of Medicine, Miami, FL, USA*

MARIO A. SAPORTA • *Department of Neurology, University of Miami Miller School of Medicine, Miami, FL, USA; Department of Human Genetics, University of Miami Miller School of Medicine, Miami, FL, USA*

GRAHAM K. SHEA • *Department of Orthopaedics and Traumatology, LKS Faculty of Medicine, The University of Hong Kong, Hong Kong, China*

DAISY K.Y. SHUM • *School of Biomedical Sciences, Li Ka Shing Faculty of Medicine, The University of Hong Kong, Hong Kong, China*

JUAN P. SOLANO • *Pediatric Critical Care, University of Miami Miller School of Medicine, Miami, FL, USA*

PAVEL STUPAKOV • *Department of Surgery, Klinikum rechts der Isar, Technische Universität München, Munich, Germany*

SABINE TASCHNER-MANDL • *Children's Cancer Research Institute, Vienna, Austria*

STEFFEN TELLER • *Department of Surgery, Klinikum rechts der Isar, Technische Universität München, Munich, Germany*

EVA C. THOMA • *Stem Core, Stem Cells Ltd, Australian Institute for Bioengineering and Nanotechnology, University of Queensland, St Lucia, QLD, Australia*

Y.P. TSUI • *School of Biomedical Sciences, LKS Faculty of Medicine, The University of Hong Kong, Hong Kong, China*

MATEUSZ M. URBANSKI • *Department of Biological Sciences, Hunter College, New York, NY, USA*

TAMARA WEISS • *Children's Cancer Research Institute, Vienna, Austria*

Part I

Protocols for Preparing Primary Cultures, Cell Lines and Engineered Schwann Cells

Chapter 1

Isolation of Schwann Cell Precursors from Rodents

Rhona Mirsky and Kristjan R. Jessen

Abstract

Schwann cell precursors are the first defined stage in the generation of Schwann cells from the neural crest and represent the glial cell of embryonic nerves. Highly pure cultures of these cells can be obtained by enzymatic dissociation of nerves dissected from the limbs of 14- or 12-day-old rat and mouse embryos, respectively. Since Schwann cell precursors, unlike Schwann cells, are acutely dependent on axonal signals for survival, they require addition of trophic factors, typically β-neuregulin-1, for maintenance in cell culture. Under these conditions they convert to Schwann cells on schedule, within about 4 days.

Key words Schwann cells, Nerve development, Neuregulin, Precursors, Neural crest

1 Introduction

In rodent embryos, the sciatic nerve invades the hind limb between embryo day (E) 14/15 (rat) and E12/13 (mouse) (Fig. 1). In these early embryonic nerve axons, glia and connective tissue relate to each other in a way which is radically different from that seen in late perinatal or older nerves. These early nerves consist of tightly packed axons, mostly without individual glial ensheathment, and sparsely distributed, flattened glial cell processes (Fig. 2). There is no significant extracellular space, matrix, or basal lamina. The glial cell bodies lie among the axons inside the nerve or at the nerve surface. These cells represent Schwann cell precursors, the first stage of the Schwann cell lineage [1–4].

Schwann cell precursors are generated from neural crest cells and, in turn, generate the immature Schwann cells of perinatal nerves (Fig. 3). Schwann cell precursors show multiple phenotypic differences from neural crest cells on the one hand and immature Schwann cells on the other. They differ from crest cells because they express glial differentiation genes, including myelin protein zero (Po), and other factors not expressed by crest cells, because they are found in association with axons in nerves, a characteristic glial feature, rather than migrating through extracellular matrix,

Paula V. Monje and Haesun A. Kim (eds.), *Schwann Cells: Methods and Protocols*, Methods in Molecular Biology, vol. 1739, https://doi.org/10.1007/978-1-4939-7649-2_1, © Springer Science+Business Media, LLC 2018

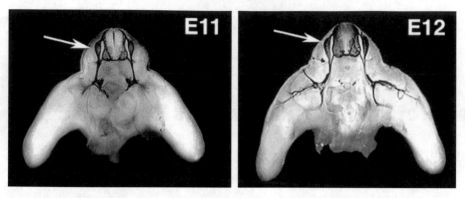

Fig. 1 Nerves in the mouse hind limb visualized using TuJ1 antibodies and immunohistochemistry. The immunostaining was carried out on whole embryos which were subsequently sectioned transversely in the region of the hind limbs. Note that nerves appear essentially confined to the trunk at E11, whereas at E12 they are present in the upper and middle part of the limb. Arrows point to DRG ×20 (Reproduced from ref. 8.)

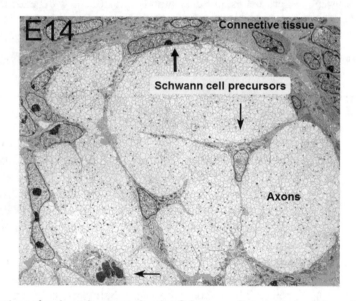

Fig. 2 The ultrastructure of embryonic nerves showing Schwann cell precursors, axons, and connective tissue. An electron microscopic image of a transverse section of a nerve in the hind limb of a rat embryo at embryonic day (E) 14. Schwann cell precursors branch among the axons inside the nerve and are also found in close apposition to axons at the nerve surface (vertical arrows). One precursor cell is undergoing mitosis (horizontal arrow). Extracellular connective tissue space (blue), which contains mesenchymal cells, surrounds the nerve but is essentially absent from the nerve itself. These nerves are also free of blood vessels, and the axons are of smaller and more uniform diameter than those seen in mature nerves. Magnification, ×2000 (Reproduced with modifications from ref. 2.)

and because they respond differently to survival, proliferation, and differentiation signals [2, 3, 5]. A large number of differences in molecular expression between Schwann cell precursors and crest cells are also revealed in gene array analysis [6, 7]. In turn, Schwann cell precursors and immature Schwann cells differ substantially in

The Schwann cell precursor lineage

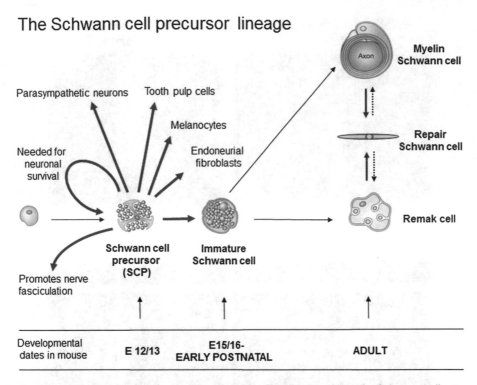

Fig. 3 The Schwann cell precursor in a developmental context. The diagram shows the Schwann cell precursor and key differentiation states in the lineage, including immature, myelin and Remak cells, and repair Schwann generated after injury. Also shown are developmental options for Schwann cell precursors and the precursor functions of promoting nerve fasciculation and the survival of developing neurons. Dates (E) refer to mouse development

molecular expression, including upregulation of S100 in immature Schwann cells [6, 7]. Further, immature Schwann cells, but not Schwann cell precursors, relate to basal lamina and connective tissue containing blood vessels. But perhaps the most striking difference concerns survival regulation, since Schwann cell precursors are acutely dependent on the axon-associated survival signal β-neuregulin-1 (βNRG1) type lll, while immature Schwann cells survive in culture without addition of survival factors due to secretion of autocrine survival signals [8–10]. Autocrine survival mechanisms are not present in Schwann cell precursors. For this reason Schwann cell precursors are routinely cultured in medium containing βNRG1 (Fig. 4).

In late embryonic nerves, Schwann cell precursors convert to immature Schwann cells. But this is not their only developmental option, since they also give rise to endoneurial fibroblasts, melanocytes in the skin, parasympathetic neurons, and tooth pulp cells [11–14]. Schwann cell 1root ganglion (DRG) neurons and motoneurons [2, 4, 10].

Rat Schwann cell precursors   Rat Schwann cells
E14   New-born

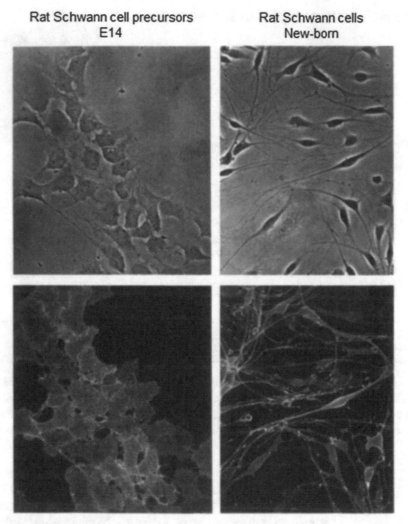

Fig. 4 Developmental changes in the morphology and cellular relationships between Schwann cell precursors and Schwann cells. The cells were dissociated from either E14 or newborn rat nerves. They were immunolabeled with p75 NTR antibodies and photographed with fluorescence optics (lower panels) and phase contrast (upper panels) 20 h after plating; the cells were maintained in neuron conditioned medium. Radical changes in shape accompany precursor to Schwann cell conversion. The precursors are flattened, whereas the immature Schwann cells from newborn nerves are bi- or tripolar and have tapered processes that often arise relatively abruptly from a rounded perinuclear area. Magnification, 1065× (Reproduced from ref. 1.)

2   Materials

For all procedures, use autoclaved, ultrapure (UP) water.

2.1   Enzyme Cocktail for Digestion of Nerves

1. 10× Krebs solution (without Ca^{2+} and Mg^{2+}): 14 g NaCl, 0.7 g KCl, 0.325 g KH_2PO_4. Make up to 200 ml with UP water. Autoclave to sterilize. Store at 4–8 °C.

2. Ca^{2+}/Mg^{2+} free medium: Mix 80 ml UP water, 10 ml 10× Ca^{2+}/Mg^{2+} free 10× Krebs solution, 2 ml 50% essential amino

acid solution, 2.5 ml 7.5% NaHCO$_3$, 0.5 ml 0.3% phenol red solution, 0.4 ml 50% glucose sterile. Filter mixture through 0.22 μm filter to sterilize and store at 4–8 °C.

3. Collagenase (Worthington collagenase type 2 LSO 041472).

4. Hyaluronidase (type IV-S: from bovine testes, Sigma-H 3631).

5. Trypsin inhibitor (Trypsin inhibitor from Glycine max (soybean) Sigma-T 6522).

6. Enzyme cocktail: Prepare 15 ml of Ca^{2+}/Mg^{2+} free medium containing 2 mg/ml collagenase (*see* **Note 1**), 1.2 mg/ml hyaluronidase, and 0.3 mg/ml trypsin inhibitors. The solution will give 30× 600 μl aliquots. Filter mixture through 0. 22 μm filter and aliquot into 1 ml Eppendorf tubes. Snap freeze in liquid nitrogen and store at −20 °C.

2.2 Defined Medium

1. DMEM/Ham's F-12 solution: Make 1:1 mixture of DMEM and Ham's F-12 plus Glutamax (Gibco life technologies 31331-028) and add penicillin/streptomycin (penicillin 100 IU/ml, streptomycin 100 IU/ml).

2. Transferrin stock solution (10 mg/ml): Weigh out 500 mg transferrin (transferrin, bovine Sigma T8027) powder and add to 50 ml 1× PBS, stir to dissolve, and then filter sterilize using a 0.22 μm filter. Make 1 ml aliquots and snap freeze in liquid nitrogen and store at −20 °C.

3. Putrescine stock solution (1.6 mg/ml): Weigh out 80 mg of putrescine powder and add to 50 ml PBS, stir to dissolve, and then filter sterilize using 0.22 μm filter. Make 1 ml aliquots, snap freeze in liquid nitrogen, and store at −20 °C.

4. Bovine serum albumin (BSA) stock solution (3.5%): Prepare 35% BSA solution in PBS then dilute 1:10 in PBS. Make 1 ml aliquots then store at 4–8 °C.

5. T4 (L-thyroxine) stock solution (400 μg/ml): Weigh out 50 mg of T4 powder, dissolve in 10 ml of 1 M NaOH to make a 5 mg/ml solution, and then dilute this 1:12.5 in 1× PBS to make a 400 μg/ml stock solution. Make 100 μl aliquots, snap freeze in liquid nitrogen, and store at −20 °C.

6. Progesterone stock solution (0.006 mg/ml): Weigh out 0.6 mg progesterone powder and dissolve in 10 ml of 100% analytic grade ethanol. Store at 4–8 °C; reseal stock tube with cling film after each use.

7. Insulin stock solution (5.7 mg/ml, 10^{-3} M): Add 2.5 ml of 10 mg/ml insulin solution (Sigma I9278-5 ml) to 1.89 ml UP water to make a total of 4.39 ml. Store both insulin solutions at 4–8 °C. Make up fresh insulin stock every 6 weeks.

8. Dexamethasone stock solution (0.05 mg/ml): Weigh out 0.5 mg dexamethasone (Sigma D4902, MW 392.46) and

dissolve in 10 ml 100% analytic grade ethanol. Keep stock at 4–8 °C; reseal with cling film after each use.

9. T3 (L-thyronine) stock solution (0.101 mg/ml): Weigh out 1.01 mg T3 (3,3′,5-Triiodo-L-thyronine sodium salt, Sigma T6397), dissolve in 10 ml of 100% analytic grade ethanol, and store at −20 °C. Do not aliquot. Store in 15 ml Falcon tube at −20 °C; reseal with cling film after each use.

10. Selenium stock solution (1.6 mg/ml): Weigh out 8′ mg of sodium selenite (Sigma S5261), dissolve in 5 ml of UP water, filter sterilize using a 0.22 µm filter, aliquot into 10 µl aliquots, snap freeze in liquid nitrogen, and store at −20 °C.

11. Defined medium: To make 100 ml of defined medium, take 96 ml of 1:1 DMEM/Ham's F-12 solution and add stock solutions of each component as the following (final concentration of each component is shown in parenthesis): 1 ml transferrin (100 µg/ml), 1 ml putrescine (16 µg/ml), 1 ml BSA (0.035%), 100 µl T4 (400 ng/ml), 100 µl progesterone (60 ng/ml), 100 µl insulin (5.7 µg/ml), 77 µl dexamethasone (38 ng/ml), 10 µl T3 (10.1 ng/ml), and 10 µl selenium (0.16 ng/ml) (Meier et al. [9]). Store defined medium at 4–8 °C and use within 2 weeks.

2.3 Defined Medium Containing βNRG1 for Culture of Schwann Cell Precursors

Since Schwann cell precursors die without added βNRG1 (or FGF plus IGF, *see* ref. [10]), we routinely resuspend the cells after centrifugation and maintain them in culture, using defined medium supplemented with 10–20 ng/ml βNRG1.

1. PBS/1% BSA solution: Dilute 1 ml of 35% BSA prepared in PBS in 34 ml of PBS.

2. High-concentration βNRG1 stock solution (50 µg/ml): Resuspend lyophilized βNRG1 EGF domain (RD systems 396-HB-050) in 1 ml of PBS/1% BSA to make a 50 µg/ml solution. Vortex thoroughly to ensure the entire pellet is resuspended, make 200 µl aliquots from this solution in 1.5 ml Eppendorf tubes, snap freeze in liquid nitrogen, and store at −80 °C (high-concentration stock).

3. Low-concentration βNRG1 stock solution (10 µg/ml): Mix 800 µl of PBS/1% BSA solution with one aliquot (200 µl) of high-concentration 50 µg/ml βNRG1 stock to obtain a 10 µg/ml solution. Make 50 µl and 10 µl aliquots of this in 0.5 ml Eppendorf tubes, snap freeze in liquid nitrogen, and store at −80 °C (low-concentration stock).

4. Defined medium supplemented with βNRG1: Mix one 10 µl aliquot of low-concentration (10 µg/ml) βNRG1 stock with 10 ml of defined medium to give a final βNRG1 concentration of 10 ng/ml.

2.4 Coating of Coverslips and Dishes for Cell Plating	1. Deep glass Petri dishes.

2.4 Coating of Coverslips and Dishes for Cell Plating

1. Deep glass Petri dishes.

2. 13 mm glass coverslips: Place approximately 100 coverslips in a deep glass Petri dish and bake in an oven for 4 h at 140 °C to sterilize. Use silica gel to keep them dry after baking unless they are to be coated immediately.

3. Tissue culture dishes.

4. 1 mg/ml poly-L-lysine prepared in UP water.

5. 1 mg/ml laminin stock solution: Store at 80 °C until ready to use, then thaw slowly on ice over several hours, aliquot into 50 µl aliquots, snap freeze in liquid nitrogen, and store aliquots at −20 °C.

2.5 Dissection of Embryonic Nerves and Schwann Cell Precursor Preparation

1. Timed pregnant rats (embryonic day [E] 14.5) or mice (E12.5): Use female rats of mice aged optimally from 2 months to 5–6 months old. Males can be older than this. Count the day that a vaginal plug is observed at noon as E0.5.

2. Sylgard-coated Petri dishes: Make up Sylgard 184 according to the manufacturer's instructions (10 parts elastomer to 1 part hardener) in a disposable plastic container with a disposable stirring stick. Pour it into several plastic 9 cm Petri dishes to approximately 0.5–1 cm depth, and then allow to set at high temperature. When set the Sylgard will be transparent and embryos can be pinned to the surface of the dish using fine dissecting pins for dissection of the embryo nerve. The dishes should be sterilized with 100% analytic grade ethanol and washed with purified water and dried before use.

3. 9 cm plastic Petri dish.

4. L15 medium.

5. 70% ethanol.

6. Coarse scissors, sterilized using 70% ethanol.

7. Forceps, sterilized using 70% ethanol.

8. Finer forceps, sterilized using 70% ethanol.

9. Fine iridectomy scissors, sterilized by 100% analytic grade ethanol.

10. Fine-tipped forceps (No. 5), sterilized by 100% analytic grade ethanol.

11. Dissecting microscope with fiber optic lighting.

12. Two 3 cm plastic Petri dishes.

13. L15 medium supplemented with 5% horse serum (for mouse dissection) or 10% fetal calf serum (for rat dissection).

14. Defined medium supplemented with βNRG1.

15. PLL and laminin-coated coverslips or dishes (*see* Subheading 3.1).

2.6 Antibodies

1. S100 polyclonal rabbit antibody (DAKO).
2. p75NTR polyclonal rabbit antibody (Merck Millipore).
3. L1 ASCS4 mouse anti-rat L1 antibody (Developmental Studies Hybridoma Bank).
4. 324 rat anti-mouse L1 antibody (Merck Millipore).
5. Alexa 488 or Cy3 secondary antibodies.

3 Methods

3.1 Preparation of Coated Coverslips and Dishes for Cell Plating

1. Poly-L-lysine coating of coverslips: Add 25 ml of 1 mg/ml poly-L-lysine in UP water to coverslips, seal with cling film (Nescofilm), put on a shaker for 24 h, remove poly-L-lysine, wash with six changes of water over 3 days on the shaker, leave to air-dry thoroughly by standing each coverslip on end around large sterile Petri dishes or in sterile mini-racks, store desiccated at room temperature in a sterile Petri dish sealed with cling film, and if possible do not use until 24 h after drying. They will keep for months.

2. Poly-L-lysine coating of dishes: Tissue culture dishes are treated with a 100 µg/ml solution of poly-L-lysine in UP water. Dilute 1 mg/ml solution 1:10 in UP water to make a 100 µg/ml solution. Treat dishes for 2 h at room temperature, then remove the solution, and leave the dishes to dry without washing. Seal them with cling film and store desiccated at room temperature. They will keep for months.

3. Laminin coating of poly-L-lysine-coated coverslips or dishes: Thaw an aliquot of stock laminin solution slowly (do not refreeze), and dilute this stock solution to a final concentration of 20 µg/ml in DMEM or L15 medium in the case of coverslips and wells and 10 µg/ml for tissue culture dishes. Do not adjust the pH of the solution as laminin binds better at an alkaline pH. Pipette the laminin solution on to the center of a PLL-coated coverslip sitting in a 4- or 24-well tissue culture plate. We routinely use 10–20 µl for rat cells and 5–10 µl for mouse cells. For culturing larger cell number in a well without coverslip or in 3 cm tissue culture dishes, use 50–100 ml of laminin. Leave the solution on for at least 1 h at room temperature, and remove immediately prior to plating cells; do not allow to dry.

3.2 Embryo Dissection and Schwann Cell Precursor Preparation

Schwann cell precursors are prepared essentially according to the methods described previously [1, 8, 15].

1. Sacrifice pregnant female rat at E14.5 or mouse at E12.5 using an approved procedure.

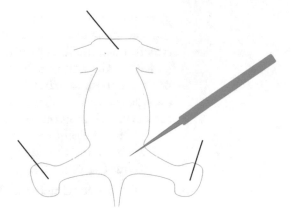

Fig. 5 Diagram showing a rodent embryo, pinned on a Sylgard surface with three dissection pins. The appropriate position of the forceps for removing the sciatic nerve without visual guidance is indicated

2. After killing the pregnant animal, pin it out on a cork dissecting board covered with aluminum foil. Spray the fur with 70% ethanol to sterilize, open up the abdominal cavity with appropriate size sterile scissors, and gently remove the uterine sac with embryos which will be clearly visible. Place this sac in a 9 mm plastic Petri dish containing L15 medium. Place the lid on the Petri dish and keep it on ice.

3. Using fine scissors carefully open up the sac to liberate one embryo to establish correct dating. This is easily judged by observing the digits of front and hind limbs. Excellent illustrations are available at: https://embryology.med.unsw.edu.au/embryology/index.php/Rat_Development. This website shows both rat and mouse embryos. Comparable illustrations are also in ref. 16.

4. Having established that the embryos are of the correct age (~E14.5 rat, ~E12.5 mouse), remove an embryo for dissection. Remove only one embryo for dissection at a time, and leave the others in the embryonic sac on ice. Remove the embryo from the amniotic sac with fine forceps and iridectomy scissors. Decapitate and pin the embryo out with the ventral side down using three metal pins on a Sylgard-coated Petri dish that has been sterilized with 100% analytic grade ethanol and then washed with UP water and briefly with L15 medium (Fig. 5).

5. The sciatic nerves are removed from the hind limb using a dissecting microscope and fine (No. 5) dissecting forceps. The forceps need to be "as new," namely, completely straight and without scratches. First, the skin needs to be removed from the hind leg and a little way up the back. L15 medium can be added to the dish as needed to prevent drying. There is a quick and perhaps surprisingly easy way to liberate the nerves (not a single nerve at this developmental stage but a delicate plexus) from the leg. To do this, insert the number 5 forceps at the

juncture of the developing limb and the torso (Fig. 5) holding the forceps at about 45° angle relative to the surface of the lab bench. Push the forceps toward, but not quite to, the midline (spine) of the embryo. Now close the forceps firmly and pull out the nerves, which, with a little practice, come out intact as a single entity, a small plexus of nerves. To decrease the risk of the nerves slipping out of the forceps as they are pulled out of the leg, it is helpful to lower the forceps a little (decrease the 45° angle) after the forceps are closed but before they are pulled out (*see* **Note 2**).

6. Place the dissected nerve in 3 cm plastic Petri dish containing L15 medium, which is kept on ice, until sciatic nerves from all embryos have been collected. When the nerve is released from the forceps into the medium, it is sometimes obvious that tissue fragments adhere to the nerve. They should be gently cleaned away. It is important to prevent the nerves from reaching the surface of the medium, where they can be pulled apart by the surface tension. The best way of doing this is to entwine each nerve, after cleaning if this is needed, with the nerve(s) already there, seeking to form one ball of tangled nerves. As a rough guide, one litter should give 100,000 cells after dissociation.

7. For enzymatic dissociation, use forceps to transfer the nerves to 600 μl of enzyme cocktail in a 3 cm Petri dish, and place it in 5% CO_2 and 95% air incubator at 37 °C for 1 h. Tilt the dish so that nerves are fully immersed. At the end of the incubation time remove the dish from the incubator and place in a tissue culture hood.

8. Triturate 5–15 times through a 1 ml blue Eppendorf-type pipette tip and then 10–15 times through a yellow 200 μl pipette tip. If the cells are not fully dissociated, re-incubate for a further 10–15 min before being triturating a further ten times with a blue 200 μl pipette tip.

9. Transfer the cell suspension to a 15 ml Falcon centrifuge tube and make the volume up to 10 ml with L15/5% HS (mouse) or L15/10% FCS (rat), in order to dilute out the enzymes. Centrifuge for 10 min at 162.4 × g in a Heraeus Primo R Centrifuge with swinging bucket rotor or equivalent.

10. Carefully remove the medium, leaving the pellet behind. Resuspend the cell pellet gently with trituration (yellow pipette tip) in 0.5–1 ml of defined medium typically supplemented with βNRG1.

11. Count the cells in a hemacytometer or other counting device. Adjust to the desired cell concentration using the resuspension medium. The number and concentration of cells to be plated

can obviously be varied. For plating on coverslips, we routinely use 2000 cells per 10–20 µl for rat cells and 2000 cells per 5–10 µl for mouse cells.

12. Plate the cells on PLL- and laminin-coated glass coverslips in wells in a 4- or 24-well tissue culture plate (or directly on PLL- and laminin-coated tissue culture well or on a 3 cm tissue culture dish) using a volume of cell suspension equal to the volume of the laminin solution used to laminin-coat the coverslip, well, or tissue culture dish (*see* Subheading 3.1, **step 3**).

13. After 1.5 h for mouse cells and 2.5 h for rat cells, top up each well with 400 µl of the same medium used for resuspension and plating. Schwann cell precursors do not express autocrine survival loops and, unlike Schwann cells, they die by apoptosis within 24 h in vitro, unless the medium is supplemented with survival factors [1]. By far the most potent of these is (βNRG1) [15], although the combination of FGFs and IGF2 also supports precursor survival [17]. For routine culturing, the defined medium we use for resuspension, plating, and top-up contains 10–20 ng/ml of βNRG1 EGF domain (Fig. 4). Schwann cell precursor purity can be monitored by immunolabeling cultures with L1 antibodies or alternatively antibodies to p75NTR which label a slightly higher percentage of cells [15] (*see* **Note 1**).

3.3 Schwann Cell Precursor Survival Assay

The survival assay described here is essentially that described previously ([1, 8, 17], *see* also [9]).

1. Schwann cell precursors are plated in drops on coverslips as described above. The choice of culture media will depend on the question under investigation (*see* Subheading 3.1, **step 13** above). After 3 h at 37 °C and 5% CO_2, one set of coverslips is fixed, without top-up, for immunolabeling with L1 or p75NTR antibodies (*see* **Note 3**). The number of immunolabeled cells on these coverslips (typically about 95% of total cells present) is counted. This number is interpreted as the number of cells that plated successfully and is used as a reference point for the quantification of survival at later time points.

2. Sister coverslips are topped up to 400 µl with appropriate medium and cultured for the desired length of time, typically 24 h. The cells are then fixed for immunolabeling with L1 or p75NTR antibodies and stained with Hoechst dye, and the number of surviving Schwann cell precursors counted. Survival percentage is the number of cells present at a certain time point, such as 24 h, as a percentage of the number of cells that had plated successfully at 3 h (*see* **Note 4**).

4   Notes

1. The concentration can vary between 2 and 4 mg/ml depending on the batch but is typically used at 2 mg/ml final concentration.

2. Note that using this method, the leg tissue is not dissected away to visualize the nerves. Rather, it relies on inserting the forceps into the leg in the correct location so that the nerves come to lie between the two tips of the forceps before they are closed. This is not as difficult as it may sound. Practicing on one litter (~10 embryos) should be sufficient to achieve a good success rate. Although the method is described here for the hind leg, with practice it can be used equally to obtain brachial plexus nerves from the forelimb. As an alternative to this approach, it is possible to dissect the nerves out under visual guidance. In our experience this is achievable, but more difficult than the method described above.

3. Schwann cell precursor purity can be monitored by immuno-labeling cultures with L1 antibodies (ASCS4 anti-L1 antibody or 324 anti-L1 antibody for rat or mouse, respectively) or alternatively antibodies to p75NTR which labels a slightly higher percentage of cells [15]. Rat precursor cultures are about 95% pure at 3 h after plating when labeled with antibodies to p75NTR [1, 15]. Mouse precursor cultures are ≥93% pure when labeled with L1 antibodies after 20 h in culture in the presence of βNRG1 [8, 15].

4. In contrast to Schwann cells [9], where cells that die during the culture period stay attached to the coverslips, Schwann cell precursors that die during the culture period detach from the coverslips and float into the medium, from which they can be collected and shown to have laddered DNA indicative of apoptotic death [1, 16]. Thus ≥98% of the cells present at 24 h under various culture conditions are living cells. It follows that the number of L1- or p75NTR-positive cells present on the coverslips at any given time point accurately reflects the number of cells that have survived up to that point.

Acknowledgment

The work in the laboratory of the authors was supported by the Wellcome Trust (grants 042257/Z/94/Z and 036963/1.5) and the Medical Research Council of Great Britain.

References

1. Jessen KR, Brennan A, Morgan L, Mirsky R, Kent A, Hashimoto Y, Gavrilovic J (1994) The Schwann cell precursor and its fate: a study of cell death and differentiation during gliogenesis in rat embryonic nerves. Neuron 12:509–527

2. Jessen KR, Mirsky R (2005) The origin and development of glial cells in peripheral nerves. Nat Rev Neurosci 6:671–682

3. Woodhoo A, Sommer L (2008) Development of the Schwann cell lineage: from the neural crest to the myelinated nerve. Glia 56:1481–1490

4. Jessen KR, Mirsky R, Lloyd AC (2015) Schwann cells: development and role in nerve repair. Cold Spring Harb Perspect Biol 7(7):a020487

5. Jessen KR, Mirsky R (2008) Negative regulation of myelination: relevance for development, injury, and demyelinating disease. Glia 56:1552–1565

6. Buchstaller J, Sommer L, Bodmer M, Hoffmann R, Suter U, Mantei N (2004) Efficient isolation and gene expression profiling of small numbers of neural crest stem cells and developing Schwann cells. J Neurosci 24:2357–2365

7. D'Antonio M, Michalovich D, Paterson M, Droggiti A, Woodhoo A, Mirsky R, Jessen KR (2006) Gene profiling and bioinformatic analysis of Schwann cell embryonic development and myelination. Glia 53:501–515

8. Dong Z, Sinanan A, Parkinson D, Parmantier F, Mirsky R, Jessen KR (1999) Schwann cell development in embryonic mouse nerves. J Neurosci Res 56:334–334

9. Meier C, Parmantier E, Brennan A, Mirsky R, Jessen KR (1999) Developing Schwann cells acquire the ability to survive without axons by establishing an autocrine circuit involving insulin-like growth factor, neurotrophin-3, and platelet-derived growth factor-BB. J Neurosci 19:3847–3859

10. Birchmeier C, Nave KA (2008) Neuregulin-1, a key axonal signal that drives Schwann cell growth and differentiation. Glia 56:1491–1497

11. Joseph NM, Mukouyama YS, Mosher JT, Jaegle M, Crone SA, Dormand EL, Lee KF, Meijer D, Anderson DJ, Morrison SJ (2004) Neural crest stem cells undergo multilineage differentiation in developing peripheral nerves to generate endoneurial fibroblasts in addition to Schwann cells. Development 131:5599–5612

12. Dyachuk V, Furlan A, Shahidi MK, Giovenco M, Kaukua N, Konstantinidou C, Pachnis V, Memic F, Marklund U, Müller T et al (2014) Parasympathetic neurons originate from nerve-associated peripheral glial progenitors. Science 345:82–87

13. Espinosa-Medina I, Outin E, Picard CA, Chettouh Z, Dymecki S, Consalez GG, Coppola E, Brunet JF (2014) Parasympathetic ganglia derive from Schwann cell precursors. Science 345:87–90

14. Kaukua N, Shahidi MK, Konstantinidou C, Dyachuk V, Kaucka M, Furlan A, An Z, Wang L, Hultman I, Ahrlund-Richter L, Blom H, Brismar H, Lopes NA, Pachnis V, Suter U, Clevers H, Thesleff I, Sharpe P, Ernfors P, Fried K, Adameyko I (2014) Glial origin of mesenchymal stem cells in a tooth model system. Nature 513:551–554

15. Dong Z, Brennan A, Liu N, Yarden Y, Lefkowitz G, Mirsky R, Jessen KR (1995) Neu differentiation factor is a neuron-glia signal and regulates survival, proliferation, and maturation of rat Schwann cell precursors. Neuron 15:585–596

16. Kaufman MH (1998) The atlas of mouse development. Elsevier, Amsterdam

17. Gavrilovic J, Brennan A, Mirsky R, Jessen KR (1995) Fibroblast growth factors and insulin growth factors combine to promote survival of rat Schwann cell precursors without induction of DNA synthesis. Eur J Neurosci 7:77–85

Chapter 2

Preparation of Neonatal Rat Schwann Cells and Embryonic Dorsal Root Ganglia Neurons for In Vitro Myelination Studies

Patrice Maurel

Abstract

The ability to understand in great details, at the molecular level, the process of myelination in the peripheral nervous system (PNS) is, in no minor part, due to the availability of an in vitro culture model of PNS myelination. This culture system is based on the ability to prepare large population of highly purified Schwann cells and dorsal root ganglia neurons that, once co-cultured, can be driven to form in vitro well-defined myelinated axon units. In this chapter, we present our detailed protocols to establish these cell cultures that are derived from modifications of procedures developed 35–40 years ago.

Key words Rat, Rodent, Schwann cell, DRG, Dorsal root ganglia, Neuron, Co-culture, Myelination

1 Introduction

The in vitro culture model of peripheral nerve system (PNS) myelination evolved from studies that were characterizing Schwann cells' responses to contact with axons from dorsal root ganglia (DRG) neurons [1–4]. These experiments benefited from recently developed techniques to isolate and culture purified Schwann cells [5, 6] and DRG neurons [6]. One of the most important findings was that under appropriate conditions, Schwann cells form compact myelin sheaths around the axons of the purified DRG neurons [3, 4]. These myelin sheaths are structurally comparable to those formed in vivo, with the formation of nodes of Ranvier, paranodes, juxtaparanodes, and internodes. Over the ensuing 35+ years of research, the Schwann cell/DRG neuron co-culture system has been central, on its own and as a necessary complement to in vivo mouse models, to most of the major discoveries pertaining to PNS myelination: from the molecular characterization and assembly of these different domains to the characterization of the molecular signals that initiate, promote, and regulate Schwann cell fate and

Paula V. Monje and Haesun A. Kim (eds.), *Schwann Cells: Methods and Protocols*, Methods in Molecular Biology, vol. 1739, https://doi.org/10.1007/978-1-4939-7649-2_2, © Springer Science+Business Media, LLC 2018

behavior. This type of myelinating culture model has been in fact so important that the lack, until recently [7], of an equivalent central nervous system (CNS) in vitro system has been considered *a longtime limitation in studying the molecular basis of CNS myelination* [8]. With the development of molecular biology tools and techniques such as lentiviral transduction (*see* Chapter 12) that allow for the easy manipulation of Schwann cells and DRG neurons, allied to the simplicity of the model (only two cell types), the Schwann cell/DRG neuron myelinating culture system is likely to continue contributing to the comprehensive dissection of the molecular mechanisms that control PNS myelin formation. Of course, the in vitro culture system cannot fully represent what happens in vivo. These molecular mechanisms need to be placed in the context of the complex morphological process that myelination is. That is where the strength of the in vivo model lies.

This chapter is divided in three main sections that provide detailed step-by-step information to prepare purified Schwann cells and DRG neuron cultures and to set up the myelinating co-culture system. Sciatic nerves from postnatal day 2 Sprague-Dawley rats are used as the source of Schwann cells, and NGF-dependent neurons are isolated from embryonic day 15 DRGs. The procedures include three basic steps: (1) dissection (i.e., the harvesting of sciatic nerves or DRGs), (2) dissociation of tissues into individual cells and plating, and (3) purification of Schwann cells or DRG neurons with antimetabolic and antimitotic agents. For Schwann cells, two additional steps, to expand and freeze the cells, are described. These protocols allow generating highly purified (>99%) Schwann cells (upward of 1×10^8 cells at third passage) and DRG neurons (upward of 300 individual cultures in 10 days).

2   Materials

2.1   Chemicals and Biochemicals (See Note 1)

2.1.1   For Schwann Cell Preparation

1. Culture-grade poly-L-lysine (MW range of 70,000–150,000): stock concentration is prepared at 1 m/ml in culture-grade water, filter sterilized, and stored at 4 °C, protected from light. Working concentrations of 100 and 10 μg/ml are made freshly at the time of use in sterile culture-grade water.

2. Dulbecco's Modified Eagle's Medium (DMEM) with 4.5 g/l of glucose and without L-glutamine.

3. Fetal bovine serum (FBS; Gibco 16000) (*see* **Note 2**), heat-inactivated; store in 50 ml aliquots at −80 °C (*see* **Note 3**).

4. Forskolin: prepare a 2 mM stock solution in 100% ethanol; store at −20 °C.

5. EGF-D: recombinant EGF domain of human neuregulin-1-ß1 (R&D Systems 396-HB): prepare a stock at 100 μg/ml in DMEM and store at −80 °C in 50 μl aliquots.

6. GlutaMAX™-I.

7. [100×] Penicillin/streptomycin solution (respectively, at 10,000 units/ml and 10,000 μg/ml).

8. Hanks' balanced saline solution (HBSS), calcium- and magnesium-free.

9. Leibovitz's L-15 medium.

10. Collagenase, type 1 (Worthington Biochemical Corp. cat # LS004194): prepare a 1% stock solution in culture-grade sterile distilled water; filter sterilize; store in 1 ml aliquots at −80 °C.

11. Cytosine-ß-arabinofuranoside hydrochloride (Ara-C): prepare a 1 mM stock solution in culture-grade sterile distilled water; filter sterilize; store 1 ml aliquots at −80 °C.

12. Anti-Thy-1.1 antibody at 1 mg/ml (AbD Serotec MCA04G).

13. 2.5% trypsin in HBSS.

14. 0.25% trypsin/2 mM EDTA solution in HBSS.

15. Dimethyl sulfoxide (DMSO), suitable for cell culture.

16. 70% ethanol.

2.1.2 For DRG Neuron Preparation

1. Matrigel, Growth Factor Reduced (GRF) (Corning, cat # 356231): upon receiving the Matrigel vial, thaw on ice and then use ice-cold Neurobasal to bring the concentration down to 2.5 mg/ml (*see* **Notes 4** and **5**). Make 1.1 ml aliquots (*see* **Note 25**) in microcentrifuge tubes that are kept on ice (*see* **Note 5**). Immediately store at −80 °C. Matrigel will be used at a final concentration of 250 μg/ml, in Neurobasal (*see* Subheading 4.1 and **Note 5**).

2. Neurobasal medium (Gibco cat # 21103).

3. B27 supplement, 50× (Gibco cat # 17504-044).

4. D(+) glucose, suitable for cell culture.

5. Mouse NGF 2.5S: prepare stock solution at 1 mg/ml in sterile culture-grade water. Keep 50 μl aliquots at −80 °C until use. Use at a final concentration of 50 ng/ml.

6. 5-Fluoro-2′-deoxyuridine (FUDR) + Uridine (U): prepare a stock solution in Neurobasal medium, at a final concentration of 10 mM each. Use at a final concentration of 10 μM each.

7. GlutaMAX™-I.

8. [100×] Penicillin/streptomycin solution (respectively, at 10,000 units/ml and 10,000 μg/ml).

9. 0.25% trypsin solution in HBSS.

10. 70% ethanol.

2.1.3 For Myelinating Schwann Cell/DRG Neuron Co-cultures

1. Minimum Essential Medium Eagle (MEM) with Earle's salts, without glutamine.

2. L-Ascorbic acid, suitable for cell culture.

3. Fetal bovine serum (FBS; Gibco 16000) (*see* **Note 1**), heat-inactivated; store in 50 ml aliquots at −80 °C (*see* **Note 2**).

4. D(+) glucose, suitable for cell culture.

5. Hanks' balanced saline solution (HBSS), calcium- and magnesium-free.

6. GlutaMAX™-I.

7. 0.25% trypsin/2 mM EDTA solution in HBSS.

2.2 Culture Dishes, Plates, and Coverslips (See Notes 6 and 7)

1. ø60 mm tissue culture dish (for Schwann cell preparation).

2. ø100 mm tissue culture dish (for Schwann cell and DRG neuron preparations).

3. ø35 mm tissue culture dish (for DRG neuron preparation).

4. 24-well plates (for DRG neuron preparation).

5. ø12 mm glass coverslips, 0.13–0.17 mm thick (for DRG neuron preparation).

2.3 Culture Media and Solutions (See Note 8)

1. *SC-BM*: culture medium to maintain Schwann cells: DMEM, 10% FBS, 2 mM GlutaMAX™-I.

2. *SC-EM*: culture medium to expand Schwann cells: SC-BM supplemented with 2 μM forskolin and 10 ng/ml EGF-D.

3. *DRG-SM*: culture medium to maintain DRG neurons: Neurobasal, 2% B27 supplement, 1% GlutaMAX™-I, 0.08% (w/v) glucose, and 50 ng/ml 2.5S NGF.

4. *DRG-FU*: culture medium to purify DRG neurons: DRG-SM supplemented with 10 μM of FUDR and 10 μM Uridine.

5. *CCM*: Schwann cell/DRG neuron co-culture medium: MEM, 10% FBS, 1% GlutaMAX™-I, 0.4% (w/v) glucose, and 50 ng/ml 2.5S NGF.

6. *MM*: myelinating medium: CCM supplemented with 50 μg/ml of ascorbic acid.

7. *Freezing media*: SC-BM supplemented with 10% DMSO.

2.4 Animals

1. *Schwann cell preparation*: postnatal day 2 (p2) Sprague-Dawley rat pups; two litters (about 20–25 pups) (*see* **Note 9**).

2. *DRG neuron preparation*: embryonic day 15 (E15) Sprague-Dawley rat embryos; 1 pregnant female (about 10–16 embryos) (*see* **Note 9**).

2.5 Dissecting Tools and Equipment

1. Stereoscopic dissecting microscope.

2. Large scissors (3), 13 cm, straight, blunt/blunt tips.

3. Extra-fine Bonn scissors (1), 8.5 cm, straight, sharp/sharp tips.

4. Spring scissors (1), 10 cm, straight, sharp/sharp tips.

5. Moria fine scissors (1), 10.5 cm, curved, sharp/sharp tips.

6. Graefe forceps (1), 10 cm, curved, serrated tips.

7. Tissue forceps (3), 12 cm, straight, 1×2 teeth.

8. Dumont forceps #5 mirror, finished forceps, straight, standard tips (0.1×0.06 mm) (*see* **Note 10**).

9. Dumont forceps #5 mirror, finished forceps, straight, biology tips (0.05×0.02 mm) (*see* **Note 10**).

10. Dumont #5/45 coverslips forceps.

11. Needles 27 $G^{1/2}$ and 23 $G^{1/2}$.

12. 15 ml conical tubes, sterile.

13. 50 ml conical tubes, sterile.

14. Hemocytometer.

15. 2 ml serological pipets.

16. Wide orifice 200 µl pipet tips (*see* **Note 11**).

17. 200 µl regular pipet tips.

18. 200 µl Pipetman or equivalent.

19. 9 in. glass Pasteur pipets, sterile.

20. 2 ml cryotubes.

21. Humidified tissue culture incubator set at 37 °C and 10% CO_2 for Schwann cell cultures.

22. Humidified tissue culture incubator set at 37 °C and 5% CO_2 for DRG neurons and myelinating co-cultures.

23. Centrifuge with swing bucket and adaptors that accommodate 15 and 50 ml conical tubes.

24. Liquid nitrogen tank, vapor phase.

25. −80 °C freezer.

26. 0.22 µm filtering devices, low-protein-binding membrane.

27. Dissecting board (*see* **Note 12**).

28. Water bath set at 37 °C.

29. Heat block set at 37 °C.

30. Microcentrifuge tubes.

31. Vacuum line.

3   Preparation of Primary Schwann Cell Cultures from Neonatal Rat Sciatic Nerves

3.1   Day 0: Preparing Culture Plates

1. Coat ø60 mm plates with 2 ml of poly-L-lysine (10 µg/ml). Incubate overnight at 37 °C in a humidified cell culture incubator (% of CO_2 is not important at this time).

   Prepare one ø60 mm plate per four pups (eight sciatic nerves collected).

3.2   Day 1: Harvesting Sciatic Nerves, Tissue Dissociation, and Cell Plating

3.2.1   Preparation

1. Sterilize the dissection tools by immersing them into 70% ethanol for 30 min (*see* **Note 13**). Then, air-dry.

   The following tools will be needed: one large (13 cm) scissors, curved Moria fine scissors, curve Graefe forceps, Dumont forceps #5 (standard tips), and spring scissors.

2. Prepare one 15 ml conical tube with 13 ml of L-15 medium; keep chilled on ice.

3. To a 15 ml conical tube, add 1 ml of 2.5% trypsin solution and 8 ml of L-15 medium. Thaw quickly (37 °C, water bath) a 1 ml aliquot of 1% collagenase and transfer to the trypsin solution. Invert the tube a few times to mix. Keep warm at 37 °C (water bath).

4. Finish preparing the ø60 mm plates. Using a Pasteur pipet connected to a vacuum line, aspirate the poly-L-lysine solution and rinse quickly once with culture-grade sterile water. Discard the water and replace with 3 ml of SC-BM media, supplemented with Penicillin/Streptomycin ([1×] final concentration). Keep the plates at 37 °C in the 10% CO_2 humidified cell culture incubator (*see* **Note 14**).

5. Wrap the dissecting board with absorbent paper towel. Spray with 70% ethanol.

3.2.2   Dissection (One Pup at a Time)

1. Euthanize one rat pup by decapitation using the large scissors (13 cm). Place the animal on the dissecting board, dorsal side up, and pin it into place with 27 $G^{1/2}$ (legs) and 23 $G^{1/2}$ (body) needles, splaying the hind legs into an inverted V shape (*see* Fig. 1a). Spray the body with 70% ethanol.

2. Using the small Graefe forceps, pull up the skin. Then using the small curved scissors, cut the skin away from the hind limbs and the lower back (*see* Fig. 1b).

3. With the #5 Dumont forceps, gently tease apart the *quadriceps femoris* (QF), medial hamstring (MH), and *biceps femoris* (BF) muscles, from the knee to the hip bone (*see* Fig. 1c).

4. Locate the sciatic nerve, which runs posterior and parallel to the femoral bone (*see* Fig. 1c). Gently grasp the sciatic nerve with one the Dumont forceps and cut the nerve at each end

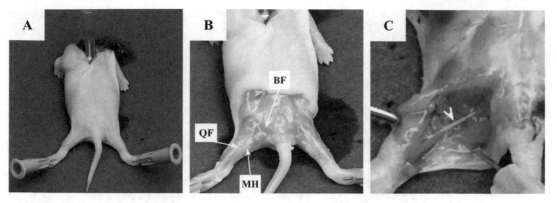

Fig. 1 Harvesting sciatic nerves from p2 neonatal rats. (**a**) A euthanized pup on the dissecting board, dorsal side up, pinned down in place with the hind legs splayed in an inverted V shape. (**b**) The skin has been removed from the lower back and hind legs. *QF* quadriceps femoris, *BF* biceps femoris, *MH* medial hamstring. (**c**) The sciatic nerve (*arrowhead*) after teasing the muscles apart

with the spring scissors. The dissected nerves will be about 1 cm (0.5 in.) in length.

5. Transfer the sciatic nerves immediately into the 15 ml conical tube that contains the ice-cold L-15 medium.

6. Repeat **steps 1** through **5** for all the remaining pups.

3.2.3 Tissue Dissociation

1. Once all the sciatic nerves have been harvested, centrifuge the 15 ml tube for 5 min at 50 × *g*, at room temperature.

2. Gently decant the supernatant and resuspend the sciatic nerves with the 10 ml of the pre-warmed trypsin-collagenase solution. Incubate at 37 °C in the water bath for 30 min, inverting the tube a couple of times every 5 min.

3. Centrifuge for 5 min at 50 × *g*, at room temperature. Decant the supernatant and gently resuspend the pelleted sciatic nerves with 10 ml of SC-BM. Repeat the centrifugation.

4. Decant the supernatant and add 2 ml of SC-BM supplemented with Penicillin/Streptomycin ([1×] final concentration) to the sciatic nerves. Using a 2 ml serological pipet, triturate the sciatic nerves until complete dissociation of the tissue; it should take about 10–20 pipettings.

5. Add SC-BM supplemented with Penicillin/Streptomycin so that you have 1 ml of dissociated tissue for every eight sciatic nerves that were collected.

6. Plate 1 ml of the cell suspension per ø60 mm plate. Incubate for 24 h at 37 °C in the 10% CO$_2$ humidified cell culture incubator. The cultures should consist of Schwann cells and fibroblasts (*see* Fig. 2a).

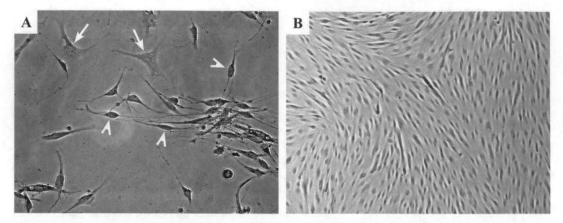

Fig. 2 Purification of rat Schwann cells. (**a**) Cells 24 h after dissection. Schwann cells have an elongated, spindle-shaped morphology (arrowheads), whereas fibroblasts are very flat and have a spread-out morphology (arrows). (**b**) Monolayer of purified Schwann cells at confluence. Note the elongated shape of the cells and swirling pattern of the monolayer

3.3 Day 2–8: Removal of Fibroblasts by Antimetabolic Treatment (See Note 15)

1. After 24 h (day 2), use a glass Pasteur pipet to aspirate the culture media, and wash the cells twice with 3 ml of HBSS.

2. After the last wash, add 3 ml of SC-BM containing Ara-C at 10 µM (1:100 dilution from stock), and incubate the cultures in the humidified cell culture incubator (37 °C, 10% CO_2) for 3 days (day 5; *see* **Note 16**).

3. Remove the culture media. Wash the cells twice with 3 ml of HBSS and replace with 3 ml of SC-EM (no antibiotics should be necessary from now on) to initiate Schwann cell proliferation (*see* **Note 17**). Place the cultures in the incubator for another 3 days (day 8).

3.4 Day 8 to About Day 12: Removal of Contaminating Fibroblasts by Complement-Mediated Killing (See Note 18)

1. On day 8, wash the cells once with 3 ml of HBSS.

2. Add 3 ml of SC-EM, supplemented with 60 µl of the anti-Thy-1.1 antibody, and incubate for 30 min in the cell culture incubator.

3. Add 400 µl of rabbit complement; mix by gently swirling the dishes.

4. Incubate for 30 min to 3 h in the cell culture incubator. Monitor the cells under a phase contrast microscope every 20 min, to assess the death of the fibroblasts as well as the health of the Schwann cells (*see* **Note 19**).

5. Remove the culture medium and wash the cells twice with 3 ml of HBSS.

6. Add SC-EM, and incubate cells in the humidified cell culture incubator (37 °C and 10% CO_2) for 2 days (day 10).

7. Observe the cultures for fibroblasts contamination and Schwann cell viability. Schwann cells should have a bipolar, spindle-shaped morphology (arrowheads in Fig. 2a), distinguishable from the fibroblasts flattened and spread-out morphology (arrows in Fig. 2a). At this stage, Schwann cell purity should be greater than 99%. If it is not, the complement-mediated killing (**steps 1** through **6**) should be repeated before expanding the cells further.

8. Otherwise, keep the cultures until they form a 100% confluent monolayer (about 3–4 more days; day 11 to day 12) (*see* **Notes 20** and **21**).

9. When Schwann cells reach confluence (Fig. 2b), coat ø100 mm plates with 5 ml of poly-L-lysine (10 μg/ml). Incubate overnight at 37 °C in the humidified cell culture incubator (% of CO_2 is not important at this time).

Prepare one ø100 mm plate per ø60 mm plate.

3.5 Day 12 to About Day 17: Expanding Schwann Cells to Passage 1

1. Finish preparing the ø100 mm plates. Remove the poly-L-lysine solution and rinse quickly once with culture-grade sterile water. Discard the water and replace with 5 ml of SC-EM media. Keep the plates at 37 °C in the humidified cell culture incubator (10% CO_2).

2. Wash the Schwann cells twice with 3 ml of HBSS. Discard HBSS washes and add 2 ml of 0.25 trypsin-EDTA solution; incubate at 37 °C for about 5 min. Observe the cells under phase contrast microscopy while gently shaking the culture dish.

3. When the cells start to come off the dish, add 2 ml of SC-BM, and collect the cells in 15 ml tubes, one tube per ø60 mm plate.

4. Centrifuge for 5 min, at $200 \times g$, at room temperature.

5. Discard the supernatants and gently resuspend the cells in 5 ml of SC-EM medium.

6. Plate cells onto the ø100 mm tissue culture plates, cells from one ø60 mm plate to one ø100 mm plate. Incubate cells in the humidified cell culture incubator (37 °C, 10% CO_2) until cells are confluent in about 5 days. This split will correspond to passage 1 (*see* **Note 22**).

7. These are the cells that we use for lentiviral transduction experiments (*see* Chapter 12) once they have reached 100% confluence (*see* **Note 23**).

3.6 Day 17 to About Day 22: Expanding Schwann Cells to Passage 2

1. When 100% confluent, the number of Schwann cells per ø100 mm plate should be about $4–5 \times 10^6$. Cells will be split 1:4. The day before splitting, prepare four ø100 mm plates for every ø100 mm plate to be split. Follow **step 9** (Subheading 3.4) and **step 1** (Subheading 3.5). In **step 1**, prepare the plates with 9 ml of SC-EM medium.

2. Then follow **steps 2** through **6** from Subheading 3.5, with the following modifications:

- **Step 2**: wash cells with 5 ml of HBSS, and use 5 ml of 0.25 trypsin-EDTA solution.

- **Steps 3** and **4** are not changed; collect cells in one tube per ø100 mm plate.

- **Step 5**: resuspend cells in 4 ml of SC-EM.

- **Step 6**: plate 1 ml of cell suspension per ø100 mm plate.

3.7 Further Expansion

1. Follow steps in Subheading 3.6 to expand Schwann cells further (*see* **Note 24**).

2. Before using Schwann cells in any experiments, replace the culture medium (SC-EM) with Schwann cell medium that does not contain forskolin and EGF-D (SC-BM) for 2–3 days.

3.8 Freezing Schwann Cells for Storage

1. Wash cells twice with 5 ml of HBSS.

2. Add 5 ml of 0.25 trypsin solution; incubate at 37 °C for 5 min.

3. Observe the cells under light microscopy while gently shaking the culture dish.

4. When cells start to come off the dish, add 2 ml of SC-BM and pool the cells in one or two 50 ml conical tubes.

5. Centrifuge for 10 min, at $200 \times g$, at room temperature.

6. Discard supernatant and gently resuspend the cells in 5–10 ml of SC-BM (adjust the volume based on the number of plates being used). Count cells using a hemocytometer.

7. Centrifuge for 10 min, at $200 \times g$, at room temperature.

8. Discard supernatant and gently resuspend the cells in ice-cold freezing medium, 1 ml per 2×10^6 Schwann cells.

9. Make 1 ml aliquots in cryotubes, and freeze overnight at −80 °C.

10. The following day, transfer frozen aliquots to a vapor phase liquid nitrogen tank for long-term storage.

3.9 Thawing Schwann Cells

1. Quickly thaw a frozen aliquot at 37 °C in a water bath.

2. Transfer the cell suspension to a 15 ml tube containing 9 ml of pre-warmed (37 °C) SC-BM.

3. Centrifuge for 10 min, at $200 \times g$, at room temperature.

4. Discard supernatant and gently resuspend the cells in 8 ml of SC-EM medium.

5. Plate cells onto two ø100 mm tissue culture plates, 4 ml per plate (prepared the day before; *see* Subheadings 3.4 **step 9** and 3.5 **step 1**).

6. Incubate cells at 37 °C and 10% CO_2 in a humidified cell culture incubator until cells are 100% confluent (about 5–7 days). Change culture medium with fresh 10 ml of SC-EM at day 3.

7. When cells are confluent, replace medium with SC-BM for 3 days before use in experiments. Alternatively, cells may also be passaged (*see* **Note 24**) following the steps in Subheading 3.6.

4   Preparation of Dorsal Root Ganglia Neurons from Rat Embryos (*See* Note 25)

4.1   Day 0: Preparing Culture Plates

1. Using the coverslips forceps, place one glass coverslip in each well of 24-well plates (*see* **Note 26**).

2. Thaw, on ice (*see* **Note 5**), one, two, or three aliquots of Matrigel (1 aliquot per 4½ 24-well plates).

3. Depending on the number of Matrigel aliquots to be thawed, prepare one 15 ml conical tube with 9.9 ml of ice-cold Neurobasal (1 aliquot) or one 50 ml conical tube with either 19.8 or 29.7 ml of ice-cold Neurobasal (2 or 3 aliquots). Keep tube on ice.

4. Transfer the Matrigel to the tube containing the Neurobasal medium, using a prechilled 2 ml serological pipet (*see* **Note 5**). Ensure thorough mixing by immediately pipetting up and down several times, keeping the tube on ice. Matrigel will be at a final concentration of 250 μg/ml.

5. Using a 200 μl wide orifice pipet tip, gently transfer 100 μl of Matrigel solution directly onto each glass coverslip. The solution must stay on the coverslip (*see* **Note 27**).

6. Slowly and without any abrupt moves, transfer the plates at 37 °C into the humidified cell culture incubator set at 5% CO_2. Incubate overnight.

4.2   Day 1: Harvesting DRGs, Tissue Dissociation, and Plating

4.2.1   Preparation

1. Sterilize the dissection tools by immersing them into 70% ethanol for 30 min (*see* **Note 13**). Then, air-dry.
   The following tools will be needed: three large scissors, three tissue forceps, curved Graefe forceps, extra-fine Bonn scissors, Dumont forceps #5 (standard tips), Dumont forceps #5 (biology tips), and spring scissors.

2. Prepare one 1.5 ml microcentrifuge tube with 1 ml of L-15 medium; keep warm in the heat block at 37 °C.

3. Prepare two ø100 plates with 30 ml of L-15 at room temperature.

4. Prepare one ø35 mm plate with 4 ml of L-15 at room temperature.

5. Warm 5 ml of DRG-BM in a 15 ml conical tube at 37 °C in water bath.

6. Warm 20–40 ml of DRG-BM in a 50 ml conical tube at 37 °C in water bath (*see* **Note 28**).

1. Euthanize the pregnant rat by CO_2 inhalation.

2. Place the rat ventral side up, on absorbent paper towels. Sterilize by spraying with 70% ethanol until the fur is thoroughly soaked (*see* **Note 29**).

3. Using one tissue forceps, lift the belly skin up and cut a buttonhole with one of the large scissors. Insert the blunt-ended tips of the scissors in between the skin and the abdominal muscles to separate both apart. Cut the skin up to the sternum and down to the pubic symphysis, then left and right. Move the skin flaps away.

4. With a clean pair of tissue forceps and large scissors, cut the abdominal muscle wall as was done for the skin and move the flaps away (*see* **Note 30**).

5. With the last clean tissue forceps, grab the cervix and lift up gently. Using the last clean large scissors, cut the cervix close to the vaginal orifice and continue lifting. The uterine horns should follow easily. Cut away any tissue that comes along (uterine artery and some fat). Transfer the horns to a sterile empty (no medium) ø100 mm plate (*see* **Note 30**).

1. Using the Graefe forceps, pinch the outside wall of the uterine horns at the level of an embryo. Use the Bonn scissors to cut a small hole through the wall of the uterus. When you see the amniotic sac, punch a whole. The embryo will naturally come out due to the internal pressure from the amniotic fluid. Cut the umbilical cord. Use the curved end of the Graefe forceps to lift (do not squeeze) the embryo and transfer to one of the ø100 mm plate with L-15 medium.

2. Repeat **step 1** until all embryos have all been collected in the same dish.

3. Swirl the dish to wash the embryos. Then transfer the embryos to a clean ø100 mm plate with L-15 medium.

1. Under magnification of the dissecting scope, hold an embryo in place with a #5 Dumont forceps (standard tips) and euthanize by cutting off the head with the spring scissors (line 1 in Fig. 3a). Repeat for all embryos.

2. Place an embryo to its side and pin it down through the belly with one of the #5 Dumont forceps (standard tips). Insert the tips of spring scissors vertically toward the center of the embryo (center along the anteroposterior axis) and ventral to the vertebral column.

3. Make a couple of sequential cuts toward the anterior end of the embryo and then a couple of sequential cuts toward the posterior end of the embryo, along line 2 shown in Fig. 3a.

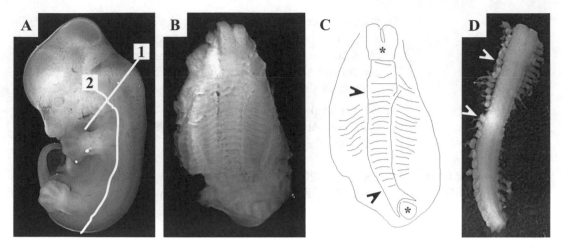

Fig. 3 Representative steps of the DRG dissection. (**a**) An E15 rat embryo. Lines 1 and 2 represent cuts that are to be made as described in Subheading 4.2.4. (**b**) The dorsal part of the embryo, ventral side up, after cutting along lines 1 and 2. (**c**) A schematic representation of panel B. The arrowheads point to the vertebral body, and the asterisks indicate visible portions of the spinal cord. (**d**) An isolated spinal cord with DRGs (arrowheads) attached

4. Remove the ventral half portion of the embryo including all the limbs. Continue until all embryos are similarly processed. Transfer the ventral portions (*see* Fig. 3b) into a clean ⌀100 mm plate with L-15 medium.

5. Lay the remaining dorsal tissue on its back so that the ventral side is facing up. Hold the tissue down with one #5 Dumont forceps (standard tips) and remove any remaining organs (mostly lungs, kidneys, and aorta at this stage) from the ventral surface with the other one. The vertebral body should be visible in the middle of the dorsal plate (*see* Fig. 3b, c).

6. While gently holding the dorsal plate in the central region (in reference to the anteroposterior axis) with the tips of one of the #5 forceps (standard tips) on either side of the vertebral column, use the other #5 forceps (standard tips) to cut the vertebrae along the anterior to posterior axis. To do so, insert one branch of the forceps in between the vertebral body and the spinal cord, the other branch remaining outside. Close and pull out to cut open the vertebral body. This will expose the underlying spinal cord.

7. Gently lift up the anterior end of the spinal cord and pull it up, freeing it from the vertebral canal. Dorsal root ganglia attached along the sides of the cord will follow (*see* **Note 31**). Transfer the spines with the attached DRG to the ⌀35 mm dish with L-15 medium (*see* Fig. 3d). A good dissection should have about 35–40 DRGs per spine.

8. Switch to using the #5 Dumont Forceps biology tips. While gently holding the spinal cord between one set of forceps, use the other one to pinch off the ganglia from the dorsal root. Be careful not to carry over any meninges from the spinal cord. Collect all the ganglia and discard the cord.

9. Repeat **step 8** until the DRGs from all the spines have been collected. All the collection is done in the ø35 mm dish, setting the DRGs aside in a pile.

4.2.5 Tissue Dissociation and Plating of the Cells

1. Using the 200 μl Pipetman setup with a wide orifice tip (*see* **Note 32**), collect all DRGs in one go and transfer to the microcentrifuge tube containing the L-15 in the heat block.

2. Most of the DRGs will settle down quickly by gravity. There is no need to spin. If a few DRGs cling to the sidewall, gently tap with you finger to dislodge them, and they will settle down.

3. Use the 200 μl Pipetman to gently remove the L-15 medium (*see* **Note 33**), and replace with 1 ml of 0.25% trypsin (*see* **Note 34**). Incubate at 37 °C in the heat block for 15 min. Gently flick the tube to resuspend the DRGs (*see* **Note 35**) and incubate for an additional 15 min.

4. Use the 200 μl Pipetman to gently collect the DRGs and transfer to 5 ml of DRG-SM in a 15 ml conical tube. Do not aspirate up and down to dissociate the DRGs just yet (*see* **Note 35**). Close the tube, gently invert a couple of time, and centrifuge at $200 \times g$ for 10 min at room temperature.

5. While **step 4** is ongoing, finish preparing the 24-well plates. To do so, place a 200 μl pipet tip (standard yellow tip) to the end of a glass Pasteur pipet connected to vacuum. This will prevent the glass coverslip to become vacuum-stuck to the end of the Pasteur pipet. Aspirate the excess of Matrigel off the coverslips.

6. Once the DRGs have been pelleted (**step 4**), use a Pasteur pipet connected to a vacuum line to aspirate the supernatant, leaving about 50 μl behind. Using a wide bore 200 μl pipet tip, resuspend the cells in 200 μl of DRG-SM. It should take no more than about 10 up-and-down pipettings. Transfer the dissociated cells into the DRG-SM prepared in the 50 ml tube (*see* note in **step 8**). Mix by inverting and swirling.

7. Seed 100 μl of the cell suspension directly on each coverslip using the wide bore 200 μl pipet tips. The drop should stay on the coverslip. Gently mix the cell suspension in the 50 ml tube after every 24-well plate to ensure an even suspension and density homogeneity across all cultures. Transfer the plates gently to a humidified cell culture incubator at 5% CO_2 and 37 °C, and incubate for 24 h.

8. *Note*: the seeding is done at a density of 1.35 DRG per coverslip. Therefore, as an example, for 100 cultures, one would need 135 dissociated DRGs in 10 ml of medium. Since we prepare an extra of 15% (*see* **Note 28**), the actual number of DRGs is 155 whose cells are resuspended in 11.5 ml of medium.

9. For transduction experiments (*see* Chapter 12), DRG cultures are used 24 h after plating.

4.2.6 DRG Neuron Purification

1. Gently shake the plates so that the 100 µl of culture media drop off to the side. Aspirate the drop with a Pasteur pipet connected to a vacuum line. Do not aspirate directly on top of the cultures.

2. Feed the cultures with 200 µl of DRG-FU, and incubate at 5% CO_2 and 37 °C, for 72 h.

3. Replace media with 200 µl of DRG-BM, and incubate at 5% CO_2 and 37 °C, for 48 h.

4. Cultures should be >99% clean. Fibroblasts do not grow in B-27-based Neurobasal media.

5. Change media two more times with fresh DRG-BM, every other day. DRG neurons will develop an extensive neurite network (Fig. 4). Cultures will be ready to use 10 days after dissection and can be kept for about a month for myelination studies if media is changed every 2–3 days (weekday/weekend schedule) with fresh DRG-BM.

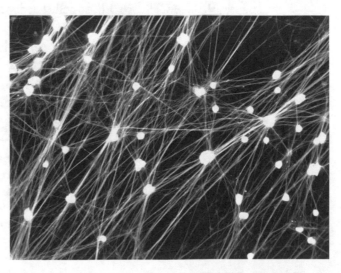

Fig. 4 Representative image of a culture of purified DRG neurons. Phase contrast image. The bright spheres are the neuron bodies; the meshing extensions are the neurite network

6. If an additional round of cleanup with antimitotic agents is necessary after **step 4**, repeat **steps 2** and **3** one more time (*see* **Note 36**).

5   Setting Up the Myelinating Schwann Cell/DRG Neuron Co-culture System

Schwann cells are seeded at a density of 1×10^5 cells per DRG neuron culture. One ø100 mm plate of Schwann cell culture should have about $4–5 \times 10^6$ cells, i.e., enough for about 40–50 DRG neuron cultures. Adapt the following steps (given for one plate) based on the specifics of your experiment.

1. Three days before use, change Schwann cell media from SC-EM to SC-BM.

2. Then wash the cells twice with 5 ml of HBSS at room temperature.

3. Discard HBSS and add 5 ml of 0.25 trypsin-EDTA solution; incubate at 37 °C for about 5 min.

4. Observe the cells under light microscopy while gently shaking the culture dish.

5. When the cells come off the dish, add 2 ml of CCM, and collect the cells in a 15 ml conical tube.

6. Centrifuge for 10 min, at $200 \times g$, at room temperature.

7. Discard supernatant and gently resuspend the cells in 5 ml of CCM.

8. Count the cells using the hemocytometer.

9. Use CCM to adjust cell density to 1×10^5 cells per 500 µl of media.

10. Gently aspirate DRG neuron culture media and add 500 µl of Schwann cell suspension per DRG culture. Incubate at 37 °C, 5% CO_2, for 24 h (*see* **Note 37**).

11. The following day, change media to remove any dead cells, replacing with fresh 200 µl of CCM. Incubate for 6 days, replacing with 200 µl of fresh CCM at day 3.

12. On the sixth day, initiate myelination by changing the CCM culture medium to MM medium, 200 µl. Change every other day during the weekdays, 3 days over the weekend. A few myelinated segments should be visible by phase contrast within 4 days (*see* **Note 38**). Figure 5 shows a representative image of a myelinated culture at day 10.

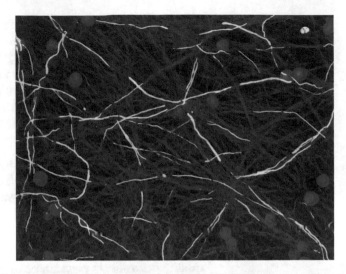

Fig. 5 Immunofluorescent visualization of a myelinated co-culture. Ten-day-old culture. The neurite network is detected by staining for neurofilaments (red). Compact myelinated segments (green) are detected by staining for myelin basic protein, a major component of the compact myelin sheath

6   Notes

1. Whenever possible, select chemical reagents that have been certified for cell culture or biology work.

2. Serum quality is critical for Schwann cell preparation and myelination. Some serum lots will induce substantial Schwann cell proliferation even in the absence of added mitogenic factor such as EGF-D. Also not all lots of serum will support myelination. Screen several lots to identify a serum that is adequate for both Schwann cell preparation and myelination. We usually test for proliferation by EdU incorporation assays (Invitrogen C10337). Basal level of incorporation in purified Schwann cells cultured in SC-BM for 48 h, with EdU added during the last 4 h should be less than 5%. Test for myelination following the protocol in Subheading 5. The Gibco catalog # 16000 for FBS is provided as an indication as to the quality of serum to look for.

3. Heat inactivate for 30 min in a water bath set at 56 °C. Have a bottle with an equivalent volume of water with a thermometer. Begin timing when the thermometer reaches 56 °C.

4. Matrigel concentration is usually around 9–13 mg/ml. Pay attention to this lot-to-lot variability to prepare the 2.5 mg/ml stock solution.

5. Keeping Matrigel on ice using ice-cold Neurobasal and pre-chilled pipets (by running ice-cold media up and down several

times) is absolutely necessary. Matrigel at high concentrations (1 mg and above) will start polymerizing at about +10 °C and will gel very quickly if brought up to room temperature. It will not gel at the final concentration of 250 μg/ml and can then be used like any other culture substrate. Do not freeze/thaw aliquots; we found that the Matrigel substrate peeled off easily when we used aliquots that went through a couple of freeze/thaw cycles.

6. Use culture plates sterilized by gamma ray irradiation. Ethylene oxide sterilization leaves residuals to which many primary cells are sensitive to, particularly Schwann cells and DRG neurons.

7. There are different types of glass material (borosilicate, soda lime), surface treated or not, to which many cell types are sensitive to. We exclusively use coverslips made from the highest-quality borosilicate glass that can be found from suppliers such as Bellco Glass, Inc. (cat # 1943-10012A) and Carolina (cat # 633029). We have used these coverslips for close to 10 years without any issues.

8. All media and solutions are sterilized using 0.22 μm, low-protein-binding membrane, filtering devices.

9. The average litter size for Sprague-Dawley rats is 10–12 pups.

10. The mirror finish of the tips will prevent adhesive tissues, particularly neural tissue, from sticking to the tips. It will facilitate the dissection but it is not an absolute requirement.

11. They will facilitate the collection and rapid transfer of intact DRGs from plate to tubes. They will also be used to dissociate DRGs by up-and-down pipetting. The shearing forces generated by the narrow orifice of regular tips will damage the neurons. While a 2 ml serological pipet can be used, the small length of the 200 μl tips also minimizes the loss of neurons that would stick to the inside wall of the serological pipet.

12. A thin Styrofoam platform (about ½ in. thick), covered with towel paper and sprayed with 70% ethanol, is adequate. The thickness is more important: sufficient to provide hold to the syringe needles, but not too thick to create focusing problems once under the dissecting microscope.

13. Most materials used in the fabrication of dissecting tools develop brownish, rusty-looking spots upon repeated autoclaving. Whether these may have some toxic effects on cells we do not know. We prefer to err toward the side of caution.

14. The 3 ml per ⌀60 mm plate may seem a lot, and once the cell suspension derived from the dissociation of the sciatic nerves is added, the final volume will be 4 ml. This will ensure an even distribution of the cells across the plates, which will facilitate the next step of fibroblasts elimination.

15. Schwann cells proliferate very little in medium containing only 10% FBS and lacking growth factors such as Neuregulin-1-ß1 (*see* **Note 2**). However fibroblasts do proliferate. Ara-C is an antimetabolic agent that, after conversion into cytosine arabinoside triphosphate, impairs DNA synthesis of rapidly dividing cells, such as fibroblasts.

16. After 3 days of treatment with Ara-C, there should be many dead cells, mainly fibroblasts, floating in the medium. Schwann cells, spindle-shaped, should remain attached to the plates.

17. Schwann cells will respond to forskolin and EGF-D by proliferating and will start to crowd out the remaining fibroblasts

18. Thy 1.1 is a murine alloantigen that is displayed at the surface of rat fibroblasts, but not of rat Schwann cells [9]. Using an anti-Thy 1.1 antibody in conjunction with serum complement, it is therefore possible to selectively kill the remaining contaminating fibroblasts by complement-mediated cytolysis.

19. The length of treatment with complement is variable from one dissection to another, even for the same combination of Thy1.1 and complement lot numbers. Fibroblasts should visibly shrivel under phase microscope observation.

20. The time it will take to reach confluence depends on many factors, such as FBS quality and lot # of EGF-D. For this particular step, it will also depend on the efficiency of the dissociation of the sciatic nerves to recover Schwann cells, and of their quality after the Ara-C treatment.

21. Note the elongated shape of the Schwann cells, and the swirly pattern of the culture, typical of confluent, non-differentiated Schwann cells. The cells should not look spread out in every direction like fibroblasts.

22. We usually have enough sciatic nerves to prepare 5–6 ø60 mm plates, i.e., 5–6 ø100 mm plates once at passage 1. When at 100% confluence, there will be about $4–5 \times 10^6$ Schwann cells per ø100 mm plates.

23. Lentiviral transduction efficiency decreases with the number of passages. While it is still possible to use Schwann cells from a later passage if the lentiviral construct is small (about 5 kb between 5′ and 3′ LTRs), it will become a problem when using larger constructs.

24. Schwann cells expanded up to passage 5 will myelinate quite well. It is not recommended to expand further as they will gradually start losing their characteristics. Their response to growth factors diminishes, and they tend to enter senescence and start to myelinate poorly. Cells from passages 1–4 can be frozen for expansion up to the fifth passage in total, i.e., if starting from a frozen passage 3 aliquot, expand only twice. Fifth passage Schwann cells are to be used and unused cells

are to be discarded. We routinely use passage 1 for lentiviral transduction experiments, expand the remaining plates up to passage 3 to use cells in experiments that do not require transduction, and freeze the remaining cells.

25. The whole dissection, from rat euthanasia to the plating of the cells from dissociated DRGs should take no more than 3 h. The longer it takes and the lower the viability of the neurons will be.

26. An average pregnancy yields about 12 embryos. A good dissection will provide spinal cords with about 35–40 DRGs, i.e., a minimum of about 420 DRGs. At 1.35 DRG per culture, one can set up 310 cultures with one dissection, i.e., 13 24-well plates. The 1.1 ml size of the Matrigel aliquots allow for maximizing the use of Matrigel with regard to the uncertainty as to the number of embryos that will be available, since the plates must be prepared the day before dissection. We usually prepare nine 24-well plates, enough in most cases for the bi-weekly needs of a team of three students.

27. Surface tension, the slight viscosity on the diluted Matrigel, and the asperities of the glass coverslip edge ensure that the 100 μl of Matrigel solution will remain on the coverslip. Make sure that the coverslips are not touching the side of the wells. Usually less than 0.5% of the coverslips will have the drop of Matrigel fall off.

28. Prepare volume amounts based on the number of cultures to set up: 100 μl per culture. Due to volume discrepancies when using serological pipet and Pipetmans, we always prepare 15% more than is needed. The number of DRGs collected is for that volume of media to ensure that all coverslips will receive cells at the proper density.

29. It is not necessary to shave the fur off.

30. The use of a succession of scissors/forceps pairs to remove the skin, muscles, and then the uterine horns decreases the chances of contamination. Make sure not to cut through the intestines, which would lead to a massive contamination of the uterine horns and jeopardize the sterility of the embryo dissection.

31. Spinal cord and DRGs are on opposite side of the vertebrae. At this embryonic stage (E15), the vertebrae are barely starting to be cartilaginous and very soft. The DRGs should come along easily. If the embryos are slightly miss-timed (E16–E16.5), it may be necessary to gently score the vertebrae with one of the forceps or too many DRGs will be left imbedded in the dorsal tissue.

32. The narrow opening of a standard tip will damage the DRGs. You can use a Pasteur pipet whose opening has been smoothed by heat. However DRGs do stick quite well to glass.

33. Do not aspirate with a Pasteur pipet connected to a vacuum line. The DRGs are loose and will be sucked up.

34. Do not use trypsin that contains EDTA. The chelation of calcium will kill the neurons.

35. Do not flick hard. The DRGs, although dissociating, remain structurally coherent as a small ball of cells. Keep it that way until it is actually needed to separate the cells away from each other.

36. In our 10-year long experience with this protocol, it has not proven necessary.

37. Do not use DRG cultures immediately after using the DRG-FU media. Always change the media at least twice within a course of 4 days. Trace amounts of FU will either kill the Schwann cells or at the very least will render then unhealthy and myelination will be quantitatively and qualitatively very poor.

38. How long to keep the cultures to myelinate will depend on one's research questions. 10–15 days provide well-myelinated cultures with thousands of segments suitable for quantitative Western blot analyses as well as immunofluorescence qualitative and quantitative analyses of the cultures.

References

1. Bunge MB, Williams AK, Wood PM (1982) Neuron-Schwann cell interaction in basal lamina formation. Dev Biol 92(2):449–460

2. Cornbrooks CJ, Carey DJ, McDonald JA, Timpl R, Bunge RP (1983) In vivo and in vitro observations on laminin production by Schwann cells. Proc Natl Acad Sci U S A 80(12):3850–3854

3. Eldridge CF, Bunge MB, Bunge RP, Wood PM (1987) Differentiation of axon-related Schwann cells in vitro. I. Ascorbic acid regulates basal lamina assembly and myelin formation. J Cell Biol 105(2):1023–1034

4. Moya F, Bunge MB, Bunge RP (1980) Schwann cells proliferate but fail to differentiate in defined medium. Proc Natl Acad Sci U S A 77(11):6902–6906

5. Brockes JP, Fields KL, Raff MC (1979) Studies on cultured rat Schwann cells. I. Establishment of purified populations from cultures of peripheral nerve. Brain Res 165(1):105–118

6. Wood PM (1976) Separation of functional Schwann cells and neurons from normal peripheral nerve tissue. Brain Res 115(3): 361–375

7. Watkins TA, Emery B, Mulinyawe S, Barres BA (2008) Distinct stages of myelination regulated by gamma-secretase and astrocytes in a rapidly myelinating CNS coculture system. Neuron 60(4):555–569. https://doi.org/10.1016/j.neuron.2008.09.011

8. Barres BA (2008) The mystery and magic of glia: a perspective on their roles in health and disease. Neuron 60(3):430–440. https://doi.org/10.1016/j.neuron.2008.10.013

9. Brockes JP, Fields KL, Raff MC (1977) A surface antigenic marker for rat Schwann cells. Nature 266(5600):364–366

Chapter 3

Isolation and Expansion of Schwann Cells from Transgenic Mouse Models

Jihyun Kim and Haesun A. Kim

Abstract

The most widely used method (Brockes' method) for preparing primary Schwann cell culture uses neonatal rat sciatic nerves as the primary source of Schwann cells. The procedure is relatively simple and yields a highly purified population of Schwann cells in a short period of time. The method has also been used to prepare Schwann cells from mice, however, with limitation. For example, Brockes' method is not applicable when the genotypes of mouse neonates are unknown or if the mouse mutants do not develop to term. We described a method ideal for preparing Schwann cells in a transgenic/knockout mouse study. The method uses embryonic dorsal root ganglia as the primary source of Schwann cells and allows preparing separate, highly purified Schwann cell cultures from individual mouse embryos in less than 2 weeks.

Key words Mouse Schwann cells, Mouse embryos, Dorsal root ganglia

1 Introduction

The study of Schwann cell development and myelination has been facilitated by the availability to isolate and establish pure population of primary Schwann cells. In addition, transgenic mouse models and naturally occurring mouse mutants serve as increasing important tools for the study of Schwann cell biology. The most widely used method of preparing Schwann cell culture was originally developed by Brockes et al. [1], which uses sciatic nerves of newborn rats as the primary source of Schwann cells. The procedure is relatively simple and yields a highly purified population of Schwann cells in a short period of time. A similar method can be used to prepare Schwann cells from mouse sciatic nerves [2–6]; however, the use of multiple neonatal nerves limits the application of the method only to wild-type mice or mouse mutants that develop to term.

In this chapter, we describe a method for establishing a purified population of Schwann cells from a single mouse embryo (E12.5–13.5). The method was originally described in studies

Paula V. Monje and Haesun A. Kim (eds.), *Schwann Cells: Methods and Protocols*, Methods in Molecular Biology, vol. 1739,
https://doi.org/10.1007/978-1-4939-7649-2_3, © Springer Science+Business Media, LLC 2018

defining abnormalities in Schwann cells derived from mice deficient in neurofibromatosis 1 (*Nf1*) gene locus [7, 8]. The *Nf1*-null mutant mice die in utero between E11 and E15 prior to the formation of mature peripheral nerves [9, 10], making it impossible to use Brockes' method. Kim's method allows to establish separate Schwann cell cultures from individual mouse embryos and generates highly purified (>99.5%) mouse Schwann cells in large quantity in less than 2 weeks [11–20].

2   Materials

2.1 Solutions and Media

1. Leibovitz's L-15 medium.

2. Trypsin, 0.25% in Hanks' balanced salt solution without Ca^{++} and Mg^{++}.

3. *L-15/10%FBS:* L-15 supplemented with 10% fetal bovine serum.

4. *DRG plating medium*: Dulbecco's Modified Eagle's Medium (DMEM) with high glucose, 10% FBS, 2 mM L-glutamine, 50 ng/mL of nerve growth factor (NGF 2.5S). Pre-warmed to 37 °C before use.

5. *N2/NGF:* 1:1 ratio of DMEM and F-12 and N2 supplement, 50 ng/mL of NGF. Pre-warmed to 37 °C before use.

6. *Trypsin-collagenase solution*: 0.25% trypsin solution supplemented with 0.1% collagenase type 1.

7. *Monoclonal anti-Thy 1.2 solution*: Dilute anti-Thy 1.2 antibody 1:1000 in DMEM, and store 1 mL aliquots at −80 °C.

8. *Rabbit complement solution*: Reconstitute lyophilized rabbit complement in 5 mL of ddH$_2$O. Add 35 mL of DMEM and filter sterilize. Store 1 mL aliquots at −80 °C.

9. *Schwann cell medium:* DMEM, 10% FBS, 2 mM L-glutamine. Pre-warm to 37 °C before use.

10. *Schwann cell growth medium*: Schwann cell medium supplemented with 10 ng/mL of recombinant human neuregulin-β-1 EGF domain and 2.5 mM forskolin. Pre-warmed to 37 °C before use.

11. *N2-Schwann cell growth medium*: 1:1 ratio of DMEM and F-12 and N2 supplement. Add 10 ng/mL of recombinant human neuregulin-β-1 EGF domain and 2.5 mM forskolin. Pre-warmed to 37 °C before use.

12. *Poly-L-lysine solution*: Poly-L-lysine, 0.05 mg/mL prepared in 0.15 M sodium borate pH 8.0, filter sterilized.

13. Phosphate buffered saline (PBS).

2.2 Equipment

1. Scissors (1), 16.5 cm, standard, straight sharp/blunt tips.

2. Forceps (1), 12 cm, narrow pattern, straight.

3. Fine iris scissors (1), 8.5 cm, straight.

4. Graefe forceps (1), 10 cm, straight.

5. Spring scissors (1), 8.5 cm.

6. Forceps (1), Dumont curved tip, #7.

7. Forceps (2), Dumont fine tip, #5.

8. Dissecting microscope.

9. 100 mm Petri dishes (3).

10. 35 mm culture dishes (three per embryo harvested).

11. Eppendorf tubes (one per embryo harvested).

12. 15 mL conical tubes (two per embryo harvested).

13. Sterile glass Pasteur pipettes, cotton plugged.

14. Air incubator, 37 °C.

15. Tissue culture incubator, 37 °C, 90% humidity, 10% CO_2.

16. 27 $G^{1/2}$ needles, sterile.

17. Sterile glass Pasteur pipettes, cotton plugged.

---

3 Methods

3.1 DRG Dissection and Plating of the Dissociated DRG (Day 1)

3.1.1 Preparation of Embryos

1. Sterilize forceps and scissors by immersing them into 70% ethanol for 20 min. Air dry. Alternatively, insert the tips of the instruments into a hot bead sterilizer for 10–20 s.

2. Euthanize the pregnant mouse by CO_2 exposure. Place the mouse on its back and clean the abdomen area with 70% ethanol.

3. Using a set of scissors (16.5 cm) and forceps (12 cm), cut across the lower abdomen through the skin and the muscle layer. While being careful not to touch the internal organs, continue cutting up the left and right side to expose the internal cavity.

4. Using a set of fine iris scissors and Graefe forceps, remove the uterine horn and transfer in a sterile 100 mm Petri dish.

5. With spring scissors and curved forceps, cut across a uterine sac and push the embryo out. Separate the embryo from the placenta and amniotic membrane, and transfer it into a 100 mm Petri dish containing L-15. Continue until all the embryos are removed.

6. Label two sets of 35 mm culture dishes according to the number of embryos harvested (e.g., label the dishes E1, E2, E3, E4, E5 for five embryos removed). Fill each dish with 3 mL of L-15. Also prepare 1.5 mL Eppendorf tubes labeled accordingly.

Step 1: DRG dissection and dissociation

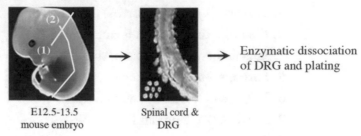

E12.5-13.5
mouse embryo

Spinal cord &
DRG

→ Enzymatic dissociation
of DRG and plating

Step 2: Separation of Schwann cell-neuron network

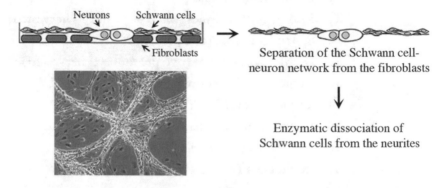

Neurons   Schwann cells

Fibroblasts

→ Separation of the Schwann cell-
neuron network from the fibroblasts

↓

Enzymatic dissociation of
Schwann cells from the neurites

Step 3: Purification and expansion of Schwann cells

Anti Thy 1.2-
complement-
mediated lysis
of fibroblasts →

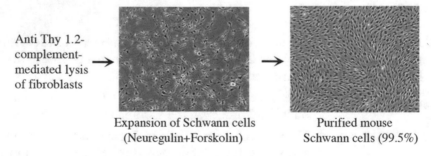

Expansion of Schwann cells
(Neuregulin+Forskolin)

Purified mouse
Schwann cells (99.5%)

Fig. 1 A schematic diagram of a procedure for isolating Schwann cells from a mouse embryo

7. Place each embryo in a separate 35 mm dish containing L-15 (*see* **Note 1**).

8. Place a dish containing an embryo under a dissecting microscope. Decapitate the embryo with spring scissors (Fig. 1 step 1, line 1), and transfer the head to a pre-labeled 1.5 mL Eppendorf tube prepared in **step 6**. Repeat the step for all the embryos. Place the Eppendorf tubes at 4 °C until ready to process the heads for preparing genomic DNA for genotyping.

9. Place the embryo on its side. Insert the tips of spring scissors vertically into the rostral end (where the head used to be), just anterior to the vertebral column, and begin making 3–4

sequential cuts from rostral to caudal (from the head to tail) direction (Fig. 1, step 1, line 2). Remove the ventral half portion of the embryo including all the limbs.

10. Lay the remaining dorsal tissue on its back so that the ventral side is facing up. Using two sets of fine forceps, remove any remaining organs (lungs, kidneys, aorta) from the ventral surface. The vertebral body should be visible in the middle of the dorsal plate.

11. Repeat **steps 9** and **10** for all the embryos (*see* **Note 2**).

3.1.2 Removal of the Spinal Cord

1. While gently holding the cord between forceps, use the other set to pinch off the ganglia from the dorsal root starting from one end of the spinal cord. Be careful not to remove any meninges from the spinal cord. Remove all the ganglia (approximately 37–42 ganglia/spinal cord/embryo) and discard the cord. Repeat the step for all the spinal cords.

3.1.3 DRG Dissociation

1. Gently swirl each dish to collect all ganglia toward the center of the dish. Under a dissecting microscope, remove the L-15 medium from the dish using a cotton-plugged glass Pasteur pipette while being careful not to remove any ganglia.

2. Add 1.5 mL of 0.25% trypsin to each dish. Place the dishes into a 37 °C air incubator for 15–20 min. Swirl dishes every 10 min. Meanwhile, label 15 mL conical tubes according to the number of embryos, and add 10 mL of L-15/10% FBS into each tube.

3. Transfer ganglia collected from individual embryo into a 15 mL conical tube containing 10 mL L-15/10% FBS.

4. Centrifuge the ganglia at $50 \times g$ for 5 min.

5. Remove the supernatant while being careful not to disturb the pellet. Add 10 mL of L-15/10% FBS. Centrifuge for 5 min as above.

6. Remove the supernatant and add 1 mL of DRG plating medium to each tube. Prepare narrow-bore glass Pasteur pipettes by flaming the tips. Triturate the pellet 10–15 times through the tip of a pipette or until an even cell suspension is obtained. Be careful not to generate too many air bubbles while triturating. Repeat the step for all the tubes and make sure to use separate pipette for each pellet.

7. Add additional 1 mL of DRG plating medium to the cell suspension.

8. Plate dissociated cells (2 mL) derived from a single embryo in an uncoated 35 mm culture dish. Place dishes in a tissue culture incubator (*see* **Note 3**).

3.2 Expansion of Schwann Cell Precursors and Separation of Schwann Cell-Neuron Network from the Fibroblasts (Day 2 to Day 8)

3.2.1 Expansion of Schwann Cell Precursors

1. Sixteen to eighteen hours after plating of the dissociated DRG, replace the medium with 2 mL of N2/NGF. Under the serum-free condition, fibroblast growth is suppressed, while axon-associated Schwann cell precursors continue to proliferate. There is no need to change the medium (*see* **Note 4**).

2. Meanwhile, prepare genomic DNA from each embryo head collected in **step 8** of the previous section, and genotype each embryo by PCR. Mark each dish with the appropriate genotype of the embryo from which the cells were derived.

3.2.2 Separation of Schwann Cell-Neuron Network from Fibroblasts (6–7 Days Following Plating of the Dissociated DRG)

1. Under a dissecting microscope, neuron-associated Schwann cells are visible as beaded structures along neurites. Axon bundles along the periphery of the dish should be visible under a dissecting scope. Starting from a corner of the plate where axonal bundles have formed, lift up the Schwann cell-neuron network using a 27 G$^{1/2}$ needle, and slowly peel it off toward the center of the plate. Usually, the neuronal network that is being peeled off can be distinguished from the fibroblast layer left behind under the microscope (Fig. 2b). If DRGs are initially dissociated well, almost complete separation of the Schwann cell-neuron network should be achieved from the underlying fibroblast layer (*see* **Note 5**).

2. After successfully detaching Schwann cell-neuron network from the plate, transfer the culture medium containing the floating neuronal network into a 15 mL conical tube. Collect all the Schwann cell-neuron networks from embryos of the same genotype into the same tube.

3.3 Mouse Schwann Cell Purification and Expansion (Day 8 to Day 14)

1. After collecting Schwann cell-neuron networks from all the genotypes, pellet the cells by centrifugation at 200 × g for 5 min.

2. Remove the supernatant, and resuspend the pellet in trypsin-collagenase solution in a volume of approximately 0.5 mL per embryo. For example, if cells from six embryos of the same genotype are pooled, 3 mL of the enzyme solution is used to dissociate Schwann cells. However, if cells from a single embryo are processed, a minimum amount of 1 mL is used.

3. Incubate cells at 37 °C for 30 min. Swirl tubes every 10 min.

4. Stop the enzyme reaction by adding Schwann cell medium. Pellet the cells as above.

5. Remove the supernatant. Resuspend the pellet in 1 mL of anti-Thy 1.2 antibody solution. Incubate for 30 min at 37 °C.

6. Pellet the cells by centrifugation and remove the supernatant. Resuspend the cells in 1 mL of rabbit complement solution. Incubate for 30 min at 37 °C.

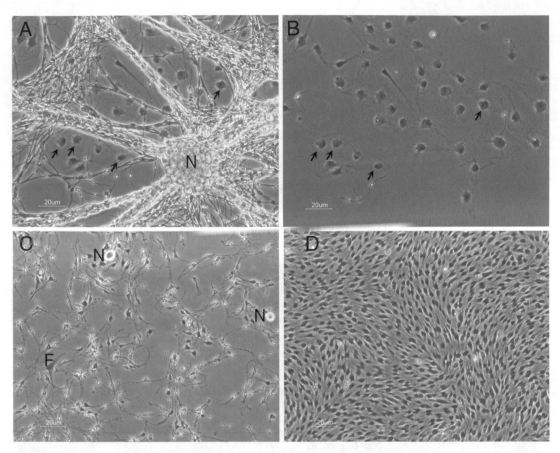

Fig. 2 Mouse Schwann cell purification. (**a**) Phase-contrast micrographs of mouse-dissociated DRG culture on Day 7. Neuronal cell bodies (N) and extended axons are seen. Schwann cell bodies are seen tightly associated with the axons. Large flat fibroblasts (arrows) are also present with no obvious association with the neurons. (**b**) Most of the fibroblasts are left behind after removal of the Schwann cell-neuron network. (**c**) Mouse Schwann cells plated after being dissociated from the neurons. Neuronal cell bodies (N) and contaminating fibroblasts (F) are also seen in the culture. (**d**) After being maintained for 2 days in a serum-free condition in the presence of neuregulin and forskolin, the culture is populated with Schwann cells with the characteristic bipolar, spindle-shaped cell bodies

7. Count the cell number. An average of 0.3×10^6 cells are obtained from a single embryo. Thus, if cells were pooled from three embryos, there should be about 1×10^6 cells. Plate cells from the same genotype onto a poly-L-lysine-coated 100 mm culture dish in Schwann cell medium. If less than 0.5×10^6 cells are obtained, plate the cells onto a 60 mm culture dish.

Preparing poly-L-lysine-coated plates:

(a) Add PBS to a tissue culture dish, just enough to wet the bottom of the dish. Remove the PBS.

(b) Add poly-L-lysine solution enough to cover the surface of the dish (3 mL for 100 mm dish). Leave at room temperature for 30 min.

(c) Remove the poly-L-lysine. Rinse the dish with PBS three times. After the final wash, leave PBS in the dish until ready to use.

8. Next day, bipolar, spindle-shaped Schwann cells should be visible with few neuronal cell bodies (Fig. 2c). Change the medium to N2-Schwann cell growth medium. Neuregulin and forskolin contained in the growth medium start to stimulate Schwann cell proliferation, while growth of any contaminating fibroblasts is suppressed under the serum-free condition. Neurons should die off within 2 days in the absence of NGF in the medium.

9. There is no need to change the medium until the cultures become confluent (6–7 days after the plating), at which point the cultures should contain 99–99.5% mouse Schwann cells (Fig. 2d). Replate the cells onto new PLL-coated plates at a density of 0.5×10^6 cells/100 mm plate. From this point on, the cells are grown in Schwann cell growth medium (*see* **Note 6**).

3.4 Freezing Cells for Storage

1. When cells are 80–90% confluent, wash cells twice with 5 mL of HBSS.

2. Add 5 mL of 0.25% trypsin-EDTA solution; incubate at 37 °C for 2–5 min.

3. Trypsinize the cells and add 5 mL of L-15/10% FBS. Collect cells in a 15 mL conical tube.

4. Centrifuge for 10 min, at $200 \times g$, at room temperature.

5. Discard supernatant and gently resuspend the cells in 10 mL of L-15/10% FBS; count cells using a hemocytometer.

6. Centrifuge and discard supernatant. Gently resuspend the cells in ice-cold freezing medium (10% DMSO, 90% FBS) with a final density of 2×10^6 Schwann cells/1 mL.

7. Make 1 mL aliquots in cryotubes, and freeze overnight at −80 °C.

8. Next day, transfer frozen aliquots to liquid nitrogen for long-term storage.

3.5 Thawing Schwann Cells

1. Quickly thaw a frozen aliquot in a 37 °C water bath.

2. Transfer the cell suspension to a 15 mL conical tube containing 9 mL of pre-warmed (37 °C) L-15/10% FBS. Centrifuge.

3. Discard supernatant and gently resuspend the cells in 8 mL of Schwann cell growth medium.

4. Plate cells onto one 100 mm poly-L-lysine-coated culture dish.

4   Notes

1. From this point on, embryos are processed individually.

2. If L-15 medium in the dish becomes too cloudy with blood, transfer the dorsal plate into a new 35 mm dish with fresh L-15.

3. Dissociated DRG cultures are composed of neurons, fibroblasts, and the Schwann cell precursors. It is important to use uncoated plastic culture dishes since the use of any substrate (collagen, Matrigel)-coated dishes makes the fibroblast become bundled up with neurons and the associated Schwann cells, making it difficult to separate the Schwann cell-neuron network from the underlying fibroblast layer.

4. Expect the following morphologies of the dissociated DRG culture after plating. *Twenty-four hours after plating*, most cells should have attached to the plate. *Forty-eight hours*, many neurons extend neurites. Schwann cell precursors beginning to make contact with neurites should be visible. However, the precursors do not show the characteristic spindle-shaped morphology of Schwann cell at this point. They look rather flat with fibroblast-like morphology. *Four days*, extensive neuronal network develops. Schwann cells greatly increase in number as they proliferate in contact with neurons. Fibroblasts appear as large flat cells with no obvious association with neurons (Fig. 2a). If everything goes well as it should, fibroblasts should remain quiescent without obvious increase in number under the serum-free condition. *Six to seven days*, virtually, the entire neurite network should be occupied by Schwann cells. Schwann cells begin to acquire spindle-shaped morphology as they mature in culture. DRG axons extend out to the periphery of the dish and begin to bundle up along the curvature of the dish.

5. If dense populations of fibroblasts are observed in the direction from which neurons are being peeled off, cut off the neuronal network at that point and start lifting the remaining network on the plate from a new point. Sometimes, fibroblasts become condensed heavily underneath neuronal cell bodies making the separation difficult at the region. In this case, the fibroblast-dense area can be excised from the plate using a surgical razor blade before proceeding with the separation procedure. It is crucial to obtain Schwann cell-neuron networks that are free of contaminating fibroblasts at this point, because removal of fibroblasts from the later Schwann cell cultures is more difficult.

6. As the passage number increases, mouse Schwann cells tend to become less responsive to growth factors and begin to senesce.

The demise of cells occurs faster in serum-free media than in serum-containing media. After switching to Schwann cell growth medium, the mouse Schwann cells can be grown up to passage 6–7 or frozen down and stored in liquid nitrogen for future use (*see* Subheading 3.4 for freezing cells).

References

1. Brockes JP, Fields KL, Raff MC (1979) Studies on cultured rat Schwann cells. I. Establishment of purified populations from cultures of peripheral nerve. Brain Res 165(1):105–118

2. Manent J, Oguievetskaia K, Bayer J, Ratner N, Giovannini M (2003) Magnetic cell sorting for enriching Schwann cells from adult mouse peripheral nerves. J Neurosci Methods 123(2): 167–173

3. Seilheimer B, Schachner M (1987) Regulation of neural cell adhesion molecule expression on cultured mouse Schwann cells by nerve growth factor. EMBO J 6(6):1611–1616

4. Shine HD, Sidman RL (1984) Immunoreactive myelin basic proteins are not detected when shiverer mutant Schwann cells and fibroblasts are co-cultured with normal neurons. J Cell Biol 98(4):1291–1295

5. Stevens B, Tanner S, Fields RD (1998) Control of myelination by specific patterns of neural impulses. J Neurosci 18(22):9303–9311

6. Zhang BT, Hikawa N, Horie H, Takenaka T (1995) Mitogen induced proliferation of isolated adult mouse Schwann cells. J Neurosci Res 41(5):648–654

7. Kim HA, Rosenbaum T, Marchionni MA, Ratner N, DeClue JE (1995) Schwann cells from neurofibromin deficient mice exhibit activation of p21ras, inhibition of cell proliferation and morphological changes. Oncogene 11(2):325–335

8. Kim HA, Ling B, Ratner N (1997) Nf1-deficient mouse Schwann cells are angiogenic and invasive and can be induced to hyperproliferate: reversion of some phenotypes by an inhibitor of farnesyl protein transferase. Mol Cell Biol 17(2):862–872

9. Jacks T, Shih TS, Schmitt EM, Bronson RT, Bernards A, Weinberg RA (1994) Tumour predisposition in mice heterozygous for a targeted mutation in Nf1. Nat Genet 7(3):353–361

10. Brannan CI, Perkins AS, Vogel KS et al (1994) Targeted disruption of the neurofibromatosis type-1 gene leads to developmental abnormalities in heart and various neural crest-derived tissues. Genes Dev 8(9):1019–1029

11. Ng A, Logan AM, Schmidt EJ, Robinson FL (2013) The CMT4B disease-causing phosphatases Mtmr2 and Mtmr13 localize to the Schwann cell cytoplasm and endomembrane compartments, where they depend upon each other to achieve wild-type levels of protein expression. Hum Mol Genet 22(8): 1493–1506

12. Ratner N, Williams JP, Kordich JJ, Kim HA (2005) Schwann cell preparation from single mouse embryos: analyses of neurofibromin function in Schwann cells. Methods Enzymol 407:22–33

13. Chen Z, Liu C, Patel AJ, Liao CP, Wang Y, Le LQ (2014) Cells of origin in the embryonic nerve roots for NF1-associated plexiform neurofibroma. Cancer Cell 26(5):695–706

14. De Vries GH, Boullerne AI (2010) Glial cell lines: an overview. Neurochem Res 35(12): 1978–2000

15. Miyamoto Y, Torii T, Takada S, Ohno N, Saitoh Y, Nakamura K et al (2015) Involvement of the Tyro3 receptor and its intracellular partner Fyn signaling in Schwann cell myelination. Mol Biol Cell 26(19):3489–3503

16. Miyamoto Y, Torii T, Kawahara K, Tanoue A, Yamauchi J (2016) Dock8 interacts with Nck1 in mediating Schwann cell precursor migration. Biochem Biophys Rep 6:113–123

17. Reuss DE, Habel A, Hagenlocher C, Mucha J, Ackermann U, Tessmer C et al (2014) Neurofibromin specific antibody differentiates malignant peripheral nerve sheath tumors (MPNST) from other spindle cell neoplasms. Acta Neuropathol 127(4):565–572

18. Torii T, Miyamoto Y, Takada S, Tsumura H, Arai M, Nakamura K et al (2014) In vivo knockdown of ErbB3 in mice inhibits Schwann cell precursor migration. Biochem Biophys Res Commun 452(3):782–788

19. Torii T, Miyamoto Y, Yamamoto M, Ohbuchi K, Tsumura H, Kawahara K et al (2015) Arf6 mediates Schwann cell differentiation and myelination. Biochem Biophys Res Commun 465(3):450–457

20. Wu J, Liu W, Williams JP, Ratner N (2017) EGFR-Stat3 signalling in nerve glial cells modifies neurofibroma initiation. Oncogene 36: 1669–1677

Chapter 4

Isolation, Culture, and Cryopreservation of Adult Rodent Schwann Cells Derived from Immediately Dissociated Teased Fibers

Natalia D. Andersen and Paula V. Monje

Abstract

Adult Schwann cell (SC) cultures are usually derived from nerves subjected to a lengthy step of predegeneration to facilitate enzymatic digestion and recovery of viable cells. To overcome the need for predegeneration, we developed a method that allows the isolation of adult rat sciatic nerve SCs immediately after tissue harvesting. This method combines the advantages of implementing a rapid enzymatic dissociation of the nerve fibers and a straightforward separation of cells versus myelin that improves both cell yield and viability. Essentially, the method consists of (1) acute dissociation with collagenase and dispase immediately after removal of the epineurium layer and extensive teasing of the nerve fibers, (2) removal of myelin debris by selective attachment of the cells to a highly adhesive poly-L-lysine/laminin substrate, (3) expansion of the initial SC population in medium containing chemical mitogens, and (4) preparation of cryogenic stocks for transfer or delayed experimentation. This protocol allows for the procurement of homogeneous SC cultures deprived of myelin and fibroblast growth as soon as 3–4 days after nerve tissue dissection. SC cultures can be used as such for experimentation or subjected to consecutive rounds of expansion prior to use, purification, or cryopreservation.

Key words Peripheral nerve, Teased fibers, Primary Schwann cell cultures, Myelin, Cryopreservation, Fibroblasts

1 Introduction

Schwann cell (SC) cultures are useful in vitro models to study SC biology associated with normal and abnormal nerve development, myelination, and regeneration [1, 2]. Procurement of large populations of SCs is relevant to therapeutic approaches aimed at promoting regeneration and myelination in the injured peripheral and central nervous systems [3–5].

Large-scale manufacturing of SC cultures was made possible after the discovery of critical soluble growth factors able to expand the cells over several passages in vitro [6–8]. Though SCs can be isolated from essentially any type of nerve at any stage of

Paula V. Monje and Haesun A. Kim (eds.), *Schwann Cells: Methods and Protocols*, Methods in Molecular Biology, vol. 1739,
https://doi.org/10.1007/978-1-4939-7649-2_4, © Springer Science+Business Media, LLC 2018

differentiation, the use of postnatal nerve tissues has been preferred due to the ease of isolation and high expansion potential of the SCs. In adult nerves, the large load of myelin and the presence of mature connective tissue layers make the separation of SCs versus myelin difficult without damaging the cells themselves. In addition, the presence of abundant connective tissue represents a substantial source of fibroblasts [6, 9] that can contaminate the SC cultures in the short or long term.

It has been shown that delaying enzymatic dissociation of the nerve explants by introducing an in vitro or in vivo pre-degeneration step results in increased cell yields and viability of the initial SC populations obtained from adult nerve digestion [6, 10–15]. However, a delayed dissociation may not be suitable for all applications, as the time needed for pre-degeneration can be considerable and the risk of fibroblast contamination can be increased. To eliminate the pre-degeneration step, we developed a reliable protocol that allows the rapid isolation of adult rat SCs by enzymatic digestion of teased nerve fibers immediately after tissue harvesting [16]. This method introduces a series of steps aimed at facilitating enzymatic dissociation, removing myelin and increasing cell yields and viability. First, the epineurium is mechanically removed, eliminating the most prevalent source of contaminating fibroblasts. Second, an extensive mechanical teasing of the nerve fibers is applied prior to incubation with proteolytic enzymes, which ensures a more complete digestion of the fibers and the release of highly viable individual cells. Third, the cells are plated in suspension as individual droplets onto a highly adhesive laminin substrate, which permits the fast separation of adherent SCs from floating myelin debris. This protocol yields highly viable populations consisting mainly of SCs with minimal fibroblast and myelin contamination. The SC cultures can be used or analyzed as soon as 3 days after plating or further expanded in medium containing chemical mitogens to increase the number of cells. The storage of the excess of cells as cryogenic stocks is made possible by following a simple protocol that does not cause detrimental effects on viability or biological activity after thawing and replating. With the proper modifications, this protocol is suitable for the isolation of SCs from different species and types of nerves. The minimally manipulated SC cultures obtained through this method can be used as such in experimentation. They can also be purified or cryopreserved at essentially any step of the process.

2   Materials

All materials, reagents, and solutions used for nerve tissue dissection and processing should be sterile and cell culture grade. Use ice-cold buffers and solutions during the steps of dissection, teasing of the nerve fibers, and maintenance of tissues at all times prior to enzymatic treatment.

2.1 Tissue Harvesting, Removal of the Epineurium, and Teasing of the Nerve Fibers

1. Female Sprague-Dawley rat, adult, ~3-month-old.
2. 70% (v/v) ethanol.
3. Surgical fine scissors with straight sharp and blunt tips.
4. Forceps with serrated tips and straight or curved ends.
5. Scalpel handle with a surgical blade.
6. Vannas spring scissors.
7. Dumont number 3, 4, and 5 forceps with straight tips.
8. 60 mm × 15 mm polystyrene cell culture dishes.
9. L-15 medium. Leibovitz's L-15 medium supplemented with 25 µg/ml gentamicin. Maintain the L-15 medium on ice prior to and during the steps of nerve isolation, cleaning and teasing of the nerve fibers.
10. Horizontal laminar flow hood.
11. Dissecting stereo microscope.
12. Fiber-optic illuminator. 150 W dual-gooseneck fiber-optic illuminator for cool light illumination.

2.2 Enzymatic Dissociation of the Nerve Fibers

1. High-glucose Dulbecco's Modified Eagle's Medium (DMEM) with phenol red, pH 7.2.
2. Enzymatic cocktail. Prepare a 10 × stock solution by dissolving 125 mg of dispase II (neutral protease) and 25 mg of type I collagenase in 5 ml of DMEM. Sterilize by filtration through 0.22 µm filters. Store the 10× stock in aliquots at −80 °C. The 10× stock can be stored frozen for at least 1 year without significant loss of enzymatic activity. At the moment of use, thaw the aliquots of the 10 × stock on ice, and prepare a working dilution in DMEM. This solution is referred to as 1× enzymatic cocktail.
3. CO_2 cell incubator. Set at 37 °C with a humid atmosphere containing 9% CO_2 (*see* **Note 1**).
4. Plastic transfer pipettes with fine tip.

2.3 Plating and Growth of SC Cultures

1. 15 ml round-bottom centrifuge tubes with snap cap, polypropylene.
2. Straight end glass Pasteur pipettes, 5 × 3 in., with a flame-narrowed tip.
3. FBS/HBSS medium: 40% (v/v) decomplemented fetal bovine serum (FBS) prepared in HBSS, pH 7.2.
4. Benchtop centrifuge. Low-speed centrifuge with swinging bucket rotor, set at 4 °C prior to use.

5. Disposable 10 ml serological pipets, polystyrene.

6. Pipet-Aid.

7. 100 mm × 20 mm polystyrene cell culture dishes.

8. Distilled water, cell culture grade.

9. Poly-L-lysine (PLL) stock and working solutions. Prepare a 100× stock solution of PLL by resuspending 500 mg of PLL (powder) in 12.5 ml of 1.9% (w/v) sodium tetraborate and 12.5 ml of 1.2% (w/v) boric acid. Prepare a 1× PLL working solution by diluting the 100× stock in distilled water. Aliquots of the 100× PLL stock solution can be stored at −20 °C for up to 1 year. The 1× PLL working solution can be prepared in advance and stored at 4 °C for up to 1 month.

10. Plastic paraffin film.

11. Dulbecco's phosphate-buffered saline (DPBS), pH 7.2.

12. Laminin stock and working solutions. The laminin stock solution consists of a commercially available sterile 1 mg/ml laminin from Engelbreth-Holm-Swarm murine sarcoma basement membrane, aliquoted and stored at −80 °C. Thaw the laminin aliquots slowly at 2–8 °C at the moment of use to avoid irreversible gelification. Prepare a working solution of laminin by diluting 55 μg of laminin stock solution in 10 ml of DPBS to coat one 100 mm cell culture dish. Scale the volume of the laminin working solution up or down according to the surface and the number of dishes used. It is recommended to use about 0.7–1 μg of laminin per cm² of coated surface.

13. PLL-/laminin-coated dishes. Add 1× PLL working solution to cover the bottom of a 100 mm cell culture dish, and incubate it for 1 h at room temperature. Wash the plate three times with distilled water, and let it dry under the laminar flow hood. Air-dried PLL-coated dishes properly sealed with plastic paraffin film can be stored for up to 1 month at 4 °C. To proceed with laminin coating, add 10 ml of laminin working solution per PLL-coated dish. Incubate the plate at room temperature for 1 h, or place it at 4 °C until use without removing the laminin solution. Prepare at least five PLL-/laminin-coated dishes for plating the cell material derived from the processing of two sciatic nerves (*see* **Note 2**).

14. HBSS. Hank's balanced salt solution, formulated without calcium or magnesium. Containing phenol red, pH 7.2.

15. Low proliferation medium. DMEM supplemented with 10% FBS, 1% (v/v) 200 mM L-alanyl-L-glutamine dipeptide in 0.85% NaCl, 1% penicillin-streptomycin, and 25 μg/ml gentamicin.

16. High proliferation medium. Low proliferation medium supplemented with 10 nM heregulin-β1177–244 and 2 μM forskolin.

17. CO_2 cell incubator.

18. Inverted phase contrast microscope.

2.4 Expansion and Passaging of SC Cultures

1. Low and high proliferation media.

2. Disposable 5 and 10 ml serological pipettes, polystyrene.

3. HBSS.

4. Trypsin/EDTA solution. At the moment of use, prepare a working solution of trypsin/EDTA in ice-cold HBSS from a 10× stock consisting of 0.5% trypsin/EDTA. Maintain the stock and working solutions of trypsin/EDTA on ice at all times.

5. Inverted phase contrast microscope.

6. 50 ml polypropylene conical-bottom centrifuge tubes.

7. Benchtop centrifuge.

8. Digital cell counter or hemocytometer.

9. Trypan blue solution: 0.4% (w/v) trypan blue in 0.81% sodium chloride and 0.06% potassium phosphate dibasic solution (optional).

10. PLL-/laminin-coated 100 mm cell culture dishes.

11. CO_2 cell incubator.

2.5 Cryopreservation Thawing, and Plating of SC Stocks

1. Low and high proliferation media.

2. Trypsin/EDTA working solution and HBSS.

3. 15- and 50-ml polypropylene conical-bottom centrifuge tubes.

4. Inverted phase contrast microscope.

5. Benchtop centrifuge.

6. Cell counter.

7. Trypan blue solution (optional).

8. Freezing medium: 10% (v/v) dimethyl sulfoxide (DMSO) prepared in FBS (*see* **Note 3**).

9. Cryogenic vials, polypropylene, 2 ml, round bottom, self-standing.

10. Freezing container. Polycarbonate container with a tube holder, filled with 100% (v/v) isopropyl alcohol (*see* **Note 4**).

11. −80 °C laboratory freezer.

12. Liquid nitrogen tank.

13. Containers with wet ice and dry ice.

14. PLL-/laminin-coated dishes.

15. CO_2 cell incubator.

3   Methods

The protocol described below has been optimized for the processing of two sciatic nerve biopsies obtained from an adult female Sprague-Dawley rat. The schematic diagram and representative images presented in Figs. 1 and 2, respectively, are provided to illustrate the methodological steps in sequential order. Figure 3 contains representative images of typical SC cultures obtained via the drop-plating method. Proper scaling up or down of this protocol may be necessary according to the experimental needs and the availability of resources. The use of tissues from rats of a different gender or strain should render similar results. Optimization of this protocol for other types of nerves and species may be needed.

3.1   Tissue Harvesting, Removal of the Epineurium, and Teasing of the Nerve Fibers

Work inside a laminar flow, and implement good practices in the use of sterile technique for the procurement, handling, and use of cells and tissues. This is imperative to reduce the time of operation and avoid contamination in this and all subsequent steps. Prepare an excess of sterile forceps and other dissection tools for the fast replacement of instrumentation while performing the steps of tissue dissection and teasing of the nerve fibers. The risk of contamination can be minimized by the use of antibiotics in all media formulations, but pertinent precautions should be taken at all times to minimize exposure to potential sources of contamination. The use of antifungal reagents such as amphotericin B is not recommended for SC cultures, as this drug impairs cell survival. Prepare all reagents, materials, and equipment beforehand, and start the procedure in the early afternoon of day 0 so as to adhere to the digestion times recommended in Subheading 3.2 (*see* Fig. 1).

1. Euthanize an adult rat according to the regulations set forth by institutional policies for animal care and use.

2. Immediately after, place the rat body with its dorsal side facing upward on a clean surface. Thoroughly douse the hind limb area with 70% ethanol to disinfect the skin giving at least 1 min of skin contact so as to minimize the risk of contaminating the cell preparations. Next, make an incision in the skin located at the end of the abdominal zone (in the posterolateral hind limb area) using surgical fine scissors.

3. Lift and grasp the skin using serrated forceps. Extend the cut forward from the site of incision parallel to the femur and toward the knee by using surgical scissors (*see* **Note 5**).

4. Using a scalpel handle with a surgical blade, make a superficial cut to divide the muscle parallel to and just inferior to the femur. This procedure will expose the sciatic nerve.

5. Place the tip of the Vannas spring scissors right below the sciatic nerve, and proceed to open both ends to detach the nerve

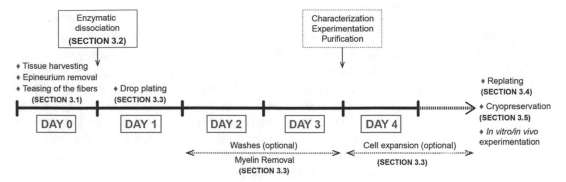

Fig. 1 Schematic diagram depicting the isolation of SCs from immediately dissociated teased nerve fibers. The timeline summarizes the sequence of events in the isolation protocol highlighting each section described in the chapter. Dotted lines represent undetermined or optional steps. This method allows the preparation of viable SC cultures with minimal myelin and fibroblast contamination. The cultures can be used for experimentation, purification, or cryopreservation prior to or after being subjected to expansion

from the surrounding muscular and adipose tissue. Next, cut the most proximal end of the nerve while holding it with a Dumont number 5 forceps. Then, cut the distal end of the nerve at the level of the knee.

6. Place the nerve segment in a 60 mm dish containing ice-cold L-15 medium immediately after dissection.

7. Proceed to dissect the sciatic nerve from the contralateral side as described above.

8. Prepare the laminar flow hood by placing a dissecting stereo microscope and fiber-optic illuminator in a conveniently located spot.

9. (Optional) While working under the laminar flow hood, rinse the nerve segments by transferring them into new 60 mm dishes containing ice-cold L-15 medium to remove blood and non-peripheral nerve tissues loosely attached to the nerve explant.

10. Remove any excess tissue from the sciatic nerves by using Dumont number 3 or 4 forceps while observing under the dissecting stereo microscope (*see* **Note 6**).

11. Transfer the clean nerve tissue into a new 60 mm dish containing ice-cold L-15 medium.

12. Mechanically remove the epineurium layer as one single sheath. Use Dumont number 4 or 5 forceps to grasp the nerve fibers at the proximal end, and subsequently pull the epineurium layer toward the distal end (*see* **Note 7**). Repeat this procedure with the contralateral nerve.

13. Discard the epineural layers, and collect the fibers from both nerves in a new 60 mm dish containing ice-cold L-15 medium.

14. Tease the nerve fibers with the aid of Dumont number 5 forceps by pulling off individual nerve fascicles prior to separating them into fibers of decreased caliber. The teasing step is repeated until all fascicles are separated as finely as possible into individual fibers regardless of the initial caliber of the fascicles (*see* Fig. 2a). This procedure is carried out while observing the tissue at the highest possible magnification under the dissection stereo microscope (*see* **Note 8**).

15. Carefully transfer the teased nerve fibers into a new 60 mm dish containing ice-cold L-15 medium using Dumont number 5 forceps for final inspection. At this stage, visually monitor the fiber preparation at high magnification under the dissecting

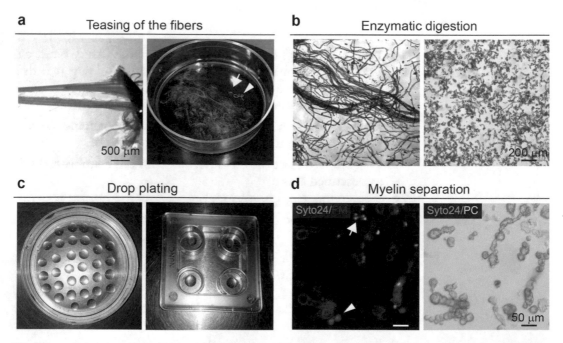

Fig. 2 Nerve teasing, dissociation, and plating of SCs obtained from immediately dissociated nerves. (**a**) Low-magnification view of the nerve teasing step (left). Macroscopic visualization of the teased fibers (*right*). An incomplete teased fiber (*arrowhead*) and segment of non-peripheral nerve tissue (arrow) were separated towards the side of the dish for further teasing and removal, respectively. (**b**) Phase contrast images showing dissociated fibers at 3 (left) and 18 h (right) after addition of the enzymatic cocktail. (**c**) Drop plating of the cell suspension onto PLL-/laminin-coated dishes. A variety of cell culture dishes can be used according to the experimental needs, such as 100 mm plates (left) for cell expansion and multiwell plates (right) for analysis. (**d**) End product obtained immediately after dissociation and plating. The fluorescence microscopy and phase contrast (PC) images show abundant cell nuclei stained with Syto24 (green) and floating myelin labeled with FluoroMyelin™ (FM, red). These preparations contain myelin-free cells (arrow) and myelin-associated cells (arrowhead)

stereo microscope to ensure the absence of non-teased fibers or excess tissue. Continue to mechanically tease and wash the fibers by transferring them to new L-15-containing dishes as many times as needed. Frequent addition or replacement of the L-15 medium is recommended to avoid warming of the media and to reduce the risk of contamination.

3.2 Enzymatic Dissociation of the Nerve Fibers

The enzymatic dissociation of the nerve fibers is the most important step that affects cell yields and viability. The temporal course of dissociation needs to be controlled and optimized for each enzymatic preparation particularly if mature nerves with fully developed connective tissue layers are used. We recommend the use of an enzymatic cocktail solution composed of type I collagenase and dispase II, as reported previously [6]. Increasing the digestion time or the concentration of the enzymes can increase cell recovery but may compromise the viability of the isolated cells. For best results, we suggest the experimenter to optimize this step by monitoring the progression of the digestion (*see* Fig. 2b) and the viability of the cells throughout the time course of digestion [16].

1. Remove the L-15 medium from the 60 mm dish containing the teased fibers by using a plastic transfer pipette being careful not to disturb the fibers.

2. Add 3 ml of 1X enzymatic cocktail to the teased fibers derived from both nerves ensuring that the fibers are completely immersed in this solution.

3. Incubate the nerve fibers with the enzymatic cocktail overnight (on average 15–18 h) in a CO_2 cell incubator (*see* **Note 9**).

4. (Optional) Monitor the progression of the enzymatic treatment and the isolation of viable cells by phase contrast (*see* Fig. 2b) and fluorescence microscopy intermittently during the digestion step (*see* **Note 9**).

3.3 Plating and Growth of SC Cultures

The drop plating of the cell suspensions onto a laminin-coated surface in the presence of chemical mitogens enhances the viability and growth rate of the cells. Plating the suspensions as small droplets allows SCs to rapidly attach to the surface and thus separate from the floating myelin. The combination of soluble heregulin and forskolin synergistically enhances SC proliferation and reduces fibroblast growth [8]. This section describes the protocol used for cell plating immediately after enzymatic dissociation of the teased nerve fibers. Proliferating SC cultures containing minimal myelin and fibroblast contamination can be obtained within 3 days post-plating.

1. Before cell plating, remove the laminin solution from each dish, wash the dish with distilled water, and let it dry in the laminar flow hood up until no traces of liquid are observed by visual inspection. The plates typically dry within 30 min.

2. Transfer the digested nerve fibers into a 15 ml round-bottom tube by using a glass Pasteur pipette. Gently rinse the dish twice by using 5 ml of FBS/HBSS medium to stop the enzymatic digestion and increase the recovery of cells.

3. Collect the cells by low-speed centrifugation at $200 \times g$ for 10 min at 4 °C.

4. Remove the supernatant by using a glass Pasteur pipette, and mechanically resuspend the cell pellet by adding 5–10 ml of low proliferation medium. Gently and smoothly pipet the cell suspension up and down using a glass Pasteur pipette until it becomes homogeneous and no clumps are visible (*see* **Note 10**).

5. Centrifuge the cell suspension at $200 \times g$ for 5 min at 4 °C (*see* **Note 11**).

6. Remove the supernatant, and resuspend the cells in 4–5 ml of high proliferation medium for plating onto PLL-laminin-coated dishes.

7. Plate the cell suspension as discrete 30 μl drops regularly distributed throughout the surface of a PLL-/laminin-coated dish. Usually, 20–30 drops are plated per 100 mm dish as shown in Fig. 2c.

8. Carefully transfer the culture dishes to a CO_2 cell incubator, and incubate the cells overnight for at least 15–18 h without disturbing the drops (*see* **Note 12**).

9. The following day, use a phase contrast microscope to confirm that the cells have attached properly. Gently and smoothly fill each 100 mm dish with 8–10 ml of high proliferation medium (*see* **Note 13**).

10. Feed the cultures with high proliferation medium three times per week. Monitor the progression of the culture on a daily basis with a phase contrast microscope. The increase in cell density should become apparent as soon as 2 days after plating.

11. (Optional) Determine the viability (*see* Fig. 3a), identity, and purity (*see* Fig. 3c–e) of the cultures by means of immunostaining using antibodies against SC-specific markers such as p75^NGFR, O4, S100, and glial fibrillary acidic protein (GFAP), alone or together with Thy-1, which is a fibroblast-specific marker (*see* **Note 14**). Representative images of SC cultures immunostained with antibodies for the abovementioned markers are presented in Fig. 3c–e. Myelin contamination can be determined by staining with FluoroMyelin™ (Fig. 3b).

12. Use the SC cultures for experimentation, or further expand the cells in high proliferation medium as described in Subheading 3.4. The use of an intermediate cell purification step is optional, as the percentage of contaminating Thy-1-positive cells usually does not exceed 10% of the population at this stage.

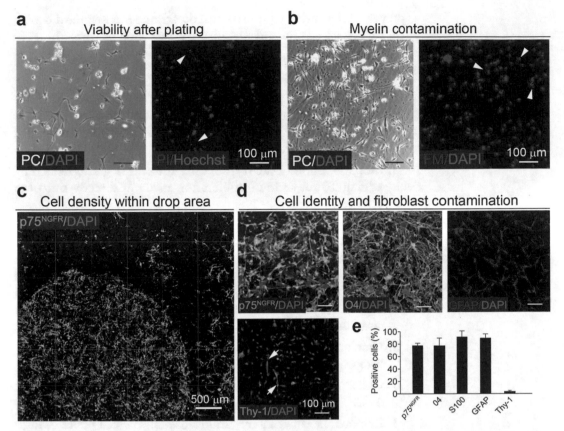

Fig. 3 Viability and phenotypic characterization of adherent SC cultures. (**a**) Viability of adherent SCs 24 h after plating. In the phase contrast (PC) image (left), note the numerous cells exhibiting a typical SC morphology. In the fluorescence microscopy image (right), note the low percentage of dead cells (arrowheads) labeled with propidium iodide (PI, red). All nuclei are counterstained with Hoechst (blue). (**b**) Myelin contamination 72 h post-plating. Staining with FluoroMyelin™ (FM, red) and DAPI (blue) reveals a high proportion of cells versus myelin debris (arrowheads). (**c–d**) Confirmation of SC phenotype through immunostaining 72 h post-plating. The low-magnification image (**c**) reveals the distribution and density of p75^{NGFR}-positive SCs within the area initially delimited by the drop. The high-magnification images (**d**) reveal the high purity and typical morphology of individual cells, as evidenced by the expression of markers specific to SCs (p75^{NGFR}, O4, and GFAP) and fibroblasts (Thy-1), respectively. (**e**) Fibroblast contamination 72 h post-plating. Quantitative results of a typical immunofluorescence analysis after staining the cultures with SC- and fibroblast-specific antibodies. Note the low percentage of fibroblasts as judged by the proportion of Thy-1-positive cells (arrows)

3.4 Expansion and Passaging of SC Cultures

The primary SC cultures obtained by this method can be passaged and further expanded to obtain sufficient numbers of cells for experimentation (*see* **Note 15**). Monitor the cultures daily by phase contrast microscopy to prevent overgrowth of the populations within the area originally restricted by the individual droplets (*see* Fig. 3c). Given the fact that the available surface area is limited and the cells actively divide in proliferation medium, the cultures can reach confluence within 4–5 days after plating. At this time the cultures need to be used for experimentation or passaged to prevent cell loss due to substratum detachment. A standard protocol for cell expansion and subculture is provided below:

1. Once the cultures reach confluence within local areas, remove the culture medium and rinse the cultures with 10 ml of HBSS.

2. Add 5 ml of trypsin/EDTA working solution to the cells. Incubate the plates for 2–5 min at room temperature while monitoring the cells under the phase contrast microscope for signs of cell detachment (*see* **Note 16**).

3. Stop the action of the trypsin once the cells begin to detach by adding 10 ml of low proliferation medium directly onto the cultures.

4. Collect the cell suspension and transfer it to a 50 ml centrifuge tube. Rinse the dish with 5 ml of low proliferation medium to help detach and collect remnant cells. Use a 10 ml plastic pipette to transfer the cells while pipetting up and down several times (not more than 6–8 times) during the collection and transfer steps. This helps to both detach the cells from the dish and dissociate the small clumps.

5. Confirm the effectiveness of trypsinization and collection by inspection of the culture plate by phase contrast microscopy.

6. Collect the cells by centrifugation at $200 \times g$ for 10 min at 4 °C.

7. Remove the supernatant, and resuspend the cell pellet in high proliferation medium for replating onto PLL-/laminin-coated dishes.

8. (Optional) Count the cells and estimate the percentage of cell viability according to the method of choice (*see* **Note 17**).

9. Plate the cells as homogeneously dispersed cells (*see* **Note 18**).

10. Place the dishes inside a CO_2 cell incubator immediately after seeding for improved attachment and growth.

11. Replace the medium two to three times per week with high proliferation medium until the cultures reach confluency.

12. Repeat **steps 1–11** in case further expansion is desired.

3.5 Cryopreservation Thawing, and Platingof SC Cultures

Primary SC cultures can be cryopreserved without losing their distinctive biochemical and functional properties [16]. In this section, the steps necessary to prepare cells for cryogenic storage and their subsequent replating after thawing are described in detail. It is recommended to use healthy SC cultures harvested in the logarithmic or exponential phase of growth for optimal results. Standard safety practices in cell cryopreservation should be used to minimize the risk of contamination and exposure of the SCs to the toxic effects of the DMSO.

*3.5.1 Preparation
and Storage
of Cryogenic Stocks*

1. Prepare a single cell suspension of SCs by harvesting the SC cultures via trypsinization and collection by centrifugation, essentially as described in Subheading 3.4 (**steps 1–6**).

2. Resuspend the cell pellet in low proliferation medium, count the cells, and determine their viability by the method of choice.

3. Spin the cells at $200 \times g$ for 10 min at 4 °C.

4. Discard the supernatant, and gently resuspend the cell pellet in ice-cold freezing medium at a density of $1–2 \times 10^6$ cells/ml. Gently resuspend the cells up and down using a 10 ml pipette once or twice to obtain a homogeneous cell suspension prior to aliquoting.

5. Aliquot 1.5 ml of the cold cell suspension directly into properly labeled cryogenic vials placed on ice. Work fast as this is a time- and temperature-sensitive step that can seriously affect the survival and recovery of the cells.

6. Immediately transfer the cryovials to an isopropanol-filled freezing container to be placed in a −80 °C freezer.

7. Twenty-four hours after, transfer the cryovials to a liquid nitrogen tank for long-term storage.

*3.5.2 Plating of Cells
from Cryogenic Stocks*

1. For thawing frozen stocks, transfer the cryogenic vials directly from the liquid nitrogen to a safe container filled with dry ice for transportation to the laminar flow hood.

2. Thaw the cells quickly by placing the vials in a 37 °C water bath up until 70% of the volume is melted.

3. Transfer the cell suspension (1.5 ml) to a 50 ml centrifuge tube containing at least 15 ml of ice-cold low proliferation medium. Diluting the freezing medium at 1:10–1:20 with low proliferation medium is recommended to reduce exposure of the SCs to the DMSO (*see* **Note 19**).

4. Collect the cells by centrifugation at $200 \times g$ for 10 min at 4 °C.

5. Remove the supernatant, and immediately add 10 ml of high proliferation medium to resuspend the cell pellet. Gently pipette the cells up and down 2–3 times using a 10 ml pipette up until no cell clumps are observed.

6. Plate the cell suspension in a 100 mm PLL-/laminin-coated dish, incubate inside a CO_2 cell incubator, and allow for expansion as described in Subheading 3.4 (*see* **Note 20**).

4   Notes

1. SCs are highly sensitive to alkalization of the culture medium with a reduction of cell viability. Incubation of the cells in an 8–9% CO_2 atmosphere is preferred to prevent pH changes

when bicarbonate-buffered media formulations such as DMEM are used. It is of utmost relevance to work fast and avoid excessive exposure to the airflow, as this can lead to media alkalization. The use of phenol red-containing media and buffers is recommended to easily monitor pH levels based on color changes.

2. Sequential coating of plates with PLL and laminin provides a better substrate for adhesion and survival of the initial populations when compared to plates coated with PLL alone. Laminin coating may not be required, but it is preferable to maximize the rate of cell proliferation.

3. Commercially available freezing media for mammalian cell cultures such as Recovery™ (Gibco) can be used with nearly identical results in cell viability after thawing.

4. Commercially available freezing containers such as Mr. Frosty™ (Nalgene) placed in a −80 °C standard freezer are useful to deliver the appropriate cooling rate for optimal cryopreservation of SCs.

5. Removing the skin in an inverted V shape is useful to expose the posterior hind limb area up until reaching the knee. This area of the knee, also known as the popliteal fossa, is where the sciatic nerve branches into the fibular or peroneal, tibial, and sural nerves.

6. Remove any muscle, fat, or blood vessels that remain attached to the nerves with Dumont number 3 or 4 forceps. These tissues can increase the load of contaminating cells into the SC cultures if they are not efficiently removed prior to enzymatic dissociation.

7. The epineurium is the outermost layer of connective tissue that surrounds the nerve and constitutes the main source of contaminant fibroblasts. The percentage of fibroblasts in the final cell preparations can be strongly decreased by mechanical removal of this layer simply by pulling it from one extreme while holding the fascicles from the opposite side. Once separated from the fibers, this layer can be recognized on the basis of its whitish color and elastic properties conferred by the high content of collagen. The epineurium layer can also be collected and placed individually into a dish containing ice-cold L-15 for independent digestion and isolation of the constituent cells, as described in [16].

8. The teasing step consists of repeated pulling off the individual nerve fascicles up until fibers of the least possible caliber are obtained. Collectively, the steps of dissection, epineurium removal, and teasing of one nerve biopsy can be completed within a time frame of 30–45 min. The processing time is determined mostly by the technical skills of the operator.

9. The enzymatic digestion can be regularly monitored under the phase contrast microscope to visually follow the progression of tissue digestion. Enzymatic treatment is typically performed overnight to maximize the effectiveness of digestion and ensure cell release without the need for mechanical dissociation. However, the action of the enzymes may be halted at an earlier stage (e.g., 3–4 h after treatment) as soon as most of the individual myelinating cells get released from their association to extracellular matrix components. A determination of cell viability by means of staining with fluorescent dyes can prove useful at this and other steps of the process. This can be achieved by means of co-staining with propidium iodide, which selectively labels the nuclei of dead cells, and Syto24 (green), which labels the nuclei of all cells (reference control) and visualization by fluorescence microscopy (*see* Fig. 2d). Typically, assessment of propidium iodide incorporation immediately after overnight dissociation renders less than 10% propidium iodide-positive cells [16].

10. Mechanical dissociation is of relevance to reduce cell clumping and produce a single cell suspension for plating. Immediately isolated SCs have a tendency to adhere to each other and form aggregates of various sizes that may not adhere to the substrate. Caution should be taken at this step given the fact that excessive mechanical dissociation can compromise the viability of the cells.

11. Floating myelin debris and cell clumps can be observed by phase contrast and fluorescence microscopy using combined staining with a general nuclear marker and a selective myelin stain, such as FluoroMyelin™ (*see* Fig. 3b). If the digested cell suspension does not look homogeneous and tissue fragments or large cell clumps are observed, the steps of mechanical dissociation and centrifugation can be repeated. An optional filtering step using 70 μm nylon filters might be performed prior to centrifugation to remove myelin debris. However, avoid excessive manipulation and mechanical dissociation so as not to compromise the viability of the cells. Cell counting prior to myelin removal is challenging due to interference with myelin autofluorescence. We have found that fluorescence staining with the nuclear stains Syto24 and propidium iodide (red) is a useful combination to estimate the yields and viability during and after dissociation [16].

12. The drop-plating method allows the recovery of high densities of adherent cells per unit area. It contributes to the survival of the adherent cells by limiting fluid shearing motion while preventing the attachment of myelin debris. Plating the cells as drops in 100 mm dishes is recommended for in vitro expansion,

but plating in other formats is preferred for other uses such as determination of purity and viability (*see* Fig. 2c) [16].

13. At this point, cells begin to attach and extend processes on the laminin substrate, while most myelin debris remains floating. The separation of myelin from the viable adherent cells represents a major advantage of the drop-plating method. Gentle washes with DMEM can be performed to remove loosely attached myelin and other cellular impurities in this and all subsequent steps. Avoid the use of vacuum to aspirate the culture medium. Instead, use a glass Pasteur pipette with a thin tip to manually remove/replace the medium. It is advisable to leave a small volume of medium in each plate so as to prevent cell detachment by the flow created with the addition of new medium.

14. At 3 days post-plating, most SCs remain viable and display a typical elongated, spindle-shaped morphology. At this stage, most of the cells are actively proliferating and express the SC markers S100, p75^{NGFR}, GFAP, and O4 (cell surface sulfatide). The percentage of Thy-1-expressing cells should be less than 5% (*see* Fig. 3d–e) [16]. See Chapter 6 for a standard immunostaining protocol using live or fixed SC cell cultures.

15. SCs typically acquire a bipolar morphology and align along each other by forming a pattern reminiscent of a fingerprint. Adult rat SC cultures can be passaged up to 5–8 times without losing their morphology, growth properties, and ability for in vitro differentiation [16]. Extended culture can lead to an increase in the percentage of fibroblasts after 2–3 rounds of expansion.

16. Mechanical tapping of the dish usually helps to accelerate cell detachment. Check the progression of the dissociation by phase contrast microscopy, and stop the trypsinization as soon as detachment is observed. Confluent SC cultures typically detach as bundles. This is a time-sensitive step that is determined empirically in each case as assessed by microscopic observation. Avoid over-trypsinization of the cells, as this reduces cell survival.

17. Once the cells are in suspension, cell counting is useful to determine an appropriate dilution for use or replating. Viability assays can be performed using trypan blue exclusion assays [16]. This method should render a percentage of live cells higher than 90–95% if the time and magnitude of trypsinization are strictly monitored.

18. It has been estimated that confluent cultures can render ~100–150 × 10³ cells/cm². Subculture the cells by replating the cell suspensions in high proliferation medium at a ratio of 1:3–1:10 according to experimental needs.

19. SCs are particularly sensitive to the presence of DMSO in the freezing medium. Therefore, it is essential to remove the freezing medium as soon as possible after thawing of the stocks. A dilution of 1:10–1:20 with low proliferation medium is recommended in order to reduce the cryoprotector concentration and minimize cytotoxicity prior to centrifugation and replating.

20. Monitor the condition of the cells by phase contrast microscopy 2–3 h after plating. At this time, it is possible to observe some cells attached to the substrate extending branched processes, particularly if PLL-/laminin-coated dishes are used. Lack of attachment may be indicative of poor viability or a problem with the substrate. It is recommended to change the medium the following day after plating in case dead cells or excessive debris are observed.

Acknowledgments

We thank the technical assistance provided by S. Srinivas, G. Piñero, B. Kuo, and K. Bacallao and the editorial assistance provided by K. Ravelo. We thank J. Guest for critically reviewing the manuscript. This work was supported by NIH-NINDS (NS084326), The Craig Neilsen Foundation (M1501061), The Miami Project to Cure Paralysis, and The Buoniconti Fund. The authors declare no conflicts of interest with the content of this article.

References

1. Jessen KR, Mirsky R, Lloyd AC (2015) Schwann cells: development and role in nerve repair. Cold Spring Harbor Perspect Biol 7(7):a020487. https://doi.org/10.1101/cshperspect.a020487

2. Brosius Lutz A, Barres BA (2014) Contrasting the glial response to axon injury in the central and peripheral nervous systems. Dev Cell 28(1):7–17. https://doi.org/10.1016/j.devcel.2013.12.002

3. Tetzlaff W, Okon EB, Karimi-Abdolrezaee S, Hill CE, Sparling JS, Plemel JR, Plunet WT, Tsai EC, Baptiste D, Smithson LJ, Kawaja MD, Fehlings MG, Kwon BK (2011) A systematic review of cellular transplantation therapies for spinal cord injury. J Neurotrauma 28(8):1611–1682. https://doi.org/10.1089/neu.2009.1177

4. Li J, Lepski G (2013) Cell transplantation for spinal cord injury: a systematic review. Biomed Res Int 2013:786475. https://doi.org/10.1155/2013/786475

5. Bunge MB (2016) Efficacy of Schwann cell (SC) transplantation for spinal cord repair is improved with combinatorial strategies. J Physiol. https://doi.org/10.1113/jp271531

6. Morrissey TK, Kleitman N, Bunge RP (1991) Isolation and functional characterization of Schwann cells derived from adult peripheral nerve. J Neurosci 11(8):2433–2442

7. Porter S, Clark MB, Glaser L, Bunge RP (1986) Schwann cells stimulated to proliferate in the absence of neurons retain full functional capability. J Neurosci 6(10):3070–3078

8. Rahmatullah M, Schroering A, Rothblum K, Stahl RC, Urban B, Carey DJ (1998) Synergistic regulation of Schwann cell proliferation by heregulin and forskolin. Mol Cell Biol 18(11):6245–6252

9. Bunge MB, Wood PM, Tynan LB, Bates ML, Sanes JR (1989) Perineurium originates from fibroblasts: demonstration in vitro with a retroviral marker. Science (New York, NY) 243(4888):229–231

10. Casella GT, Bunge RP, Wood PM (1996) Improved method for harvesting human Schwann cells from

mature peripheral nerve and expansion in vitro. Glia 17(4):327–338. https://doi.org/10.1002/(SICI)1098-1136(199608)17:4<327::AID-GLIA7>3.0.CO;2-W

11. Verdu E, Rodriguez FJ, Gudino-Cabrera G, Nieto-Sampedro M, Navarro X (2000) Expansion of adult Schwann cells from mouse predegenerated peripheral nerves. J Neurosci Methods 99(1–2):111–117

12. Mauritz C, Grothe C, Haastert K (2004) Comparative study of cell culture and purification methods to obtain highly enriched cultures of proliferating adult rat Schwann cells. J Neurosci Res 77(3):453–461. https://doi.org/10.1002/jnr.20166

13. Haastert K, Mauritz C, Chaturvedi S, Grothe C (2007) Human and rat adult Schwann cell cultures: fast and efficient enrichment and highly effective non-viral transfection protocol. Nat Protoc 2(1):99–104. https://doi.org/10.1038/nprot.2006.486

14. Pietrucha-Dutczakv M, Marcol W, Francuz T, Golka D, Lewin-Kowalik J (2014) A new protocol for cultivation of predegenerated adult rat Schwann cells. Cell Tissue Bank 15(3):403–411. https://doi.org/10.1007/s10561-013-9405-x

15. Niapour N, Mohammadi-Ghalehbin B, Golmohammadi MG, Amani M, Salehi H, Niapour A (2015) Efficacy of optimized in vitro predegeneration period on the cell count and purity of canine Schwann cell cultures. Iran J Basic Med Sci 18(3):307–311

16. Andersen ND, Srinivas S, Pinero G, Monje PV (2016) A rapid and versatile method for the isolation, purification and cryogenic storage of Schwann cells from adult rodent nerves. Sci Rep 6:31781. https://doi.org/10.1038/srep31781

Chapter 5

Detailed Protocols for the Isolation, Culture, Enrichment and Immunostaining of Primary Human Schwann Cells

Tamara Weiss, Sabine Taschner-Mandl, Peter F. Ambros, and Inge M. Ambros

Abstract

This chapter emphasizes detailed protocols for the effective establishment of highly enriched human Schwann cell cultures and their characterization via immunostaining. The Schwann cells are isolated from immediately dissociated fascicle tissue and expanded prior to purification. Two purification methods are described that use either fluorescence-activated cell sorting for the Schwann cell marker TNR16 (p75^{NTR}) or a less-manipulative two-step enrichment exploiting the differential adhesion properties of Schwann cells and fibroblasts, which is especially useful for low Schwann cell numbers. In addition, a method to determine Schwann cell purity via stained cytospin slides is introduced. Together with an immunofluorescence staining procedure for the combined analysis of extra- and intracellular markers, this chapter provides a solid basis to study human primary Schwann cells.

Key words Human Schwann cell culture, Schwann cell isolation, Schwann cell enrichment, Cytospin preparation, Immunofluorescence analysis

1 Introduction

Several protocols to culture primary Schwann cells have been introduced over the last decades. However, these predominantly involved the use of Schwann cells obtained from rats. To address the increasing interest in human Schwann cells, this chapter describes refined protocols for the fast generation and immunofluorescence-based characterization of highly pure human Schwann cell cultures. These include procedures for (1) Schwann cell isolation and (2) enrichment, (3) determination of purity, and (4) immunostaining [1, 2] alongside with representative images of expected results.

An ex vivo or in vivo predegeneration phase, usually 8–14 days, of nerve tissue has been shown to increase the cell yield by allowing myelin clearance, dedifferentiation and proliferation of Schwann cells to take place [3–6]. The here presented protocol omits this

Paula V. Monje and Haesun A. Kim (eds.), *Schwann Cells: Methods and Protocols*, Methods in Molecular Biology, vol. 1739,
https://doi.org/10.1007/978-1-4939-7649-2_5, © Springer Science+Business Media, LLC 2018

time-consuming predegeneration step as recent studies demonstrated that viable and proliferative primary Schwann cell cultures can be obtained after immediate dissociation of both rat and human nerve tissue [1, 7, 8]. Another reason supporting an immediate processing of human nerve tissue is the early onset of senescence in human Schwann cells [1, 9]. Alongside with viability, highly pure Schwann cell cultures are essential to achieve robust results from downstream applications. This need encouraged different approaches for the separation of Schwann cells from contaminating fibroblasts that would otherwise progressively outnumber the Schwann cells with increased culture time. Methodological advances replaced the common fibroblast elimination procedures using antimitotic chemicals [10, 11] or complement-mediated cytolysis against fibroblast marker THY1 [10, 12] by more selective and effective approaches. One of them takes advantage of the capacity of rat Schwann cells to metabolize D-valine [8]. As fibroblasts lack the D-amino acid oxidase (DAO) [13], D-valine-substituted medium was demonstrated to successfully select for rat Schwann cells, while most fibroblasts died because of the depletion of this essential amino acid [8]. The attempt to apply this method for human Schwann cell cultures was abandoned as our expression data indicated that DAO is barely expressed in human primary Schwann cells [1]. The "cold-jet" technique is another purification method reported to significantly increase rat and human Schwann cell purity [14, 15]. Yet, personal experience showed that this method was not efficient in detaching human Schwann cells under the tested experimental conditions (unpublished observations). Hence, the first enrichment protocol described here represents a straightforward, cost- and time-effective alternative for the enrichment of human Schwann cells. This two-step procedure exploits the differential adhesion properties of Schwann cells and fibroblasts and results in Schwann cell cultures of about 96% purity without causing substantial cell loss [1]. In addition, targeted enrichment methods by magnetic- or fluorescence-activated cell sorting (MACS or FACS) for Schwann cell-specific proteins such as TNR16 (NGFR or p75[NTR]) or CADM3 (NECL1) have been shown to achieve highly pure Schwann cell cultures obtained from embryonic and adult rat tissue [7, 16, 17]. These procedures are suitable when high Schwann cell numbers are available because the effectiveness of MACS and FACS comes at the expense of increased cell loss. Thus, the second Schwann cell enrichment protocol described below involves a fast, FACS-based purification using labeled antibodies for p75[NTR] that yields in over 99% pure human Schwann cells.

After enrichment, information on Schwann cell purity is commonly obtained by flow cytometry, which requires the sacrifice of a high number of cells. Another strategy is the immunofluorescence staining analysis of sister cultures, but they may differ in purity. In addition, Schwann cells tend to grow upon fibroblasts, which impede proper discrimination of the two cell types particularly in areas of high cell density. To overcome these obstacles, this chapter involves the preparation and immunofluorescence staining of cytospin slides enabling the determination of Schwann cell purity on a single-cell level [1]. This method is of special benefit for approaches with limited cell numbers as well as routine applications dependent on a consistent readout procedure. Moreover, cytospins can be prepared after the exposure of Schwann cells to different experimental conditions and stored at −20 °C, which allows to perform further immunofluorescence analyses at a later time point.

Finally, an immunofluorescence staining protocol for the combined analysis of intra- and extracellular markers/proteins is described. This procedure not only saves precious Schwann cells but also enables the discovery of cell subpopulations by the generation of multidimensional protein expression and localization data on single cells. Of note, the herein provided protocols may be readily implemented with Schwann cells from other species.

2   Materials

2.1   Isolation and Dissociation of Human Nerve Fascicle Tissue

2.1.1   Fascicle Isolation

1. Tissue culture hood.
2. Supplemented MEMα: MEMα (Gibco Life Technologies, #32561-029) containing 10 U/ml penicillin, 0.1 mg/ml streptomycin, 25 mM 4-(2-hydroxyethyl)-1-piperazineethanesulfonic acid (HEPES), 1 mM sodium pyruvate, and 10% (w/v) fetal bovine serum (FBS).
3. Sterile, disposable plastic forceps (Cole-Parmer, #DF8088S).
4. Glass petri dishes, sterilized.
5. Safety scalpel.

2.1.2   Dissociation of Fascicle Tissue

1. Sterile, disposable syringe filters (0.2 μm).
2. Standard CO_2 incubator: 37 °C with 5% CO_2 in a humidified atmosphere.
3. Dissociation solution: supplemented MEMα containing 0.125% (w/v) collagenase type IV, 1.25 U/ml dispase II, and 3 mM calcium chloride ($CaCl_2$). The addition of $CaCl_2$ increases collagenase enzyme activity.

2.2   Culture of Dissociated Fascicle-Derived Cells

2.2.1   Coating of Wells

1. Standard CO_2 incubator.
2. Tissue culture hood.
3. 1× Dulbecco's phosphate-buffered saline (PBS) without calcium and magnesium.
4. Laminin working solution: Thaw the laminin stock solution overnight at 4 °C and aliquot it to avoid repeated freezing-

thawing cycles. Prepare a 4 µg/ml laminin working solution in ice-cold PBS. Store the laminin working solution at 4 °C up to 14 days.

5. Poly-L-lysine (PLL) working solution: Prepare a 0.01% (w/v) PLL solution by reconstituting PLL hydrobromide (70–150 kDa) in sterile bi-distilled water. Store at 4 °C for up to 6 months.

2.2.2 Seeding of Dissociated Fascicle Cells

1. Schwann cell expansion medium (SCEM): MEMα medium containing 20 mM HEPES, 10 U/ml penicillin, 0.1 mg/ml streptomycin, 1 mM sodium pyruvate, 1% (w/v) FBS, 0.5× (w/v) N-2 supplement, 10 ng/ml heregulinβ-1 (PeproTech, #100-03), 10 ng/ml FGF-basic, 5 ng/ml PDGF-AA, and 2 µM forskolin. Store SCEM at 4 °C for up to 7 days.

2. Benchtop centrifuge with 15 ml conical centrifuge tube inserts set up at room temperature (RT).

3. Tissue culture hood.

4. Standard CO_2 incubator.

2.3 Enrichment and Seeding of Schwann Cells

2.3.1 Two-Step Schwann Cell Enrichment Procedure

1. Tissue culture hood.

2. Standard CO_2 incubator.

3. Phase contrast microscope.

4. Sterile tissue culture-treated 6-well plates.

5. Accutase™, 4 °C. Ready-to-use proprietary formulation that contains proteolytic and collagenolytic enzymes. Recommended for use in replacement of trypsin.

6. 1000 µl micropipette.

7. Supplemented MEMα.

8. Benchtop centrifuge.

2.3.2 Fluorescence-Activated Cell Sorting for the Schwann Cell Marker p75[NTR]

1. Accutase™, pre-warmed to 37 °C.

2. Supplemented MEMα.

3. AF647-labeled p75[NTR] antibody: Prior to use, label 1 µg p75[NTR] antibody (Cell Signaling, #8238) with Zenon Rabbit IgG Labeling Kit Alexa Fluor 647 (Life Technologies, #Z-25308) according to the manufacturer's protocol.

4. FACS buffer: PBS containing 0.1% (w/v) BSA and 0.05% (w/v) sodium azide. Store at 4 °C for up to 1 month.

5. 5 ml round-bottom tubes with cell strainer snap cap (Corning, #352235).

6. Cell sorter. Standard cell sorter equipped with a 640 nm laser excitation line. Used to sort and detect AF647-labeled Schwann cells in the allophycocyanin (APC) channel (670 ± 15 nm).

7. Tissue culture hood.

8. Benchtop centrifuge with 15 and 50 ml conical centrifuge tube inserts, set up at 4 °C.

2.3.3 Seeding
of Enriched Schwann Cells

1. μ-Slides 8 Well (Ibidi, #80826). These 8-well chamber slides provide excellent optical quality for (confocal) immunofluorescence analysis.

2. PLL/laminin-coated culture dishes. Prepared as needed for experimentations. Coated culture dishes can be stored covered with PBS at 37 °C for at least 2 days.

3. Cell counter.

4. SCEM.

5. Tissue culture hood.

6. Standard CO_2 incubator.

2.4 Preparation and Immunofluorescence Staining of Cytospin Slides for the Determination of Schwann Cell Purity

2.4.1 Preparation of Cytospin Slides

1. Microscope slides.

2. Cytocentrifuge (Shandon Cytospin 2 or equivalent model for the preparation of cytospins with a diameter of 5 mm).

3. Cytofunnels (Thermo Scientific, #A78710018).

4. Filter cards (Thermo Scientific, #5991022).

5. Microscope slide boxes.

6. Sterile, disposable syringe filter (0.4 μm).

7. Sodium chloride (NaCl) solution: Prepare a 0.9% (w/v) NaCl solution in bi-distilled water.

2.4.2 Immunofluorescence Staining of Cytospin Slides

1. 4% stabilized paraformaldehyde (PFA) solution, phosphate-buffered, pH 7,4 °C.

2. Bovine serum albumin (BSA) working solution: Prepare a 1% (w/v) BSA solution in PBS. Ensure complete dissociation. Filter-sterilize through a 0.4 μm syringe filter and store in aliquots at −20 °C for up to 6 months.

3. Swine and goat serum. It is recommended to use serum derived from the host species which served to generate the secondary antibody.

4. Blocking solution 1: BSA working solution containing 1.5% (w/v) swine serum and 1.5% (w/v) goat serum.

5. 0.1% Triton-X100 solution: BSA working solution containing 0.1% (w/v) Triton-X100. After adding Triton-X100 to the BSA working solution, rotate the tube for 30 min. Store at −20 °C for up to 6 months.

6. S100 plus vimentin primary antibody mix: Prepare a 1:200 dilution of rabbit anti-human S100 (Dako, #Z0311) and a 1:300 dilution of chicken anti-human vimentin (Millipore, #AB5733) in 0.1% Triton-X100 solution containing 0.5% (w/v) swine serum and 0.5% (w/v) goat serum.

7. FITC plus AF647 secondary antibody mix: Prepare a 1:50 dilution of fluorescein (FITC)-labeled swine anti-rabbit (Dako, #F0205) and a 1:300 dilution of AF647-labeled goat anti-chicken (Life Technologies, #SA5-10073) in 0.1% Triton-X100 solution containing 0.5% (w/v) swine and 0.5% (w/v) goat serum.

8. 4′,6-Diamidino-2-phenylindole (DAPI) working solution: Prepare a 2 μg/ml DAPI solution in PBS. Store at 4 °C for up to 6 months.

9. Mounting medium (VECTASHIELD, Vector Laboratories).

10. Coplin jars.

11. Humidity chamber.

12. Cover glasses.

13. Blotting paper.

14. Liquid repellent pen.

15. Glue.

16. Fluorescence microscope with light source and filters enabling detection of FITC (Ex495/Em519), A647 (Ex650/Em665) and DAPI (Ex358/Em461).

2.5 Immuno-fluorescence Staining of Schwann Cells Grown in 8-Well Chamber Slides

1. PFA solution.

2. PBS.

2.5.1 Fixation of Cells

2.5.2 Simultaneous Staining for Intracellular S100 and Vimentin

1. Permeabilization solution 1: BSA working solution containing 0.3% (w/v) Triton-X100, 1.5% (w/v) swine serum, and 1.5% (w/v) goat serum.

2. S100 plus vimentin primary antibody mix.

3. FITC plus AF647 secondary antibody mix.

4. DAPI working solution.

5. Mounting medium (Fluoromount-G, SouthernBiotech).

6. Confocal microscope.

2.5.3 Sequential Staining for Extracellular p75^{NTR} and Intracellular Vimentin

1. Blocking solution 2: BSA working solution containing 3% (w/v) swine serum.

2. p75^{NTR} primary antibody solution: Dilute rabbit anti-human p75^{NTR} (Cell Signaling, #8238S) at 1:300 in BSA working solution containing 1% (w/v) of swine serum.

3. FITC secondary antibody solution: Dilute FITC-labeled swine anti-rabbit at 1:50 in BSA working solution containing 1% (w/v) of swine serum.

4. Permeabilization solution 2: BSA working solution containing 0.3% (w/v) Triton-X100 and 3% (w/v) goat serum.

5. Vimentin primary antibody solution: Dilute chicken anti-human vimentin at 1:300 in intracellular antibody solution containing 1% (w/v) goat serum.

6. AF647 secondary antibody solution: Dilute AF647-labeled goat anti-chicken at 1:300 in intracellular antibody solution containing 1% (w/v) goat serum.

7. DAPI working solution.

8. Mounting medium (Fluoromount-G, SouthernBiotech).

9. Confocal microscope allowing excitation and detection of FITC, AF647 and DAPI (e.g., a laser scanning confocal microscope equipped with a 405 nm diode laser and a white-light laser).

3 Methods

The sections below provide a detailed description of isolation and purification methods previously used to generate highly enriched human Schwann cell cultures for transcriptomic and proteomic analyses [1]. Alongside, the cytospin preparation and immunostaining protocol allow for a reliable and fast determination of Schwann cell purity, which requires only a small number of cells. Furthermore, the provided immunofluorescence staining procedure is specifically adapted to but not exclusive for human Schwann cells grown on microscope chamber slides.

3.1 Isolation and Dissociation of Human Nerve Fascicle Tissue

All steps are carried out under sterile conditions in a tissue culture hood.

3.1.1 Fascicle Isolation

1. Transfer a fresh piece of human peripheral nerve into a sterile glass petri dish and cover it with supplemented MEMα (*see* **Note 1**).

2. Cut the nerve into 3-cm-long pieces and use disposable plastic forceps to pull the fascicles out of the epineurium (*see* **Note 2**).

3. Collect the isolated fascicles in a separate glass dish containing supplemented MEMα (*see* Fig. 1a).

3.1.2 Dissociation of Fascicle Tissue

1. Place 4–5 fascicles per well of a 6-well plate and add 2 ml of dissociation solution per well.

2. Cut the fascicles into about 5 mm fragments and incubate them in a standard CO_2 incubator for about 20 h.

3. Proceed with coating of the wells (*see* Subheading 3.2.1).

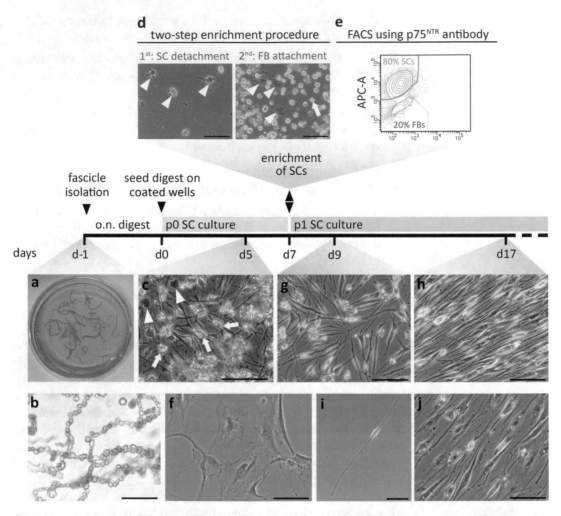

Fig. 1 Scheme for human Schwann cell cultivation and enrichment. (**a**) Isolated fascicles from the human nerve to be digested immediately after harvesting. (**b–d**) and (**f–j**) show phase contrast images. (**b**) Dissociated fascicle tissue after overnight incubation with dissociation enzymes. (**c**) A p0 culture consisting of fibroblasts (arrowheads) and Schwann cells with bi- to multipolar processes and myelin inclusions (arrows). (**d**) Two-step enrichment procedure that includes (1) a short Accutase™ treatment, which primarily detaches Schwann cells while most fibroblasts remain attached (arrowheads), and (2) the removal of remaining fibroblasts by selective attachment to plastic (arrowheads), while Schwann cells stay floating in the medium (arrow). (**e**) Enrichment via FACS using the Schwann cell marker p75NTR. A representative FACS plot shows that the p0 culture consists of 80% Schwann cells (SCs) and 20% fibroblasts (FBs). (**f**) Cultured p1 fibroblasts displaying a broad and flattened morphology. (**g–h**) p1 Schwann cells exhibiting the typical parallel alignment after 10 days in culture. (**i**) A single Schwann cell displaying the expected spindle-shaped morphology. (**j**) A culture of senescent Schwann cells. Note the enlarged and granulated cytoplasm and dark appearance of the nuclei. Black bars represent 100 µm

4. The next day, mechanically dissociate the fascicle tissue into a suspension (*see* Fig. 1b) using a 1000 µl micropipette (*see* **Note 3**).

5. Proceed to seed the fascicle-derived cells as described in Subheading 3.2.2.

3.2 Culture of Dissociated Fascicle-Derived Cells

Coating of culture dishes with PLL and laminin is widely used to ensure the attachment, survival, and growth of Schwann cells [1, 7, 8]. PLL provides positively charged attachment sites and therefore enhances binding of the negatively charged cell membranes. Laminins are recognized by Schwann cell-expressed integrins [18]. The cells are grown in medium that contains growth factors and additives known to promote the survival and proliferation of Schwann cells [3, 15, 19–22], while a minimal serum concentration decelerates fibroblast growth [1, 23, 24].

All the following steps are carried out under sterile conditions in a tissue culture hood.

3.2.1 Coating of Wells

1. Prepare as many wells of a 6-well plate as used during the step of fascicle tissue dissociation.

2. Add 800 µl of PLL working solution per well for 10 min at RT (*see* **Note 4**).

3. Wash the wells carefully with sterile bi-distilled water and let them dry in the tissue culture hood for 2 h.

4. Add 1 ml of laminin working solution per well and incubate the plate in a standard CO_2 incubator overnight or for at least 6 h.

5. Wash the PLL/laminin-coated wells twice with PBS.

6. Add 1.5 ml of SCEM to each well and store the plate in a standard CO_2 incubator until the cells are ready for seeding.

3.2.2 Seeding of Fascicle-Derived Cells

1. Transfer the fascicle suspension of each well into a separate 15 ml conical centrifuge tube. Add 8 ml of supplemented MEMα to stop the activity of dissociation enzymes.

2. Centrifuge the tubes at $200 \times g$ for 5 min at RT, aspirate the supernatant, and resuspend each cell pellet in 1 ml SCEM. Proceed to count the cells if an initial cell number is required (*see* **Note 5**).

3. Add the resuspended cells from each tube to a separate PLL/laminin-coated well preincubated with 1.5 ml SCEM (*see* **Note 6**). Transfer the plate into a standard CO_2 incubator.

4. Do not disturb the culture for 3 days before checking the cells. The primary cell culture is referred to as "passage zero" (p0).

5. Replace half of the medium every 3 days and remove as much floating myelin debris as possible (*see* **Note 7**).

6. Proceed with Schwann cell enrichment when the p0 culture has reached about 90% confluency (*see* Subheading 3.3). Each p0 well is expected to contain $4–7 \times 10^5$ Schwann cells. Consider to coat the 8-well chamber slides with PLL/laminin for immunofluorescence analysis the day before (as described in Subheading 3.2.1). Coat additional culture dishes as needed for experimentation. Scale up or down the volume of coating solution according to the size of the culture surface used.

3.3 Enrichment and Seeding of Schwann Cells

The choice of any given Schwann cell enrichment method is dependent on the number and purity of the cells required for experimentation. In Subheadings 3.3.1 and 3.3.2, two alternative methods of Schwann cell enrichment are described that have been optimized for the use with human Schwann cells. The two-step enrichment procedure (Subheading 3.3.1) results in Schwann cell cultures of about 96% purity and causes no substantial cell loss (*see* **Note 8**). The FACS-based enrichment procedure (Subheading 3.3.2) achieves a Schwann cell purity of over 99%, but yields are substantially lower (approximately 60% recovery rate).

3.3.1 Two-Step Schwann Cell Enrichment Procedure

All steps are carried out under sterile conditions in a tissue culture hood.

1. Transfer the conditioned SCEM of a p0 well to a 15 ml conical centrifuge tube and keep it at RT.

2. Wash the cells carefully with PBS.

3. Add 1 ml of ice-cold Accutase™ to the p0 well and incubate the cells for 2 min at RT.

4. Perform the first enrichment step as follows. First, tilt the plate to about 20°, absorb the Accutase™ with a 1000 μl micropipette, and apply a constant stream of Accutase™ to wash off primarily the Schwann cells (*see* **Note 9**). Second, repeat this procedure until all Schwann cells are detached from the surface. Use a phase contrast microscope to monitor the process and to confirm that most fibroblasts are still attached (*see* Fig. 1d, arrowheads). Third, transfer the cell suspension into a 15 ml conical centrifuge tube containing 8 ml of supplemented MEMα. Wash the well with 1 ml of supplemented MEMα and add it to the tube.

5. Repeat the first enrichment step for the remaining p0 wells. Collect the conditioned SCEM from each p0 well in one tube and keep it at RT until used for seeding; *see* Subheading 3.3.3.

6. Centrifuge the tube containing the cell suspension at $200 \times g$ for 5 min at RT.

7. Remove the supernatant and resuspend each cell pellet in 2 ml of supplemented MEMα.

8. Perform the second enrichment step as follows. First, transfer each cell suspension into a separate well of a 6-well plate (*see* **Note 6**). Second, incubate the cells for 30 min in a standard CO_2 incubator. Use a phase contrast microscope to monitor that the fibroblasts have attached to the surface (*see* Fig. 1d arrowheads) while most Schwann cells are still in suspension (*see* Fig. 1d arrow) (*see* **Note 10**).

9. Harvest and pool the medium with floating Schwann cells in a 15 ml conical centrifuge tube.

10. Add 1 ml of supplemented MEMα to each well with attached fibroblasts. Use a phase contrast microscope to monitor weakly attached Schwann cells that appear as small, round cells with a shiny white rim (*see* Fig. 1d, arrow). Absorb the medium with a 1000 µl micropipette and apply a constant stream of medium to wash off the weakly attached Schwann cells. Transfer them to the tube containing the Schwann cell suspension (*see* **Note 11**).

11. For seeding of the enriched Schwann cells, proceed as indicated in Subheading 3.3.3.

3.3.2 Fluorescence-Activated Cell Sorting for the Schwann Cell Marker p75^{NTR}

1. Transfer the conditioned SCEM of p0 cultures to a 15 ml conical centrifuge tube and keep it at RT until used for seeding; *see* Subheading 3.3.3.

2. Wash the cells carefully with PBS.

3. Add 1 ml of pre-warmed Accutase™ per p0 well and incubate the cells for 5 min at RT.

4. Transfer the detached p0 cells from each well in a separate 15 ml conical centrifuge tube and complete the volume to 10 ml with supplemented MEMα.

5. Centrifuge the tubes at $200 \times g$ for 5 min at RT and aspirate the supernatant.

6. Resuspend each cell pellet in 1 ml ice-cold PBS and pool the cell suspensions in a 15 ml conical centrifuge tube.

7. Centrifuge the tube at $300 \times g$ for 5 min at 4 °C.

8. Aspirate the supernatant and resuspend the pellet in the suitable volume of ice-cold FACS buffer (*see* **Note 12**).

9. Dilute the p75^{NTR}-AF657 antibody at 1:200 in the cell suspension. Vortex and incubate the tube in the dark for 30 min at 4 °C.

10. Wash the cells with 2 ml of ice-cold FACS buffer and centrifuge the tube at $300 \times g$ for 5 min at 4 °C.

11. Aspirate the supernatant and resuspend the cell pellet in the suitable volume of FACS buffer (*see* **Note 12**).

12. Absorb the cell suspension using a 1000 µl pipette and gently press the pipette tip onto the nylon mesh of a cell strainer snap cap attached to a 5 ml round-bottom tube. Slowly but constantly pass the cell suspension through the cell strainer snap cap to collect a single-cell suspension in the 5 ml round-bottom tube.

13. Transfer the 5 ml round-bottom tube into a 50 ml conical centrifuge tube and spin down the single-cell suspension at $400 \times g$ for 1 min (*see* **Note 13**).

14. Resuspend the cell pellet and use a cell sorter to purify the p75^{NTR}-AF647-positive Schwann cells. Collect the sorted Schwann cells in a sterile 5 ml round-bottom tube filled with 2 ml of FBS. A representative FACS plot in Fig. 1e shows the

partitioning of p75^{NTR}-AF647-positive Schwann cells and p75^{NTR}-AF647-negative fibroblasts.

15. For seeding of the enriched Schwann cells, proceed as indicated in Subheading 3.3.3.

3.3.3 Seeding
of Enriched Schwann Cells

1. Centrifuge the tube containing the enriched Schwann cell suspension and the tube containing the conditioned SCEM at $300 \times g$ for 5 min at RT.

2. Aspirate the supernatant of the Schwann cell-containing tube and resuspend the cell pellet in the suitable volume of conditioned SCEM (*see* **Note 14**).

3. Calculate the cell number of the enriched Schwann cell suspension using a cell counter.

4. Take an aliquot containing 3000 cells. Add conditioned SCEM to complete the volume to 100 µl and store on ice. This sample is to be used for the preparation of cytospin slides.

5. Add conditioned SCEM and fresh SCEM at a 1:1 ratio to PLL/laminin-coated dishes and proceed to seed the Schwann cells. The following seeding scheme is recommended, but the number of cells may be adapted according to the planned experiments. Seed 4×10^4 cells per well of an 8-well chamber slide (in 250 µl SCEM) for immunofluorescence analysis. Alternatively, seed 1×10^5 cells per well of a 12-well plate (in 1.5 ml SCEM) and 2.5×10^5 cells per well of a 6-well plate (in 3 ml SCEM) for other experimental approaches.

6. Use the aliquot of enriched Schwann cells to prepare two cytospin slides for the determination of Schwann cell purity as described in Subheading 3.4.

7. Replace half of the medium of p1 Schwann cells every 2–3 days (*see* **Note 15**).

3.4 Preparation and Immunofluorescence Staining of Cytospin Slides for the Determination of Schwann Cell Purity

The preparation of cytospins allows to distinguish between different cell populations within a sample on a single-cell level. Here, Schwann cells are discriminated from fibroblasts on the basis of their specific expression of S100. Vimentin is used as a dual marker for Schwann cells and fibroblasts as it is expressed at high levels in both cell types.

3.4.1 Preparation
of Cytospin Slides

The following steps have been established with the Shandon Cytospin 2 cytocentrifuge. If equivalent models are used, adapt the procedure according to the cytocentrifuge user guide.

1. Assemble two cytofunnels with a filter card and a microscope slide each.

2. Pre-wet the microscope slide by adding 50 µl of the NaCl solution to each cytofunnel and run the cytocentrifuge at $41 \times g$ for 1 min.

3. Load each cytofunnel with 50 μl of the aliquot of enriched Schwann cells. Then, add 50 μl of the NaCl solution. Run the cytocentrifuge at $23 \times g$ for 7 min.

4. Carefully detach the cytofunnel and the filter card from the microscope slide without damaging the fresh cytospin.

5. Label the cytospin slides and let them air-dry for 30 min.

6. Perform the immunofluorescence staining for S100 and vimentin as described in Subheading 3.4.2. Alternatively, store the cytospin slides in a microscope slide box at −20 °C and stain them at a later time point.

3.4.2 Immunofluorescence Staining of Cytospin Slides

The following steps are carried out at RT unless noted otherwise. The PBS wash involves two 5-min incubation steps in a Coplin jar filled with fresh PBS. For incubation steps with blocking and antibody solutions, apply a volume of 100 μl per cytospin slide.

1. If using frozen slides, thaw these inside the closed microscope slide box for 25 min. Proceed directly to **step 2** if fresh slides are used.

2. Immerse the slides for about 5 s in a PBS-filled Coplin jar. From now, do not allow the cytospins to dry out.

3. Remove the excess PBS by tapping the slides onto a paper towel.

4. Transfer the slides to a Coplin jar filled with ice-cold PFA solution for fixation. Close the lid and incubate them at 4 °C for 10 min.

5. Wash cytospin slides with PBS and remove the excess of liquid.

6. Transfer the slides into a humidity chamber. Use a liquid repellent pen to draw a circle with 5 mm distance around the area of cytospinned cells. This barrier will confine the applied solutions to the area of cytospinned cells. Immediately add the blocking solution 1 and incubate the slides in a closed humidity chamber for 20 min.

7. Remove the blocking solution and incubate the cells with the S100 plus vimentin primary antibody mix in the closed humidity chamber for 1 h at RT or overnight at 4 °C.

8. Wash the slides with PBS and remove the excess of liquid.

9. Incubate the cells with the FITC plus AF647 secondary antibody mix for 1 h in the closed humidity chamber.

10. Wash slides with PBS and remove the excess liquid.

11. Add 100 μl of the DAPI working solution per cytospin slide and incubate cells for 2 min.

12. Wash the slides three times with PBS and remove the excess liquid.

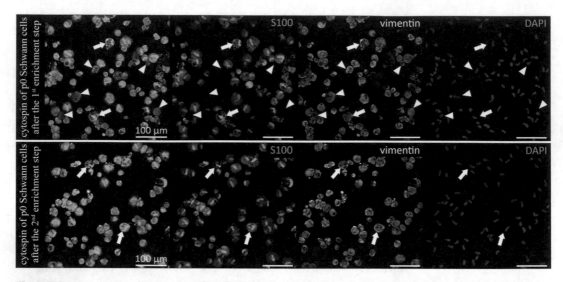

Fig. 2 Cytospins of p0 cultures stained for S100 and vimentin. The upper panel shows a representative cytospin of a p0 culture after the first enrichment step. Schwann cells are positive for S100 and vimentin, whereas fibroblasts are negative for S100 and positive for vimentin (arrowheads). The lower panel illustrates that Schwann cells are highly enriched after the second enrichment step. Note the typical elongated Schwann cell nuclei of p0 cultures. S100 is strongly expressed in the cytoplasm and nuclei of Schwann cells. Vacuolated Schwann cells (arrows) usually contain myelin inclusions or other cellular debris

13. Add 30 μl of mounting medium onto the cells and carefully place a coverslip on top (*see* **Note 16**).

14. Gently press a blotting paper on the coverslip to absorb the excess of mounting medium and seal the coverslip with glue.

15. Use a fluorescence microscope to count Schwann cells (S100-positive, vimentin-positive cells) and fibroblasts (S100-negative, vimentin-positive cells), (*see* Fig. 2). Alternatively, take images of a minimum of 500 cells and count at a later time point.

3.5 Immunofluorescence Staining of Schwann Cells Grown in 8-Well Chamber Slides

The combined immunofluorescence staining of intra- and extracellular markers is very useful to analyze the presence of a marker of interest in addition to a cell-specific marker. Here, an immunostaining protocol of intracellular vimentin combined with Schwann cell-specific intracellular S100 (Subheading 3.5.2) or extracellular p75^{NTR} (Subheading 3.5.3) is provided. This protocol may be used for other combinations of antibodies for extra- and intracellular markers. However, the optimal incubation time, dilution, and species of primary and secondary antibodies need to be adapted to each particular antibody combination.

The following steps are carried out at RT unless noted otherwise. The PBS wash involves two incubation steps with fresh PBS for 5 min each. Close the lid of the 8-well chamber slide during incubation times to protect the samples from drying out. For incubation steps with blocking and antibody solutions, apply a volume of 150 µl per well of an 8-well chamber slide.

3.5.1 Fixation of Cells

1. Carefully remove the medium from an 8-well chamber slide containing adherent Schwann cells using a 1000 µl micropipette (*see* **Note 17**).

2. Wash the cells with PBS.

3. Fix the cells by adding 300 µl of ice-cold PFA solution per well. Close the lid and incubate the cells for 15 min at 4 °C.

4. Remove the fixative and wash the cells with PBS. Cells can be stored in PBS at 4 °C for 24 h.

5. Proceed to Subheading 3.5.2 for simultaneous staining of intracellular S100 and vimentin or to Subheading 3.5.3 for sequential staining of extracellular p75^{NTR} and intracellular vimentin.

3.5.2 Simultaneous Staining for Intracellular S100 and Vimentin

1. Remove the PBS and incubate the cells with permeabilization solution 1 for 10 min.

2. Remove the permeabilization solution 1 and incubate the cells with the S100 plus vimentin primary antibody mix for 1 h at RT or overnight at 4 °C.

3. Remove the S100 plus vimentin primary antibody mix and wash the cells with PBS.

4. Incubate the cells with the FITC plus AF647 secondary antibody mix for 1 h.

5. Remove the FITC plus AF647 secondary antibody mix and wash with PBS.

6. Proceed to Subheading 3.5.3, **step 13**.

3.5.3 Sequential Staining for Extracellular p75^{NTR} and Intracellular Vimentin

1. Remove PBS and incubate the cells with the blocking solution 2 for 30 min.

2. Remove blocking solution 2 and incubate cells with the p75^{NTR} primary antibody solution overnight at 4 °C.

3. The next day, remove the p75^{NTR} primary antibody solution and wash the cells with PBS.

4. Add the FITC secondary antibody solution and incubate the cells for 1 h.

5. Remove the FITC secondary antibody solution and wash cells with PBS.

6. To fix the anti-p75^{NTR}-FITC staining, add 300 µl of the PFA solution per well and incubate the cells for 10 min at 4 °C.

7. Remove the fixative and wash the cells with PBS.

8. Incubate the cells with the permeabilization solution 2 for 10 min.

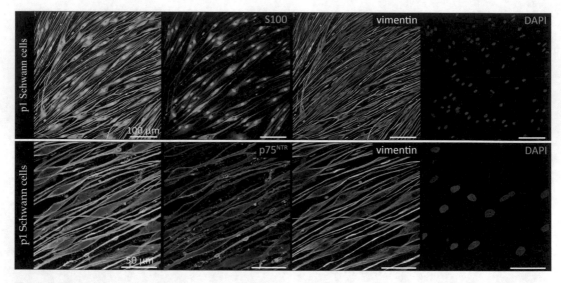

Fig. 3 Stained p1 Schwann cells grown in chamber slides. The upper panel shows Schwann cells after combined staining for intracellular S100 and vimentin. The lower panel depicts Schwann cells sequentially stained for extracellular p75^NTR and intracellular vimentin. Note that the vimentin staining displays a filamentous network within the Schwann cell cytoplasm and within the long processes

9. Remove the permeabilization solution 2 and incubate the cells with the vimentin staining solution for 1 h at RT or overnight at 4 °C.

10. Remove the vimentin staining solution and wash the cells with PBS.

11. Incubate the cells with the AF647 secondary antibody solution for 1 h.

12. Remove the AF647 secondary antibody solution and wash the cells with PBS.

13. Add the DAPI working solution and incubate the cells for 2 min.

14. Wash the cells with PBS and add 300 μl of mounting medium per well. When stored at 4 °C, the immunofluorescence should be stable for about 1 week.

15. Use a confocal microscope to take pictures of the stained cells (*see* **Note 18**). Representative images are illustrated in Fig. 3.

4   Notes

1. This protocol was developed by using human peripheral sensory or mixed sensory and motor nerves from amputated extremities. Nerve pieces should be kept in sterile saline solution to avoid drying up during the transfer.

2. Use one forceps to fix the nerve piece and another one to grasp and pull out the fascicle endings. Although human fascicles can be identified by the naked eye, a stereomicroscope may be of advantage to exclusively harvest dense, white fascicles. The careful removal of the epineurium and loosen connective tissue is essential to reduce the initial number of fibroblasts.

3. If the fascicle tissue failed to disaggregate properly, add additional 0.125% (w/v) collagenase type IV, 1.25 U/ml dispase II, and 3 mM $CaCl_2$ into the wells and incubate the fascicles for 2–6 h. A representative image of dissociated fascicle tissue is shown in Fig. 1b.

4. Repeated redistribution of the solution over the whole surface of the well is required during the incubation period to ensure homogenous coating.

5. Take a 10 μl aliquot of cells for counting and store the 15 ml conical centrifuge tube containing the dissociated fascicles in a standard CO_2 incubator until further use. Add 90 μl of supplemented MEMα to the aliquot and prepare two cytospin slides as indicated in Subheading 3.4.1. Fix the cells as described in Subheading 3.4.2, **steps 2–5**, and proceed to the DAPI staining as indicated in Subheading 3.4.2, **steps 11–14**. Count intact, DAPI-positive cell nuclei of both slides using a fluorescence microscope and calculate the total number of cells. This procedure takes up to 50 min but is advantageous over counting viable cells using a counting chamber as cells may not be well identifiable within the abundant myelin and cellular debris.

6. Drench a paper with 70% ethanol and put on the bench. Gently tab the 6-well plate three times onto the drenched paper. This procedure enables equal seeding by stopping the rotation of the medium and reducing static electricity.

7. Cultures at p0 are primarily composed of Schwann cells and fibroblasts with floating myelin and cellular debris in the medium (*see* Fig. 1c). Several p0 Schwann cells usually possess myelin inclusions of different sizes that appear round and shiny in the phase contrast microscope (*see* Fig. 1c). The amount and size of the myelin inclusions decrease during time in culture due to ongoing myelinophagy, i.e., the autophagocytosis and degradation of myelin debris [25]. Most p0 cultures reach 90% confluency at around day 7 post-plating. If reduced cell density is observed, continue to expand the cells for another week. When cultures have not reached 90% confluency after that time but show the typical bi- to multipolar and no senescent morphology (*see* Fig. 1c,g vs j), proceed with the two-step Schwann cell enrichment procedure as

described in Subheading 3.3.1. A longer cultivation time of p0 cultures is not recommended as older cultures may behave differently, which could interfere with the reproducibility of follow-up experiments.

8. The higher number of fibroblasts after the two-step enrichment procedure may be of advantage for immunostaining-based characterizations as the cell-type specificity of antibodies can be directly compared between Schwann cells and fibroblasts (e.g., fibroblasts may serve as internal negative control for proteins exclusively expressed by Schwann cells).

9. If Schwann cells do not detach easily, incubate p0 cultures with Accutase™ for another 30–60 s. Avoid air bubbles during the procedure.

10. The time of fibroblast attachment may vary if cells derived from different species are used.

11. To further cultivate the attached p1 fibroblasts, add 3 ml supplemented MEMα to each well and incubate the cells in a standard CO_2 incubator for 3–4 days. Figure 1f illustrates the morphology of cultured fibroblasts at p1.

12. It is recommended to resuspend the cells harvested per p0 well of a 6-well plate in 100 μl FACS buffer.

13. This step is essential to collect the single-cell suspension still adhering underneath the nylon mesh.

14. It is recommended to resuspend the enriched Schwann cells harvested per p0 well of a 6-well plate in 500 μl conditioned SCEM before cell counting.

15. The p1 Schwann cells show a bi- to multipolar morphology 1 day after seeding. During culture time, the Schwann cells proliferate and get aligned, and the amount and size of myelin inclusions are reduced. After about 10 days post-plating, the Schwann cell culture has formed the typical swirled parallel alignment and most cells have acquired a spindle-shaped morphology (see Fig. 1h,i). A reduced proliferative activity and senescent Schwann cells are commonly observed at p3. Senescent Schwann cells show an enlarged, granulated cytoplasm and dark-appearing cell nuclei (see Fig. 1j). In rare cases, Schwann cell cultures undergo senescence at earlier passages. Consider to terminate these cultures as they are likely to diverge in behavior and marker expression from viable Schwann cells.

16. Avoid air bubbles within the mounting medium by gently placing the coverslip using a curved tip forceps.

17. Tilt the chamber slide before pipetting. All manipulations should be carried out with a 1000 μl micropipette while avoiding vigorous pipetting so as not to detach the Schwann cells.

It is recommended to reduce the flow rate by cutting about 2 mm off the 1000 µl pipette tip. Place the pipette tip on the same well edge at all times. Do not allow the cells to dry out.

18. While performing confocal imaging, make sure to cover the whole cell volume. Start the scanning at the substrate level to image the adherent processes and move up to the elevated cell body by taking z-stack images at about 1 µm distance each. Typically, four to five stacks are needed to scan a whole Schwann cell. Save the images as maximum projection of total z-stacks to obtain a sharp picture representative of the culture (*see* Fig. 3).

Acknowledgments

We wish to thank Dieter Printz (FACS core unit, Children's Cancer Research Institute, Vienna, Austria) for cell sorting and Prof. Dr. Reinhard Windhager (Department of Orthopedic Surgery, Medical University of Vienna, Austria), Dr. Hugo B. Kitzinger and Prof. Dr. Chieh-Han Tzou (Division of Plastic and Reconstructive Surgery, Medical University of Vienna, Austria), and their patients for providing human nerve tissue samples. The research leading to these results has received funding from the St. Anna Kinderkrebsforschung (Vienna, Austria) and an FFG (TisQuant, EraSME, by the Austrian Research Promotion Agency) grant to Peter F. Ambros.

References

1. Weiss T, Taschner-Mandl S, Bileck A, Slany A, Kromp F, Rifatbegovic F, Frech C, Windhager R, Kitzinger H, Tzou CH, Ambros PF, Gerner C, Ambros IM (2016) Proteomics and transcriptomics of peripheral nerve tissue and cells unravel new aspects of the human Schwann cell repair phenotype. Glia 64(12): 2133–2153. https://doi.org/10.1002/glia.23045

2. Ambros IM, Attarbaschi A, Rumpler S, Luegmayr A, Turkof E, Gadner H, Ambros PF (2001) Neuroblastoma cells provoke Schwann cell proliferation in vitro. Med Pediatr Oncol 36(1): 163–168. https://doi.org/10.1002/1096-911x(20010101)36:1<163::aid-mpo1040>3.0.co;2-2

3. Casella GT, Bunge RP, Wood PM (1996) Improved method for harvesting human Schwann cells from mature peripheral nerve and expansion in vitro. Glia 17(4):327–338. https://doi.org/10.1002/(SICI)1098-1136(199608)17:4<327::AID-GLIA7>3.0.CO;2-W

4. Keilhoff G, Fansa H, Schneider W, Wolf G (1999) In vivo predegeneration of peripheral nerves: an effective technique to obtain activated Schwann cells for nerve conduits. J Neurosci Methods 89(1):17–24

5. Morrissey TK, Kleitman N, Bunge RP (1991) Isolation and functional characterization of Schwann cells derived from adult peripheral nerve. J Neurosci 11(8):2433–2442

6. Liu HM, Yang LH, Yang YJ (1995) Schwann cell properties: 3. C-fos expression, bFGF production, phagocytosis and proliferation during Wallerian degeneration. J Neuropathol Exp Neurol 54(4):487–496

7. Andersen ND, Srinivas S, Piñero G, Monje PV (2016) A rapid and versatile method for the isolation, purification and cryogenic storage of Schwann cells from adult rodent nerves. Sci Rep 6. https://doi.org/10.1038/srep31781

8. Kaewkhaw R, Scutt AM, Haycock JW (2012) Integrated culture and purification of rat

Schwann cells from freshly isolated adult tissue. Nat Protoc 7(11):1996–2004. https://doi.org/10.1038/nprot.2012.118

9. Rutkowski JL, Kirk CJ, Lerner MA, Tennekoon GI (1995) Purification and expansion of human Schwann cells in vitro. Nat Med 1(1):80–83

10. Brockes JP, Fields KL, Raff MC (1979) Studies on cultured rat Schwann cells. I Establishment of purified populations from cultures of peripheral nerve. Brain Res 165(1):105–118

11. Wood PM (1976) Separation of functional Schwann cells and neurons from normal peripheral nerve tissue. Brain Res 115(3):361–375

12. Assouline JG, Bosch EP, Lim R (1983) Purification of rat Schwann cells from cultures of peripheral nerve: an immunoselective method using surfaces coated with anti-immunoglobulin antibodies. Brain Res 277(2):389–392

13. Gilbert SF, Migeon BR (1975) D-Valine as a selective agent for normal human and rodent epithelial cells in culture. Cell 5(1):11–17

14. Jirsova K, Sodaar P, Mandys V, Bar PR (1997) Cold jet: a method to obtain pure Schwann cell cultures without the need for cytotoxic, apoptosis-inducing drug treatment. J Neurosci Methods 78(1-2):133–137

15. Haastert K, Mauritz C, Chaturvedi S, Grothe C (2007) Human and rat adult Schwann cell cultures: fast and efficient enrichment and highly effective non-viral transfection protocol. Nat Protoc 2(1):99–104

16. Vroemen M, Weidner N (2003) Purification of Schwann cells by selection of p75 low affinity nerve growth factor receptor expressing cells from adult peripheral nerve. J Neurosci Methods 124(2):135–143

17. Spiegel I, Peles E (2009) A novel method for isolating Schwann cells using the extracellular domain of Necl1. J Neurosci Res 87(15):3288–3296. https://doi.org/10.1002/jnr.21985

18. Previtali SC, Feltri ML, Archelos JJ, Quattrini A, Wrabetz L, Hartung H (2001) Role of integrins in the peripheral nervous system. Prog Neurobiol 64(1):35–49

19. Rutkowski JL, Tennekoon GI, McGillicuddy JE (1992) Selective culture of mitotically active human Schwann cells from adult sural nerves. Ann Neurol 31(6):580–586. https://doi.org/10.1002/ana.410310603

20. Davis JB, Stroobant P (1990) Platelet-derived growth factors and fibroblast growth factors are mitogens for rat Schwann cells. J Cell Biol 110(4):1353–1360

21. Lemke GE, Brockes JP (1984) Identification and purification of glial growth factor. J Neurosci 4(1):75–83

22. Rahmatullah M, Schroering A, Rothblum K, Stahl RC, Urban B, Carey DJ (1998) Synergistic regulation of Schwann cell proliferation by heregulin and forskolin. Mol Cell Biol 18(11):6245–6252

23. Lopez TJ, De Vries GH (1999) Isolation and serum-free culture of primary Schwann cells from human fetal peripheral nerve. Exp Neurol 158(1):1–8. https://doi.org/10.1006/exnr.1999.7081

24. Needham LK, Tennekoon GI, McKhann GM (1987) Selective growth of rat Schwann cells in neuron- and serum-free primary culture. J Neurosci 7(1):1–9

25. Gomez-Sanchez JA, Carty L, Iruarrizaga-Lejarreta M, Palomo-Irigoyen M, Varela-Rey M, Griffith M, Hantke J, Macias-Camara N, Azkargorta M, Aurrekoetxea I, De Juan VG, Jefferies HB, Aspichueta P, Elortza F, Aransay AM, Martinez-Chantar ML, Baas F, Mato JM, Mirsky R, Woodhoo A, Jessen KR (2015) Schwann cell autophagy, myelinophagy, initiates myelin clearance from injured nerves. J Cell Biol 210(1):153–168. https://doi.org/10.1083/jcb.201503019

Magnetic-Activated Cell Sorting for the Fast and Efficient Separation of Human and Rodent Schwann Cells from Mixed Cell Populations

Kristine M. Ravelo*, Natalia D. Andersen*, and Paula V. Monje

Abstract

To date, magnetic-activated cell sorting (MACS) remains a powerful method to isolate distinct cell populations based on differential cell surface labeling. Optimized direct and indirect MACS protocols for cell immunolabeling are presented here as methods to divest Schwann cell (SC) cultures of contaminating cells (specifically, fibroblast cells) and isolate SC populations at different stages of differentiation. This chapter describes (1) the preparation of single-cell suspensions from established human and rat SC cultures, (2) the design and application of cell selection strategies using SC-specific (p75[NGFR], O4, and O1) and fibroblast-specific (Thy-1) markers, and (3) the characterization of both the pre- and post-sorting cell populations. A simple protocol for the growth of hybridoma cell cultures as a source of monoclonal antibodies for cell surface immunolabeling of SCs and fibroblasts is provided as a cost-effective alternative for commercially available products. These steps allow for the timely and efficient recovery of purified SC populations without compromising the viability and biological activity of the cells.

Key words Peripheral nerve, Schwann cells, Primary cultures, Fibroblasts, Magnetic bead separation, p75[NGFR], Thy-1, O1, O4, Cell sorting

1 Introduction

Dissociated peripheral nerve fascicles typically give rise to heterogeneous cultures that contain varying proportions of Schwann cells (SCs) and contaminating cells, which are mainly fibroblasts derived from the epi-, peri-, and endoneurial layers of connective tissue. Further adding to the heterogeneity of SC cultures from peripheral nerve and various other sources is the usual presence of SC subpopulations that display distinct phenotypic features or stages of differentiation.

Kristine M. Ravelo, Natalia D. Andersen contributed equally to this work.

Paula V. Monje and Haesun A. Kim (eds.), *Schwann Cells: Methods and Protocols*, Methods in Molecular Biology, vol. 1739, https://doi.org/10.1007/978-1-4939-7649-2_6, © Springer Science+Business Media, LLC 2018

Nearly all methods to purify SC populations exploit specific biochemical and functional differences between SCs and fibroblasts. Fibroblasts are highly proliferative cells, and their growth can be attenuated by a variety of methods, including (1) the addition of antimitotic drugs in the culture medium [1–3], (2) the use of selective medium containing a low serum concentration [4] or lacking some fibroblast essential nutrients [5], and (3) the incorporation of mitogenic factors such as heregulin and forskolin, a cAMP-stimulating agent that thwarts fibroblast contamination by providing a competitive growth advantage to the SCs [6, 7]. Differential attachment and detachment procedures [8–12] along with immunological methods such as complement-mediated cell lysis [1, 13], and magnetic cell separations relying on differential expression of cell surface antigens [14–17], are also available to reduce fibroblast growth.

Magnetic-activated cell sorting (MACS) is a versatile technique that can be easily modified to overcome the challenges that arise in the isolation of cell types in mixed populations. For MACS, cells are collected as a single-cell suspension and labeled with an antibody against a cell surface marker of choice. Conjugating these cells to magnetically tagged antibodies allows for the desired cell population to be sorted from the contaminating cells under the presence of a magnetic field. In addition to its capacity to purify a given cell type in a series of simple steps, MACS allows for the recovery of both the magnetically labeled and unlabeled fractions as independent, viable populations. Both the magnetic labeling and the separation are gentle and have minimal influence on cell viability and function. Implementation of MACS does not require expensive or sophisticated equipment. It consistently provides a high efficiency of purification when cells are selectively labeled on their surface. Arguably, the most important prerequisite for an efficient isolation using the MACS system is a thorough understanding of cell surface marker(s) that are suitable to achieve specific and selective cell labeling. Scalability of the MACS procedure is possible by using columns of larger capacity without significantly increasing the time of operation or reducing the efficiency of purification.

Here, optimized MACS protocols are described to overcome two main challenges in the development of purification strategies for SC cultures: (1) the separation of SCs from contaminating cells (specifically, fibroblast-like cells) and (2) the isolation of SCs at different stages of differentiation. To separate SCs from contaminating cells, positive and negative cell selection protocols are provided based on the differential expression of the low-affinity neurotrophin receptor p75[NGFR] and the adhesion molecule Thy-1 in SCs and fibroblasts, respectively. Positive SC selection based on the expression of early (O4) and late (O1) markers of SC differentiation is also provided to separate SCs exhibiting distinct maturation stages. In this chapter, the following procedural steps are described in detail: (1) the magnetic immunolabeling of surface antigens using cells in suspension, followed by (2) the magnetic separation, and finally (3) the characterization, identification, and determination of viability in the pre- and post-sorting cell populations (*see* Fig. 1). Importantly, pertinent notes are provided with suggested adaptations for use with cultured human and rat SCs, along with recommendations on how to improve cell purity and recovery in the post-MACS cellular products. An optional protocol for the preparation of monoclonal antibodies suitable for cell labeling and immunological characterization of the resultant populations is included. These cell sorting strategies can be customized to generate the desired population(s) through positive selection, depletion, and/or multiparameter sorting with a separation efficacy of 95% or higher [17] irrespective of the proportion of contaminating cells in the mixed culture.

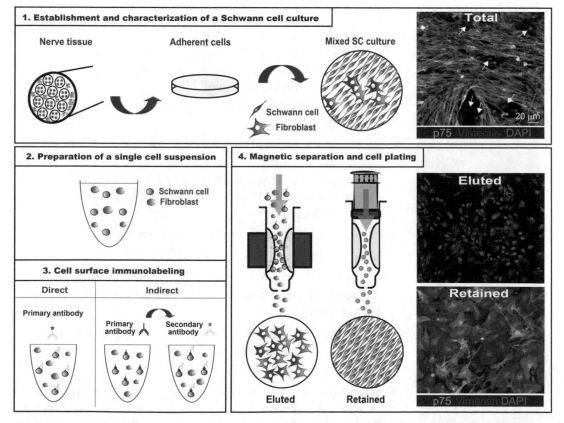

Fig. 1 Schematic diagram representing the main steps used for MACS of mixed SC cultures. A representation of direct and indirect strategies to separate magnetically labeled SCs (gray) from non-labeled fibroblasts (red) is depicted to highlight the steps used in a generic MACS procedure. A representative immunofluorescence microscopy analysis of human SC cultures deprived of fibroblasts by indirect p75NGFR immunolabeling followed by MACS is provided to illustrate one of the potential uses of this protocol (right panels). Human SCs are characterized by their p75NGFR-positive, vimentin-positive phenotype, whereas fibroblasts are characterized by their p75NGFR-negative, vimentin-positive one (arrows). The strategy of choice needs to be selected based on the composition of the starting population, the goal of the MACS experiment, and the availability of reagents

2   Materials

The materials described below are recommended to maximize the effectiveness of MACS separation using heterogeneous human or rat SC cultures. All reagents, media, and buffers should be sterile and cell culture grade. Suitable alternatives are provided in Subheading 4 in the event certain materials are not available to the user. Information on commercial suppliers (such as product information for the MACS system) is provided and should be used as recommended. The use of alternative magnetic cell surface labeling products (outside the MACS system and products listed in this protocol) has not been tested and may affect the purification outcome. This protocol uses research grade materials, reagents, and cells. Preclinical or clinical purification of SC cultures is feasible but has not been tested yet.

2.1 Establishment and Characterization of Human and Rat SC Cultures

1. An established, healthy culture of human or rat origin consisting of a monolayer of SCs. Cultures can be derived from essentially any source, passage, or stage of differentiation. The use of confluent cell cultures is recommended to maximize initial cell yields per isolation experiment.

2. Laminar airflow hood, preferably with vertical airflow for use with biological substances.

3. Sterile disposable plasticware. This may include but is not limited to serological pipettes, polystyrene cell culture bottles, and conical centrifuge tubes of assorted volumes for the preparation and storage of cells, cell culture media, antibiotics, and other supplements.

4. Sterile Pasteur pipette connected to a vacuum line aspirator.

5. CO_2 incubator. Standard humidified cell incubator set to 37 °C with 8–9% CO_2.

6. Refrigerated benchtop centrifuge equipped with swinging-bucket rotor.

7. Bottle-top filter (low-binding poly-ether sulfone membrane, 0.22 μM) for cell culture media sterilization.

8. Distilled cell culture grade water.

9. Heregulin-β1 stock solution. 133 μM recombinant human heregulin-β1 (PeproTech, Catalog # 100-03) prepared in sterile water and aliquoted and stored at −80 °C. It is important to avoid repeated freezing-thawing cycles of heregulin stocks.

10. Molecular biology grade 200 proof absolute ethanol.

11. Forskolin stock solution. 20 mM forskolin (Sigma-Aldrich, Catalog # F6886) prepared in 95% ethanol and aliquoted and stored at −20 °C.

12. DMEM. Dulbecco's Modified Eagle's Medium with high glucose and phenol red, pH 7.2.

13. De-complemented fetal bovine serum (FBS). To de-complement the FBS, thaw the stock at room temperature (RT), prepare aliquots in 50 mL tubes, and heat-inactivate at 56 °C for 30 min. Proceed to freezing the aliquots at −20 °C for long-term storage.

14. 100× GlutaMAX™ supplement. 200 mM L-alanyl-L-glutamine dipeptide in 0.85% NaCl (Gibco, Catalog # 35050-061).

15. 100× penicillin-streptomycin stock solution containing 10,000 units/mL penicillin and 10,000 µg/mL streptomycin.

16. Low proliferation medium. DMEM supplemented with 10% (v/v) FBS, 1% GlutaMAX™, 1% (v/v) penicillin-streptomycin, and 0.1% (v/v) gentamicin. Sterile filtered. May be kept at 4 °C for up to 1 month.

17. Proliferation medium. Low proliferation medium supplemented with 10 nM heregulin-β1 and 2 µM forskolin. Sterile filtered. May be kept at 4 °C for up to 1 month without significant loss of mitogenic activity.

18. Hanks' Balanced Salt Solution (HBSS). Formulated without calcium and magnesium and with phenol red, pH 7.2.

2.2 Preparation of Viable Single-Cell Suspensions

1. 10× trypsin stock solution. 0.5% trypsin and 0.02% ethylenediaminetetraacetic acid (EDTA) prepared in phosphate buffer saline (PBS). Store in aliquots at −20 °C.

2. HBSS.

3. 1× dissociation buffer. 0.05% trypsin and 0.002% EDTA prepared in chilled HBSS, pH 7.2.

4. Low proliferation medium as prepared in Subheading 2.1.

5. Proliferation medium as prepared in Subheading 2.1.

6. Inverted phase contrast microscope for routine inspection of cell cultures.

7. Automated cell counter or hemocytometer.

8. Sterile disposable plasticware as described in Subheading 2.1.

9. Sterile Pasteur pipette connected to a vacuum line aspirator as described in Subheading 2.1.

10. Trypan blue solution. 0.4% (w/v) trypan blue prepared in 0.81% sodium chloride (NaCl) and 0.06% potassium phosphate dibasic solution.

11. (Optional) Sterile pre-separation 70 µm cell strainers (Miltenyi, Catalog # 130-095-823) to remove large cell clumps or debris and prevent obstruction of the column.

12. (Optional) Dead Cell Removal Kit (Miltenyi, Catalog # 130-090-101) to remove dead or apoptotic cells in suspension prior to separation.

2.3 MACS for Human and Rat SC Cultures

SC cultures may consist of an assortment of contaminating fibroblast-like cells in addition to a variable mixture of SC sub-populations, each identifiable by the antigenic dissimilarities of cell surface markers indicative of the developmental stage. The protocols described in the following sections use stage-specific markers to separate SC populations at increasing stages of differentiation via positive cell selection. The suggested SC-specific markers are the following: (1) p75^{NFGR}, the low-affinity neurotrophin receptor characteristic of neural crest-derived cells; (2) O4 (sulfatide), a general surface lipid antigen that serves as a precursor-to-immature SC transition marker; and (3) O1 (galactocerebroside), a lipid antigen sensitive to cAMP and axon contact stimulation that is an indicator of SC differentiation into a myelinating phenotype. It is important to understand that high levels of expression of p75^{NFGR} are maintained throughout the SC lineage with the exception of the myelinating phenotype [18], whereas O4 expression levels remain high in both mature myelinating and nonmyelinating SCs. Non-SC populations such as fibroblasts, macrophages, and stem cells can also be present in various proportions in SC cultures derived from the dorsal root ganglia and peripheral nerves. Fibroblasts can be recognized and depleted of the SC populations by positive cell selection based on their high content of fibronectin and cell surface expression of membrane anchored glycoproteins such as Thy-1 (CD90) that are absent in SCs. Table 1 summarizes the different labeling strategies that are recommended based on our previous findings [17, 22] and unpublished data.

2.3.1 Indirect Magnetic Cell Labeling

1. Primary antibodies. Antibodies can be in the form of commercially available purified monoclonal antibodies or conditioned medium generated from the supernatant of a hybridoma cell culture (as described in Subheading 3.5). The primary antibodies should recognize and bind with high affinity and specificity to cell surface SC-specific antigens (e.g., p75^{NGFR}, O4, and O1) or fibroblast-specific antigens (e.g., Thy-1 in rat cells or equivalent antigen in human cells). Primary antibodies should be tested for their activity and specificity of live cell labeling prior to their use in MACS (*see* Table 1).

2. Secondary antibodies conjugated to magnetic microbeads. Secondary antibodies may be used in both positive and negative selection protocols (*see* Table 1).

3. 30% (w/v) bovine serum albumin (BSA) fraction V stock solution. Prepared in Dulbecco's modified phosphate buffer saline (D-PBS), pH 7.2. Use ultrapure, cell culture grade BSA for all applications.

4. Ultrapure 0.5 M EDTA stock solution, pH 8.0.

5. 10 × D-PBS with calcium and magnesium, pH 7.2.

Table 1

Reagents and strategies available for MACS of human and rat SC cultures. To increase the efficiency of separation, it is suggested to customize the protocol based on the composition of the initial cell population. Though direct labeling strategies are faster, no major differences in the efficiency of separation have been found with respect to indirect cell labeling. SC purification can be achieved by positive selection (e.g., using p75[NGFR], O4, or O1 immunolabeling) or negative selection (e.g., using fibroblast-specific immunolabeling) with nearly identical results [17]. The indirect labeling of human SC cultures with human-specific p75[NGFR] (clone 8737) provides purities of 80–99% depending on the initial proportion of contaminating cells. Labeling of rat cultures with p75[NGFR], Thy-1, O4, and O1 antibodies allows for an enrichment of >95% in cells expressing any given marker. The efficiency of SC purification using fibroblast-specific microbeads is variable and dependent on the nature of the contaminating human cells. Primary and secondary antibodies conjugated to magnetic microbeads may be purchased directly from the manufacturer. As per our knowledge, the hybridoma cell lines for 192-p75[NGFR], O4, and O1 are not currently commercially available. These cell lines may be obtained from private repositories or requested to the laboratories credited for the original publications (*see* References listed in the Table)

Strategy	Sorting antigen	Immunoreactivity	Primary antibody	Secondary antibody
Indirect	p75[NGFR]	Rat only	192-IgG Hybridoma [19]	Anti-mouse IgG MicroBeads
		Human only	8737-IgG Hybridoma ATCC # HB8737	Miltenyi # 130-047-102
	O4	Human, rat, other	O4-IgM Hybridoma [20, 21]	Anti-mouse IgM MicroBeads
	O1	Human, rat, other	O1-IgM Hybridoma [20, 21]	Miltenyi # 130-047-501
	Thy-1	Rat only	Thy-1.1-IgM Hybridoma ATCC # TIB-103	
Direct	p75[NGFR]	Human, other	Anti-CD271 MicroBeads Miltenyi # 130-099-023	–
	O4	Rat, other	Anti-O4 MicroBeads Miltenyi # 130-094-543	–
	Myelin-specific	Rat, other	Myelin Removal Beads Miltenyi # 130-096-731	–
	Fibroblast-specific	Human only	Anti-Fibroblast MicroBeads Miltenyi # 130-050-601	–

6. Indirect sorting buffer. 0.5% (w/v) BSA, 2 mM EDTA prepared in D-PBS, pH 7.2. Sterile filtered and degassed prior to use.

2.3.2 Direct Magnetic Cell Labeling

1. Primary antibodies conjugated to magnetic microbeads. Anti-CD271 (also known as p75[NGFR]) microbeads, fibroblast-specific microbeads, and O4 microbeads (Miltenyi, Catalog # 130-096-670), (*see* Table 1).

2. Direct sorting Buffer: 0.5% (w/v) BSA prepared in D-PBS, pII 7.2. Sterile filtered and degassed prior to use.

2.3.3 Magnetic
Separation

1. DMEM.

2. MS columns (Miltenyi, Catalog # 130-042-201) positioned on the MACS separator. Use the columns according to the recommendations provided by the manufacturer unless specified in Subheading 4.

3. (Optional) Large cell columns (Miltenyi, Catalog # 130-042-202).

4. (Optional) LS columns (Miltenyi, Catalog # 130-042-401).

5. Direct or indirect sorting buffer as determined by the type of separation. Prepared as indicated in Subheadings 2.3.1 and 2.3.2.

6. MACS separator of choice as determined by the column of choice. MiniMACS Separator (Miltenyi, Catalog # 130-042-102) for single separations or OctoMACS Separator (Miltenyi, Catalog # 130-042-109) for simultaneous separation of up to eight samples. The separators are magnetically attached to the MACS multistand (Miltenyi, Catalog # 130-042-303) and positioned inside a laminar flow hood. (Optional) The MidiMACS Separator (Miltenyi, Catalog # 130-042-302) is needed for single separations using LS columns.

7. Round-bottom, sterile polystyrene tubes (5 mL) and tube rack for the collection of eluates.

2.3.4 Cell Plating

1. 100× poly-L-lysine (PLL) stock solution. A 2% (w/v) PLL stock solution is obtained by dissolving 500 mg of PLL (powder) in 25 mL of borate buffer. The stock solution should be sterile filtered prior to aliquoting and storing at −20 °C. A 1× working solution of PLL is prepared by diluting the stock solution in chilled distilled water prior to coating the substrate.

2. Mouse laminin stock solution. 1 mg/mL laminin from Engelbreth-Holm-Swarm murine sarcoma sterile filtered in TRIS-buffered NaCl (Sigma-Aldrich, Catalog # L2020). Aliquoted and stored at −80 °C.

3. PLL-laminin-coated cell culture dishes. 24-well (for immunofluorescence assays) and 10 cm plates (for cell expansion) sequentially coated with PLL and mouse laminin. It is recommended to incubate the plates for 1 h at RT with 0.18 mL of PLL working solution per cm^2 and air dry the plates prior to overnight incubation at 4 °C with mouse laminin. The laminin is prepared in chilled D-PBS using 1 μg laminin per cm^2. Double coating with PLL and laminin is preferred for enhanced SC adhesion and survival pre- and post-separation.

2.4 Assessment of SC Purity and Viability Via Fluorescence Microscopy Analysis

2.4.1 Immunolabeling and Propidium Iodide Incorporation in Live, Adherent Cells

1. 1× D-PBS.

2. Primary antibodies recognizing the SC- or fibroblast-specific marker of choice. Antibodies can consist of undiluted conditioned medium from hybridoma culture supernatant (*see* Subheading 3.5) or commercially available antibodies diluted in D-PBS.

3. Propidium iodide (PI) stock solution diluted to 1 μg/mL in DMEM or D-PBS.

4. Hoechst 33342 (bis-benzimide) stock solution diluted to 1 μg/mL in DMEM or D-PBS.

5. Fixation solution. 4% (w/v) paraformaldehyde in D-PBS, pH 7.2.

6. Blocking solution. 5% (v/v) normal goat serum (NGS) prepared in D-PBS.

7. Green fluorescent secondary antibodies. Alexa Fluor™ (488)-conjugated secondary antibodies (or equivalent) prepared at 1:300 in blocking solution.

2.4.2 Immunolabeling of Fixed Cells

1. 100% (v/v) methanol kept at −20 °C.

2. 1× D-PBS.

3. Blocking solution.

4. Primary antibodies recognizing the cell marker of choice. Antibodies against intracellular SC markers such as S100 (a calcium binding protein) can be used to selectively label the SC population regardless of the stage of differentiation.

5. Fluorescent secondary antibodies.

6. 4′,6-diamidino-2-phenylindole dihydrochloride (DAPI) diluted to 1 μg/mL in D-PBS. Hoechst may also be used for nuclear staining of fixed cultures.

2.4.3 Fluorescence Microscopy, Mounting, and Long-Term Preservation of Fixed Cells

1. Inverted fluorescence microscope with fluorescein/rhodamine/UV filters and phase contrast lenses.

2. Post-fixation solution. 1% (w/v) paraformaldehyde prepared in D-PBS, pH 7.2.

3. Mounting reagent. 2.5% (w/v) 1,4-diazabicyclo [2.2.2] octane (DABCO) and 0.1% (w/v) sodium azide prepared in 90% glycerol/10% PBS. Stored at 4 °C protected from light.

4. Sterile microplate sealing film.

5. Aluminum foil.

2.5 (Optional) Culture of Hybridoma Cell Lines and Preparation of Conditioned Medium

1. Cryopreserved stocks of hybridoma cell lines producing mouse monoclonal antibodies. Protocols have been optimized using conditioned medium derived from the hybridoma cell lines listed in Table 1. Information on the respective sources, providers, and specificity of immunoreaction for each antibody are included in the table and its legend.

2. Iscove's Modified Dulbecco's Medium (IMDM) with phenol red, pH 7.2.

3. Hybridoma culture medium. IMDM supplemented with 10% (v/v) FBS, 1% GlutaMAX™, 1% (v/v) penicillin-streptomycin, and 0.1% (v/v) gentamicin.

4. T-75 cell culture-treated flasks with filter screw caps containing a 0.2 μm hydrophobic membrane. Used for initial plating and expansion of hybridoma cells.

5. CO_2 incubator.

3 Methods

The methods described below outline (1) general guidelines on the preparation and identification of heterogeneous SC cultures suitable for MACS, (2) the collection of viable cell suspensions obtained by enzymatic dissociation, (3) the immunolabeling of the resultant cell suspensions followed by magnetic separation, and (4) the assessment of purity, viability, and identity of the resultant populations as a means to determine the efficiency of the MACS protocol (*see* Fig. 2). An optional protocol for the preparation, storage, and use of conditioned medium from hybridoma cell lines is included as a cost-effective source of monoclonal antibodies suitable for cell labeling and immunofluorescence microscopy. Each purification strategy should be designed and customized according to the population used as starting material and the type of separation desired. The information presented in Table 1 is provided to assist the user with the overall experimental design. Results from a representative MACS experiment are presented in Fig. 2.

3.1 Establishment and Characterization of Human and Rat SC Cultures

1. Establish a culture of human or rat origin containing a proportion >10% of cells identifiable as SCs (*see* **Note 1**).

2. Maintain the culture in proliferation medium until it reaches a confluent or sub-confluent state and passage into new dishes until the desired number of cells is obtained (*see* **Notes 2** and **3**).

3. Perform an immunofluorescence microscopy analysis of an unpurified cell sample to justify the strategy used for purification. Refer to Subheading 3.4 and Table 1 for suggested methods and antibodies, respectively, suitable in the identification of SCs and contaminating cells.

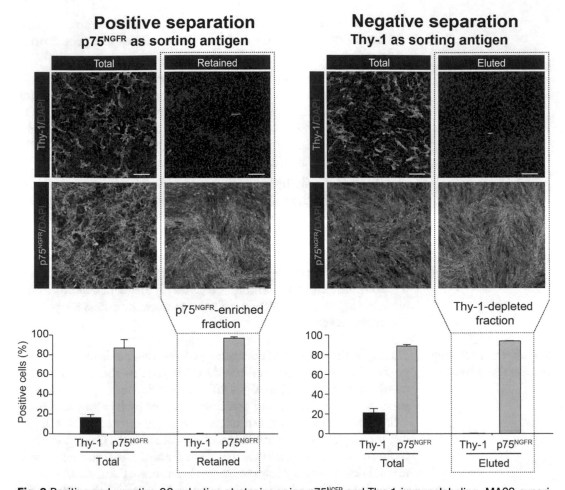

Fig. 2 Positive and negative SC selection strategies using p75^{NGFR} and Thy-1 immunolabeling. MACS experiments using SC enrichment (positive selection) and fibroblast depletion (negative selection) were based on indirect immunolabeling with SC-specific (anti-p75^{NGFR}) and fibroblast-specific (anti-Thy-1) antibodies, respectively. Adult rat SC cultures were detached by trypsinization (Subheading 3.2) and labeled with conditioned medium enriched in p75^{NGFR} (derived from 192-IgG hybridoma cells) or Thy-1 antibodies (derived from Thy1.1-IgM hybridoma cells) prior to magnetic labeling and separation (Subheading 3.3). Cells were plated for analysis as described in Subheading 3.3.4. A live cell immunofluorescence microscopy analysis of the total (mixed) and SC-containing (pure) fractions was performed using the same antibodies used for MACS (Subheading 3.4.1). Note the high efficiency of purification obtained with both positive and negative separation strategies, as evidenced by the image data (upper panels) and the quantitative analysis of results (lower panels). The purified SC populations are highly viable and can be used in experimentation immediately after collection. No SC toxicity has been found in association with magnetic labeling. Scale bars, 100 μm

3.2 Preparation of Viable Single-Cell Suspensions

1. Replace the proliferation medium from the culture dishes of interest with an equivalent volume of HBSS to remove cellular debris (*see* **Notes 4** and **5**).

2. Remove the HBSS by gentle aspiration and replace it with 1 volume of dissociation solution. Monitor the detachment of the cells by phase contrast microscopy and if needed, tap the side of the dish to assist in the dislodging of the cells (*see* **Note 6**).

3. Once the cells are fully detached from the substrate, stop the action of the trypsin by adding 2 volumes of low proliferation medium.

4. Collect the trypsinized cell suspension and transfer them to a sterile conical tube of the appropriate volume. When collecting the cells gently, pipette the solution up and down (e.g., two to three times) using a 10 mL serological pipette to ensure a single-cell suspension is obtained. Collect any remaining cells on the substrate by using 1 volume of low proliferation medium.

5. Pellet the cells by centrifugation ($188 \times g$, 8 min at 4 °C). Aspirate the supernatant and gently resuspend the pellet in low proliferation medium to obtain a single-cell suspension.

6. Proceed to determine cell numbers by the counting method of choice. Estimate the proportion of viable cells by mixing an aliquot of the cell suspension at a 1:1 ratio with the trypan blue solution prior to counting. *See* **Note 7** for management of the cell suspensions in case dead cells, clumps, or debris are present at this stage.

3.3 MACS for Human and Rat SC Cultures

Having a thorough understanding of the composition of the cell population to be subjected to MACS is vital to the optimization, troubleshooting, and analysis of the purified end product. Cell sorting can be done using positive selection by labeling and separating the cell population of interest or negative selection by labeling and depleting the population of unwanted cells. Of importance, different selection methods may be combined to provide experimental flexibility for multiparameter isolations of different cell subpopulations. Table 1 summarizes the various cell labeling strategies that have been optimized for cultures of human and rat SCs. The protocol for indirect magnetic cell labeling entails (1) surface labeling using a carefully selected primary antibody followed by (2) the magnetic tagging of these surface antigens with a secondary antibody conjugated to virus-sized (50 nm) magnetic microbeads. Direct cell labeling uses a single step of incubation with commercially available primary antibodies conjugated with magnetic microbeads (*see* Fig. 1). Indirect and direct cell labeling protocols are described in Subheadings 3.3.1 and 3.3.2, respectively. The conditions for direct or indirect cell surface labeling are similar regardless of the primary antibody used but may be changed if other antibodies not presented in Table 1 are chosen.

The following is a generic protocol describing the use of MS columns, as these are suitable to purify cells for a variety of downstream applications. It is important to estimate the expected proportion of labeled cells based on previous knowledge of the composition of the original population and use this information to estimate the number of columns needed, as each MS column has a maximum retention capacity of up to 1×10^7 labeled cells.

These columns should be handled according to the instructions provided by the manufacturer. After purification, the cells may be used in experimentation as suspension cells (*see* Subheading 3.3.3) or plated for continued expansion or analysis (*see* Subheading 3.3.4).

3.3.1 Indirect Magnetic Cell Labeling

1. Label a cell suspension containing up to 2×10^7 cultured cells by incubating the cells with the desired purified antibody or conditioned medium for 20 min at 4 °C (*see* Table 1). To efficiently label the suspension with undiluted conditioned media, gently mix the cells at a ratio of 0.5–1 mL of medium per 1.0×10^6 cells. Refer to the manufacturer's instructions for the suggested dilutions of commercial antibodies.

2. Regularly "flick" the tube to assure distribution of the cells while incubating.

3. Collect the cells by centrifugation ($188 \times g$, 8 min at 4 °C) to remove unbound antibodies. Resuspend the pelleted cells in 1–2 mL of DMEM and repeat the centrifugation step.

4. Resuspend the cell pellet in 80 μL of indirect sorting buffer and add 20 μL of microbead-conjugated secondary antibodies per 10^7 cells (Table 1). Mix by gently pipetting up and down twice with a micropipette (*see* **Notes 8–10**).

5. Incubate the cells at 4 °C for 15 min and "flick" occasionally to assure even distribution of the cell suspension.

6. Without removing the secondary antibody mixture, add 3–5 mL of cold DMEM to the tube. Proceed to collect the cells by centrifugation ($188 \times g$, 8 min at 4 °C).

7. Follow the description provided in Subheading 3.3.3.

3.3.2 Direct Magnetic Cell Labeling

1. Label a cell suspension containing up to 2×10^7 cells by incubation with the desired magnetically conjugated primary antibody for 20 min at 4 °C (*see* Table 1). When using O4 microbeads or Myelin Removal Beads, a mixture of 90 μL of direct sorting buffer and 10 μL of antibody microbeads are sufficient to capably label 10^7 SCs. For anti-fibroblast beads, 80 μL of buffer with 20 μL of microbeads is recommended. Regularly "flick" the tube to assure distribution while incubating.

2. Without removing the antibody mixture, add 3–5 mL of DMEM to the tube. Proceed to collect the cells by centrifugation ($188 \times g$, 8 min at 4 °C).

3. Follow description provided in Subheading 3.3.3.

3.3.3 Magnetic Separation

1. While centrifuging the cell samples, prepare the MACS separation system by placing the magnetic separator on the stand in a laminar flow hood. Position the MS column(s) on the magnetic separator and wash the MS column twice with 500 μL of

direct or indirect sorting buffer according to the type of separation. If using an LS column, use 3 mL of sorting buffer (*see* **Notes 11** and **12**).

2. Resuspend the pelleted cells in 1 mL of sorting buffer to ensure a homogeneous suspension. Proceed to separation by transferring the cells to the MS column and collecting the eluted fractions. Be sure to visually monitor the progression of the flow and collect the eluates in the 5 mL sterile tubes placed directly below the column.

3. Once the initial suspension has been eluted from the column, proceed to wash the column twice with 500 μL of sorting buffer. Be sure to collect this elution and combine with the elution obtained in the previous step.

4. As soon as all washes have been eluted, cap the tube and maintain the cells at 4 °C until centrifugation. This portion of cells is herein referred to as the "eluted fraction". Once the cells are separated, it is recommended to immediately add a large volume of low proliferation medium to better support the viability of the cells up until centrifugation.

5. Remove the column from the magnetic separator and place it on top of an open, sterile 5 mL collection tube. Add 1 mL of sorting buffer to the column.

6. Recover the cells of interest from the column by applying the plunger evenly and slowly. This portion of cells collected is herein referred to as the "retained fraction". Cap the conical tube containing the retained fraction and immediately proceed to collect by centrifugation. It is also suggested to dilute the sorting buffer with low proliferation medium immediately after recovery of the retained cells.

7. Centrifuge both the eluted and retained fractions ($188 \times g$, 8 min at 4 °C).

8. Resuspend the resultant pellets in 1 mL of proliferation medium. Proceed to counting the cells in each fraction and determine viability by trypan blue staining to assess the recovery of viable cells (*see* **Note 13**).

9. Use the cells as such for experimentation or plate them for continued growth or analysis, as described in Subheading 3.3.4. Refer to **Note 14** for strategies used to improve purity.

3.3.4   Cell Plating

1. Plate the cells obtained in the eluted and retained fractions in PLL-laminin-coated cell culture dishes in proliferation medium. If analysis via fluorescence microscopy is desired, plate the cells in coated 24-well plates at a density of 50,000–80,000 cells/well. If expansion of each population is desired, plate the cells in coated 10 cm dishes at a density of $1–2 \times 10^6$ cells/dish.

2. Incubate the cells in the CO_2 incubator for a minimum of 24 h to guarantee adhesion to the substrate prior to live cell labeling, fixation, or collection (*see* **Note 15**).

3. Analyze the eluted and retained fractions to assess for efficiency of separation as described in Subheading 3.4.

3.4 Assessment of Purity and Viability Via Fluorescence Microscopy Analysis

Following cell sorting, replicates of all fractions collected may be assessed in multi-well dishes for the determination of purity and viability. A sample of unpurified, unlabeled cells should be used as a reference control to reveal the composition of the initial cell population. Different choices for measuring SC viability and metabolic activity are possible [23]. The effectiveness of separation can be determined by immunofluorescence microscopy analysis or other methods [17]. Simple generic protocols for purity determination in live (*see* Subheading 3.4.1) and fixed cells (*see* Subheading 3.4.2) are provided below. Antibodies used for immunostaining can be the same ones used for MACS. A description of recommended antibodies to use in immunofluorescence microscopy of rat SCs from adult and postnatal nerves can be found in our previous publication [17]. A live/dead assay using PI incorporation can be performed together with live cell labeling with SC- or fibroblast-specific antibodies to gauge the efficiency of separation and the viability in the same samples (*see* Subheading 3.4.1).

3.4.1 Immunolabeling and PI Incorporation in Live, Adherent Cells

1. Gently aspirate the culture medium and wash the cells with D-PBS to remove debris (*see* **Note 15**).

2. Proceed to fill the multi-wells with a solution containing the primary antibody of choice (or 500 μL of undiluted conditioned medium) and incubate for 20 min at RT (*see* **Note 16**).

3. Wash the wells with D-PBS to remove unbound antibodies. Repeat this wash step twice.

4. Dilute the green fluorescent secondary antibodies (1:300) in D-PBS together with PI (1:1000) and Hoechst (1:1000). Add this preparation to the cells and incubate them for 20 min at RT.

5. Gently aspirate the secondary antibody solution and wash two or three times with D-PBS.

6. Observe by fluorescence microscopy and take images of the live cells using the appropriate filters for each label making sure not to exceed a total time of 30–45 min to avoid capping of antibodies on the cell surface (*see* **Note 17**).

7. Add fixation solution to all wells and incubate for 20 min at RT. Cautiously remove the fixation solution and dispose appropriately. Wash the wells two or three times with D-PBS.

8. Proceed by performing post-fixation, analysis, and storage of fixed cultures as described in Subheading 3.4.3.

3.4.2 Immunolabeling
of Fixed Cells

1. (Optional) Gently aspirate the culture medium and wash the cells with D-PBS.

2. Fix the cells as described in Subheading 3.4.1, **step 7**.

3. (Optional) Double-fix and permeabilize the cells by filling the well with cold methanol for 10 min. Aspirate the methanol solution quickly and wash the cells three times with D-PBS to remove traces of methanol. This step is only needed for the detection of intracellular antigens such as S100.

4. Add 500 μL/well of blocking solution for 30 min at RT.

5. Remove the blocking solution and replace it with a solution containing 300–500 μL/well of the primary antibody of choice (or 500 μL of undiluted conditioned medium).

6. Incubate overnight at 4 °C with gentle agitation to maximize immunolabeling.

7. Collect the primary antibody mixture and perform three consecutive washes using D-PBS. Follow the washes with the addition of 300–500 μL/well of fluorescent secondary antibodies (1:300) for 1 h at RT, shielded from light. A general nuclear stain, such as DAPI or Hoechst 33,342, may be incorporated into the secondary antibody solution at this time or added independently at a following step.

8. Remove the secondary antibody solution and conduct three additional washes with D-PBS.

9. Proceed as described in Subheading 3.4.3.

3.4.3 Fluorescence
Microscopy Analysis,
Mounting, and Long-Term
Preservation of Fixed Cells

By comparing fluorescent extra-/intracellular staining of SC-specific (or fibroblast-specific) markers to a general nuclear marker, one can discern the purity of the resultant cell populations from the MACS procedure. Examples of such image analysis can be found in our recent publication [17]. For long-term preservation of immunostained samples, post-fixation is required to maintain the immunoreaction with fluorescent antibodies.

1. Post-fix the immunostained cells by adding enough post-fixation solution to coat the bottom of the well. Incubate at RT for 10 min.

2. Remove the post-fixation solution, wash once with D-PBS, and cover the wells with 500 μL of mounting reagent.

3. Observe the fluorescence under an inverted fluorescence microscope using the appropriate filters and take digital images for analysis and documentation. *See* **Note 18** for a comment on data analysis.

4. For optimal storage seal the plate with a sterile adhesive plate cover to prevent evaporation and wrap the dishes with aluminum foil. Store at 4 °C (*see* **Note 19**).

3.5 (Optional) Culture of Hybridoma Cell Lines and Preparation of Conditioned Medium

The use of hybridoma supernatant as a source of monoclonal antibodies for live cell labeling prior to MACS and subsequent immunostaining may be used as an economical alternative to commercially available purified antibodies. The following protocol has been used to produce monoclonal antibodies for the immunological detection and sorting of rat and human SCs [17, 24].

1. Retrieve a stock of the hybridoma cell line of choice from the liquid nitrogen tank (*see* Table 1).

2. Thaw the cells quickly in a 37 °C water bath, gently shaking the vial to accelerate the thawing step.

3. Monitor the thawing process by eye inspection and immediately transfer the cell stock to a conical tube containing ice-cold low proliferation medium to dilute the cryopreservant (DMSO). Use at least 20 mL of medium per 2 mL of cell stock.

4. Pellet the cells by centrifugation ($188 \times g$, 8 min at 4 °C) and resuspend them in 10 mL of hybridoma culture medium per 1×10^6 cells.

5. Transfer the cell suspension into a T-75 flask (up to 20 mL/flask) and observe the cells by phase contrast microscopy to visually assess their density at time zero.

6. Place the flask in a CO_2 incubator for expansion (*see* **Note 20**).

7. Monitor the cultures daily by phase contrast microscopy to assess the state of health of the cells and their rate of expansion. Continue to culture the cells up until obtaining a sufficiently dense population (i.e., a culture exhibiting at least three to four times higher density than the one initially plated).

8. Passage the cells in hybridoma culture medium by splitting the suspensions at 1:10 ratio in T-75 flasks. Subculture the cells up until obtaining the desired final volume of medium. As a reference, a volume of 100–300 mL is sufficient for most routine applications.

9. Place the flasks in a CO_2 incubator for expansion and monitor the cultures daily by phase contrast microscopy up until the cultures get sufficiently dense and the medium turns acidic with a yellowish color (*see* **Note 21**).

10. Harvest the hybridoma conditioned medium enriched in monoclonal antibodies by transferring the cells in suspension into 50 mL conical tubes, pelleting the cells by centrifugation ($188 \times g$, 8 min, at 4 °C), and collecting the supernatant.

11. Transfer the obtained supernatant from all culture flasks to a bottle-top filtering system and filter to remove any remaining hybridoma cells and debris

12. Test the supernatant (conditioned media) for antibody activity and specificity by live cell immunostaining of cultured SCs (*see* Subheading 3.4) prior to use in experimentation.

13. Store the conditioned medium in aliquots at −80 °C for long-term use.

4 Notes

1. The population to be used as starting material can be derived from essentially any source or species if SC identity is confirmed by means of specific immunological detection of cell surface p75[NGFR], O4, and/or O1 in live cells. *See* Chapter 4 for a description of a detailed protocol for establishing a primary SC culture from adult rat sciatic nerve.

2. The use of the chemical mitogens heregulin and forskolin in the culture medium has been known to reduce the growth of fibroblasts while synergistically enhancing SC proliferation [6, 25]. The addition of these mitogens is recommended but not essential for the maintenance and survival of SC cultures before and after purification. Medium containing 10% FBS can be used for plating of the cell suspensions in case chemical mitogens are not available, or is undesirable experimentally.

3. A standard MACS purification experiment should use a minimum of 1.5×10^6 total cells for optimal separation and recovery of sufficient numbers of viable cells in all fractions. It is advised to provide regular media changes (e.g., on a 2–3 day feeding schedule) prior to cell isolation to maximize the viability of the cell populations. The use of confluent cultures is recommended to maximize initial cell yields. A confluent 10 cm plate of human and rat SCs contains 4–5×10^6 and 7–9×10^6 cells, respectively. By daily monitoring of the progression of the cultures by phase contrast microscopy, one can determine the arrival to confluency by the concurrent detection of pattern formation and cell alignment along with the disappearance of figures of cell division in the population.

4. The use of a single-cell suspension of highly viable cells is instrumental for the success of the MACS procedure. Dead cells will nonspecifically bind to antibodies and distort the purification results. Removal of dead cells or debris can be easily carried out by successive washes with HBSS prior to enzymatic dissociation of adherent cell cultures. Purifying cells of low viability may negatively affect the specificity of the immunoreaction and the efficiency of purification.

5. Avoid the use of alkaline medium at all times during the culture, trypsinization, and purification steps. The viability of cultured SCs is highly impaired when exposed to media or

buffers exhibiting pH levels at or above 8, even for short periods of time. Changes in pH levels are visually evident when medium/buffers containing phenol red are used. Phenol red-containing media is highly recommended.

6. This protocol uses mild trypsinization using trypsin/EDTA to enzymatically produce a single-cell suspension containing SCs. Mechanical dissociation is not recommended for SC cultures. Attempts using cell scrappers or other forms of mechanical detachment often lead to poor cell recovery and viability. The dissociation buffer needs to be prepared immediately before use and maintained on ice. Dissociation time for SC cultures needs to be determined empirically by the experimenter. Time required for trypsinization depends on the cell type, confluency of the culture, and potency of the trypsin solution. Always monitor the course of detachment by observation under the phase contrast microscope to avoid reduced viability due to over-trypsinization. Confluent human SCs usually detach as layers and may be difficult to isolate into single cells. When cultures are heavily populated or strongly attached to the substrate, a 1.5× concentration of dissociation solution may be used with caution to facilitate detachment. The action of the trypsin is inhibited by protease inhibitors that are present in the serum; therefore, any DMEM-based medium supplemented with at least 10% FBS may be used as an alternative to stop the action of trypsin.

7. Cell suspensions containing clumps (SC aggregates), dead cells, or myelin debris may affect the specificity of antibody labeling and decrease the effectiveness of cell sorting due to obstruction of the column. If clumps of SCs are observed it is recommended to filter the cell suspensions through 70 μm filters prior to separation. SCs (human or rat) tend to adhere to one another in suspension and form aggregates after trypsinization if medium without serum is used. Avoid the problem of spontaneous cell aggregation by working fast at all steps and using serum in all media formulations. Removal of dead cells and myelin debris can be accomplished using commercially available Dead Cell and Myelin Removal kits in addition to cell washes with HBSS (when adhered) prior to trypsinization and purification.

8. All solutions used for MACS, including the sorting buffer, must be precooled prior to use. This will prevent the internalization of antibodies and reduce the risk of nonspecific antibody binding. The sorting buffer can be used within 1–2 months of preparation.

9. DMEM may be used in all washing steps in the MACS procedure to maintain cell viability.

10. FBS may be used instead of BSA in the sorting buffers for highly sensitive cells such as SCs. If reduced cell viability is observed in indirect labeling protocols, this can be due to the presence of EDTA in the sorting buffer. EDTA helps prevent cell clumping in suspension during the steps of double immunolabeling. To prevent cell death, DMEM containing 10% FBS can be used as an alternative sorting buffer. The concentration of EDTA can be reduced by half. It is recommended to monitor the suspensions of labeled cells throughout the process to ensure cell viability while avoiding clump formation.

11. The type and number of columns used in each purification needs to be determined empirically by the user according to the species and number of cells. MS columns are suitable for use with rat or human SC populations not exceeding 1×10^7 labeled cells for any given specific antigen. We recommend using no more than 2×10^7 total SCs per MS column for optimal results. It is possible that a larger load of cells can be used based on the expected capacity for each MS column, which is estimated to be up to 2×10^8 total cells by the manufacturer. However, the upper limit for each column has not been tested for SCs and needs to be determined by the experimenter.

12. Large-scale purifications can be accomplished by increasing the number of MS columns per sample of labeled cells or using LS columns designed for increased retention capacity. Although the purification of human SCs via MS columns is feasible, a decreased flow rate through the column and a reduced total cell recovery have been observed. This is likely due to the larger size of the human SCs with respect to those of the rat. Large cell columns are a suitable alternative if cultures of human SCs are used; these columns are designed for the separation of cells of 14–30 μm in diameter. The large cell column's metallic mesh contains larger beads. This improves the contact and retention of the labeled cells, allowing for a faster flow of non-labeled cells through the column.

13. As with any purification procedure, it is expected that a proportion of cells will not be recovered at the end of the procedure. This proportion can range from 10–50% depending on a variety of factors. Therefore, it is recommended to use an excess of cells from the outset as determined by the demands of the downstream work.

14. The efficiency of separation can be enhanced by immediately applying the eluted or retained cell fractions over a second column. This procedure can help eliminate contaminating cells and readily enhance the purity of the resultant final populations. This is reasonable when an excess of cells is used as starting material, noting that passing through multiple columns

diminishes total cell yields. An alternative method to improve purity is to use an intermediate step of recovery and expansion in vitro prior to subjecting the cells to a second round of labeling and purification by MACS.

15. It is imperative to assure optimal cell adhesion and extension of processes throughout the course of live cell labeling prior to fixation. Reducing the manipulation of the cells outside the CO_2 incubator and time of operation in the laminar flow hood facilitates cell retention. Repeated observations by phase contrast microscopy can help identify and prevent cell loss due to detachment. The appearance of rounded cells attached via short processes close to the border of the culture well is an early sign of detachment. If not treated in due time, detachment can progress throughout the population and lead to irreversible cell loss. Note that SCs typically retract their processes and detach from their dishes if alkaline media and/or media without calcium and magnesium are used during washes or incubation with antibodies. The use of a laminin substrate reduces but does not prevent the loss of cells due to detachment.

16. SC immunostaining with O4 and O1 antibodies should be carried out exclusively in live cells, as fixation impairs the specificity of antibody binding to membrane lipid antigens. Immunostaining with all other antibodies described in Table 1 can be carried out in live or fixed cell cultures with equal efficiency of staining.

17. Cell surface staining of SCs with antibodies against p75[NGFR], O4, and O1 should be smooth and homogeneous throughout the SC's plasma membrane. A granulated staining on the cell surface due to antibody capping (i.e., cross-linking of antigens by a specific antibody reaction) can be seen when fixation or documentation of cell surface fluorescence is delayed. Avoid this artifact by working fast and placing the cells on ice up until fixation.

18. See Andersen et al. [17] for a quantitative assessment of the efficiency of separation based on cell surface expression of O4, p75[NGFR], and Thy-1 by immunofluorescence microscopy analysis. Quantification of intracellular antigens such as S100 and glial fibrillary acidic protein (GFAP) is also possible. We have found that Western blot analysis is a sensitive tool to confirm the results and provide a quantitative estimation of the efficiency of purification based on detection of alternative SC markers such as cyclic nucleotide phosphodiesterase (CNPase) and nestin, which are absent in the fibroblast populations [17].

19. Cultures that have been preserved as recommended remain intact even several years after the immunoreaction was performed. Post-fixation is essential for the long-term preservation of the immunoreaction with fluorescent antibodies. Paraformaldehyde and/or methanol fixation does not alter the fluorescence of Alexa Fluor™ conjugates.

20. Note that most hybridoma cell lines grow in suspension and cells will appear round and shiny under phase contrast microscopy.

21. To maximize antibody titer, proceed to collect the conditioned medium once the cultures become heavily populated and begin displaying a proportion of crenated cells suspended in yellowish (acidic) medium.

Acknowledgments

We are grateful for the technical assistance provided by Dr. Ketty Bacallao and Ms. Blanche Kuo. We thank Dr. James Guest for critically reviewing the manuscript. The work presented in this chapter was generously supported by the NIH-NINDS (NS084326), The Craig Neilsen Foundation (339576), The Miami Project to Cure Paralysis, and The Buoniconti Fund. The authors declare no conflicts of interest with the contents of this article.

References

1. Brockes JP, Fields KL, Raff MC (1979) Studies on cultured rat Schwann cells. I. Establishment of purified populations from cultures of peripheral nerve. Brain Res 165(1):105–118

2. Calderon-Martinez D, Garavito Z, Spinel C, Hurtado H (2002) Schwann cell-enriched cultures from adult human peripheral nerve: a technique combining short enzymatic dissociation and treatment with cytosine arabinoside (Ara-C). J Neurosci Methods 114(1):1–8

3. Lehmann HC, Chen W, Mi R, Wang S, Liu Y, Rao M, Hoke A (2012) Human Schwann cells retain essential phenotype characteristics after immortalization. Stem Cells Dev 21(3):423–431. https://doi.org/10.1089/scd.2010.0513

4. Komiyama T, Nakao Y, Toyama Y, Asou H, Vacanti CA, Vacanti MP (2003) A novel technique to isolate adult Schwann cells for an artificial nerve conduit. J Neurosci Methods 122(2):195–200

5. Kaewkhaw R, Scutt AM, Haycock JW (2012) Integrated culture and purification of rat Schwann cells from freshly isolated adult tissue.

Nat Protoc 7(11):1996–2004. https://doi.org/10.1038/nprot.2012.118

6. Levi AD BR, Lofgren JA, Meima L, Hefti F, Nikolics K, Sliwkowski MX (1995) The influence of heregulins on human Schwann cell proliferation. J Neurosci 15(2):1329–1340

7. Casella GT, Bunge RP, Wood PM (1996) Improved method for harvesting human Schwann cells from mature peripheral nerve and expansion in vitro. Glia 17(4):327–338

8. Jirsova K, Sodaar P, Mandys V, Bar PR (1997) Cold jet: a method to obtain pure Schwann cell cultures without the need for cytotoxic, apoptosis-inducing drug treatment. J Neurosci Methods 78(1-2):133–137

9. Pannunzio ME, Jou IM, Long A, Wind TC, Beck G, Balian G (2005) A new method of selecting Schwann cells from adult mouse sciatic nerve. J Neurosci Methods 149(1):74–81. https://doi.org/10.1016/j.jneumeth.2005.05.004

10. Jin YQ, Liu W, Hong TH, Cao Y (2008) Efficient Schwann cell purification by differential cell detachment using multiplex collage-

nase treatment. J Neurosci Methods 170(1):140–148. https://doi.org/10.1016/j.jneumeth.2008.01.003

11. Wang HB, Wang XP, Zhong SZ, Shen ZL (2013) Novel method for culturing Schwann cells from adult mouse sciatic nerve in vitro. Mol Med Rep 7(2):449–453. https://doi.org/10.3892/mmr.2012.1177

12. Haastert K, Mauritz C, Chaturvedi S, Grothe C (2007) Human and rat adult Schwann cell cultures: fast and efficient enrichment and highly effective non-viral transfection protocol. Nat Protoc 2(1):99–104. https://doi.org/10.1038/nprot.2006.486

13. Tao Y (2013) Isolation and culture of Schwann cells. Methods Mol Biol 1018:93–104. https://doi.org/10.1007/978-1-62703-444-9_9

14. Manent J, Oguievetskaia K, Bayer J, Ratner N, Giovannini M (2003) Magnetic cell sorting for enriching Schwann cells from adult mouse peripheral nerves. J Neurosci Methods 123(2):167–173

15. Vroemen M, Weidner N (2003) Purification of Schwann cells by selection of p75 low affinity nerve growth factor receptor expressing cells from adult peripheral nerve. J Neurosci Methods 124(2):135–143

16. van Neerven SG, Krings L, Haastert-Talini K, Vogt M, Tolba RH, Brook G, Pallua N, Bozkurt A (2014) Human Schwann cells seeded on a novel collagen-based microstructured nerve guide survive, proliferate, and modify neurite outgrowth. Biomed Res Int 2014:493823. https://doi.org/10.1155/2014/493823

17. Andersen ND, Srinivas S, Pinero G, Monje PV (2016) A rapid and versatile method for the isolation, purification and cryogenic storage of Schwann cells from adult rodent nerves. Sci Rep 6:31781. https://doi.org/10.1038/srep31781

18. Jessen K, Mirsky R (2005) The origin and development of glial cells in peripheral nerves. Nat Rev Neurosci 6(9):671–682

19. Chandler C, Parsons L, Hosang M, Shooter E (1984) A monoclonal antibody modulates the interaction of nerve growth factor with PC12 cells. J Biol Chem 259(11):6882–6889

20. Sommer I, Schachner M (1981) Monoclonal antibodies (O1 to O4) to oligodendrocyte cell surfaces: an immunocytological study in the central nervous system. Dev Biol 83(2):311–327

21. Sommer I, Schachner M (1982) Cell that are O4 antigen-positive and O1 antigen-negative differentiate into O1 antigen-positive oligodendrocytes. Neurosci Lett 29(2):183–188

22. Soto J, Monje P (2017) Axon contact-driven Schwann cell dedifferentiation. Glia 65:864–882. https://doi.org/10.1002/glia.23131

23. Piñero G, Berg R, Andersen N, Setton-Avruj P, Monje P (2016) Lithium reversibly inhibits Schwann cell proliferation and differentiation without inducing myelin loss. Mol Neurobiol 10(2). https://doi.org/10.1007/s12035-016-0262-z

24. Bacallao K, Monje P (2015) Requirement of cAMP signaling for Schwann cell differentiation restricts the onset of myelination. PLoS One 10(2). https://doi.org/10.1371/journal.pone.0116948

25. Monje P, Bartlett Bunge M, Wood P (2006) Cyclic AMP synergistically enhances neuregulin-dependent ERK and Akt activation and cell cycle progression in Schwann cells. Glia 53(6):649–659

Chapter 7

Culture and Expansion of Rodent and Porcine Schwann Cells for Preclinical Animal Studies

Adriana E. Brooks, Gagani Athauda, Mary Bartlett Bunge, and Aisha Khan

Abstract

Cell-based therapies have become a major focus in preclinical research that leads to clinical application of a therapeutic product. Since 1990, scientists at the Miami Project to Cure Paralysis have generated extensive data demonstrating that Schwann cell (SC) transplantation supports spinal cord repair in animals with spinal cord injury. After preclinical efforts in SC transplantation strategies, efficient methods for procuring large, essentially pure populations of SCs from the adult peripheral nerve were developed for rodent and pig studies. This chapter describes a series of simple procedures to obtain and cryopreserve large cultures of highly purified adult nerve-derived SCs without the need for additional purification steps. This protocol permits the derivation of ≥90% pure rodent and porcine SCs within 2–4 weeks of culture.

Key words Peripheral nerve, Schwann cells, Fibroblasts, Preclinical research, Spinal cord injury, Clinical trials, Regenerative medicine, Food and Drug Administration (FDA)

1 Introduction

There is ample evidence that Schwann cells (SCs) support repair of injured peripheral nerves [1] and injured spinal cords [2–4]. In the rat, transplanted SCs reduce tissue loss and cyst formation, promote axon regeneration into the implant, and provide myelin for those axons. The SCs can be obtained from adult rat peripheral nerves, specifically the sciatic or sural nerves, isolated and purified in culture and given a combination of mitogenic factors for expansion in number. All of these findings contributed to the approval by the Food and Drug Administration (FDA) of a clinical trial testing the safety of transplantation of autologous SCs. An advantage of choosing SCs for cell therapy is that they can be obtained from a piece of peripheral nerve from the subject who is spinal cord injured, thereby avoiding the prospect of immune rejection following SC transplantation. The first clinical trial on autologous SC

Paula V. Monje and Haesun A. Kim (eds.), *Schwann Cells: Methods and Protocols*, Methods in Molecular Biology, vol. 1739, https://doi.org/10.1007/978-1-4939-7649-2_7, © Springer Science+Business Media, LLC 2018

transplantation conducted in subacutely injured subjects has been completed and found to be safe [5].

Although many spinal cord injury studies had been done over the course of 20 years [2, 6], the FDA required many additional ones to be performed in order to answer questions about SC survival for 6 months, migration from the implant, dose and toxicity (including minimal toxic dose), tumor formation, and method of injection into the spinal cord injury site [7]. Assessment of sensory changes was also performed after transplantation. Because so many SCs were required for these additional animals, different methods were explored for their efficient isolation, expansion, and purification in vitro. One of the challenges was to culture SCs for several passages without being overrun by fibroblasts using clinically relevant methods and procedures. This overgrowth of fibroblasts was suppressed simply by a combination of pre-degeneration of the nerve explants and the continued use of two chemical mitogens, neuregulin and forskolin, in the culture medium. The SCs obtained by these methods showed no signs of toxicity or genetic alterations in cultures expanded up to passage 5, even when the cells were subjected to cryopreservation.

The protocol described in this chapter differs from that used to generate SCs for the rat transplantation studies conducted over many years [2, 6]. The goal of the preclinical SC team was to prepare rat SCs in ways similar to those employed for preparing human SCs for clinical trials [7, 8]. For the rat transplantation studies, pituitary extract, heregulin, and forskolin were added as mitogens, and the cells were grown on polylysine. Only the two latter mitogens and laminin-coated culture dishes were used to generate SCs for the preclinical investigations [8]. The time required to generate the rat SCs for the preclinical studies was 3 weeks shorter than that for preparing the cells for transplantation, mainly due to different timing of mitogen addition.

This chapter presents the resulting protocol developed to obtain and cryopreserve highly purified cultures of rat and porcine SCs for preclinical studies carried out in animal models of spinal cord injury. This protocol introduces a series of modifications over existing protocols for adult SC culture [9, 10] that helped remove undefined factors from the culture medium, such as pituitary extract, without reducing the potential for cell expansion. Because the rat spinal cord was too small to test injection methods appropriate for transplantation into humans, a pig spinal cord injury model was developed [11–14] in order to successfully devise injection methods into a larger spinal cord. The protocol for obtaining highly purified populations of rat SCs was adequate for preparing the porcine cells with only minor modifications. The completion of these preclinical studies demonstrated that the final protocol for SC culture and expansion was effective, efficient, and reproducible to obtain large quantities of SCs from a modest segment of peripheral nerve tissue [5]. Both the porcine and rodent SC products had high viability and stability as established by the series of tests required by the FDA for product release [15]. The SC product derived from the use of these protocols was potent because the expanded cells retained their capability to respond to SC-specific mitogenic factors in vitro.

2 Materials

2.1 Equipment

1. Dissecting microscope. Stereo microscope with 6.3:1 zoom.
2. Fiber optic illuminator. High intensity 150 W halogen lamp.
3. Biological safety cabinet (BSC). A standard Class II Type A2 cabinet is recommended.
4. Horizontal laminar flow hood.
5. Forceps, 14.5–17 cm long.
6. Surgical scissors. Use two different types of Metzenbaum dissecting scissors, one straight with a sharp tip and the other curved with a blunt tip.
7. Fine forceps. Use Dumont forceps, style: #3, #4, or #5, approximately 11–12 cm with a standard straight tip.
8. Ice bucket with ice.
9. Benchtop centrifuge. Use a centrifuge that allows spinning the cells at a relative centrifugal force of $300 \times g$ and $450 \times g$. Place the settings at 4 °C for 5 min, with high brakes "*on*".
10. Pipette aid.
11. Micropipettes. Use assorted sizes varying from P10 to P1000.
12. Cell incubator. Use a humidified incubator at 37 °C, with 8% CO_2.
13. Freezer, −80 °C.
14. Liquid nitrogen cell storage tank.
15. Refrigerator, +4 °C.
16. Hemocytometer and coverslip.
17. Water bath, 37 °C.

2.2 Supplies

1. Liquid waste aspiration system. Use a collection flask attached to a suction pump or a similar device to aspirate the wash supernatants.
2. Conical tubes, 50 mL.
3. Pipettes of various sizes, varying from 1 to 50 mL, including aspirating pipettes.
4. Round tissue culture treated petri dishes, two sizes, 10 and 15 cm in diameter.

5. Micropipette tips. Use various sizes ranging from 10 to 1000 µL.

6. Tissue culture flasks, T-75 cm² flasks made of polyethylene with a canted neck and vented cap.

7. 6-well sterile tissue culture plates made of polystyrene.

8. Cryopreservation vials. The use of cryogenic vials with 1.5 mL capacity is recommended, but 5 mL vials have also been validated.

9. Freezing containers. This protocol utilizes Nalgene® Mr. Frosty for 1 or 5 mL vials. These containers require filling with isopropyl alcohol following the instructions provided by the manufacturer.

10. Bottle-top vacuum filter systems. Use systems with a cellulose nitrate membrane and a pore size of 0.22 µm. Filters of various sizes (e.g., 50 mL, 500 mL, and 1 L) are advised but not required.

11. Cell chamber slides with glass coverslips. The use of polystyrene 4-well chamber slides with a surface area of 1.7 cm² /well is recommended.

2.3 Reagents

1. 40 mg/mL gentamicin sulfate stock solution.

2. Leibovitz's L-15 Medium. Use L-15 Medium containing L-glutamine and sodium pyruvate, but no glucose or phenol red.

3. Sterile distilled water. Use cell culture quality water distilled and filtered, with no added substances.

4. Recombinant human Heregulinβ-1 (HRG1-B1). Prepare the working solution fresh each time of use by adding 500 µL of sterile distilled water to a 50 µg vial of HRG1-B1.

5. Forskolin. Powder from *Coleus forskohlii*, ≥98% pure (HPLC). Prepare a 15 mM forskolin stock solution by reconstituting 10 mg of forskolin powder in 1.6 mL of dimethyl sulfoxide. Aliquot this stock solution into 100 µL individual aliquots and store them at −20 °C for up to 6 months.

6. 200 nM (100×) L-glutamine stock solution. Prepare 13 mL aliquots of L-glutamine stock solution and store them at −20 °C.

7. Fetal bovine serum (FBS), heat inactivated and gamma irradiated.

8. Dulbecco's Modified Eagle's Medium (DMEM), high glucose, no L-glutamine, and no phenol red.

9. Neutral protease NB. From *Clostridium histolyticum*, chromatographically purified with a neutral protease activity of ≥0.50 U/mg. Prepare a stock solution by diluting the lyophilized powder in sterile distilled water to achieve 20 DMCU/ mL. Calculate the volume of sterile distilled water needed

depending on the DCM units of each specific lot of neutral protease. Prepare 1 mL aliquots of this stock solution and store frozen at −80 °C for up to 1 year.

10. Collagenase NB1. From *Clostridium histolyticum*, each vial contains not less than 2000 PZ units of collagenase. Prepare a stock solution by dissolving the lyophilized powder in 40 mL of a 50 mM Tris/10 mM calcium acetate buffer prepared in sterile distilled water, with an adjusted pH of 7.1. It is recommended that both reagents be ultrapure grade. Prepare 1 mL aliquots of this stock solution and store frozen at −80 °C for up to 1 year.

11. Calcium acetate hydrate, 99.99% trace metal basis.

12. ≥99.9% Tris, (hydroxymethyl) aminomethane, ultrapure grade.

13. 100 mg/mL calcium chloride, United States Pharmacopeia (USP) grade.

14. Dimethyl sulfoxide (DMSO). Use 99% DMSO with a maximum water content of 0.5%.

15. Laminin stock solution. Use laminin from Engelbreth-Holm-Swarm murine sarcoma basement membrane. The stock solution is distributed in vials containing 1–2 mg/mL in Tris-buffered NaCl.

16. 1× Dulbecco's phosphate-buffered saline (DPBS), no calcium and no magnesium.

17. 70% ethanol in distilled water.

18. 1× TrypLE™ Select Enzyme, no phenol red.

19. 1× Hank's balanced salt solution (HBSS), no calcium, no magnesium, no phenol red.

20. 0.4% trypan blue stain solution.

21. Belzer UW® Cold Storage Solution.

2.4 Preparation of Media, Solutions, and Plates

1. *Dissection medium*: Add 0.6 mL of gentamicin to 499.4 mL of Leibowitz's L-15 Medium and mix well. Prepare 50 mL aliquots which may be stored at +4 °C for up to 1 month.

2. *Schwann cell growth (SCG) medium*: 10 nM HRG1-B1, 2 μM forskolin, 4 mM L-glutamine, 51 μg/mL gentamicin, and 10% (v/v) FBS, prepared in DMEM medium. Mix 548.6 mL of DMEM medium (at room temperature), 62.5 mL of FBS, 83 μL of forskolin stock solution, 0.8 mL of gentamicin, 12.5 mL of L-glutamine, and 0.5 mL of the reconstituted HRG1-B1 stock solution. Filter the solution through a 1 L, 0.22 μm vacuum filter system. This medium may be stored at +4 °C for up to 1 month.

3. *D10 medium*: 10% (v/v) FBS in DMEM medium. Add 50 mL of FBS to 450 mL of DMEM medium. Mix well, and then

filter the solution through a 500 mL vacuum filter system. Store the medium at +4 °C for up to 1 month.

4. *Dissociation enzyme working solution*: Prepare this solution fresh, and keep cold (+4 °C) until ready to use. Collect one aliquot of the collagenase NB1 stock solution and one aliquot of the neutral protease NB stock solution. Allow both vials to thaw slowly on ice. Concurrently prepare a dilution buffer with 3.1 mM $CaCl_2$ in 25 mL of DMEM medium. Remove 8.9 mL of this buffer to a clean 15 mL conical tube. In this second conical tube, mix the 1 mL aliquot of neutral protease NB stock solution and 100 μL of the collagenase NB1 stock solution.

5. *Cryopreservation solution*: 10% (v/v) DMSO prepared in undiluted FBS. Prepare this solution fresh, do not store. Determine the total volume of cryopreservation solution needed to resuspend the SCs at a concentration of 2×10^6 cells/mL.

6. *Laminin-coated T-75 cm² flasks*: Use 1 μg of laminin stock per cm² surface area. In a 50 mL conical tube, add 75 μg of the laminin stock for every 10 mL of 1× DPBS. Resuspend the solution using a pipette. Immediately add the 10 mL of the solution to each T-75 cm² flask to be coated. Ensure that the entire surface is covered with the laminin solution, and then place the flask(s) in the cell incubator for 2 h. After the incubation period, aspirate the laminin solution and wash the flask two times with 10 mL of 1× DPBS. Do not aspirate the last wash until the cells are ready to be plated. The flask may also be stored for up to 48 h at +4 °C (*see* **Note 1**).

2.5 Tissue Procurement

The protocol described uses nerve tissue derived from the following animal species:

1. The anterior cutaneous branch of the femoral nerve is obtained from Yucatan miniature pigs.

2. The sciatic nerve is from Fischer 344 adult rats.

3 Methods

3.1 Manual Separation of Nerve Fascicles from the Epineural Connective Tissue

This is a generic protocol to produce SCs from pigs and rats although it may be easily adaptable to other animal species. There are minor differences between these two species, which are noted throughout each step. Through this chapter, we refer to the SCs originating from porcine and rats as pSCs and rSCs, respectively. Once harvested, the nerve segments may be stored and transported to the cell processing facility in Belzer UW® Cold Storage Solution or DMEM medium with antibiotics. The procedure should begin

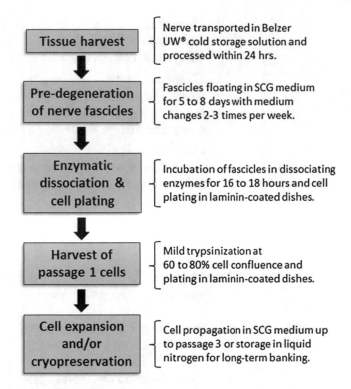

Fig. 1 Process time line from the receipt of the nerve tissue through expansion and cryopreservation of SCs

immediately upon receipt at the processing laboratory (*see* **Note 2**). The schematic diagram presents a synopsis of the procedural steps described in the sections below (Fig. 1).

1. Wipe all working surfaces and equipment thoroughly with 70% ethanol. Set up all reagents and supplies in the BSC and the horizontal laminar flow hood.

2. Prepare the dissection medium and keep it cold until ready to be used. Transfer approximately 7 mL of this dissection medium to each of the 10 cm tissue culture dishes (*see* **Note 3**).

3. Set up the dissecting microscope and fiber optic illuminator inside the horizontal laminar flow hood to aid in the visualization of the nerve fascicles.

4. Place the transportation container containing the nerve(s) in the horizontal laminar flow hood as well. Use sterile forceps to remove the nerve from the transportation container and carefully place it in a dry 15 cm tissue culture dish.

5. Using surgical scissors, trim off the excess fat tissue attached to the nerve and cut the nerve approximately into 0.5–1 cm segments (*see* **Note 4**).

6. Place three to five segments of peripheral nerve tissue into each tissue culture dish containing 7 mL of dissection medium. Keep these dishes on ice for the remainder of the process.

7. Clasp the nerve tissue segment, at the opposite end from where the fascicles will be pulled, using #3 forceps. Start pulling each fascicle away from the ensheathing connective tissue using fine forceps #4 or #5 (*see* **Note 5**).

8. Only those fascicles which are free of perineurial connective tissue should be kept in the culture. Discard any fascicles with indistinct and irregular boundaries (*see* **Note 6**).

9. Once all clean fascicles have been collected in one 10 cm tissue culture dish containing dissection medium, transfer the dish to the BSC to prepare for culture. Replace the dissection media with SCG medium (*see* **Note 7**).

10. Maintain the tissue culture dish inside the cell incubator for approximately 1 week.

3.2 Pre-degeneration of Nerve Fascicles

The period of pre-degeneration of the nerve tissue allows Wallerian degeneration to take place within the nerve explants, inducing Schwann cell dedifferentiation and proliferation before tissue dissociation. Additionally, studies have found that pre-degeneration in vitro reduces the outgrowth of fibroblasts from the cultured tissue [16]. During this period (5–8 days), the SCG medium is replenished every other day or twice during the week at a minimum. Follow the steps described below.

1. Wipe all surfaces of the BSC thoroughly with 70% ethanol.

2. Remove the 10 cm tissue culture dish with the nerve fascicles from the cell incubator and place it inside the BSC.

3. Aspirate the existing SCG medium without disturbing the fascicles. Add 7–10 mL of room temperature SCG medium using a sterile pipette, and place the fascicles back to the cell incubator.

4. Repeat **steps 2** and **3** at least twice a week until tissue dissociation (Subheading 3.3).

3.3 Dissociation of Nerve Fascicles to Isolate Schwann Cells

Enzymatic dissociation of pre-degenerated nerve fascicles is essential to yield a single cell suspension suitable for plating and propagation (*see* **Note 8**).

1. Prepare the dissociation enzyme working solution as described in Subheading 2.4.

2. Transfer the fascicles to 1 well of a 6-well plate using #3 fine forceps.

3. Add approximately 2 mL of dissociation enzyme working solution to the well with fascicles. If necessary add up to 3 mL to ensure that all the fascicles are submerged in the enzyme solution.

4. Place the 6-well plate in the cell incubator for 16–18 h.

5. After the tissue dissociation period, neutralize the enzyme solution by adding FBS, 10% (v/v) of the total volume of the dissociation working solution to the nerve fascicles. Essentially add 0.2 mL of FBS to 1 well of the 6-well plate (*see* **Note 9**).

6. Mechanically dissociate the fascicles by gently pipetting a few times and then transferring the resulting cell suspension to a 50 mL conical centrifuge tube (*see* **Note 10**).

7. Rinse the well of the 6-well plate in which the fascicles were dissociated with 2 mL of room temperature D10 medium, and then add this wash to the same centrifuge tube.

8. Increase the total volume to 40 mL by adding D10 medium. Centrifuge the pSCs at $300 \times g$ and rSCs at $450 \times g$. Observe the centrifuge settings described in Subheading 2.1 (*see* **Note 11**).

9. Repeat the previous step to wash the SCs once more. However, before increasing the volume to 40 mL, collect 100 μL of the cell suspension to obtain a viable cell count using a traditional hemocytometer (*see* **Note 12**).

3.4 Plating of Dissociated Schwann Cells at Passage Zero

Passage zero (P0) SCs are plated at a density of 13.3×10^4 viable cells per cm^2. Based on several validation studies conducted during the development of this protocol, this seeding density was found to be optimal for the adhesion and propagation of the freshly dissociated SCs. The steps described below indicate the use of T-75 cm^2 flasks, but smaller tissue culture dishes, such as T-25 cm^2 flasks or 6-well plates, may also be utilized.

1. Gently resuspend the SC suspension at a concentration of 0.1×10^6 viable cells/mL in SCG medium.

2. Aspirate the 1× DPBS from the laminin-coated T-75 cm^2 flask(s) immediately before plating the cells. Wash the flask(s) twice with 10 mL of 1× DPBS to remove any excess laminin solution.

3. Add 10 mL of the cell suspension to each laminin-coated T-75 cm^2 tissue culture flask.

4. Place the flask(s) in the cell incubator for approximately 6–10 days.

5. Replace the SCG medium every other day (*see* **Note 13**).

6. Observe the culture under an inverted microscope after performing each medium change to estimate the cell confluency, and assess the possibility of microbial contamination (*see* **Note 14**).

**3.5 Harvesting
and Replating
of Passage 1 (P1)
Schwann Cells**

P1 SCs are plated at a seeding density of 6.6×10^3 viable cells per cm² in SCG medium. This is the recommended minimum seeding density. Several validation studies conducted at our cell processing facility have confirmed that this density supports SC attachment, proliferation, and cell-to-cell contact for optimal expansion of the cell population (*see* Fig. 2a, b).

1. Wipe all working surfaces especially inside the BCS with 70% ethanol. In addition, wipe the outer surface of each tissue culture flask with 70% ethanol before they are placed inside the BSC. Use caution not to wet the paper filter inside the cap of the flask.

2. Aspirate all the SCG medium in the T-75 cm² tissue culture flask. Then, add a small volume of 1× HBSS to rinse out the

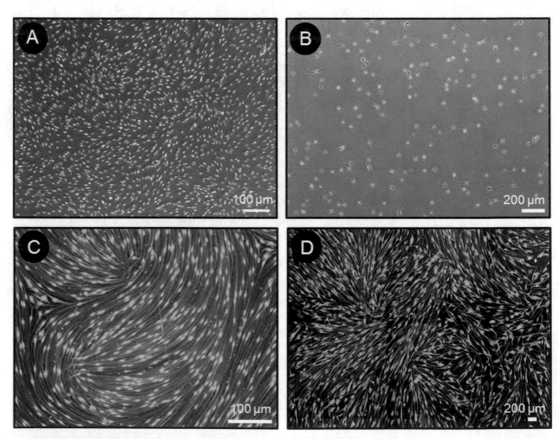

Fig. 2 (**a**) Phase contrast image of a confluent passage-3 rSC culture observed at 10×. (**b**) Phase contrast image of rSCs plated at a low density (approximately 4×10^3 viable cells per cm²) for expansion. (**c**) S100 protein immunostaining (green) of passage-2 rSCs forming a pattern that characterizes the high confluence of the cells in vitro, observed at 10×. Nuclei were counterstained with Hoechst (blue). (**d**) pSCs isolated from the anterior cutaneous branch of the femoral nerve obtained from a Yucatan miniature pig. The pSCs were obtained following the protocol described in this chapter and expanded in vitro until harvest passage-2. Cultures consisted of 94% pure SCs based on immunostaining with GFAP (green) and fibronectin (red)

serum from the SCG medium. Aspirate quickly and add 3 mL of undiluted TrypLE™ Select.

3. Place the cultures for 3–5 min inside the BSC at room temperature. Visually inspect the cultures under the phase contrast microscope for detachment of the cells from the culture surface (*see* **Note 15**).

4. Once the cells are completely detached from the dish surface, neutralize the TrypLE™ Select reagent by adding approximately 7 mL of D10 medium to stop the enzymatic reaction.

5. Collect all the cell suspension into 50 mL conical tube(s) and centrifuge the pSCs at $300 \times g$ and the rSCs at $450 \times g$. Use the settings described in Subheading 2.1.

6. Remove the tube(s) from the centrifuge and aspirate the supernatant. Resuspend the pellet in approximately 3–5 mL of D10 medium.

7. Increase the cell suspension volume to approximately 10 mL. From this volume collect 100 μL. Follow the steps described in Subheading 3.4 to obtain a viable cell count.

8. Increase the volume of the remaining cell suspension to 40 mL also with D10 medium, and centrifuge once more. Use the centrifugation settings described in Subheading 2.1.

9. Resuspend the cells at a concentration of 1×10^6 viable cells/mL in SCG medium, and plate them into tissue culture flasks freshly coated with laminin at a seeding density of 6.6×10^3 viable cells per cm² in SCG medium. Alternatively, the cells may be cryopreserved for future use (*see* Subheading 3.6).

10. For cell purity assessment by immunostaining, plate the SC suspension into a 4-well chamber slide at a seeding density of 50,000 cells per well.

11. Fix the slides with a 4% paraformaldehyde solution in 1× DPBS within 24 h (*see* **Note 16**).

3.6 Cryopreservation

Cells can be cryopreserved at passage 1 or any subsequent passage thereafter for future use (*see* **Note 17**). We have found that 2×10^6 viable cells per mL in the cryopreservation solution is the optimal cell concentration for long-term banking. We recommend the use of a 10% final concentration of DMSO as cryoprotectant. Because our cell processing facility often prepares large batches of cells for banking, over 60×10^6 cells in some cases, we have modified slightly the preparation of a typical cryopreservation solution. To avoid a prolonged exposure of the SCs to the toxic effects of the DMSO, the cells are separated into independent small batches while being resuspended in FBS only. The DMSO is added at a final concentration of 10% (v/v) directly to the cells suspended in FBS in a stepwise manner. This procedure allows the experimenter to comfortably finish processing one batch at a time, while the remaining batches are maintained on ice in the absence of DMSO.

1. Resuspend the cell pellet in a small volume of FBS. It is recommended to use approximately 2–5 mL of FBS. Then, add the remaining volume of FBS to achieve the 90% of the total volume in which the cells will be resuspended for cryopreservation. Mix gently.

2. Add the DMSO gradually not to produce an osmotic shock in the cells.

3. Mix gently and quickly proceed to aliquot the cell suspension into the cryovials. The use of 2 mL cryovials is recommended.

4. Place the cryovials inside freezing containers.

5. Transfer the freezing containers with the cells to a −80 °C freezer immediately, and store them for 24–72 h. This will allow the water to move out of the cells while freezing slowly to avoid the formation of large water crystals with potential to destroy organelles and membranes.

6. Transfer the cryovials to a liquid nitrogen storage tank for long-term storage (*see* **Note 18**).

4   Notes

1. There is significant variation in cell adherence when the laminin-coated surfaces dry out. It becomes evident in areas where there is minimum or no cell growth. If the flask(s) will be stored at +4 °C (up to 24 h), they should be put in a ziplock or equivalent sealed plastic bag. Ensure there is enough 1× DPBS to cover the entire surface area and that the flask(s) is placed on an evenly flat surface in the refrigerator.

2. The tissue is typically processed immediately upon receipt. Nonetheless, comparable results can be obtained from nerves processed within 24 h of harvest. We do not recommend storing the nerve tissue for more than 36 h, as the tissues become softer and gluey which makes the process of separating the fascicles more difficult.

3. Adding more than 7 mL of dissection medium to the petri dish makes the nerve segments float and difficult to keep in place for fascicle extraction. In addition, it increases the risk of medium spillage or loss of nerve tissue. Adding less than 7 mL increases the risk of dehydration of the nerve segments, especially when it takes a longer time for processing. The nerve segments should be placed on ice to minimize tissue degradation, while the separation of fascicles is performed.

4. Straight and sharp Metzenbaum scissors are recommended to cut the nerve into smaller segments. They allow a clean cut with well-delineated edges to facilitate a precise selection of fascicles during separation. The blunt curved scissors are recommended to trim the excess fat tissue that typically surrounds the nerve. The nerve segments should be cut at a length of approximately 1 cm long. Longer segments will make the task of extracting the fascicles difficult, especially in the case of bifurcated fascicles. By contrast, cutting segments shorter than 1 cm will increase the duration of the process unnecessarily.

5. Observe the forceps tips under the dissecting microscope prior to use. Confirm that they are sharp and not bent before initiating the pulling of fascicles; otherwise, the fascicles may be damaged during processing. The tips of the forceps should be perfectly aligned without any gaps or overlap. If gaps or overlaps exist, it is hard to grip the fascicles to separate them from the epineural connective tissue. It is recommended to use #3 fine forceps to hold the nerve and #4 or #5 to separate the fascicles from the epineural connective tissue.

6. The fascicles have a smooth and shiny surface with clearly defined borders. When a tail of connective tissue adheres to the departing fascicle, an attempt should be made to remove it by pulling it away from the end of the fascicle with the aid of forceps #4. If the perineurium fails to separate, that fascicle must be excluded from the culture in order to preserve the highest purity possible in the resultant SC populations.

7. Fascicles often stick to the plastic pipettes. Therefore, special care should be taken to not aspirate or touch the fascicles with the pipette tip. This process is enabled by tilting the petri dish slightly, allowing the accumulation of dissection medium and fascicles to one side.

8. Taking into consideration the extended incubation period of the fascicles in dissociation enzymes (16–18 h), it is recommended to initiate this step toward the end of a working day. This will allow the staff member to return the next morning and proceed with the dissociation of the tissue.

9. Prepare the tissue culture flasks or slides by coating them with laminin before initiating this process because a 2-h incubation period is required. If the laminin-coated flasks are not ready to be used, place the cell suspension on ice.

10. The tissue incubation in dissociation enzymes enhances the SC's sensitivity to shear stress-induced mechanical damage. However, gentle pipetting is essential for the isolation of primary cells. Cell destruction and lower viability may result if this is done too aggressively. Use a slow rate to aspirate and release. Intermittent pipetting is also recommended to lessen the mechanical damage.

11. SCs derived from porcine nerve tissue are larger and denser cells than those obtained from rats and require lower centrifugal forces for pelleting. The centrifugation speed was optimized through various validation studies developed at our cell processing facility based on the size of the cells, the percentage of cell recovery, and the viability post centrifugation.

12. Prepare a 1:1 dilution by adding 100 μL of trypan blue to the collected cell sample. Use a pipette to take 20 μL of the trypan blue-treated SC sample and add it to the hemocytometer. Gently fill both chambers underneath the coverslip. Allow the cell suspension to be drawn out by capillary action rather than pushing it in. The use of a tally counter is recommended but not required. Count the live (unstained) cells.

13. Pre-warm the SCG medium for each change of medium. A 60–80% cell confluence is recommended before trypsinization. Cell confluence is determined by the ratio between the total surface areas occupied by the cells versus the available space. Healthy proliferating SCs will reach such confluence within 6–10 days in culture. At that point they will begin making "swirl" patterns (*see* Fig. 2c).

14. Healthy SCs should have long cell processes giving them the unique spindle-shaped morphology that characterizes them. Debris should not be attached to the cells, and the medium should be clear without discoloration or turbidity after passage 1. Only at P0 will the medium appear cloudy with pale floating particles due to myelin and debris.

15. During cell harvesting, the cells should completely detach from the surface of the flask(s). This trypsinization step should not last longer than 3–5 min. If necessary tap the side of the flasks several times to accelerate the process. When the SCs become detached, the medium will become turbid.

16. To assess the purity of rSCs through immunostaining, we recommend using an anti-rat S100 antibody raised in rabbits. The S100 protein is present in the cytoplasm of SCs but not fibroblasts. To assess the purity of pSCs, we recommend using a polyclonal rabbit anti-glial fibrillary acidic protein (GFAP) antibody. Staining with an anti-fibronectin antibody is recommended for the detection of fibroblasts for both the rSCs and the pSCs (*see* Fig. 2d).

17. SCs should not be expanded further than passage 4. Subsequent passages have shown slower than normal growth until eventually they cease to replicate. Also, signs for morphological changes have been observed in the latest passages.

18. We have stored cryovials for up to 10 years in liquid nitrogen without notable decline in either the viability or potential to proliferate after thawing.

Acknowledgment

We thank Risset Silvera, Maxwell Donaldson, and Yelena Pressman for their technical assistance. We acknowledge the outstanding contributions of Linda White in the development of these preclinical protocols. We also are immensely grateful to Patrick Wood and Paula Monje for their critical review and valuable comments that greatly improved the manuscript. Key participants supporting this project were Drs. W. Dalton Dietrich, Allan Levi, James Guest, Damien Pearse, and Kim Anderson. This work was funded from the NIH-NINDS, the Miami Project to Cure Paralysis, the Buoniconti Fund, and the Christopher and Dana Reeve Foundation International Research Consortium.

References

1. Quintes S, Goebbels S, Saher G, Schwab MH, Nave KA (2010) Neuron-glia signaling and the protection of axon function by Schwann cells. J Peripher Nerv Syst 15(1):10–16. https://doi.org/10.1111/j.1529-8027.2010.00247.x

2. Fortun J, Hill CE, Bunge MB (2009) Combinatorial strategies with Schwann cell transplantation to improve repair of the injured spinal cord. Neurosci Lett 456(3):124–132. https://doi.org/10.1016/j.neulet.2008.08.092

3. Oudega M, Xu XM (2006) Schwann cell transplantation for repair of the adult spinal cord. J Neurotrauma 23(3–4):453–467. https://doi.org/10.1089/neu.2006.23.453

4. Tetzlaff W, Okon EB, Karimi-Abdolrezaee S, Hill CE, Sparling JS, Plemel JR, Plunet WT, Tsai EC, Baptiste D, Smithson LJ, Kawaja MD, Fehlings MG, Kwon BK (2011) A systematic review of cellular transplantation therapies for spinal cord injury. J Neurotrauma 28(8):1611–1682. https://doi.org/10.1089/neu.2009.1177

5. Anderson KD, Guest JD, Dietrich WD, Bunge MB, Curiel R, Dididze M, Green BA, Khan A, Pearse DD, Saraf-Lavi E, Widerstrom-Noga E, Wood P, Levi AD (2017) Safety of autologous human Schwann cell transplantation in subacute thoracic spinal cord injury. J Neurotrauma. https://doi.org/10.1089/neu.2016.4895

6. Bunge MB (2016) Efficacy of Schwann cell transplantation for spinal cord repair is improved with combinatorial strategies. J Physiol Lond 594(13):3533–3538. https://doi.org/10.1113/Jp271531

7. Bunge MB, Monje PV, Khan A, Wood PM (2017) From transplanting Schwann cells in experimental rat spinal cord injury to their transplantation into human injured spinal cord in clinical trials. Prog Brain Res 231:107–133

8. Bastidas J, Athauda G, De la Cruz G, Chan WM, Golshani R, Berrocal Y, Henao M, Lalwani A, Mannoji C, Assi M, Otero A, Khan A, Marcillo AE, Norenberg M, Levi A, Wood PM, Guest JD, Dietrich JD, Bunge MB, Pearse DD (2017) Human SC exhibit long-term cell survival, a lack of tumorigenicity and promote repair when transplanted into the contused spinal cord. Glia. https://doi.org/10.1002/glia.23161

9. Kanno H, Pearse DD, Ozawa H, Itoi E, Bunge MB (2015) Schwann cell transplantation for spinal cord injury repair: its significant therapeutic potential and prospectus. Rev Neurosci 26(2):121–128. https://doi.org/10.1515/revneuro-2014-0068

10. Morrissey TK, Kleitman N, Bunge RP (1991) Isolation and functional characterization of Schwann cells derived from adult peripheral nerve. J Neurosci 11(8):2433–2442

11. Benavides F, Santamaria J, Solano J, Levene H, Guest J (2011) Utility of monitoring spinal cord conduction during transplantation in incomplete SCI (Abstract). Paper presented at the 50th Anniversary Annual Scientific Meeting of The International Spinal Cord Society (ISCoS), Washington, DC, USA, June 4–8, 2011

12. Guest J, Benavides F, Padgett K, Mendez E, Tovar D (2011) Technical aspects of spinal cord injections for cell transplantation. Clinical and translational considerations.

Brain Res Bull 84(4–5):267–279. https://doi.org/10.1016/j.brainresbull.2010.11.007

13. Guest J, Santamaria AJ, Benavides FD (2013) Clinical translation of autologous Schwann cell transplantation for the treatment of spinal cord injury. Curr Opin Organ Transplant 18(6):682–689. https://doi.org/10.1097/MOT.0000000000000026

14. Guest J, Levene H, Padgett K, Garcia-Canet C, Benavides F, Santamaria J, Solano J (2011) Tolerance of three volumes of injected SC into the contused porcine spinal cord. Maximum tolerated dose study. Paper presented at the 29th Annual Mantional Neurotrauma Symposium, Hollywood Beach, FL, July 10–13, 2011

15. Food and Drug Administration, Department of Health and Human Services, 21 C.F.R. § 312 (2016).

16. Mauritz C, Grothe C, Haastert K (2004) Comparative study of cell culture and purification methods to obtain highly enriched cultures of proliferating adult rat Schwann cells. J Neurosci Res 77(3):453–461. https://doi.org/10.1002/jnr.20166

Chapter 8

Chemical Conversion of Human Fibroblasts into Functional Schwann Cells

Eva C. Thoma

Abstract

Direct conversion of one somatic cell type into another represents a promising approach to obtain patient-specific cells for numerous applications. Here, we describe a method allowing the transdifferentiation of human postnatal fibroblasts into functional Schwann cells via a transient progenitor stage. The conversion process is solely based on chemical treatment and does not require the overexpression of ectopic genes. The resulting induced Schwann cells (iSCs) can be characterized by expression of Schwann cell-specific proteins and neuro-supportive and myelination capacity in vitro. This strategy allows to obtain mature Schwann cells from human fibroblasts under chemically defined conditions without the introduction of ectopic genes.

Key words Transdifferentiation, Human Schwann cells, Small molecules, Direct conversion, In vitro model

1 Introduction

Schwann cells are the major glia cell type of the peripheral nervous system where they play an important role in development, homeostasis, and disease [1]. Relevant in vitro models are an important prerequisite to analyze glia cell biology and to model Schwann cell affecting disorders. Here, we present a novel approach to obtain human Schwann cells by directly converting fibroblasts into Schwann cells solely by chemical treatment and without the expression of ectopic genes [2]. The protocol consists of two main steps: First, fibroblasts are transdifferentiated into a transient neural precursor stage using a novel multikinase inhibitor. Transient precursors are passaged once and subsequently differentiated into induced Schwann cells (iSCs). These cells display Schwann cell-specific protein and gene expression patterns and neuroprotective and myelination capacity in vitro and thus represent a useful tool to study Schwann cell functions and pathophysiology.

Chemical conversion represents a promising approach to obtain Schwann cells in a cost-effective and highly controllable and standardized way. Furthermore, this method does not require the

Paula V. Monje and Haesun A. Kim (eds.), *Schwann Cells: Methods and Protocols*, Methods in Molecular Biology, vol. 1739,
https://doi.org/10.1007/978-1-4939-7649-2_8, © Springer Science+Business Media, LLC 2018

introduction of transgenes or the use of embryonic or postmortem tissues as it is the case with Schwann cells derived from pluripotent stem cells [3] and primary cells [4], respectively.

2   Materials

2.1   Cells

This protocol was optimized using commercially available postnatal human foreskin fibroblasts and embryonic stem cell-derived neural stem cells (ESC-NSCs). ESC-NSCs can be derived from human embryonic stem cells by following the protocol described previously [5].

1. Cryopreserved fibroblasts (Millipore, SCC058) (*see* **Note 1**).
2. Cryopreserved ESC-NSCs.

2.2   Cell Culture Media

Prepare all media under sterile conditions and maintain them refrigerated at 4 °C. If not indicated otherwise, media are stable for 1 month at 4 °C. Growth factors are usually prepared as 1000× stock solutions in appropriate solvents, as recommended by the manufacturer's instructions. Growth factors have to be added right before use in the preparation of culture media. For compounds dissolved in dimethyl sulfoxide (DMSO) (e.g., Compound B and CP21), stock solutions have to be pre-warmed to 37 °C directly before use and added to pre-warmed medium.

1. Fibroblast medium. FibroGRO™-LS Complete Media Kit (Millipore) is used as fibroblast medium. Complete medium consists of FibroGRO™ Basal Medium supplemented with 5 ng/ml basic fibroblast growth factor (bFGF), 50 µg/ml ascorbic acid (Sigma), 7.5 mM L-glutamine, 1 µg/ml hydrocortisone hemisuccinate, 5 µg/ml insulin, and 2% fetal bovine serum (FBS). Complete medium is light sensitive; prolonged exposure to light has to be avoided.

2. Neural induction medium. NeuroCult NS-A proliferation medium (Stemcell Technologies) supplemented with 1% penicillin/streptomycin (Gibco), 20 ng/ml bFGF, 20 ng/ml epidermal growth factor (EGF), 20 ng/ml brain-derived neurotrophic factor (BDNF), 2 µg/ml heparin (Stemcell Technologies), 500 ng/ml delta-like 4 (Dll4), 500 ng/ml Jagged1, 500 ng/ml sonic hedgehog (SHH) (PeproTech), 0.2 mM ascorbic acid, 100 ng/ml FGF8a, 10% NSC-CM (*see* **item 4**), and 2 µM Compound B (Roche) (*see* **Note 2**).

3. Neural induction medium supplemented with inhibitor mix. Neural induction medium supplemented with 500 ng/ml Noggin (Peprotech), 10 µM SB431452 (Tocris), and 1 µM CP21 (Roche, GSK3 inhibitor) [6].

4. Neural differentiation medium (N2B27). DMEM/F12 and Neurobasal (1:1) supplemented with 50 μM β-mercaptoethanol, 2% B27 supplement without vitamin A (Invitrogen), 1% N_2 supplement (Invitrogen), 1% penicillin/streptomycin, 20 ng/ml BDNF, 20 ng/ml GDNF, 1 μg/ml laminin (Roche), 0.2 mM ascorbic acid, and 0.5 mM dibutyryl-cAMP (Sigma).

5. Neural differentiation medium supplemented with inhibitor mix. Neural differentiation medium supplemented with 500 ng/ml Noggin (Peprotech), 10 μM SB431452 (Tocris), and 1 μM CP21 (Roche, GSK3 inhibitor) [6].

6. NSC-CM (medium conditioned by ESC-NSCs). ESC-NSCs are cultured on plates coated with poly-ornithine (PO) and laminin (POL) in N2B27 supplemented with 10 ng/ml bFGF, 10 ng/ml EGF, and 20 ng/ml BDNF. Medium is changed every 2–3 days, centrifuged for 5 min at $200 \times g$, and subsequently filtered through a 0.2 μm filter to remove residual cells or cell debris. NSC-CM can be stored in aliquots at −20 °C. After thawing, NSC-CM can be stored at 4 °C for up to 1 week.

2.3 Coated Cell Culture Dishes and Other Cell Culture Reagents

1. POL-coated cell culture dishes. Sterile culture dishes are covered with PO solution (Sigma, diluted 1:6 in phosphate buffer saline (PBS)) and incubated overnight at 37 °C in a 5% CO_2 incubator. Plates are washed twice with sterile PBS, covered with laminin coating solution (Invitrogen, diluted 1:500 in PBS), and incubated overnight at 37 °C in a 5% CO_2 incubator. PO-laminin-coated plates can be stored at 37 °C in a 5% CO_2 incubator for up to 2 days. After laminin coating, plates are stored at 37 °C in a 5% CO_2 incubator for up to 7 days.

2. 1× PBS (Gibco).

3. 0.05% Trypsin-EDTA (Gibco).

4. 0.25% Trypsin-EDTA (Gibco).

5. Valproic acid (VPA) stock solution. VPA (Sigma) prepared in water at a concentration of 1 mM and aliquoted and stored at −20 °C. Upon thawing, valproic acid stock solution is stored at 4 °C and used within 1 week.

6. 10 mg/ml polybrene stock solution (Millipore).

7. Trypan blue stock solution (Invitrogen).

2.4 Equipment and Supplies

2.4.1 General Equipment

1. Biosafety laminar flow cabinet class II.

2. Inverted microscope for light and fluorescence microscopy with 10×, 20×, and 40× objectives.

3. Centrifuge with insets for plate centrifugation.

4. Automated cell counter (e.g., Countess II FL, Thermo Fisher) or Neubauer chamber.

5. 37 °C incubator with humidity and gas control to maintain >95% humidity and an atmosphere of 5% CO_2 in air.

6. Water bath.

7. Vortex mixer.

8. Pipetboy.

2.4.2 *Cell Culture Consumables*

1. Sterile tissue culture-treated cell culture dishes (Corning).

2. Ultralow attachment cell culture dishes (6-well plates, Corning).

3. Serological pipet (sterile, 5–25 ml).

4. 70 μm cell strainers.

3 Methods

Perform all cell culture work in a biosafety cabinet, class II, unless indicated otherwise. Aliquot media to the required volume and pre-warm to 37 °C before use to avoid loss of activity of temperature-sensitive media components.

3.1 Thawing and Plating of Fibroblasts

Before starting the transdifferentiation protocol, human fibroblasts are thawed and cultured for a few days by following the protocol described below (*see* **Note 3**).

1. Remove vial with cryopreserved fibroblasts from the liquid nitrogen tank, and wipe the vial with disinfectant reagent (e.g., 70% isopropanol).

2. In a biosafety cabinet, unscrew the top of the cryovial for pressure equalization.

3. Place the cryovial in 37 °C water bath and turn it slowly for about 2 min until only a small clump of ice is left.

4. Wipe the vial with disinfectant reagent and open it under a biosafety cabinet.

5. Transfer the cell suspension to a 15 ml tube containing a tenfold volume (usually 10 ml) of pre-warmed fibroblast medium.

6. Centrifuge for 5 min at $200 \times g$.

7. Remove the supernatant and resuspend the cells in 1–2 ml of pre-warmed fibroblast medium.

8. Determine the number of total and dead cells by using the trypan blue exclusion assay.

9. Seed fibroblasts in a 10 cm dish at 5000–10,000 cells/cm² in fibroblast medium. Place the dish at 37 °C in a 5% CO_2 incubator.

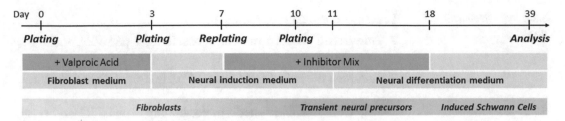

Fig. 1 Flowchart describing the main steps and the timing of fibroblast conversion into Schwann cells

3.2 VPA Treatment and Sphere Formation

In the first part of the protocol, fibroblasts are treated with VPA, a histone deacetylase inhibitor that has previously been shown to improve cellular reprogramming [7]. In this protocol, VPA treatment is used to erase epigenetic signatures and render cells more amenable for cell-fate changing signals. Subsequently, cells are treated with Compound B to induce a neural precursor stage. Compound B is a multikinase inhibitor that has been shown to enhance the proliferation of neural precursors while inhibiting their differentiation [2]. During neural induction, cells are cultured in suspension to allow the formation of sphere-like structures. The ability to form spheres in suspension is a characteristic typical of various progenitor cells. Thus, this step helps to select the cells that have converted into neural precursors. The resulting primary spheres are then dissociated and seeded for secondary sphere formation to allow a more stringent selection of those cells exhibiting precursor characteristics. During secondary sphere formation, cells are treated with compounds known to inhibit TGF-β, BMP, and GSK3 signaling, as they have been shown to promote gene transfer-dependent neural conversion of fibroblasts [8] (Fig. 1).

3.2.1 Replating of Fibroblasts (Day 0)

1. Remove medium and wash the cells once with PBS.

2. Add pre-warmed 0.05% Trypsin-EDTA and incubate for 3–4 min at 37 °C in a 5% CO_2 incubator.

3. Check cell detachment under a microscope. When cells start to detach, add pre-warmed fibroblast medium (1:1 with respect to the volume of Trypsin solution), and rinse the plate to detach remaining cells.

4. Collect the cell suspension and centrifuge at $200 \times g$ for 5 min.

5. Remove the supernatant and resuspend the cells in 1–2 ml of pre-warmed fibroblast medium.

6. Determine the number of viable cells after trypan blue staining (as described in Subheading 3.1).

7. Seed the cells at a density of 7300 cells/cm² in fibroblast medium containing 1 μM VPA. Incubate the cells for 24 h at 37 °C in a 5% CO_2 incubator.

3.2.2 VPA Treatment (Day 1)

1. Remove the culture medium.

2. Add fresh fibroblast medium containing 1 μM VPA and 6 μg/ml polybrene.

3. Incubate the cells for 24 h at 37 °C in a 5% CO_2 incubator.

3.2.3 VPA Treatment (Day 2)

1. Add fresh fibroblast medium containing 1 μM VPA and 6 μg/ml polybrene without replacing the medium.

2. Incubate the cells for 24 h at 37 °C in a 5% CO_2 incubator.

3.2.4 Seeding Fibroblasts for Primary Sphere Formation (Day 3)

1. Detach the cells by trypsinization using 0.05% Trypsin-EDTA as described in Subheading 3.2.1 (*see* **Note 4**).

2. After centrifugation, resuspend the cells in neural induction medium and determine cell number.

3. Seed the cells in ultralow attachment plates for sphere formation at a density of 5×10^5 cells/well in 2 ml medium/well.

4. Place the plates at 37 °C in a 5% CO_2 incubator. Do not move the plates for the next 4 days to allow sphere formation.

3.2.5 Dissociation of Primary Spheres and Seeding for Secondary Sphere Formation (Day 7)

1. Collect the medium containing spheres in 50 ml tubes using a 5 ml tip (*see* **Note 5**).

2. Wash the plates with 1× PBS to collect residual spheres and add to the sphere suspension collected in the 50 ml tube.

3. Wait for 10–15 min for spheres to sink to the bottom of the tube.

4. Carefully remove the supernatant and add 10 ml of 1× PBS. Invert the tubes and wait for 10–15 min for the spheres to sink to the bottom of the tube.

5. Remove the 1× PBS and add 0.25% Trypsin-EDTA by using 0.5 ml of Trypsin-EDTA to the suspension culture collected from a 6-well plate. Scale up accordingly up until reaching a maximum of 5 ml. Incubate the suspension at 37 °C for 15–30 min. Occasionally and carefully move the tube to prevent sphere aggregation at the bottom.

6. Use a 1 ml pipet to dissociate the spheres. Pipet up and down for about 1 min until no more spheres are visible (*see* **Note 6**). Then, add neural induction medium at a ratio of 1:1 (v/v) with respect to the volume of Trypsin-EDTA, and centrifuge at $200 \times g$ for 5 min.

7. Remove the supernatant, resuspend the cells in neural induction medium, and filter through a 70 μm cell strainer to remove any remaining cell aggregates.

8. Determine the cell number and seed the cells in ultralow attachment plates at a density of 3.5×10^5 cells/well using 1.5 ml/well of neural induction medium supplemented with inhibitor mix.

9. Place the plates at 37 °C in a 5% CO_2 incubator. Do not move the plates for the next 3 days to allow for sphere formation (*see* Fig. 2).

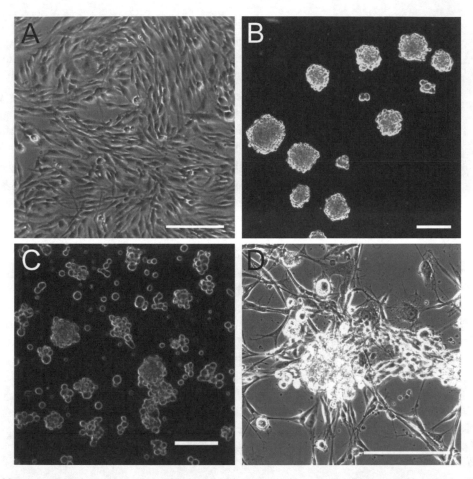

Fig. 2 Morphological changes during early stages of the conversion process. (**a**) Human fibroblasts at day 0. Cells display typical fibroblast morphology when they are at 80–90% confluency. (**b**) Primary spheres at day 6. (**c**) Secondary spheres at day 10. These structures are slightly smaller and show a slightly less regular shape compared to primary spheres. (**d**) Secondary sphere 24 h after plating on POL-coated plates (day 11). Bipolar cells have started to migrate out of the spheres. Scale bars, 200 μm

3.3 Plating of Neural Precursors and Differentiation into Schwan Cells

Cells have now converted into a transient neural precursor stage. For further differentiation into Schwann cells, transient neural precursors are seeded on POL-coated plates in a neural differentiation medium.

3.3.1 Plating of Secondary Spheres (Day 10)

1. Collect the medium containing spheres using a 5 ml tip (*see* **Note 5**).

2. Wait for 10–15 min for the spheres to sink to the bottom of the tube.

3. Remove the supernatant and collect in 50 ml tube (this supernatant is referred to as conditioned medium). Leave a small amount of medium so that the spheres remain covered with medium.

4. Add fresh neural induction medium supplemented with inhibitor mix to conditioned medium to a final ratio of 1:2 (e.g., 1 ml of fresh medium per 2 ml of conditioned medium). Carefully resuspend the spheres in the conditioned/fresh medium mix.

5. Transfer the sphere suspension to POL-coated dishes at a ratio of 1:1, e.g., spheres collected from one 6-well ultralow attachment plate to one 6-well POL-coated plate. Make sure that the spheres are evenly distributed in the plate.

6. Centrifuge the plates at $200 \times g$ for 3 min to enhance sphere attachment. Carefully place the plates at 37 °C in a 5% CO_2 incubator. Do not disturb the plates for the next 4 h to allow for sphere attachment.

3.3.2 Switch to Neural Differentiation Medium (Day 11)

1. Carefully remove the culture medium from the plates (*see* **Note 7**).

2. Carefully add neural differentiation medium supplemented with inhibitor mix.

3.3.3 Differentiation into iSCs (Days 12–18)

1. Every other day exchange 50% of the culture medium with fresh neural differentiation medium supplemented with inhibitor mix (*see* **Note 8**).

2. Proceed to Subheading 3.3.4.

3.3.4 Switch to Neural Differentiation Medium Without Inhibitor Mix (Day 18)

1. Exchange complete medium with fresh neural differentiation medium without inhibitor mix.

2. Proceed to Subheading 3.3.5.

3.3.5 Continued Differentiation of iSCs (Days 19–39)

1. Every other day exchange 50% of the culture medium with fresh neural differentiation medium (*see* **Note 8**).

2. Analyze the expression of Schwann cell markers (*see* **Note 9** and Fig. 3).

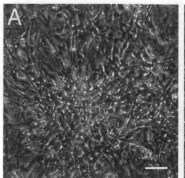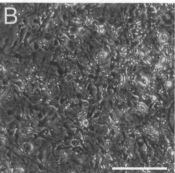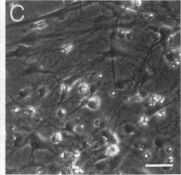

Fig. 3 Formation of iSCs. (**a**) iSCs at day 25. Cells have formed a monolayer culture, and sphere-like structures can only be detected occasionally. (**b** and **c**) iSCs at day 34. Sphere structures are not visible anymore. (**c**) Close-up image showing the typical morphology of iSCs. Scale bars, 200 μm (**a** and **b**) and 50 μm (**c**)

4   Notes

1. Other fibroblast lines can be used. In that case, it is recommended to adjust culture conditions, fibroblast media, and cell density according to the fibroblast line of choice (*see* Subheadings 3.1–3.2.3).

2. Compound B is a multikinase inhibitor that enhances the proliferation of neural precursors while inhibiting their differentiation [2]. Small amounts of Compound B can be obtained from Roche through a Material Transfer Agreement for research purposes only. The use of this compound in this method is protected by patent rights [9].

3. When starting the conversion protocol, fibroblasts should have recovered from the thawing procedure and be proliferating actively. For most fibroblast lines, 3 days are needed for full recovery. Cell confluency should be around 80–90%. If fibroblasts display an abnormal morphology, performing one more passage might be helpful. However, it is recommended to use primary fibroblasts at the lowest passage possible. Prolonged culture typically leads to a decreased rate of cell proliferation, and this negatively affects the efficiency of conversion.

4. Until day 3, cells display fibroblast morphology and do not express neural marker proteins. A reduced proliferation rate and slight signs of cellular stress (i.e., flattened cell shape and/or increased granularity) are normal and due to VPA treatment. If cell density is very low or a high rate of cell death is observed, it is recommended to increase initial seeding numbers at day 0.

5. Sphere transfer works best using a 0.5–5 ml pipet with corresponding tips. Using pipets with smaller tips might result in disruption or loss of spheres. Especially at day 10, when spheres are not supposed to be dissociated, do not use pipets with small tips (<0.5–5 ml) and avoid excessive pipetting.

6. The step of sphere dissociation is technically challenging and requires practice. If spheres are difficult to dissociate, perform a second PBS washing step before the treatment with Trypsin-EDTA. If necessary, Trypsin-EDTA treatment can be prolonged for up to 45 min. As an optional step, the Trypsin-EDTA can be removed after 20 min and replaced by fresh Trypsin-EDTA solution followed by another 20-min incubation time. Prolonged Trypsin-EDTA treatment is not recommended if a flow cytometric analysis of surface marker expression is to be performed as this may result in degradation of cell surface antigens. The use of a more gentle dissociation reagent such as Accutase is recommended for antigen preservation.

7. Spheres can still easily detach from the plates at 24 h after seeding. To maintain the spheres attached, it is recommended to perform medium changes carefully making sure not to touch the bottom of the dishes.

8. Twenty-four hours after attachment of the secondary spheres (day 11), bipolar cells start to migrate out of the spheres. These bipolar cells express the neural plate markers Sox1 and Nestin but do not express neural crest markers. At day 18, the neural crest markers SNAI1, Sox10, FOXD3, Pax3, and CD271 can be detected. Detection of CD271 can be carried out by flow cytometry, while detection of the other markers is usually carried out by immunostaining.

9. Expression of Schwann cell markers can be detected from day 30 onward starting with proteolipid protein (PLP) expression. At day 39, the cells express galactocerebroside (GalC), Krox-20, and S100B. For follow-up experiments such as coculture with neurons, it is recommended to use cells collected at around days 21–25.

Acknowledgments

This work was supported by the Roche Postdoctoral Fellowship Program.

References

1. Bhatheja K, Field J (2006) Schwann cells: origins and role in axonal maintenance and regeneration. Int J Biochem Cell Biol 38(12): 1995–1999. https://doi.org/10.1016/j.biocel.2006.05.007

2. Thoma EC, Merkl C, Heckel T et al (2014) Chemical conversion of human fibroblasts into functional Schwann cells. Stem Cell Rep 3(4):539–547. https://doi.org/10.1016/j.stemcr.2014.07.014

3. Liu Q, Spusta SC, Mi R et al (2012) Human neural crest stem cells derived from human ESCs and induced pluripotent stem cells: induction, maintenance, and differentiation into functional Schwann cells. Stem Cells Transl Med 1(4):266–278. https://doi.org/10.5966/sctm.2011-0042

4. Casella GT, Bunge RP, Wood PM (1996) Improved method for harvesting human Schwann cells from mature peripheral nerve and expansion in vitro. Glia 17(4):327–338. https://doi.org/10.1002/(SICI)1098-1136(199608)17:4<327::AID-GLIA7>3.0.CO;2-W

5. Chambers SM, Fasano CA, Papapetrou EP et al (2009) Highly efficient neural conversion of human ES and iPS cells by dual inhibition of SMAD signaling. Nat Biotechnol 27(3):275–280. https://doi.org/10.1038/nbt.1529

6. Patsch C, Challet-Meylan L, Thoma EC et al (2015) Generation of vascular endothelial and smooth muscle cells from human pluripotent stem cells. Nat Cell Biol 17(8):994–1003. https://doi.org/10.1038/ncb3205

7. Huangfu D, Maehr R, Guo W et al (2008) Induction of pluripotent stem cells by defined factors is greatly improved by small-molecule compounds. Nat Biotechnol 26(7):795–797. https://doi.org/10.1038/nbt1418

8. Ladewig J, Mertens J, Kesavan J et al (2012) Small molecules enable highly efficient neuronal conversion of human fibroblasts. Nat Methods 9(6):575–578. https://doi.org/10.1038/nmeth.1972

9. Graf M, Iacone R, EC Thoma (2015) Small molecule based conversion of somatic cells into neural crest cells. WO2015011031 A1, 29 Jan 2015

Chapter 9

Derivation of Fate-Committed Schwann Cells from Bone Marrow Stromal Cells of Adult Rats

Y.P. Tsui, Graham K. Shea, Y.S. Chan, and Daisy K.Y. Shum

Abstract

Our goal is to derive phenotypically stable Schwann cells from bone marrow stromal cells (BMSCs) for use in transplantation studies of central/peripheral nerve injuries. With the adult rat as model, here we describe steps that foster (1) expansion of the BMSC subpopulation of neural progenitors as neurosphere cells, (2) differentiation of the progenitors into Schwann cell-like cells in adherent culture supplemented with soluble factors, and (3) cell-intrinsic switch of Schwann cell-like cells to the Schwann cell fate following co-culture with sensory neurons purified from dorsal root ganglia. The derived Schwann cells retain marker expression despite withdrawal of supplements and neuronal cues, survive passaging and cryopreservation, and, importantly, show functional capacity for myelination.

Key words BMSC-derived Schwann cells, Nerve regeneration, Bone marrow stromal cells, Myelination, Cell-based therapy

1 Introduction

In nerve injury, the affected axons often fail to regrow and reinnervate target tissues or reconnect with the existing neural circuit. The main cause of failure is the absence of physical/biological guidance of axonal regrowth in the injured environment. Schwann cells [1] have been exploited to guide regrowing axons in implants across critical gaps between stumps of the injured nerve in both PNS and CNS [2–6]. The supply of donor Schwann cells is however limited by the low proliferative capacity of cells recoverable from adult peripheral nerves which are often harvested at the expense of function loss and morbidity at the donor tissue. BMSCs represent an accessible alternative and potentially autologous source of adult precursors that promise in vitro expansion and differentiation into functional Schwann cells [7, 8]. Earlier efforts of in vitro derivation of Schwann cell-like cells from BMSCs utilized a growth factor cocktail [9–11]. However, the Schwann cell-like cells lack fate commitment and revert to a fibroblast-like phenotype

Paula V. Monje and Haesun A. Kim (eds.), *Schwann Cells: Methods and Protocols*, Methods in Molecular Biology, vol. 1739, https://doi.org/10.1007/978-1-4939-7649-2_9, © Springer Science+Business Media, LLC 2018

upon withdrawal of differentiation-inducing factors. This instability limits the yield of Schwann cells and presents risk of in vivo dedifferentiation causing tumorigenesis or differentiation to myofibroblasts [12–15]. Our recent works provide evidence that these hurdles can be overcome, yielding phenotypically stable Schwann cells and progeny without further need for exogenous glia-inducing factors or neurons [16]. The derived Schwann cells are functionally capable of myelination and neurotrophic support of neurite growth both in in vitro assays and in bridging a critical gap in the sciatic nerve of a rat model [17].

2   Materials

2.1   Cell Culture Media

1. Growth factors: Basic fibroblast growth factor (bFGF, Peprotech), epidermal growth factor (EGF, Peprotech), nerve growth factor (NGF, Millipore), β-heregulin (Millipore), and platelet-derived growth factor-AA (PDGF-AA, Peprotech).

2. Ascorbic acid (Sigma-Aldrich).

3. Growth medium for rat BMSC: Supplement Minimum Essential Medium alpha modification (α-MEM) with 2 mM L-glutamine, 15% fetal bovine serum (FBS), and 1% penicillin/streptomycin (P/S).

4. Cryopreservation medium for rat MBSC: Supplement α-MEM with 2 mM L-glutamine, 20% FBS, 1% P/S, and 10% dimethyl sulfoxide (DMSO).

5. Medium for expansion of neural progenitors in suspension culture of BMSCs: Supplement Dulbecco's Modified Eagle's Medium/F12 (DMEM/F12) with 2 mM L-glutamine, 2% B27 supplement, 20 ng/ml bFGF, 20 ng/ml EGF, and 1% P/S.

6. Medium for DRG neuron maintenance: Supplement Neurobasal medium with 1% GlutaMAX, 2% B27, 20 ng/ml NGF, and 1% P/S.

7. Medium for DRG neuron purification: Supplement medium for DRG neuron maintenance with 10 μg/ml 5-fluoro-2'-deoxyuridine (FDU).

8. Medium for inducing Schwann cell-like cells (SC-LCs) as derived from neural progenitors: Supplement α-MEM with 200 ng/ml β-heregulin, 10 ng/ml bFGF, 5 ng/ml PDGF-AA, 10% FBS, and 1% P/S.

9. Medium for co-culture of SC-LCs and DRG neurons: Mix equal volume of Neurobasal media and α-MEM and supplement the mixture with 100 ng/ml β-heregulin, 5 ng/ml bFGF, 2.5 ng/ml PDGF-AA, 5% FBS, 0.5% GlutaMAX, 1% B27 supplement, 10 ng/ml NGF, and 1% P/S.

10. Maintenance medium for fate-committed Schwann cells as derived from BMSCs: Supplement DMEM/F12 with 10 ng/ml β-heregulin, 5% FBS, and 1% P/S.

11. Cryopreservation medium for fate-committed Schwann cells as derived from BMSCs: Supplement DMEM/F12 with 10 ng/ml β-heregulin, 10% FBS, 1% P/S, and 10% DMSO.

12. Medium for inducing myelination in Schwann cell/DRG neuron co-cultures: Supplement Neurobasal medium with 1% GlutaMAX, 1% B27 supplement, 20 ng/ml NGF, 50 µg/ml ascorbic acid, 5% FBS, and 1% P/S.

13. Poly-D-lysine (PDL) solution: 10 µg/ml PDL prepared in Dulbecco's phosphate-buffered saline (DPBS).

14. Laminin solution: 5 µg/ml of laminin prepared in DPBS.

15. 1× Tryple Express.

2.2 Cell Culture Apparatus and Related Materials

1. Tissue culture-treated 10- and 6-cm dishes.
2. Tissue culture-treated 6-well plates.
3. Ultra Low® 6-well nonadherent culture plates.
4. Tissue culture-treated 4-well plates.
5. 35-mm µ-dishes.
6. 40-µm cell strainer.
7. CryoTubes 1.5-ml.

2.3 Tissue Harvesting from Rats

1. Sprague Dawley (SD) rats.
   (a) Young adults (200–250 g body weight), for the bone marrow.
   (b) Timed-pregnant dams on gestational day 14–15, for dorsal root ganglia (DRG).
2. Pentobarbital sodium for euthanasia of rats.
3. 70% Ethanol.
4. DPBS supplemented with 1% P/S.
5. 21-gauge needles and 10-ml syringes for marrow extraction.

2.4 Immunocytochemistry and Flow Cytometry

2.4.1 Solutions

1. Cell fixation: 4% paraformaldehyde prepared in DPBS.
2. Blocking buffer: Supplement 1× DPBS with 2% bovine serum albumin (BSA) and 0.1% Triton X-100.

2.4.2 Primary Antibodies for Flow Cytometric Analysis of BMSCs

1. STRO-1 (mouse vs. human/rat STRO-1, R&D Systems, 1:50).
2. CD90 (mouse vs. rat CD90, BD Pharmingen, 1:200).
3. CD73 (mouse vs. rat CD73, BD Pharmingen, 1:200).
4. CD45 (mouse vs. rat CD45, BD Pharmingen, 1:200).

5. Nestin (mouse vs. rat nestin, BD Pharmingen, 1:300).

6. Mouse isotype control (mouse IgG, Thermo Fisher Scientifics, 1:300).

2.4.3 Primary Antibodies for Markers of Neural Progenitors vs. Stromal Cell Marker

1. Nestin (mouse vs. rat nestin, BD Pharmingen, 1:300).

2. Sox10 (mouse vs. rat/human nestin, R&D Systems, 1:500).

3. STRO-1 (mouse vs. human/rat STRO-1, R&D Systems, 1:50).

2.4.4 Primary Antibodies for Markers of Schwann Cell

1. p75 NGF receptor (mouse vs. rat p75 LNGFR, Millipore, 1:200).

2. S100β (rabbit vs. S100β, DAKO, 1:500).

3. SOX10 (mouse vs. rat/human S, R&D Systems, 1:500).

2.4.5 Primary Antibodies for Markers of Myelin and Axon in Myelin-Forming Culture

1. Myelin basic protein (MBP) (mouse vs. rat MBP, Millipore, 1:500).

2. Neurofilament 200 kDa (NF200) (rabbit vs. rat NF200, Sigma-Aldrich, 1:300).

3   Methods

3.1   Schedule for Derivation of Schwann Cells from BMSCs of Adult Rat

Establishment of BMSC cultures from whole bone marrow extract usually requires 18 days. Subsequent derivation of fate-committed Schwann cells requires approximately another 30 days. Conduct the procedure at 22 °C unless otherwise specified. Figure 1 outlines the schedule for the entire process.

3.2   BMSC Culture

1. Autoclave all dissection tools at 180 °C for at least 2 h.

2. Sacrifice SD rat with pentobarbital overdose (i.p., ≥200 mg/kg body weight). Spray the animal with 70% ethanol.

3. Make a cut in the skin of each hind limb and pull skin back, exposing the underlying muscles. Free the femur from the pelvic joint using sterilized scissors, making sure that the epiphysis remains intact. Similarly, free the tibia from the ankle joint and then from the knee joint. Transfer the hind limb to DPBS-P/S(1%) in a sterile petri dish placed in a microbiological safety cabinet. Perform likewise for the other hind limb.

4. Dissect the femur from surrounding muscle, keeping the ends of the bone intact.

5. Fill a 10-ml syringe with DPBS-P/S(1%). Attach a 21-gauge needle to the syringe. Carefully cut off the ends of a femur and insert the needle into the proximal end to flush the bone marrow into a sterile tube. For each femur, use 5 ml of DPBS. Filter off bone fragments or debris from the marrow extract of each rat with use of a 40-μm cell strainer. Transfer

Step 1

Day 0 – 10: Adherent culture of BM extract
Day 0 denotes the day of plating bone marrow extract onto tissue culture dishes (day 0). by day By 6~7, colonies of BMSCs are visible (A). Passage culture on day 10.

Step 2

Day 10 – 18: Establishment of BMSC culture
BMSCs in culture (B) for at most 8 passages; >90% cells are positive for STRO-1/CD90/CD73, ≥5% cells are positive for nestin and <1% cells are positive for CD45.

Step 3

Day 18 – 30: Neurosphere suspension culture
BMSCs in suspension culture foster expansion of neural progenitors in spheres reaching ≥50 μm in diameter (C); >60% sphere cells are positive for nestin.

Step 4

Day 30 – 37: From neural progenitors to Schwann cell-like cells (SC-LCs)
Sphere cells in adherent culture supplemented with soluble, glia-inducing factors progressively acquire Schwann cell-like phenotype (D).

Step 5

Day 37 – 50: SC-LC/DRG neuron co-culture
By day 45~50, SC-LCs acquire bi-/tri-polar morphology (E) typical of Schwann cells in culture.

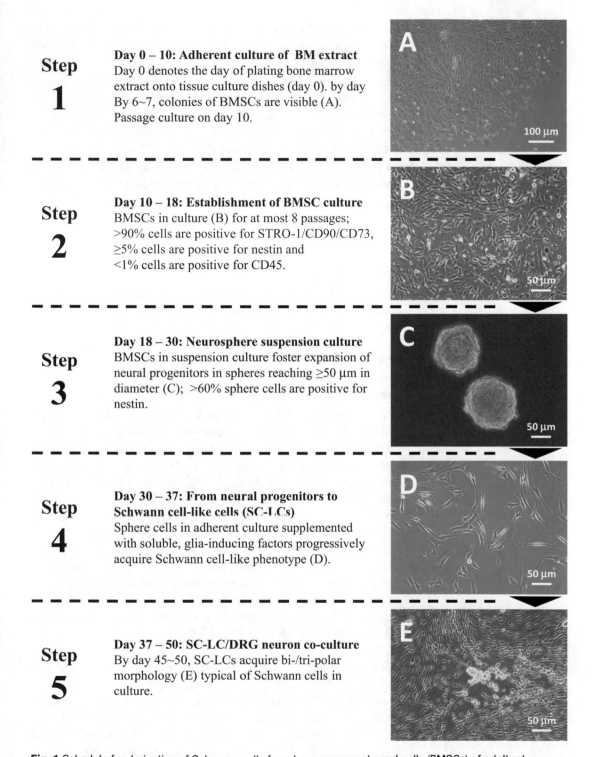

Fig. 1 Schedule for derivation of Schwann cells from bone marrow stromal cells (BMSCs) of adult rat

extract into a sterile 15-ml centrifuge tube. To avoid cross contamination, *do not* use the same syringe or cell strainer for marrow samples from multiple rats.

6. Centrifuge the tubes at $250 \times g$ for 5 min at 4 °C. Remove supernatant and resuspend the cell pellet in BMSC growth medium. Dissociate the pellet by mild trituration using a 1-ml pipetman. Plate the cell suspension recovered from one rat onto a tissue culture-treated 10-cm dish. Add 10 ml of medium to each 10-cm dish. Register the day of plating of whole marrow extract as *day 0* of the Schwann cell derivation schedule (Fig. 1).

7. Remove the medium on *day 2*. Wash the culture with 5 ml of DPBS to remove cells that remain unattached (*see* **Note 1**). Add 10 ml of fresh growth medium to the culture. Refresh growth medium every 72 h. Colonies of BMSCs are visible by day 6–7 (Fig. 1).

8. Passage BMSCs on *day 10*. Remove medium and wash the culture with 3 ml of DPBS. Remove DPBS and add 1 ml of Tryple Express and incubate at 37 °C for 5 min. Collect detached cells by washing the dish with 3 ml of growth medium. Centrifuge at $250 \times g$ for 5 min at 4 °C and resuspend in growth medium. Seed cells into new 10- or 6-cm dishes at 40,000 cells/cm². In case of 6-cm dishes, add 5 ml of medium to each dish.

9. After passaging, BMSCs will reach 80–90% confluence within 2 days. After three passages, hematopoietic cells should be mostly removed (*see* **Note 2**). Limit passaging to no more than eight times. Monitor cultures in passages 3–8 for increasing proportion of cells immunopositive for (a) BMSC markers (STRO-1, CD90, and CD73) and (b) neural progenitor marker nestin. Also monitor for decline to basal level in the proportion of cells immunopositive for the hematopoietic progenitor marker, CD45. Cultures that show >90% of cells positive for the BMSC markers, >5% of cells positive for nestin, and <1% of cells positive for CD45 can be kept in storage under cryopreservation or subjected to neurosphere suspension culture to foster neural progenitor cell expansion. For flow cytometry and immunocytochemistry of the BMSCs (*see* Subheadings 3.8 and 3.9).

3.3 BMSC Cryopreservation

1. Prepare BMSCs at the end of passage 3, 4, or 5 for cryopreservation by treatment with Tryple Express (*see* **step 8** in Subheading 3.2). Resuspend cells in cryopreservation medium. Aliquot 10^6 cells/ml medium into each CryoTube at 1 ml per tube. Keep CryoTubes at −20 °C for 2 h, then at −80 °C for 24 h, and finally at −198 °C in a liquid nitrogen tank.

2. To return cells to the culture environment, remove a CryoTube from storage in liquid nitrogen. Immediately transfer the tube to a 37 °C water bath and gently swirl the tube content to thaw cells. Dilute the content with pre-warmed growth medium and

seed cells onto tissue cultured-treated 6-cm dishes or 6-well plates at a density of 80,000 cells/cm^2. Refresh medium 2 h after seeding to remove DMSO.

3.4 Neurosphere Suspension Culture

1. On *day 18* of the Schwann cell derivation schedule, detach BMSCs by treatment with Tryple Express (*see* **step 8** in Subheading 3.2). After centrifugation and removal of supernatant, wash the cell pellet once with DMEM/F12 medium, and then resuspend in neural progenitor derivation medium. Seed BMSCs into Ultra Low® 6-well nonadherent culture plates at 1000 cells/cm^2.

2. Refresh half of the medium every 2 days. Cell clusters/spheres are visible by *day 24*. On *day 30*, sediment spheres (Fig. 1) by centrifugation at $100 \times g$ for 2 min at 4 °C.

3. Sample BMSC-derived spheres and monitor them for the proportion of cells positive for (a) the neural progenitor marker, the nestin, and (b) the neural crest progenitor marker, Sox10. By this stage, the sphere cells should be negative for the BMSC marker, STRO-1. Partially dissociate spheres by treatment with 0.5 ml of Tryple Express at 37 °C for 5–10 min followed by mild trituration. Cytospin the partially dissociated spheres onto a SuperFrost microscope slide. Perform immunocytochemistry on the immobilized sphere cells (*see* Subheading 3.9). Cultures with >60% of sphere cells positive for nestin will be taken to the next stage of the Schwann cell derivation schedule.

3.5 Preparation of Purified DRG Neuron Cultures

3.5.1 Coating Culture Substratum with PDL and Laminin

1. Coating improves cell attachment to the culture substratum. Dispense PDL solution at 1 ml/well for 6-well plates (0.3 ml/well both for 4-well plates and 35-mm μ-dishes). Incubate overnight (~16 h) at 4 °C. Wash each well with DPBS at the volumes recommended. Remove as much of the DPBS from the wells as possible.

2. Dispense laminin solution to the PDL-coated wells at 1 ml/well for 6-well plates (0.3 ml/well both for 4-well plates and 35-mm μ-dishes). Incubate overnight at 4 °C. Wash the wells with DPBS at the volumes recommended. Do not allow the coated wells to dry. Do not remove the last DPBS wash until ready to plate the cells.

3. The PDL/laminin-coated wells are to be used within the day of completing the coating procedure.

3.5.2 DRG Cultures

1. In parallel to neurosphere culture, prepare cultures of purified neurons from DRGs.

2. Autoclave all dissection tools at 180 °C for at least 2 h.

3. Sacrifice timed-pregnant rat dams on gestational days 14–15 with pentobarbital overdose. Spray extremities and abdomen with 70% ethanol.

4. Cut medially through the skin and muscles of the lower abdomen with a pair of sterile scissors to expose the uterus and embryos.

5. Harvest all embryos and place them in a sterile 10-cm culture dish filled with DPBS-P/S(1%) on ice. Using fine scissors, cut down the dorsal length of the embryo exposing the vertebral column. Then insert blade of scissors through the neck opening and cut along either side of the midline down to the tail stub; gently lift the strip of midline bone exposing the underlying spinal cord. With use of fine forceps, gently lift the spinal cord with attached DRGs from the vertebral cavity, starting from the most rostral aspect. Perform this procedure in a microbiological safety hood if possible. Transfer DRGs into 1.5-ml sterile centrifuge tubes containing DPBS.

6. Treat DRGs in each 1.5-ml tube with 0.25 ml of Tryple Express at 37 °C for 5 min. To stop digestion, add DRG neuron maintenance medium into the tubes. Centrifuge at $250 \times g$ for 5 min at 4 °C. Remove the supernatant, add fresh DRG neuron maintenance medium, and dissociate the pellet by mild trituration using a 200-µl pipetman. Seed cells onto PDL/laminin-coated 6-well plates at a density of 5000 cells/cm².

7. To purify for DRG neurons, treat DRG cell culture with antimitotic containing medium to remove proliferating cells (*see* **Note 3**). Two days after initial plating, switch to purification medium. For each purification cycle, treat DRG cultures with purification medium for 2 days, followed by 1 day of incubation in maintenance medium. After three to four cycles, removal of all proliferating cells that originate from the DRGs is expected. The resulting culture is expected to test positive for the neuron marker, TuJ1, and negative for glial cell marker, S100β.

3.6 Schwann Cell Derivation and Culture Expansion

1. On *day 30* of the Schwann cell derivation schedule, collect BMSC-derived neural progenitors (BM-NPs) as described above and resuspend in SC-LC induction medium. Plate BM-NP clusters onto PDL/laminin-coated 6-well plates (*see* Subheading 3.5.1) at a density of 5–10 clusters/cm². Refresh medium every 2 days (*see* **Note 4**).

2. On *day 37*, treat cells with Tryple Express (0.5 ml/well) at 37 °C for 5 min. Collect cell by washing with 3 ml of SC-LC induction medium, and then centrifuge at $250 \times g$ for 5 min. Resuspend the cell pellet in SC-LC/DRG neuron co-culture medium. Seed SC-LCs onto purified DRG neuron culture at a density of 1000 cells/cm². Refresh medium every 2 days.

3. During co-culture, fibroblast-like SC-LCs gradually acquire the spindle-like morphology of Schwann cells in culture (Fig. 1). By *day 50*, all SC-LCs are expected to have made the switch to the Schwann cell fate. These committed Schwann cells are referred to as BMSC-derived Schwann cells (BMSC-SCs).

4. To harvest BMSC-SCs, treat the cultures with Tryple Express (0.5 ml/well) at 37 °C for 5 min. Collect cells by washing the dish with 3 ml of Schwann cell maintenance medium. Centrifuge at $250 \times g$ for 5 min and resuspend in Schwann cell maintenance medium. Plate BMSC-SCs onto PDL/laminin-coated 6-well plates at a density of 10,000 cells/cm². The BMSC-SCs can be kept in cryopreservation or culture-expanded/passaged for up to 1 month.

3.7 In Vitro Myelination Assay

1. Prepare purified DRG neuron cultures (*see* Subheading 3.5). Seed BMSC-SCs onto DRG neuron culture at a density of 5000 cells/cm² in Neurobasal medium supplemented with B27 (2% v/v), NGF (10 ng/ml), and 10% FBS. In about 7 days when alignment of BMSC-SC with neurites becomes detectable, induce myelination by exposure to myelin-inducing medium for 15–20 days. Refresh medium every 3 days. Fix culture in 4% paraformaldehyde (PFA) in preparation for immunostaining procedure.

2. To view immunofluorescence under epi-fluorescence microscopy, set up the co-culture on PDL/laminin coating of tissue culture-treated 6-well or 4-well plates. To view immunofluorescence under confocal microscopy, set up the co-culture on PDL/laminin coating of 35-mm μ-dishes.

3.8 Flow Cytometric Analysis of BMSCs

1. Treat BMSC culture at 80–90% confluence with Tryple Express (*see* Subheading 3.2, **step 8**). Collect detached cells in the growth medium and centrifuge at $250 \times g$ for 5 min. Remove supernatant and wash the cell pellet with DPBS. Resuspend cells in 0.1 ml of DPBS and then mix in 2 ml of 4% PFA. After 10 min, centrifuge at $250 \times g$ for 5 min. Wash the cell pellet in DPBS for two to three times to remove residual PFA. Resuspend fixed cells in blocking buffer and incubate for 30 min.

2. Incubate fixed BMSCs with selected primary antibody in blocking buffer for 2 h. Centrifuge at $250 \times g$ for 5 min, remove the supernatant, and then resuspend the cell pellet in blocking buffer. Incubate the cells with the appropriate secondary antibody in blocking buffer for 30 min. Centrifuge the cells and resuspend in DPBS for flow cytometric analysis.

3.9 Immunocyto-chemistry

1. Cells/cultures to be analyzed by immunocytochemistry include bone marrow-derived neural progenitors (BM-NPs cytospun onto SuperFrost slides as described in Subheading 3.4), BMSC-SCs, DRG neurons, and myelin-forming cultures. Fix the sample cells/cultures in 4% PFA for 15 min. Wash the samples with DPBS for 3 × 5 min to remove residual PFA. Incubate samples in blocking buffer for 30 min.

2. Incubate sample with selected primary antibodies in blocking buffer at 4 °C for 16 h. Remove the solution primary antibodies,

wash with DPBS three times, wash each for 5 min, and then incubate with appropriate secondary antibodies in blocking buffer for 1 h. Remove the solution of secondary antibodies, apply 10 µg/ml Hoechst stain prepared in DPBS, and then incubate for 15 min to label cell nuclei/DNA. Wash cells with DPBS three times; wash each for 5 min to remove residual secondary antibodies and Hoechst stain. Examples of immunolabeled BM-NPs and BMSC-SCs are shown in Fig. 2.

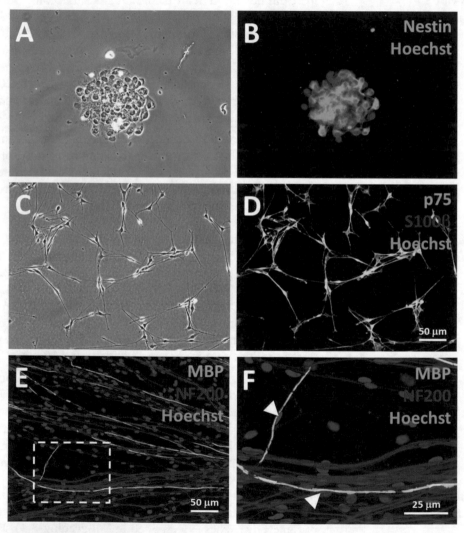

Fig. 2 Neural progenitors and Schwann cells derived from BMSCs of adult rat. BMSCs in suspension culture foster formation of spheres (**a**) in which cells are positive for the neural progenitor marker, nestin (**b**). Fate-committed Schwann cells derived from BMSCs show bi-/tripolar morphology (**c**) and immunopositivities for the Schwann cell markers, p75 and S100β (**d**), these being features typical of Schwann cells in culture. In myelin-inducing medium, co-culture of the BMSC-derived Schwann cells with purified DRG neurons shows myelin (MBP-positive) segments along NF200-positive neurites (**e**). Zoom-in view (**f**) of the boxed area (**e**) shows Schwann cell nuclei (blue with Hoechst stain, white arrowhead) in alignment with a MBP-positive segment

4   Notes

1. The bone marrow extract consists of diverse cell types. By washing the culture of bone marrow extract 48 h after initial plating, BMSC cells remain attached to the substratum, whereas non-BMSCs can be removed in the wash.

2. Hematopoietic progenitors in the bone marrow extract may not be completely removed in the first wash performed 48 h after initial plating. These however can be removed with passaging of the culture for three to four times.

3. The key to achieving the Schwann cell fate lies in contact-mediated signaling between SC-LCs and DRG neurons in co-culture. Proliferating glial progenitors endogenous in DRGs can be removed by multiple rounds of treatment with the antimitotic FDU, yielding purified DRG neurons for the co-culture.

4. Despite immunopositivities for the Schwann cell markers, p75 and S100β among SC-LCs, they lose the phenotype when the SC-LC induction medium is withdrawn. Co-culture of SC-LCs with DRG neurons provides for juxtacrine signaling that accomplishes the cell-intrinsic switch to the Schwann cell fate. The BMSC-SCs and progeny are no longer dependent on exogenous glia-inducing factors or cues from DRG neurons for maintaining phenotypic stability.

Acknowledgment

This work was supported by the Strategic Research Theme of The University of Hong Kong, Croucher Foundation MBBS/PhD Scholarship to GKHS, and the Hong Kong Research Grants Council GRF 777810 to D.K.Y.S.

References

1. Jessen KR, Mirsky R, Lloyd AC (2015) Schwann cells: development and roles in nerve repair. Cold Spring Harb Perspect Biol 7:a020487

2. Rodriguez FJ, Verdu E, Ceballos D, Navarro X (2000) Nerve guides seeded with autologous Schwann cells improved nerve regeneration. Exp Neurol 161:571–584

3. Chau CH, Shum DK, Li H, Pei J, Lui YY, Wirthlin L, Chan YS, Xu XM (2004) Chondroitinase ABC enhances axonal regrowth through Schwann cell-seeded guidance channels after spinal cord injury. FASEB J 18: 194–196

4. Pearse DD, Pereira FC, Marcillo AE, Bates ML, Berrocal YA, Filbin MT, Bunge MB (2004) cAMP and Schwann cells promote axonal growth and functional recovery after spinal cord injury. Nat Med 10:610–616

5. Bachelin C, Lachapelle F, Girard C et al (2005) Efficient myelin repair in the macaque spinal cord by autologous grafts of Schwann cells. Brain 128:540–549

6. Saberi H, Moshayedi P, Aghayan HR, Arjmand B, Hosseini SK, Emami-Razavi SH, Rahimi-Movaghar V, Raza M, Firouzi M (2008) Treatment of chronic thoracic spinal cord injury patients with autologous Schwann cell

transplantation: an interim report on safety considerations and possible outcomes. Neurosci Lett 443:46–50

7. Nagoshi N, Shibata S, Kubota Y et al (2008) Ontogeny and multipotency of neural crest-derived stem cells in mouse bone marrow, dorsal root ganglia, and whisker pad. Cell Stem Cell 2:392–403

8. Cai S, Shea GK, Tsui AY, Chan YS, Shum DK (2011) Derivation of clinically applicable Schwann cells from bone marrow stromal cells for neural repair and regeneration. CNS Neurol Disord Drug Targets 10:500–508

9. Dezawa M, Takahashi I, Esaki M, Takano M, Sawada H (2001) Sciatic nerve regeneration in rats induced by transplantation of in vitro differentiated bone-marrow stromal cells. Eur J Neurosci 14:1771–1776

10. Caddick J, Kingham PJ, Gardiner NJ, Wiberg M, Terenghi G (2006) Phenotypic and functional characteristics of mesenchymal stem cells differentiated along a Schwann cell lineage. Glia 54:840–849

11. Keilhoff G, Goihl A, Langnäse K, Fansa H, Wolf G (2006) Transdifferentiation of mesenchymal stem cells into Schwann cell-like myelinating cells. Eur J Cell Biol 85:11–24

12. Keilhoff G, Goihl A, Stang F, Wolf G, Fansa H (2006) Peripheral nerve tissue engineering: autologous Schwann cells vs transdifferentiated mesenchymal stem cells. Tissue Eng 12:1451–1465

13. Cogle CR, Theise ND, Fu D, Ucar D, Lee S, Guthrie SM, Lonergan J, Rybka W, Krause DS, Scott EW (2007) Bone marrow contributes to epithelial cancers in mice and humans as developmental mimicry. Stem Cells 25:1881–1887

14. Takashima Y, Era T, Nakao K, Kondo S, Kasuga M, Smith AG, Nishikawa S (2007) Neuroepithelial cells supply an initial transient wave of MSC differentiation. Cell 129:1377–1388

15. Pawelek JM, Chakraborty AK (2008) Fusion of tumour cells with bone marrow-derived cells: a unifying explanation for metastasis. Nat Rev Cancer 8:377–386

16. Shea GSK, Tsui YP, Chan YS, Shum DK (2010) Bone marrow-derived Schwann cell achieved fate commitment – a prerequisite for remyelination therapy. Exp Neurol 224:248–258

17. Ao Q, Fung CK, Tsui YP, Cai S, Zuo HC, Chan YS, Shum DK (2011) The regeneration of transected sciatic nerves of adult rats using nerve conduits seeded with bone marrow stromal cell-derived Schwann cells. Biomaterials 32:787–796

Chapter 10

Human Induced Pluripotent Stem Cell-Derived Sensory Neurons for Fate Commitment of Bone Marrow Stromal Cell-Derived Schwann Cells

Sa Cai, Daisy K.Y. Shum, and Ying-Shing Chan

Abstract

Here we describe the in vitro derivation of sensory neurons for use in effecting fate commitment of Schwann cell-like cells derived from human bone marrow stromal cells (hBMSCs). We adopt a novel combination of small molecules in an 8-day program that induces the differentiation of human induced pluripotent stem cells into sensory neurons. In co-cultures, the derived sensory neurons present contact-dependent cues to direct hBMSC-derived Schwann cell-like cells toward the Schwann cell fate. These derived human Schwann cells survive passaging and cryopreservation, retain marker expression despite withdrawal of glia-inducing medium and neuronal cues, demonstrate capacity for myelination, and therefore promise application in autologous transplantation and re-myelination therapy.

Key words Human BMSC-derived Schwann cells, Induced pluripotent stem cells, Small-molecule inhibitors, Sensory neurons, Fate commitment

1 Introduction

Schwann cells transplanted into the injured environments of peripheral nerve and spinal cord switch on an axon-supportive program that improves nerve regeneration and functional recovery [1–4]. To take advantage of this unique plasticity of Schwann cells in autologous transplantation for therapy, sufficient numbers of such cells are required, ideally without having to sacrifice a peripheral nerve for the graft. This need is in part addressed by in vitro derivation of Schwann cell-like cells (SCLCs) from human bone marrow stromal cells (hBMSCs) with a cocktail of glia-inducing factors [5]. The derived human SCLCs however tended to be phenotypically unstable and demonstrated limited capacity for re-myelination [6]. In another chapter of this book, we demonstrated that commitment to the Schwann cell fate can be acquired by rat SCLCs in co-culture with rat DRG neurons [7, 8]. However, the

Paula V. Monje and Haesun A. Kim (eds.), *Schwann Cells: Methods and Protocols*, Methods in Molecular Biology, vol. 1739, https://doi.org/10.1007/978-1-4939-7649-2_10, © Springer Science+Business Media, LLC 2018

limiting source of human DRG neurons presents as a barrier to translation. In this chapter, we demonstrate the use of a human induced pluripotent stem cell (hiPSC) line to bypass the translational barrier [9]. We selected five small-molecule inhibitors of key signaling pathways in an 8-day program to induce differentiation of hiPSCs into sensory neurons, reaching ≥80% yield in terms of marker proteins and electrophysiological properties. Importantly, expression of NRG1 type III, a key to Schwann cell specification and maturation into myelinating glia, was consistent among the derived human neurons (*see* ref. 10). The derived neurons fulfill the need for a surrogate of human DRG neurons in the translation to a protocol, whereby hBMSC-derived Schwann cells achieve fate commitment and meet safety requirement for autologous transplantation and re-myelination therapy. This is a breakthrough toward cell-based therapy for nerve injury.

2   Materials

2.1   hiPSC Culture and Passage

1. 6-well culture plates.

2. DMEM/F12.

3. 0.4 mg/ml BD Matrigel™ hESC-qualified matrix (BD Matrigel™, Stem Cell Technologies) prepared in DMEM/F12.

4. Maintenance medium (mTeSR™1) for hiPSCs: mTeSR™1 Basal Medium (400 ml) mixed with mTeSR™1 5× Supplement (100 ml) (Stemcell Technologies).

5. hiPSC (IMR90) clone (#1) (WiCell Research Institute).

6. 1 mg/ml Dispase.

7. Cell scrapers.

2.2   Derivation of Sensory Neurons from hiPSCs

1. Sensory neuron differentiation medium: DMEM/F12 supplemented with 10% KnockOut Serum Replacement, 0.3 μM LDN-193189, 2 μM A83-01, 6 μM CHIR99021, 2 μM RO4929097, 3 μM SU5402, 0.3 μM retinoic acid, and 1% penicillin/streptomycin (P/S).

2. Sensory neuron maintenance medium: neural basal media supplemented with 10 ng/ml NT-3, 20 ng/ml BDNF, 20 ng/ml NGF, 20 ng/ml GDNF, and 1% P/S.

2.3   hBMSC Culture and Passage

1. Fresh bone marrow.

2. Dulbecco's phosphate-buffered saline (DPBS).

3. Sterile glass Pasteur pipette.

4. 70-μm pore-sized cell strainer.

5. 1.077 g/ml Ficoll-Paque.

6. hBMSC maintenance medium: DMEM/F12 supplemented with 10% FBS and 1% P/S.

7. T75 culture flasks.

8. TrypLE™ Express.

2.4 Derivation of SCLCs from hBMSCs

1. TrypLE™ Express.

2. Neural progenitor induction medium: mix DMEM/F12 and Neurobasal medium at 1:1 (v/v) and then supplement with 40 ng/ml bFGF, 20 ng/ml EGF, 2% B27, and 1% P/S.

3. Ultra low 6-well nonadherent culture plates.

4. Superfrost microscope slides.

5. SCLC induction medium: supplement DMEM/F12 with 10% FBS, 1% P/S, 5 µM forskolin (FSK), 5 ng/ml PDGF-AA, 10 ng/ml bFGF, and 200 ng/ml β-HRG.

6. PDL/laminin-coated 6-well plates.

2.5 Directing SCLCs to the Schwann Cell Fate

1. TrypLE™ Express.

2. SCLC/sensory neuron co-culture medium: mix α-MEM and neural basal medium at 1:1 (v/v), and then supplement with 2.5 µM FSK, 2.5 ng/ml PDGF-AA, 5 ng/ml bFGF, 100 ng/ml β-HRG, 5 ng/ml NGF, 1% B27, 5% FBS, and 1% P/S.

3. Schwann cell maintenance medium: supplement DMEM/F12 with 10% FBS and 1% P/S.

4. PDL/laminin-coated 6-well plates.

5. Cryopreservation medium: supplement DMEM/F12 with 10% FBS, 1% P/S, and 10% DMSO.

6. Cryotubes.

2.6 Immunocyto-chemistry and Flow Cytometry

2.6.1 Solutions

1. DPBS.

2. Cell fixation: 4% paraformaldehyde in DPBS.

3. Blocking buffer: supplement DPBS with 2% bovine serum albumin (BSA) and 0.1% Triton X-100.

4. Fluoroshield mounting medium.

5. TrypLE™ Express or 1 mg/ml dispase.

2.6.2 Primary Antibodies for Identifying hiPSCs

1. OCT3/4 (mouse monoclonal, BD Biosciences, 1:200).

2. NANOG (mouse monoclonal, eBioScience, 1:200).

3. SSEA3 (rat monoclonal, BD Biosciences, 1:100).

4. SSEA4 (mouse monoclonal, BD Biosciences, 1:100).

2.6.3 Primary Antibodies for Identifying Sensory Neurons	1. TUJ1 (mouse monoclonal, Covance, 1:500). 2. Neurofilament 200 (mouse monoclonal, Covance, 1:500). 3. Peripherin (mouse monoclonal, Abcam, 1:500). 4. BRN3A (mouse monoclonal, Santa Cruz Biotechnology, 1:200). 5. NeuN (mouse monoclonal, Millipore, 1:500). 6. Islet-1 (mouse monoclonal, Abcam, 1:400).
2.6.4 Primary Antibodies for Identifying hBMSCs	1. CD 29 (mouse monoclonal, Santa Cruz Biotechnologies, 1:200). 2. CD 44 (mouse monoclonal, Santa Cruz Biotechnologies, 1:200). 3. CD 73 (mouse monoclonal, BD Biosciences, 1:500). 4. CD 90 (mouse monoclonal, BD Biosciences, 1:500).
2.6.5 Primary Antibodies for Identifying Neural Progenitors	1. Nestin (mouse monoclonal, BD Pharmingen, 1:300). 2. GFAP (mouse monoclonal, Abcam, 1:100).
2.6.6 Primary Antibodies for Identifying Schwann Cells	1. p75^{NTR} (rabbit polyclonal, Abcam, 1:500). 2. S100β (rabbit polyclonal, Abcam, 1:200).
2.6.7 Secondary Antibodies as Stain for Cell Nuclei	1. Goat anti-mouse/rat IgG secondary antibody, Alexa Fluor 488 conjugate (1:1000). 2. Goat anti-rabbit IgG secondary antibody, Alexa Fluor 594 conjugate (1:1000). 3. 4,6-diamidino-2-phenylindole (DAPI) stain for identifying cell nuclei.

3   Methods

3.1   Schedule for Derivation of Schwann Cells from hBMSCs

Establishment of adherent cultures of BMSCs from human bone marrow extract requires 10 days. Transfer to nonadherent culture fosters expansion of hBMSC-derived neuroprogenitors into neurospheres in 14 days. Under glia-inducing environment, cells that emigrate from the neurospheres in adherent culture undergo differentiation into SCLCs in 8 days. In parallel, we harness small-molecule inhibitors of cell signaling to direct the differentiation of hiPSCs into sensory neurons. Henceforth, co-culture of the sensory neurons with hBM-SCLCs for 14 days provides for the cell-intrinsic switch of SCLCs to the Schwann cell fate. An outline of the schedule for the entire process is shown in Fig. 1.

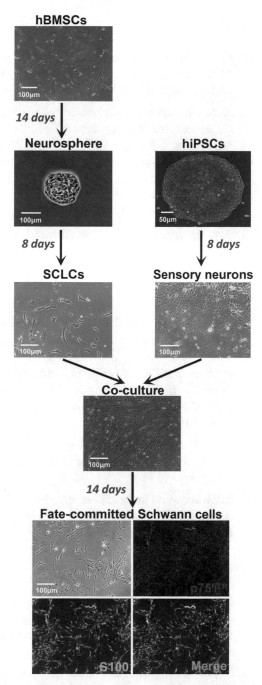

Fig. 1 Flow diagram showing the timeline for deriving fate-committed Schwann cells from hBMSCs. Left: hBMSCs are established in adherent culture. Transfer to nonadherent culture fosters expansion of hBMSC-derived neuroprogenitors into neurospheres in 14 days. Then, with neurospheres in adherent culture for 8 days under glia-inducing environment, cells that emigrate from spheres undergo differentiation into SCLCs. Right: In parallel, hiPSCs differentiate into sensory neurons in an 8-day program as directed by a cocktail of small-molecule inhibitors of key signaling pathways. Co-culture of the hiPSC-derived sensory neurons and hBMSC-derived SCLCs for 14 days mediate the cell-intrinsic switch of SCLCs to the Schwann cell fate. As shown, the fate-committed Schwann cells maintained Schwann cell morphology and retained marker expression (p75^{NTR} (red)/S100 (green) and colocalization (yellow) in the merged image) even in subcultures where glia-inducing factors were withdrawn and neuronal cues were no longer supplied

3.2 hiPSC Culture and Passage

1. Coat tissue culture-treated 6-well plates evenly with BD Matrigel™ in DMEM/F12 (1:83, v/v). Place the plate in a CO_2 incubator (37 °C) for at least 30 min.

2. Remove residual BD Matrigel™ solution from the culture plate, and immediately add mTeSR™1 (*see* **Note 1**). Seed hiPSCs at 2×10^5 per well (6-well plate), taking care to maintain cells as aggregates of 50–60 μm in diameter. Maintain the cells in culture at 37 °C, under 5% CO_2 and 95% humidity, and with medium refreshed daily. By 5–7 days after initial plating, colonies of undifferentiated cells are ready for passage.

3. Aspirate medium from the culture and rinse with DMEM/F12 (2 ml/well). Add 1 ml per well of dispase to reach 1 mg/ml. Incubate at 37 °C for 5 min. Remove dispase, and gently rinse off residual dispase with 2 ml of DMEM/F12 per well. Repeat the rinse 2–3 times.

4. To each well, add 2 ml of mTeSR™1 and scrape colonies off with use of a cell scraper. Transfer the detached cell aggregates to a 15-ml tube; rinse the well and residual aggregates with another 2 ml of mTeSR™1, and add the rinse mixture to the same 15-ml tube. Carefully triturate the suspension with use of a P1000 micropipetter (four times) making sure that the cells remain as aggregates.

5. Plate the cell aggregates onto new plates coated with BD Matrigel™ at dilutions of 1:4 to 1:8. Replace the medium every day.

6. hiPSCs can be maintained indefinitely in maintenance medium. Alternatively, proceed to the next step when cells are 60% confluent. hiPSCs maintained in these conditions should be positive for pluripotent markers such as OCT4, NANOG, SSEA3, and SSEA4.

3.3 Derivation of Sensory Neurons from hiPSCs

1. Passage hiPSCs at 60% confluence (*see* Subheading 3.2, **steps 3** and **4**), except that cells should be resuspended in DMEM/F12. Centrifuge the 15-ml tube containing cells at $160 \times g$ for 4 min at 4 °C.

2. Aspirate the supernatant. Resuspend cells with sensory neuron differentiation medium and seed at a density of (1 to 5×10^5) cells onto BD Matrigel™-coated plates (*see* **Note 2**). Incubate the cells at 37 °C, under 5% CO_2 and 95% humidity. Refresh the medium every 2–3 days.

3. Morphology changes should be noticeable by day 3 and neuronal morphology by day 6–8 in culture.

4. On day 8, replace the sensory neuron differentiation medium with the sensory neuron maintenance medium. Incubate the cells at 37 °C, under 5% CO_2 and 95% humidity. Refresh the medium every 2–3 days.

5. By day 14 in culture, assess the cells for sensory neuron identity by immunocytochemistry for the neuron-specific markers, TUJ1 and neurofilament 200, and the sensory lineage markers, peripherin, BRN3A, NeuN, and Islet-1. By flow cytometry, cells can be monitored for the sensory lineage markers, peripherin and BRN3A.

3.4 hBMSC Culture and Passage

1. Dilute 5–8 ml of fresh bone marrow to 30 ml with DPBS in a 50-ml tube and mix by inversion. Allow the tubes to stand for 15 min at room temperature for the separation of mineral fragments and fat from solution. Carefully remove the floating layer (fat) with a sterile Pasteur pipette. Then filter the cell suspension through a 70-μm pore-size cell strainer with care not to disturb the mineral pellet at the bottom.

2. Carry out density gradient centrifugation (1.077 g/ml Ficoll-Paque) at $250 \times g$ for 20 min at 4 °C. Collect the whitish ring of cells from the interphase with use of a sterile Pasteur pipette, and transfer it to a 15-ml tube. Wash cells with hBMSC maintenance medium, and centrifuge at $250 \times g$ for 5 min at 4 °C.

3. Aspirate supernatant. Resuspend the cell pellet with 15 ml of hBMSC maintenance medium, and plate onto a fresh T75 flask at a density of 100,000 cells per cm^2. Incubate at 37 °C under 5% CO_2 and 95% humidity for 48 h (*see* **Note 3**). Aspirate and discard medium together with nonadherent cells.

4. Add 15 ml of fresh hBMSC maintenance medium, and incubate cells (passage 0) at 37 °C under 5% CO_2 and 95% humidity. Maintain cultures for 8–10 days, changing medium every 2 days. A flat morphology characteristic of fibroblasts in culture should be noticeable after cell attachment to plastic plates. Proliferative colonies should be obvious as the cells approach confluence.

5. Passage hBMSCs when they reach 80–90% confluence. Aspirate and discard medium from the flask and then wash the culture with fresh DPBS. Remove the wash and then add 3 ml of TrypLE™ Express and incubate the cells at 37 °C for 3–5 min (avoid prolonged incubation). Neutralize trypsin by adding 3 ml of hBMSC maintenance medium. Aspirate the medium together with detached cells, and centrifuge at $250 \times g$ for 5 min at 4 °C. Aspirate and discard the supernatant. Resuspend the pellet, and replate the cell suspension onto fresh tissue culture T75 flasks at dilutions of 1:2 to 1:4. Passage cultures once cells reach 80–90% confluence. Subject hBMSCs between passages 4 and 8 to the following induction protocol (*see* **Note 4**).

6. Check hBMSCs for cell identity with use of immunocytochemistry and flow cytometry. The hBMSCs will be positive for such markers as CD29, CD44, CD90, and CD73 but negative for such hematopoietic stem cell markers as CD34 and CD45.

3.5 Derivation of SCLCs from hBMSCs

1. When hBMSC cultures are confluent, detach the cells with TrypLE™ Express (1 ml, at 37 °C, 5 min), and then centrifuge at 250 × g for 5 min at 4 °C.

2. Aspirate the supernatant. Resuspend the pellet in neuroprogenitor induction medium, and plate onto Ultra Low 6-well nonadherent culture plates at a density of 100,000 cells/ml. Refresh half of the medium every 72 h. Cell clusters will be noticeable after 3–5 days. Floating spheres will be visible by day 8–10 and expandable to ≥150 μm in diameter by day 12–14.

3. Collect spheres of neural progenitors by centrifugation at 160 × g for 5 min at 4 °C in a 15-ml tube.

4. Partially dissociate the neurospheres with use of TrypLE™ Express (37 °C, 5 min), and cytospin the cells (3000 cells per cm²) onto Superfrost microscope slides for immunocytochemistry. hBMSC-derived neural progenitors will be immunopositive for the neural progenitor markers nestin and GFAP but negative for the hBMSC markers CD29, CD44, CD90, and CD73. Only sphere cultures that show immunopositivities for nestin and GFAP will be subjected to the following Schwann cell derivation protocol.

5. Collect hBMSC-derived neural progenitors as described above. Resuspend cells in SCLC induction medium, and seed them onto PDL/laminin-coated 6-well plates at a density of 8–10 spheres per cm². Maintain in adherent culture for 7–8 days with medium change every 2–3 days. Spindle-like cells derived from the spheres should be detectable by 3–4 days in adherent culture. By 6–8 days in culture, SCLCs with extended processes should become obvious; these cells will be immunopositive for the glial cell marker, S100β (*see* **Note 5**).

3.6 Directing SCLCs to the Schwann Cell Fate

1. Dissociate SCLCs with TrypLE™ Express (1 ml, 37 °C, 5 min), and centrifuge at 250 × g for 5 min at 4 °C. Aspirate the supernatant, and then resuspend the pellet in medium for SCLC/sensory neuron co-culture.

2. Seed the cells onto hiPSC-derived sensory neurons (prepared as described in Subheading 3.3) at a density of 3000 cells per cm², and maintain in co-culture for 14 days. Replace the medium every 2–3 days. Typical bipolar morphology of Schwann cells should be detectable after 7 days and networks of Schwann cells in alignment with neurons should become obvious by 14 days of co-culture.

3. To harvest the Schwann cells, passage the co-culture with use of TrypLE™ Express (1 ml, 37 °C, 5 min). Add an equal volume of Schwann cell maintenance medium to neutralize the trypsin. Aspirate the detached cells, and centrifuge at 250 × g

for 5 min at 4 °C. Aspirate and discard supernatant. Resuspend the pellet in Schwann cell maintenance medium, and replate onto PDL/laminin-coated 6-well plates at a density of 10,000 cells per cm^2. hiPSC-derived sensory neurons will not survive the passage. The surviving cells not only stain positive for the Schwann cell markers, p75^{NTR} and S100, but they remain so following recovery from cryopreservation and subsequent culture expansion/passage (*see* **Note 6**).

4. For cryopreservation, detach fate-committed Schwann cells with use of TrypLE™ Express (1 ml, 37 °C, 5 min), and then resuspend in medium for Schwann cell cryopreservation. A maximum of 1×10^6 cells can be stored in each 1.5-ml CryoTube. Subject the vials to serial cooling at 4 °C for 15 min, at −20 °C for 2 h, and at −80 °C for 16 h, before storage in a liquid nitrogen tank.

5. To thaw the cryopreserved Schwann cells, remove the selected CryoTube from liquid nitrogen, and quickly transfer to a water bath (37 °C) with gentle and continuous shaking until the frozen pellet is thawed. Remove the vial from the water bath and wipe with 70% ethanol. Transfer the contents of the vial to a 15-ml tube; add 3–5 ml of medium for Schwann cell maintenance (37 °C) dropwise to the tube, gently mixing as the medium is added. Centrifuge cells at $250 \times g$ for 5 min at 4 °C. Aspirate the medium, and then gently triturate the cell pellet in an appropriate volume of Schwann cell maintenance medium before seeding onto to PDL/laminin-coated plates at 10,000 cells/cm^2. Maintain in culture at 37 °C under 5% CO_2 and 95% humidity. Change the medium every 2–3 days. Passage cells when they reach 80–90% confluence.

3.7   Cell Staining

3.7.1   Immunocyto-chemistry

1. When the cells reach the desired density, remove the medium from each well, and wash twice with DPBS. Add cell fixation solution to each well, and incubate for 20 min at room temperature.

2. Wash the wells three times with DPBS. Block nonspecific staining by incubating in blocking buffer for 30 min at room temperature.

3. Remove blocking buffer. Incubate cells at 4 °C for 16 h with selected primary antibodies in blocking buffer at the recommended dilution (*see* Subheading 2.6).

4. Remove primary antibodies, and wash the cells three times in DPBS. Incubate with appropriate secondary antibodies in blocking buffer at the recommended dilution (*see* Subheading 2.6), and incubate cells (room temperature) for 2 h in the dark.

5. Remove secondary antibodies, and rinse the cells three times in DPBS. Add Fluoroshield mounting medium together with DAPI to each well, and incubate for 10–15 min at room temperature to stain cell nuclei.

6. View stained cells under a fluorescence microscope. Stained preparations can also be stored in the dark at −20 °C for later examination.

3.7.2 Flow Cytometry

1. Dissociate cultures with use of TrypLE™ Express (for hiPSCs, use dispase). Transfer suspension of detached cells and subsequent rinses into a 15-ml tube. Then centrifuge at $160 \times g$ for 5 min at 4 °C. Aspirate and discard supernatant. Wash the cell pellet with ice-cold DPBS, centrifuge, and repeat the wash.

2. Aliquot $0.5–1.0 \times 10^6$ cells into each Eppendorf tube. Resuspend cells in 100 μl of selected primary antibody in blocking buffer, and incubate for 1 h at room temperature in the dark.

3. Wash the cells three times in blocking buffer; between washes, centrifuge the cell suspension at $250 \times g$ for 5 min at 4 °C. Dilute the fluorochrome-labeled secondary antibody in blocking buffer at the optimal dilution, and resuspend the cells in this solution. Incubate for 20–30 min at room temperature in dark.

4. Wash the cells three times before analysis by flow cytometry. Follow each wash with centrifugation at $250 \times g$ for 5 min, and then resuspend in ice-cold DPBS.

4 Notes

1. Do not allow the Matrigel-coated plates to dehydrate. Coated plates should be left at 37 °C for at least 30 min before use. If not used immediately, the plates must be sealed to prevent dehydration (e.g., with Parafilm®). The sealed plates can be stored at 4 °C for up to 7 days. Immediately before cell seeding, remove residual BD Matrigel™ solution on the surface of the gelled coat by gently tilting the plate onto one corner and allowing the excess BD Matrigel™ solution to collect in that corner. Remove the solution using a serological pipette. Take care that the tip of the pipette does not scratch the coated surface. Immediately apply a thin layer of mTeSR™1, and follow with seeding of hiPSCs in mTeSR™1 medium.

2. The cell density at the time of initiation of neural differentiation determines the viability and yield of neurons. A relatively high confluency biases toward differentiation along the peripheral neural cell lineage. Initial seeding densities should be empirically determined to reach ideal starting conditions.

3. Avoid frequent or unnecessary moving of the flasks in the first 2 days of culture to allow proper adhesion of hBMSCs to the substratum. Rapid change of medium may cause hBMSCs to detach together with the less adherent hematopoietic cell lineages in the BM sample. The medium should be changed slowly.

4. Cells of the hematopoietic lineage are often contaminants in BMSC cultures. Density gradient centrifugation can be performed to minimize this possibility. Alternatively, frequent medium change and passaging of BMSC cultures can serve to remove the less adherent cells of the hematopoietic lineage.

5. The hBMSC-derived SCLCs are phenotypically unstable. Following withdrawal of glia-inducing factors from cultures, SCLCs undergo rapid reversion to fibroblast-like morphology and lose not only expression of Schwann cell markers but also the capacity for axonal myelination.

6. Co-culture of the SCLCs with hiPSC-derived sensory neurons facilitates contact-mediated signaling that accomplishes the switch to fate-committed Schwann cells. Following withdrawal of glia-inducing factors and loss of hiPSC-derived sensory neurons with passage, the fate-committed Schwann cells demonstrate persistence of spindle-shaped morphology, immunopositivities for Schwann cell markers p75^{NTR} and S100, and the capacity for axonal myelination. Therefore, the human bone marrow-derived Schwann cells so derived hold promise for clinical application in cell-based therapy for nerve injury.

Acknowledgments

This work was supported in part by the HKRGC-General Research Fund 777810, NSFC/RGC-Joint Research Scheme N_HKU741/11, Innovation and Technology Fund (100/10), SK Yee Medical Research Fund, National Natural Science Foundation of China (81000011, 81272080), and the Strategic Research Theme on Neuroscience of The University of Hong Kong.

References

1. Rodríguez FJ, Verdú E, Ceballos D, Navarro X (2000) Nerve guides seeded with autologous Schwann cells improve nerve regeneration. Exp Neurol 161:571–584

2. Bachelin C, Lachapelle F, Girard C, Moissonnier P, Serguera-Lagache C, Mallet J et al (2005) Efficient myelin repair in the macaque spinal cord by autologous grafts of Schwann cells. Brain 128:540–549

3. Cai S, Shea GK, Tsui AY, Chan YS, Shum DK (2011) Derivation of clinically applicable Schwann cells from bone marrow stromal cells for neural repair and regeneration. CNS Neurol Disord Drug Targets 10:500–508

4. Jessen KR, Mirsky R, Arthur-Farraj P (2015) The role of cell plasticity in tissue repair: adaptive cellular reprogramming. Dev Cell 34:613–620

5. Caddick J, Kingham PJ, Gardiner NJ, Wiberg M, Terenghi G (2006) Phenotypic and functional characteristics of mesenchymal stem cells differentiated along a Schwann cell lineage. Glia 54:840–849

6. Brohlin M, Mahay D, Novikov LN, Terenghi G, Wiberg M, Shawcross SG et al (2009) Characterisation of human mesenchymal stem cells following differentiation into Schwann cell-like cells. Neurosci Res 64:41–49

7. Shea GK, Tsui AY, Chan YS, Shum DK (2010) Bone marrow-derived Schwann cells achieve fate commitment—a prerequisite for remyelination therapy. Exp Neurol 224:448–458

8. Ao Q, Fung CK, Tsui AY, Cai S, Zuo HC, Chan YS et al (2011) The regeneration of transected sciatic nerves of adult rats using chitosan nerve conduits seeded with bone marrow stromal cell-derived Schwann cells. Biomaterials 32:787–796

9. Cai S, Chan YS, Shum DK (2014) Induced pluripotent stem cells and neurological disease modeling. Acta Physiol Sinica 66:55–66

10. Cai S, Han L, Ao Q, Chan YS, Shum DK (2017) Human iPS cell-derived sensory neurons for fate commitment of bone marrow-derived Schwann cells - implication for re-myelination therapy. Stem Cells Transl Med 6:369–381

Chapter 11

Generation and Use of Merlin-Deficient Human Schwann Cells for a High-Throughput Chemical Genomics Screening Assay

Alejandra M. Petrilli and Cristina Fernández-Valle

Abstract

Schwannomas are benign nerve tumors that occur sporadically in the general population and in those with neurofibromatosis type 2 (NF2), a tumor predisposition genetic disorder. NF2-associated schwannomas and most sporadic schwannomas are caused by inactivating mutations in Schwann cells in the *neurofibromatosis type 2* gene (*NF2*) that encodes the merlin tumor suppressor. Despite their benign nature, schwannomas and especially vestibular schwannomas cause considerable morbidity. The primary available therapies are surgery or radiosurgery which usually lead to loss of function of the compromised nerve. Thus, there is a need for effective chemotherapies. We established an untransformed merlin-deficient human Schwann cell line for use in drug discovery studies for NF2-associated schwannomas. We describe the generation of human Schwann cells (HSCs) with depletion of merlin and their application in high-throughput screening of chemical libraries to identify compounds that decrease their viability. This NF2-HSC model is amenable for use in independent labs and high-throughput screening (HTS) facilities.

Key words Human Schwann cell line, Schwannoma, High-throughput viability assay, Neurofibromatosis type 2 (NF2), Drug discovery, Merlin

1 Introduction

Schwannomas originate from Schwann cells and can be sporadic or associated with neurofibromatosis type 2 (NF2). NF2 is caused by mutations in the *NF2* tumor suppressor gene that encodes a protein called merlin or schwannomin [1, 2]. Over 60% of sporadic schwannomas have mutations in the *NF2 gene* [3, 4]. Loss of merlin function causes deregulation of numerous signaling pathways controlling cell size, morphology, cell contact inhibition of growth, and survival [5]. Although merlin is ubiquitously expressed, loss of merlin function in Schwann, arachnoid, and ependymal cells selectively leads to their tumorigenesis. A hallmark and diagnostic feature of NF2 is the development of bilateral vestibular schwannomas (VS), also known as acoustic neuromas. VS commonly lead to

Paula V. Monje and Haesun A. Kim (eds.), *Schwann Cells: Methods and Protocols*, Methods in Molecular Biology, vol. 1739, https://doi.org/10.1007/978-1-4939-7649-2_11, © Springer Science+Business Media, LLC 2018

hearing loss, tinnitus, and imbalance. Compression of the brainstem by these tumors can be fatal. NF2 patients also develop substantial morbidity due to coexisting meningiomas and ependymomas [6, 7]. Currently recommended treatment includes monitoring tumor growth followed by surgical resection or radiosurgery if the tumor shows accelerated growth and is accessible. Resection typically leads to loss of function of the compromised nervous tissue [6]. The FDA has yet to approve a drug for NF2-associated schwannomas or meningiomas.

Because the merlin tumor suppressor lacks enzymatic activity and modulates numerous signaling pathways, rational selection of a target for drug development is challenging. In order to adopt a chemical genomics screening approach for NF2-associated schwannomas, a suitable merlin-deficient SC line is needed which is amenable to growth within restricted parameters. These include growth in 384- and 1526-well plates and doubling times within 24–72 h in standard growth medium. Cell-based chemical screens with a human Schwann cell line or human schwannoma cells would be the ideal cell model for drug discovery efforts. However, it is widely recognized that adult human Schwann cells (HSCs) and NF2 schwannoma cells are difficult to propagate in vitro, do not proliferate rapidly and survive only a few passages before senescence, and moreover are in scarce supply [8–11]. Our experience with cultured adult HSC gene editing for *NF2* knockout using RNA-guided clustered regularly interspaced short palindrome repeats (CRISPR)-associated nuclease Cas9 or *NF2* silencing by shRNA technology has been unfruitful because adult HSC stop dividing, regardless of merlin expression level. A common approach to overcome these limitations is to genetically immortalize human Schwann/schwannoma cells. Overexpression of human *telomerase reverse transcriptase* (*hTERT*) alone is not sufficient to immortalize adult HSC or schwannoma cells. Dual overexpression of *hTERT* and mouse cyclin-dependent kinase 4 (*Cdk4*) has been attempted and successfully established an adult HSC line [12]. Additionally, dual overexpression of *hTERT* and *SV40 large T antigen* has produced immortalized fetal HSC lines [13]. However, the molecular changes introduced by the immortalization procedure can mask and/or interfere with true merlin drug targets. For example, the HEI193 Schwann cell line is a human schwannoma cell line created from an NF2 patient schwannoma by immortalization with human papillomavirus *E6–E7* genes [14–16]. Continuous passaging of this cell line has led to chromosomal instability rendering them unsuitable for cell-based drug screening campaigns [17]. HSC of fetal origin can be expanded multiple times and generate sufficient cells for high-throughput screening because they have a faster division rate than their adult counterparts. We introduced *NF2* loss-of-function mutation by RNA-guided CRISPR/Cas9 into these cells. Despite the use of two validated *NF2*_sgRNA constructs, as soon as HSC underwent *NF2* knockout, they displayed senescence similar to the merlin-deficient meningioma cells [18, 19]. However, by reducing merlin expression level by approximately 98% of control levels with shRNA technology, the cells can be propagated in vitro up to 20 passages before they began to senesce.

Here we describe the generation of merlin-deficient HSCs of fetal origin using lentiviral delivery of NF2 shRNA. The NF2 shRNA construct is commercially available and is specific and highly effective; merlin levels remain low through multiple passages. The merlin-deficient HSCs are highly proliferative in culture and can be expanded to generate sufficient cells to screen large compound libraries at low passage number. Additionally, we describe their application in a 384-well plate format viability assay to generate ten-point semilog dose-response curves for the screening of candidate compounds that can be performed in individual laboratories.

2 Materials

2.1 Cell Culture

1. HSC isolated from fetal human spinal nerve (ScienCell, Cat. No. 1700).

2. 100 and 35 mm dishes with advanced treated surface such as CellBIND (Corning, Cat. No. 07-202-516) (*see* **Note 1**).

3. Schwann cell growth medium (ScienCell, Schwann Cell Medium Cat. No. 1701): basal medium containing 5% fetal bovine serum (FBS), 1× Schwann cell growth supplement, and 100 U/mL penicillin and 100 µg/mL streptomycin. Warm at 37 °C in a water bath before use.

4. Hank's balanced salt solution (HBSS).

5. 0.05% trypsin-EDTA.

6. D10 medium is used for trypsin inactivation when harvesting cells. D10 medium: Dulbecco's modified Eagle medium (DMEM, Gibco), 10% heat inactivated FBS, 100 U/mL penicillin, and 100 µg/mL streptomycin.

7. Automatic cell counter such as Countess (Invitrogen).

8. Trypan blue solution.

9. CO_2 incubator for all cell work set at 37 °C in a humidified atmosphere of 5–7% CO_2.

2.2 Lentiviral Transduction and Selection of Merlin Knockdown Cells

1. Human *NF2* gene-specific shRNA lentiviral particles (GenBank accession no. NM_000268; TRCN0000237845, Mission Sigma-Aldrich) (*see* **Note 2**).

2. Polybrene (hexadimethrine bromide) stock solution: Prepare 2 mg/mL polybrene by dissolving 20 mg in 10 mL of double distilled water, filter sterilize, and store at 4 °C.

3. Puromycin dihydrochloride stock solution: prepare a 10 mg/mL solution in double distilled water, filter sterilize, aliquot, and store at −20 °C.

4. (Optional) Merlin and β-actin antibodies. Primary antibodies used in Western blotting are recommended as follows. From cell signaling: rabbit monoclonal anti-Merlin (D1D8, Cat #6995), anti-Merlin (D3S3W, Cat #12888), and mouse monoclonal anti-β-actin (clone 8H10D10, Cat #3700). From Abcam: rabbit monoclonal anti-NF2/Merlin [clone EPR2573(2), Cat #ab109244].

5. (Optional) Fluorescent secondary antibodies. Goat anti-rabbit IgG conjugated with DyLight 800 4×-PEG (cell signaling, Cat #5151) and goat anti-mouse IgG conjugated with DyLight 680 (cell signaling, Cat #5470).

2.3 Dose-Response Viability Assay

1. 384-well plates. The use of 384-well plates with advanced treated surface such as CellBIND Corning #3770 (black, clear bottom) for fluorescence detection assays is recommended. The clear bottom allows to examine the attachment of cells and follow the progress of the experiment by phase or bright-field microscopy.

2. Dimethyl sulfoxide (DMSO).

3. Fluorescence plate reader such as Synergy H1 Hybrid (BioTek).

4. Stock compounds to be tested (e.g., kinase inhibitors, reductase inhibitors, histone deacetylase inhibitors) prepared at 10 mM in DMSO.

5. Stock of positive control compound in DMSO. A 25 mM rapamycin stock is recommended. Dilute the rapamycin to a final concentration of 50 μM in the assay, which is a high concentration that produces by general toxicity 100% loss of cell viability in 48 h.

6. Phenol red free Schwann cell medium (ScienCell cat. No. 1701-prf): phenol red free Schwann cell base, 5% FBS, 1× Schwann cell growth supplements, 100 U/mL penicillin, and 100 μg/mL streptomycin. This is the same Schwann cell growth medium used to culture the cells but lacks the phenol red indicator. For optimal detection of cell-based fluorescence, the use of phenol red-free medium is recommended to reduce nonspecific background fluorescence.

7. 12-well V-bottom reservoirs.

8. Twelve-multichannel pipette.

9. CellTiter-Fluor Cell Viability Assay kit (Promega).

10. Microplate dispenser. High-performance dispensing system for 384-well plates such as MultiFlo dispenser (BioTek).

11. Centrifuge with microplate swinging bucket rotor.

12. Orbital shaker.

13. GraphPad Prism or similar software for data organization, graphing, comprehensive curve fitting (nonlinear regression), and statistics.

3   Methods

3.1   Generation and Culture of Merlin-Deficient Human Schwann Cells

1. Thaw the cryovial containing frozen HSC in a 37 °C water bath with gentle agitation for 1–2 min. Wipe the outside of the vial with 70% ethanol, and add 1 mL of Schwann cell growth medium, prior to resuspending the cells.

2. Distribute the cell suspension dropwise into a 100 mm CellBIND dish with 7 mL equilibrated Schwann cell growth medium, and gently rock the dish to seed the cells evenly. Incubate in a CO_2 incubator.

3. Next day replace the medium to remove the DMSO, dead cells, and debris.

4. Harvest the cells by trypsinization when the culture reaches 90% confluency.

5. Resuspend the cells in Schwann cell growth medium, and mechanically dissociate them with a medium-coated Pasteur pipette.

6. Count the cells using an automatic cell counter.

7. Seed the HSCs into two 100 mm CellBIND plates at 600,000 cells per plate in Schwann cell growth medium and 100,000 cells into one 35 mm CellBIND plate to be used as puromycin selection control. Incubate the plates overnight in a CO_2 incubator.

8. Next day, count the cells of one 100 mm plate (this cell suspension can be split 1:2 for expansion and/or frozen as a cell stock). Calculate the volume of *NF2* gene-specific shRNA lentiviral particles needed for a multiplicity of infection (MOI) of 5 according to the virus titer (obtained from the Certificate of Analysis) and the number of cells (*see* **Note 3**).

9. For lentiviral transduction, mix the calculated volume of lentiviral particles for a MOI of 5 with Schwann cell growth medium supplemented with 8 μg/mL of polybrene (transduction medium).

10. Remove the medium from the remaining 100 mm plate, and add transduction medium to the cells. Incubate in a CO_2 incubator for 18 h (*see* **Note 4**). Change the medium of the 35 mm dish with fresh Schwann cell growth medium.

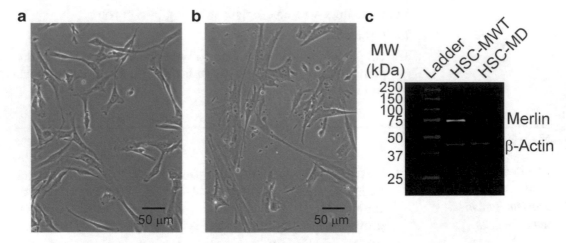

Fig. 1 Representative images of wild-type merlin HSC and merlin-deficient HSC. Phase contrast images of wild-type merlin HSC (**a**) and merlin-deficient HSC (**b**) after NF2 silencing by shRNA. Scale bar = 50 μm. (**c**) Representative image of merlin and β-actin expression by means of near-infrared fluorescence Western blot in control HSC and merlin-deficient HSC lysates. Anti-merlin (Abcam) and anti-β-actin antibodies were used at 1:1500 and 1:30,000, respectively. Anti-rabbit and anti-mouse secondary antibodies were used at 1:25,000 and 1:30,000, respectively

11. Replace transduction medium to Schwann cell growth medium, and incubate in a CO_2 incubator for another 24 h.

12. Start puromycin selection by changing the medium to Schwann cell growth medium containing 0.55 μg/mL puromycin to the transduced 100 mm dish and the 35 mm sister dish that will serve as a selection control. Replenish the growth medium containing puromycin every other day for a total period of 5 days of selection or until all the cells in the selection control dish have died (*see* **Note 5**).

13. Culture and expand the selected merlin-deficient HSCs in Schwann cell growth medium to use in drug screens or freeze for future use.

14. Confirm merlin status of the HSC line by Western blotting using a tested and validated merlin antibody (Fig. 1) (*see* **Note 6**).

3.2 Dose-Response Viability Assay

A viability assay only assesses live cells present in the well after an incubation period with compounds. A reduction of cell viability compared to vehicle-treated cells (control) means that only a lesser number of viable cells are present. It is not possible to determine with this measurement if the loss of viability is due to cell death or cell cycle arrest. This protocol tests four compounds per plate in nine semilog dose dilutions. The design of a standard plate is presented in Fig. 2. This design allows for 32 replicates for the assay and plate controls (untreated, 0.1% DMSO, and 50 μM rapamycin) to assess assay and plate quality. To perform dose-response

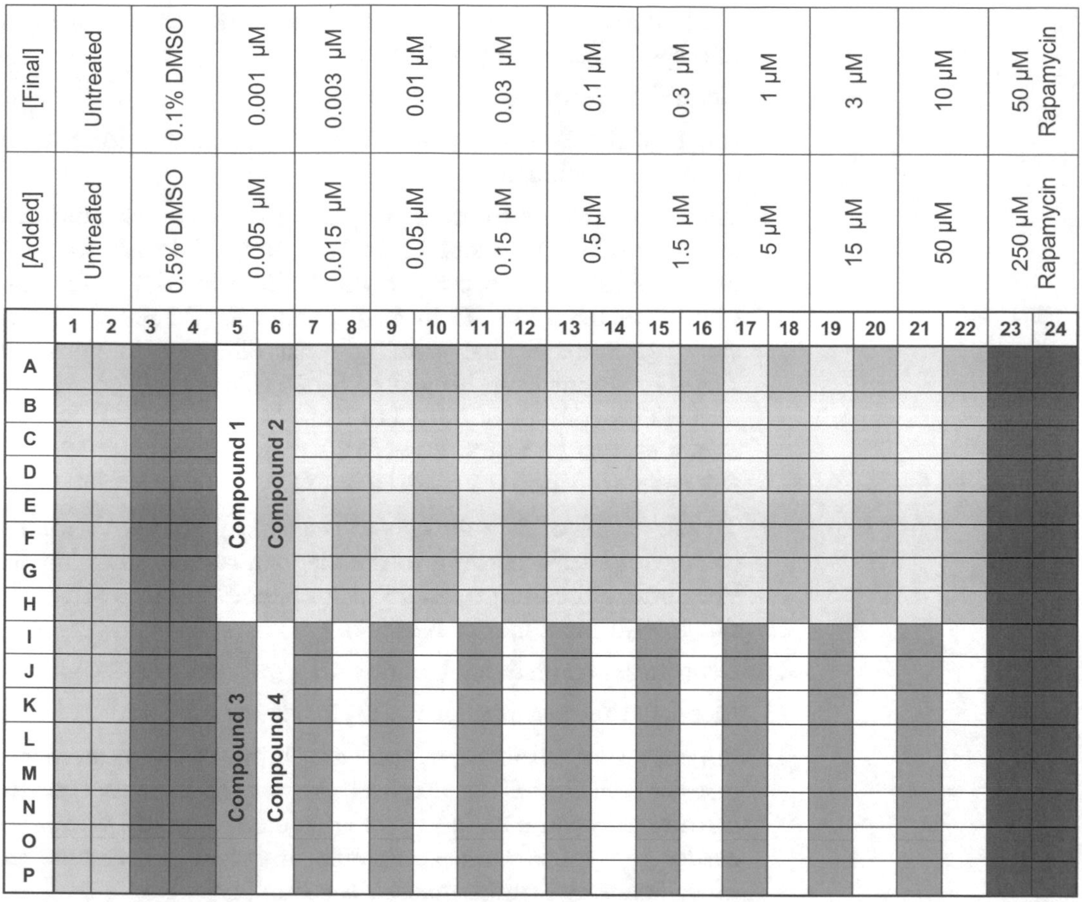

Fig. 2 Typical four drugs assay format in 384-well plates. Columns 1 and 2 contain 32 replicates of untreated cells, columns 2 and 3 contain 32 replicates of 0.1% DMSO as control, columns 23 and 24 contain 32 replicates of 50 μM rapamycin as a positive control, and columns 5–22 contain 8 replicates for each concentration of the compounds tested in a nine-concentration semilog serial dilutions ranging from 10 μM to 1 nM

screening, compounds are tested in a nine-concentration semilog serial dilution ranging from 10 μM to 1 nM with eight replicates for each concentration.

1. Harvest merlin-deficient HSCs by trypsinization.

2. Gently resuspend and thoroughly mechanically dissociate the cells in phenol red free Schwann cell growth medium by using a tissue culture medium-coated glass Pasteur pipette.

3. Count viable cells using trypan blue staining and an automatic cell counter.

4. Prepare cell suspension in phenol red free Schwann cell growth medium at a density of 1.25×10^5 cells/mL. For each 384-well CellBIND plate, prepare 8 mL of cell suspension plus the dead volume of the automatic cell dispenser. For a microplate dispenser MultiFlow model, the dead volume needed is 5 mL.

5. Seed 20 µL/well of cell suspension into 384-well plate using a microplate dispenser (*see* **Note 7**).

6. Centrifuge plate for 1 min at $40 \times g$.

7. Incubate the 384-well plates in a CO_2 incubator for 5 h to allow cell attachment.

8. Prepare treatment solutions in phenol red free Schwann cell growth medium into four 12-well reservoirs. Add 450 µL phenol red free Schwann cell growth medium in well 1. Prepare 450 µL of 0.5% DMSO in well 2 and 250 µM rapamycin in well 12. Prepare nine semilog serial dilutions in constant DMSO content from 50 µM to 5 nM (five times the desired final concentrations of 10 µM to 1 nM in 0.1% DMSO) of each compound to be tested in wells 11 through 3 of the 12-well reservoirs to match the plate layout (Fig. 2) (*see* **Note 8**).

9. Transfer 5 µL of 5× solutions to the assay plate with a multichannel pipette according to the diagram presented in Fig. 1. For untreated columns, add 5 µL of phenol red free Schwann cell growth medium (*see* **Note 9**).

10. Centrifuge the plate for 1 min at $40 \times g$.

11. Incubate the assay plate in a CO_2 incubator for 48 h.

12. Prior to assay termination, thaw and bring to room temperature the CellTiter-Fluor reagents. Prepare the 2× assay reagent mixture as recommended by the manufacturer, and vortex to ensure the substrate is completely dissolved. A volume of 10 mL of assay reagent mixture is needed per plate.

13. Terminate the assay after 48 h of compound treatment by adding 25 µL of 2× CellTiter-Fluor reagent mixture to each well using a microplate dispenser (*see* **Note 10**).

14. Mix briefly by placing the plates on an orbital shaker at 300–500 RPM for 1 min.

15. Incubate the plates in the dark at 37 °C for 45 min to allow fluorescence development.

16. Measure fluorescence intensity at 390 nm excitation and 505 nm emission using a fluorescence plate reader.

17. Calculate the coefficient of variation (CV) and the Z-factor for each plate to assess assay quality using the following calculations:

$$CV = \frac{SD \; replicates}{Mean \; replicates}$$

$$Z - factor = 1 - \frac{3 \times \left(SD_{DMSO \; ctrl} + SD_{Rapa \; ctrl}\right)}{\left(Mean_{DMSO \; ctrl} - Mean_{Rapa \; ctrl}\right)}$$

SD stands for standard deviation. Z-factor values between 0.5 and 1.0 are considered excellent [20, 21].

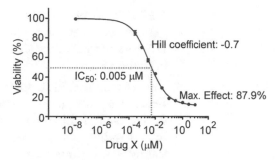

Fig. 3 Example of a dose-response viability curve. Representative dose-response curve, absolute half-maximum effective concentration IC_{50}, Hill coefficient, and maximum effect using a nonlinear regression, log [drug] vs. response, and four-parameter analysis generated with GraphPad Prism

18. Calculate the percentage of viable cells for each compound by normalizing the raw data to the mean of DMSO control wells (100% viability).

19. Analyze the dose-response data to identify promising compounds by plotting the response to drug treatment versus drug concentration in a logarithmic scale to generate a dose-response curve. Calculate the absolute half-maximum effective concentration IC_{50} (also known as half-maximum inhibition of growth IG_{50}), the Hill coefficient, and the maximum effect for each compound in the dose-response screens. The dose-response data analysis can be performed using a log [inhibitor] versus response, nonlinear regression, and four-parameter analysis included in the GraphPad Prism software (Fig. 3).

4   Notes

1. Tissue culture (TC) dishes with advanced treated surfaces facilitate the attachment and spreading of cells compared to standard TC-treated surfaces. The fetal HSCs can form foci and detach. The advanced polystyrene surface incorporates oxygen-containing functional groups which yields a net negative charge that is hydrophilic. The 100 mm advanced treated surface plates can be substituted with poly-D-lysine- or poly-L-lysine-coated standard TC-treated surfaces.

2. Mission NF2 shRNA lentiviral transduction particles are commercially available, ready to use, and self-inactivating viral particles. Lentiviral particles should be aliquoted in small volumes at first use and stored at −80 °C because repeated freeze-thawing cycles decrease their titer. They can be produced in packaging cells (e.g., HEK293T) by co-transfection of the

NF2 shRNA plasmid DNA (TRCN0000237845) with third-generation packaging plasmids. The NF2 shRNA plasmid DNA incorporates the shRNA insert into a TRC2-pLKO-puro backbone vector.

3. The Mission shRNA lentiviral particles are considered Biosafety Level 2 agents despite being replication-deficient and produced using a third-generation packaging system. Therefore, their handling, waste decontamination, and disposal should follow institutional safety rules and regulations and NIH guidelines. The MOI represents the number of infectious viral particles per cell and depends on the cell type, so the optimal MOI should be titrated in advance using reporter particles such as MISSION TurboGFP Control Transduction Particles.

4. Polybrene is a cationic molecule that increases viral transduction efficiency enhancing the virus adsorption on the cell membrane by neutralizing/reducing the negatively charged virus and cell membrane repulsive electrostatic forces. Polybrene enhances lentiviral transduction efficiency but may be toxic to some cell types; therefore, always test polybrene toxicity in a range of 0–10 μg/mL in advance in the cells to be infected. In HSCs, polybrene is routinely used at a final concentration of 8 μg/mL.

5. Puromycin selection should not begin earlier than 24 h after transduction to give cells time to express sufficient puromycin-resistance gene product. The optimal puromycin concentration for the selection of puromycin-resistant cells should be determined in advance when a new cell lot number is used. We recommend performing a puromycin kill curve ranging from 0.25 to 3 mg/mL to determine the minimal concentration necessary to kill all non-transduced cells after 3–5 days of culture. A puromycin control plate should be included in the experiment to determine selection effectiveness.

6. Fetal HSCs expressing wild-type merlin grow in monolayer and many have a typical bipolar morphology. In contrast, merlin-deficient HSCs are typically large, flat, or multipolar cells (Fig. 1a, b). When grown to confluency, merlin-deficient HSCs do not exhibit contact-dependent inhibition of growth and form multiple layers. Western blot detection of merlin in HSC expressing wild-type merlin shows a band at ~70 kDa. This band is not detected in merlin-deficient HSC (Fig. 1c).

7. For screening, cells should have been passaged at least once after thawing. Obtaining a single cell suspension is critical to ensure evenly seeded wells and a monolayer culture. The cell suspension of 1.25×10^5 cells/mL yields the optimum 2500 live cells/well in 20 μL. The optimal cell density at seeding was empirically determined during the assay optimization phase for a 48-h drug treatment. Optimal cell density can be determined

by seeding HSC in 384-well plates at multiple densities (e.g., 0–5000 cells/well in 20 μL of medium) followed by analysis of viability according to the manufacturer's recommendation. Briefly, after cell attachment half of the wells are treated with 20 μg/mL digitonin to lyse cells. At 48 h, examine untreated wells with a microscope to identify the cell density that yields a homogeneous cell monolayer. Assess cell viability using the CellTiter-Fluor viability assay. Calculate signal-to-noise ratio, signal-to-background ratio, and Z-factor for each cell density. The condition that renders the highest signal-to-noise ratio, signal-to-background ratio, and Z-factor is the optimal cell density. Do not use cells at high passage number, beyond passage 20, to avoid selection of a subpopulation of cells or spontaneously transformed cells.

8. The concentration of DMSO in the DMSO control and in the compound dilutions is maintained at 0.1%. In a 48-h DMSO tolerance test, we determined that cell viability in 0.1% DMSO was reduced only by ~1% with respect to untreated controls. The maximum compound dose that can be tested using a 10 mM stock is 10 μM so as not to exceed the limit of 0.1% DMSO. Compound dilutions are prepared 5× the final desired concentration in order to add 5 μL of compound and 20 μL of medium in each well for a total volume of 25 μL. To maintain a constant DMSO concentration, prepare the 5× 10 μM (50 μM) dose by diluting the 10 mM stock in phenol red free Schwann cell growth medium and the serial dilutions in 0.5% DMSO in phenol red free Schwann cell growth medium.

9. The positive control (50 μM rapamycin) was chosen based on a dose-response study that caused more than 98% decrease in cell viability after 48 h incubation. This control can be replaced with other pharmacological compound at a dose that become toxic and produce complete or nearly complete loss of viability after 48 h of treatment.

10. CellTiter-Glo luminescent cell viability assay (Promega) can be used instead of the CellTiter-Fluor cell viability assay (Promega). It can also be used as an orthogonal assay (use white 384-well CellBIND plates if available for the luminescent assays). Optimal cell seeding density for this assay should also be determined.

Acknowledgment

This work was supported in part by a Drug Discovery Initiative award from the Children's Tumor Foundation and with the assistance from Dr. Layton Smith, Director, Drug Discovery Florida, Sanford Burnham Prebys Medical Discovery Institute.

References

1. Rouleau GA, Merel P, Lutchman M, Sanson M, Zucman J, Marineau C, Hoang-Xuan K, Demczuk S, Desmaze C, Plougastel B et al (1993) Alteration in a new gene encoding a putative membrane-organizing protein causes neuro-fibromatosis type 2. Nature 363(6429):515–521. https://doi.org/10.1038/363515a0

2. Trofatter JA, MacCollin MM, Rutter JL, Murrell JR, Duyao MP, Parry DM, Eldridge R, Kley N, Menon AG, Pulaski K et al (1993) A novel moesin-, ezrin-, radixin-like gene is a candidate for the neurofibromatosis 2 tumor suppressor. Cell 75(4):826

3. Agnihotri S, Jalali S, Wilson MR, Danesh A, Li M, Klironomos G, Krieger JR, Mansouri A, Khan O, Mamatjan Y, Landon-Brace N, Tung T, Dowar M, Li T, Bruce JP, Burrell KE, Tonge PD, Alamsahebpour A, Krischek B, Agarwalla PK, Bi WL, Dunn IF, Beroukhim R, Fehlings MG, Bril V, Pagnotta SM, Iavarone A, Pugh TJ, Aldape KD, Zadeh G (2016) The genomic landscape of schwannoma. Nat Genet 48(11):1339–1348. https://doi.org/10.1038/ng.3688

4. Seizinger BR, Martuza RL, Gusella JF (1986) Loss of genes on chromosome 22 in tumorigenesis of human acoustic neuroma. Nature 322(6080):644–647. https://doi.org/10.1038/322644a0

5. Petrilli AM, Fernandez-Valle C (2016) Role of Merlin/NF2 inactivation in tumor biology. Oncogene 35(5):537–548. https://doi.org/10.1038/onc.2015.125

6. Blakeley JO, Plotkin SR (2016) Therapeutic advances for the tumors associated with neurofibromatosis type 1, type 2, and schwannomatosis. Neuro-Oncology 18(5):624–638. https://doi.org/10.1093/neuonc/nov200

7. Evans DG (2009) Neurofibromatosis type 2 (NF2): a clinical and molecular review. Orphanet J Rare Dis 4:16. https://doi.org/10.1186/1750-1172-4-16

8. Hayflick L (1965) The limited in vitro lifetime of human diploid cell strains. Exp Cell Res 37:614–636

9. Nair S, Leung H, Collins A, Ramsden R, Wilson J (2007) Primary cultures of human vestibular schwannoma: selective growth of schwannoma cells. Otol Neurotol 28(2):258–263. https://doi.org/10.1097/01.mao.0000247811.93453.6a

10. Rutkowski JL, Kirk CJ, Lerner MA, Tennekoon GI (1995) Purification and expansion of human Schwann cells in vitro. Nat Med 1(1):80–83

11. Turnbull VJ (2005) Culturing human Schwann cells. Methods Mol Med 107:173–182

12. Keng VW, Watson AL, Rahrmann EP, Li H, Tschida BR, Moriarity BS, Choi K, Rizvi TA, Collins MH, Wallace MR, Ratner N, Largaespada DA (2012) Conditional inactivation of Pten with EGFR overexpression in Schwann cells models sporadic MPNST. Sarcoma 2012:620834. https://doi.org/10.1155/2012/620834

13. Lehmann HC, Chen W, Mi R, Wang S, Liu Y, Rao M, Hoke A (2012) Human Schwann cells retain essential phenotype characteristics after immortalization. Stem Cells Dev 21(3):423–431. https://doi.org/10.1089/scd.2010.0513

14. Hung G, Li X, Faudoa R, Xeu Z, Kluwe L, Rhim JS, Slattery W, Lim D (2002) Establishment and characterization of a schwannoma cell line from a patient with neurofibromatosis 2. Int J Oncol 20(3):475–482

15. Lepont P, Stickney JT, Foster LA, Meng JJ, Hennigan RF, Ip W (2008) Point mutation in the NF2 gene of HEI-193 human schwannoma cells results in the expression of a merlin isoform with attenuated growth suppressive activity. Mutat Res 637(1–2):142–151. https://doi.org/10.1016/j.mrfmmm.2007.07.015

16. Sainio M, Jaaskelainen J, Pihlaja H, Carpen O (2000) Mild familial neurofibromatosis 2 associates with expression of merlin with altered COOH-terminus. Neurology 54(5):1132–1138

17. Prabhakar S, Messerli SM, Stemmer-Rachamimov AO, Liu TC, Rabkin S, Martuza R, Breakefield XO (2007) Treatment of implantable NF2 schwannoma tumor models with oncolytic herpes simplex virus G47Delta. Cancer Gene Ther 14(5):460–467. https://doi.org/10.1038/sj.cgt.7701037

18. James MF, Lelke JM, Maccollin M, Plotkin SR, Stemmer-Rachamimov AO, Ramesh V, Gusella JF (2008) Modeling NF2 with human arachnoidal and meningioma cell culture systems: NF2 silencing reflects the benign character of tumor growth. Neurobiol Dis 29(2):278–292. https://doi.org/10.1016/j.nbd.2007.09.002

19. Shalem O, Sanjana NE, Hartenian E, Shi X, Scott DA, Mikkelsen TS, Heckl D, Ebert BL, Root DE, Doench JG, Zhang F (2014) Genome-scale CRISPR-Cas9 knockout screening in human cells. Science 343(6166):84–87. https://doi.org/10.1126/science.1247005

20. Petrilli AM, Fuse MA, Donnan MS, Bott M, Sparrow NA, Tondera D, Huffziger J, Frenzel

C, Malany CS, Echeverri CJ, Smith L, Fernandez-Valle C (2014) A chemical biology approach identified PI3K as a potential therapeutic target for neurofibromatosis type 2. Am J Transl Res 6(5):471–493

21. Zhang JH, Chung TD, Oldenburg KR (1999) A simple statistical parameter for use in evaluation and validation of high throughput screening assays. J Biomol Screen 4(2):67–73

Part II

Protocols for Studying Schwann Cell Behavior and Myelination In Vitro

Chapter 12

Lentiviral Transduction of Rat Schwann Cells and Dorsal Root Ganglia Neurons for In Vitro Myelination Studies

Corey Heffernan and Patrice Maurel

Abstract

Lentiviral transduction is a gene delivery method that provides numerous advantages over direct transfection and traditional retroviral or adenoviral delivery methods. It facilitates for the transduction of primary cells inherently difficult to transfect, delivers constructs of interest to nondividing as well as dividing cells, and permits the long-term expression of sizable DNA inserts (e.g., <7 kb). The study of peripheral nerve myelination at the molecular level has long benefited from the Schwann cells/dorsal root ganglia (DRG) neurons myelinating co-culture system. As this culture system takes about a month to develop and perform experiments with, lentiviral-delivered constructs can be used to manipulate gene expression in Schwann cells and DRG neurons, primary cells that are otherwise resilient to direct transfection. Here we present our protocol for lentiviral production and purification and subsequent infection of large numbers of Schwann cells and/or DRG neurons for the molecular study of peripheral nerve myelination in vitro.

Key words Lentivirus, Schwann cell, Dorsal root ganglia neuron

1 Introduction

Since the mid-1970s [1–3], recombinant DNA technologies have been developed with a view to human gene therapies. A beneficial byproduct of the development of such technologies has been the emergence of efficient delivery systems, based on RNA (retroviruses, lentiviruses) and DNA (adenovirus, adeno-associated virus) viruses, to deliver constructs into mammalian cell for the purpose of in vitro researches. Viral transduction provides an effective approach to modify cells (either by knockdown or by expression of a protein of interest), including cells that are otherwise difficult to modify by more classical techniques such as calcium phosphate, lipofection, or electroporation. Viral transduction of mammalian cells in basic research has been in routine use since the mid-1980s [4].

RNA- and DNA-based systems each come with their advantages and disadvantages. While retroviruses provide long-term expression of the gene of interest due to their integration in the

Paula V. Monje and Haesun A. Kim (eds.), *Schwann Cells: Methods and Protocols*, Methods in Molecular Biology, vol. 1739, https://doi.org/10.1007/978-1-4939-7649-2_12, © Springer Science+Business Media, LLC 2018

genome of the host cell, their main caveat is that they do not infect quiescent cells, requiring the breakdown of the nuclear membrane during the cell cycle of dividing cells for transgenes to gain access and integrate into the genomic DNA. Although adenoviruses and adeno-associated viruses infect nondividing cells as well as dividing cells, they do not integrate into the host cell genome and thus are limited in their long-term expression. While adeno-associated viruses persist as episomal chromatin, they are small (4.7 kb) and therefore limited as to the insert size that can be transduced (about 2.5 kb).

Lentiviruses are a subtype of retroviruses. As such their integrative nature provides the advantage of long-term expression. Lentiviruses however, unlike other types of retroviruses, have also the unique capability to infect nondividing cells as well as dividing cells, thereby providing the advantage conferred by transduction systems derived from adenoviruses and adeno-associated viruses. Like any of the other type of recombinant viruses used in transduction experiments, lentiviruses can be pseudotyped to modify the tropism toward host cells. The most common pseudotyping is with the vesicular stomatitis Indiana virus glycoprotein G (VSV-G), which provides lentiviruses with a broad range of host cells [5]. Lentiviruses can also be easily pseudotyped with various glycoproteins to provide a more selective tropism toward cells of interest [6], a particularly important feature for in vivo applications. Finally, most current recombinant lentiviral vectors are based on the human immunodeficiency virus 1 (HIV-1) lentivirus. Once the required viral *cis*-acting sequences kept in the lentiviral vectors are taken into account (about 20% of the 9 kb of a wild-type lentivirus), lentiviral-based vectors can theoretically accommodate inserts up to about 7 kb in size, almost as much as that of adenovirus-based vectors (about 8 kb). Importantly, note that insert size does not refer only to the cDNA of interest (shRNA or protein encoding). It also includes mammalian promoters and sequences encoding for resistance genes and/or fluorescent proteins. The extent of these sequences will vary with the choice of lentiviral vector system and will restrict the actual size of the cDNA that can be cloned.

With their many useful characteristics, lentiviral-based vectors have been actively developed over the past 20 years. These studies have however highlighted the difficulty of developing routine protocols for the production of viral stocks that are of high quality and titer. One of the reason stems from the fact that the development of efficient packaging cell lines is proving difficult due to the cytotoxicity of the HIV Gag and Pol elements [7, 8] and of the VSV-G [9]. All viral *trans*-acting sequences are expressed from two to three separate non-viral mammalian expression vectors that need to be transfected at the same time as the lentiviral vector. Not only does the total amount of plasmid DNA have an impact on the final titer so does the relative amount of each of the plasmids [10]. Another major impediment is related to the overall size of the vector and to the size of the RNA that is to be packaged into the viral particle. All other transfection conditions being equal, an 11 kb lentiviral vector can be expected to produce viruses with an effective infectious titer (*see* **Note 1**) 100-fold lower than a vector of 7.5 kb [11, 12]. Similarly, an RNA of about 7.5 kb to be packaged will decrease the viral titer by a 1000-fold compared to an insert of about 5.5 kb, although both lentiviral constructs may have a similar overall size [11]. Several additional factors have been shown to influence the effective viral titer. Culture media, serum, and lipids [10, 13] are additional factors that can positively or negatively impact viral titer. Viral titer can be increased by the inclusion of additives such as sodium butyrate, DMSO, and chloroquine [14–17] during the transfection of the viral packaging cells. Finally cationic molecules such as protamine and polybrene have been shown to enhance the transduction efficiency of viral stocks [18–20] in certain cell types. Also, the adoption of a lentiviral purification step to the protocol outlined herein provides numerous advantages, namely, the concentration of purified viral particles (if required), resuspension of viral particles in medium ideal for target cells (instead of spent medium from the packaging cells), and the ability to freeze viral stock for later use. Needless to say all protocols need to be tailored to each lentiviral constructs and cells to be transduced.

The study of peripheral nerve myelination at the molecular level has long benefited from the Schwann cells/dorsal root ganglia (DRG) neurons myelinating co-culture system (*see* Chapter 2). As this culture system takes about a month to develop and perform experiments, the lentiviral-mediated gene expression is well suited to manipulate Schwann cells and DRG neurons, primary cells that are resistant to conventional transfection techniques. Using many of the findings of the past 20 years, we have developed a protocol that allows us to produce upward of 12×10^6 Schwann cells that are efficiently transduced, thereby abrogating the need of antibiotic selection and large-scale expansion of the Schwann cells that negatively affect their behavior. This protocol does not require large-scale viral production (only 7 ml per construct) and can also be used to transduce up to 100 cultures of DRG neurons. Depending on the nature of the lentiviral constructs (small hairpin RNAs or protein encoding), one can manipulate the expression of many target and assess their effects on the myelination process very quickly before moving to in vivo models.

2   Note on Biosafety

Lentiviral transduction systems have been designed to substantially reduce the risk that a wild-type virus will be reconstituted through recombination. They are replication-deficient and self-inactivating and do not require the TAT transactivator and any of the viral genes associated with virulence. The lentiviral transfer vector itself does not express any of the HIV-1 viral genes; the viral genes necessary for virus production and packaging are expressed in *trans* on two (second generation systems) to three (third generation systems) separate non-viral mammalian expression vectors.

Recombinant lentiviruses are however pseudotyped with VSV-G, which provides them with a wide range of host cells. More importantly, they integrate their genetic material into the host cell genome. Therefore the risk of mutagenesis and gene activation or silencing associated with integration is significant.

Finally the risks inherent to the shRNA or transgenes inserted by the lentiviral vector must also be taken into consideration.

To limit the risks of exposure while working with lentiviruses, always follow the guidelines for work at Biosafety Level (BSL-2). Your Institutional Biosafety Committee will provide such guidelines. Additional information can be found on the National Institutes of Health (http://osp.od.nih.gov) and the Center for Disease Control and Prevention (https://www.cdc.gov/biosafety) websites.

3   Materials

3.1   Chemicals, Biochemicals, and Culture Plates

1. Chloroquine diphosphate salt: prepare 82.5 mM in culture-grade water, filter-sterilize and store in 100 μl aliquots at 4 °C, and protect from light (*see* **Note 2**).

2. HEPES buffered saline (HBS): 280 mM NaCl, 100 mM HEPES, 1.5 mM Na_2HPO_4, pH 7.05 (*see* **Note 3**).

3. $CaCl_2$ solution at 2 M.

4. Cholesterol, water-soluble (Sigma cat # C4951): 30 mM in Dulbecco's PBS; store in 100 μl aliquots at −20 °C (*see* **Note 4**).

5. Lenti-X concentrator from Clontech (cat # 631231).

6. Protamine sulfate salt: from herring, grade III (Sigma cat # P4505); prepare stock at 10 mg/ml in culture-grade water, filter-sterilize, and store in 1 ml aliquots at +4 °C.

7. Forskolin: stock prepared in 100% ethanol at 2 mM and kept at −20 °C.

8. EGF-D: recombinant EGF domain of human neuregulin-1-β1 (R&D Systems 396-HB); prepare a stock at 1 mg/ml in DMEM and store at −20 °C in 50 μl aliquots.

9. Culture-grade poly-L-lysine (MW range of 70,000–150,000): stock concentration is prepared at 100 μg/ml in culture-grade water, filter-sterilized and stored at 4 °C, and light-protected. Working concentration of 10 μg/ml is made freshly in sterile culture-grade water.

10. 0.25% trypsin/1 mM EDTA.

11. Hanks' balanced saline solution (HBSS), calcium- and magnesium-free.

12. ⌀100 mm and ⌀60 mm cell culture dishes (*see* **Note 5**). Before use, plates are coated with poly-L-lysine overnight at 37 °C in a humidified cell culture incubator.

3.2 Mammalian Cells

1. HEK 293T/17 cells (ATCC® CRL-11268™) (*see* **Note 6**).

2. Schwann cells are derived from postnatal day 2 Sprague-Dawley rats. The protocol to purify Schwann cells is described in Chapter 2, Subheading 3.

3. Dorsal root ganglia (DRG) neurons are prepared from embryonic day16 Sprague-Dawley rats. The protocol to prepare DRG neuron cultures is described in Chapter 2, Subheading 4.

3.3 Culture Media

1. 293-SM, standard culture media to propagate 293T/17 cells: Dulbecco's modified Eagle medium (DMEM, Gibco 11995), 10% (v/v) fetal bovine serum (FBS, Gibco 16000), 2 mM GlutaMAX™-I (Gibco 35050), 0.1 mM nonessential amino acids (Gibco 11140), and 1 mM sodium pyruvate (Gibco 11360).

2. 293-LPM, culture media used during the production of lentiviruses by the 293T/17 cells: advanced DMEM (Gibco 12491), 2% FBS (Gibco 16000), 2 mM GlutaMAX™-I, and 0.03 mM cholesterol (*see* **Note 7**).

3. SC-SM, standard culture media to propagate Schwann cells: DMEM, 10% FBS (Gibco 16000), 2 mM GlutaMAX™-I, 2 μM Forskolin, and 10 ng/ml EGF-D.

4. DRG-SM, standard culture media for DRG neurons: Neurobasal media (Gibco 21103), 2% B27 supplement (Gibco 17504), 1% GlutaMAX™-I, 0.08% glucose, and 50 ng/ml 2.5S NGF.

3.4 Freezing Media

293-SM media supplemented with 10% DMSO.

3.5 DNA Vectors (See Note 8)

1. Packaging plasmids: psPAX2 and pMD2.G plasmids were a gift from Didier Trono (Addgene plasmids #12260 and #12259, respectively).

2. Lentiviral transfer vector: pLL3.7 was a gift from Luk Parijs (Addgene plasmid #11795; [21]).

4 Methods

4.1 Expansion of 293T/17 Viral Packaging Cells

1. If the cells are frozen, thaw an aliquot quickly by placing in a water bath at 37 °C. Otherwise jump to **step 6** if starting from an ongoing culture.

2. Quickly transfer the cells to 10 ml of pre-warmed (37 °C) 293-SM media in a 15 ml conical tube. Mix gently by inverting the tube three to four times.

3. Pellet the cells by centrifugation at $200 \times g$ for 10 min at room temperature (*see* **Note 9**).

4. Gently resuspend in 2 ml of 293-SM media and count the cells.

5. Plate in ø100 mm culture dishes (pre-coated overnight with poly-L-lysine) in 12 ml of pre-warmed (37 °C) 293-SM media at a density of 1×10^4 cells/cm^2 (*see* **Notes 10** and **11**).

6. Start expanding the cells after 4 days of culture at 37 °C and 10% CO$_2$ in a humidified incubator. Cells should be looking as shown in Fig. 1a for quality and density at time of splitting (*see* **Note 12**).

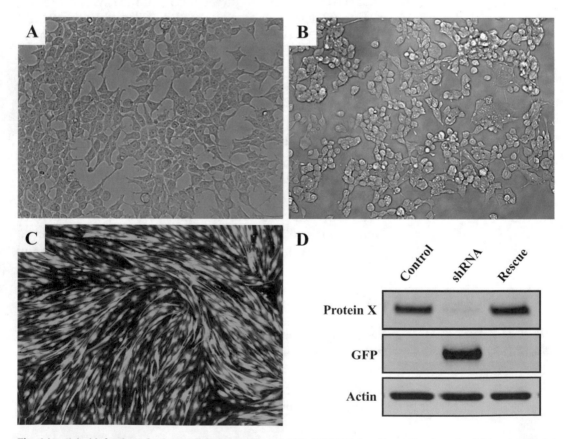

Fig. 1 Lentiviral infection of primary Schwann cells. (**a**) HEK-293T/17 density on the day of calcium phosphate transfection. Cells were seeded at a density of 8.5 x 10^4 / cm^2 in a ø100 mm culture plate 24 h earlier. (**b**) An example of unhealthy 293T/17 cells not suitable for transfection experiments. (**c**) Schwann cells 5 days post-transduction with a lentiviral shRNA construct co-expressing GFP at an MOI = 2. Note the rate of transduction efficiency at ≥95%. (**d**) Western blot analysis of protein lysate extracted from the transduced Schwann cells shown in **c** that confirms almost complete knockdown of the protein of interest by the shRNA construct

7. Remove the culture media, and wash the cells once with 5 ml of pre-warmed (37 °C) HBSS per culture plate. Do not add HBSS directly onto the cells but gently in the angle formed by the bottom and side of the dish.

8. To detach the cells, replace HBSS with 5 ml of pre-warmed 0.25% trypsin/1 mM EDTA solution, and incubate at 37 °C. Monitor visually on a light microscope for detachment of cells; it should not take more than 5 min (*see* **Note 13**).

9. Collect into a 15 ml conical tube containing 2 ml of pre-warmed 293-SM media. Mix gently by inverting the tube three to four times.

10. Go to **step 3** and repeat **steps 3** through **9** (*see* **Note 14**).

11. To freeze the cells, count the cells when at **step 9**. Then pellet the cells by centrifugation at $200 \times g$ for 10 min at room temperature. Resuspend in freezing media at a density of 2×10^6 cells/ml. Prepare 1 ml aliquots in cryotubes and transfer at −80 °C for 24 h. Then transfer into liquid nitrogen (vapor phase) for storage.

4.2 Schwann Cells and DRG Neurons

1. A detailed protocol for the isolation and purification of Schwann cells and DRG neurons is presented in Chapter 2. Time both cell preparation and viral production so that Schwann cells and DRG neurons are at the proper stage when virus is collected and purified.

2. Schwann cells are used at passage 1, DRG neurons 24 h after plating.

4.3 Lentivirus Production

4.3.1 Calcium Phosphate Transfection

1. Plate 293T/17 cells 24 h before transfection (*see* **Note 15**). To use the cells for lentivirus production, follow **steps 7–9** and then **step 3**, as described in Subheading 4.1 (Expansion of 293T/17 viral packaging cells). Differences occur in **steps 4** and **5** and are as follows:

(a) At **step 4**, cells are resuspended in 293-LPM media, not 293-SM media (*see* **Note 16**).

(b) At **step 5**, cells will be seeded at a density of 8.5×10^4 cells/cm² into ø100 mm plates. The volume of culture medium per plate is 10 ml. Twenty-four hours later (*see* **Note 17**), 293T/17 cells should look like as shown in Fig. 1a.

2. Three hours before transfection, change the media of the 293T/17 cultures with 6 ml of fresh 293-LPM medium to boost proliferation (*see* **Note 18**).

3. For each transfection, you will need to prepare two sterile 15 ml conical tubes, A and B.

Tubes labeled A will contain 300 µl of HBS.

Tubes labeled B will contain the mixture of DNA to be transfected in the following 3/2/1 proportions (*see* **Note 19**):

(a) 12 µg lentiviral transfer vector pLL3.7 (*see* **Note 8**).

(b) 8 µg packaging plasmid psPAX2.

(c) 4 µg VSV-G plasmid pMD2.G.

(d) 36 µl 2 M $CaCl_2$.

(e) Add sterile water to a final volume of 300 µl.

Vortex for 30 s at room temperature to ensure proper mixing of the $CaCl_2$ with the plasmid DNA.

4. Add the content of tube B (DNA/$CaCl_2$) to the content of tube A (HBS), 25 µl at a time. Vortex for 5 s after each addition. The whole process takes about a couple of minutes and should not be hurried (*see* **Note 20**).

5. Incubate the mix for 20 min at room temperature.

6. Add 2 µl of the 82.5 mM chloroquine stock solution, and mix by vortexing for 5 s (*see* **Note 21**).

7. Using a 1-ml Pipetman, add the mix dropwise to the 293T/17 cultures. Mix by shaking the culture disk; do not swirl (*see* **Note 22**). Final volume will be 6.6 ml, and chloroquine should be at a final concentration of 25 µM.

8. After 5 h of incubation at 37 °C, 10% CO_2, in a humidified incubator, change the culture media with fresh 7 ml of 293-LPM medium.

9. Incubate the transfected 293T17 cells for another 36 h before collecting the supernatants that should contain the lentiviral particles (*see* **Note 23**).

4.3.2 Concentration of Lentiviruses

1. Collect viral supernatants (7 ml per transfection) into 15 ml conical tubes, and pellet down all cell debris by centrifugation at room temperature for 20 min at $1600 \times g$.

2. Transfer 6 ml (*see* **Note 24**) of the clarified supernatants into a clean sterile 15 ml conical tube, and add 2 ml of Lenti-X concentrator (ratio of 3:1 (ml:ml), virus:Lenti-X). Mix gently by inverting the tube and incubate at 4 °C overnight.

3. After centrifuging the samples at 4 °C for 45 min, at $1600 \times g$, an off-white pellet should be visible.

4. Remove the supernatant very carefully, and resuspend the pellet gently with 1 ml of culture media. The choice of media will depend on the cells to be transduced. Use SC-SM for Schwann cells or DRG-SM for DRG neurons.

5. We do not routinely freeze the viral production. Therefore timing Schwann cell and DRG neurons cultures to be at the appropriate stage for transduction (*see* Subheading 4.2, **step 2**) is

essential. Freezing the viral stock results in a drastic decrease in the number of transduced Schwann cells. There is however no noticeable impact in the efficiency to transduce DRG neurons.

4.4 Transduction of Schwann Cells

1. Twenty-four hours before transduction, take one or two ⌀100 mm culture plate(s) of confluent Schwann cells that are at passage 1 (*see* Chapter 2, Subheading 3).

2. Aspirate the culture media, and wash the cells once with 5 ml of pre-warmed (37 °C) HBSS. Do not add HBSS directly on to the cells but gently in the angle formed by the bottom and side of the dish.

3. Replace HBSS with 5 ml of pre-warmed 0.25% trypsin/1 mM EDTA solution, and incubate at 37 °C. It should take about 5–10 min.

4. Collect into a 15 ml conical tube containing 2 ml of pre-warmed SC-SM media. Mix gently by inverting the tube three to four times.

5. Pellet the cells by centrifugation at $200 \times g$ for 10 min at room temperature.

6. Gently resuspend in 2 ml of SC-SM media and count the cells.

7. For each viral construct that will be used for transduction, plate Schwann cells in all 6 wells of a 6-well plate, in 3 ml of pre-warmed (37 °C) SC-SM media per dish, at a density of 3×10^5 cells per well (*see* **Note 25**).

8. Incubate at 37 °C, 10% CO_2, in a humidified cell culture incubator, for 24 h.

9. On the day of transduction (which is on the day of lentiviral concentration), bring the 1 ml of resuspended viral particle (Subheading 4.3.2, **step 4**) to 6 ml with SC-SM media. Add 6 μl of protamine (*see* **Note 26**) to a final concentration of 10 μg/ml. Mix gently by inverting the tube. We also use this preparation for virus titration assays (*see* Subheading 4.6).

10. Remove the culture media from Schwann cell cultured in 6-well plates, and replace with 1 ml of the lentivirus/protamine suspension, 1 ml per plate, one full plate (6 wells) per lentiviral construct. Incubate at 37 °C and 10% CO_2 for 24 h.

11. After 48 h, change the culture media with fresh SC-SM, and keep the cultures at 37 °C and 10% CO_2 for an additional 5 days. Observe the cells live with an inverted epifluorescent microscope to assess the expression of GFP. We routinely obtain close to 100% of the cells expressing GFP (Fig. 1c) (*see* **Note 27**).

12. Use one well per plate for Western blot analysis to determine the extent of knockdown or expression of the transgene of interest, as well as lack of effects from the control construct.

If you have replaced the GFP with a tagged transgene of interest, do not lyse cells directly from the plate, as you will need to assess the % of cells expressing the construct by immunocytochemistry for the tag. Proceed as follows:

For each construct, trypsinize cells from one well. Wash cells with 2 ml of HBSS and then replace with 1 ml of pre-warmed 0.25% trypsin/1 mM EDTA solution and incubate at 37 °C. It should not take more than 5 min. Collect into a 15 ml conical tube containing 1 ml of pre-warmed SC-SM media. Mix gently by inverting the tube three to four times. Pellet the cells by centrifugation at $200 \times g$ for 10 min at room temperature. Resuspend in 1 ml of SC-SM. Count the cells, and seed about 25,000 cells onto a PLL-coated, ø10 mm glass coverslip, in 600 µl of SC-SM. Proceed with your immunocytochemistry protocol 48 h later. Pellet the remaining cells, lyse the pellet with your preferred lysis buffer, and proceed with protein quantitation and Western blotting. An example is shown in Fig. 1d (*see* **Note 28**).

13. For each construct, trypsinize the cells from the remaining five wells as described in **step 12**. Pool the cells and split across four ø100 mm plates in 10 ml of SC-SM per plate. Grow at 37 °C and 10% CO_2 until 100% confluent (about 5–6 days). This step will provide about 1.6×10^7 cells per construct (*see* **Note 29**).

4.5 Transduction of DRG Neurons

1. Transduction is performed 24 h (*see* **Note 30**) after the DRG dissection and the plating of the DRG neurons, as described in Chapter 2, Subheading 4.2.5, **step 9**. Cells are at a density of about 25,000 cells (neurons, Schwann cells, fibroblasts) per coverslips.

2. Determine how many DRG cultures you need to transduce per lentiviral construct. Per culture/construct, we use 10 µl of the 1 ml lentiviral preparation (Subheading 4.3.2, **step 4**) to 200 µl of DRG-SM media. Multiply these amounts (10 µl/200 µl) by the number of cultures you need to transduce, plus 10%, and mix into a 15 ml conical tube. Add 1:1000 volume of protamine to a final concentration of 10 µg/ml.

3. Remove the culture media from the DRG cultures, and replace with 200 µl of the lentivirus/protamine suspension prepared in **step 2**. Incubate at 37 °C, 5% CO_2 for 24 h.

4. Then proceed from **step 2** onward as described in Chapter 2, Subheading 4.2.6.

5. As mentioned, we usually do not freeze viruses, and we usually use most of our production for Schwann cells. However we do freeze the virus in 50 µl aliquots in 500 µl microcentrifuge tubes for use with DRG transductions (*see* Subheading 4.3.2, **step 5**).

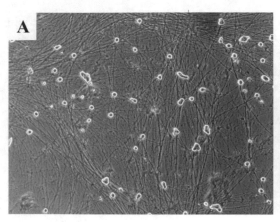 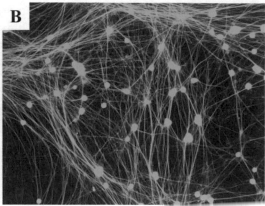

Fig 2 Lentiviral infection of primary DRG neurons. (**a**) Phase contrast image of a 10-day-old culture of DRG neurons that was transduced with the pLL3.7 vector 24 h after setting up the cultures, i.e., 9 days prior this image. (**b**) Same field as in **a** imaged for the expression of GFP. Note the rate of transduction at 100% efficiency

6. Figure 2 shows purified DRG neuron cultures 10 days post-transduction. We have so far found transduction efficiency to be 100%.

4.6 Lentivirus Titration

Viral titer is determined using Schwann cells. Since our constructs either carry a fluorescence reporter gene (GFP) or express a tagged version of our protein of interest, we perform a fluorescent titration assay.

1. On the day the viral supernatant is collected (Subheading 4.3.2, **step 1**), seed 15 wells of a 24-well plate with 5000 Schwann cells per well, in at least 600 μl of SC-SM (*see* **Note 31**). Incubate the cells overnight at 37 °C, 10% CO_2.

2. Bring the 1 ml of resuspended viral particle (Subheading 4.3.2, **step 5**) to 6 ml with SC-SM media. Add 6 μl of protamine to a final concentration of 10 μg/ml. Mix gently by inverting the tube. Prepare dilutions with SC-SM containing 10 μg/ml of protamine. We usually prepare five dilutions ranging from 1:10 to 1:1000, 1 ml of each.

3. Remove media from the Schwann cell cultures, and replace with 300 μl of the diluted virus, three cultures per dilutions, and incubate for 48 h.

4. If your construct expresses a fluorescence reporter gene, change culture media with 300 μl of dPBS containing Hoechst 33342 at a final concentration of 1 μg/ml, and incubate for 10 min at room temperature. Quickly rinse with dPBS and keep cultures in 300 μl of dPBS.

If your construct does not express a fluorescent protein (e.g., GFP) but a tagged protein (or a protein not normally present in Schwann cells), process the cultures for immunofluorescence staining following the protocol that is appropriate for the antigen to be detected, and counterstain with Hoechst 33342. Keep the cells in 300 µl of dPBS.

In either case, observe with an inverted epifluorescent microscope under 10× magnification. Determine the percentage of transduction by averaging the ratio of fluorescent protein (e.g., GFP)/Hoechst (or tag/Hoechst) from ten random fields.

5. Only dilutions that result in a transduction rate of 10% or less after 36 h are used for determining the viral titer (*see* **Note 32**), which is calculated as follows:

$$\frac{5000 \times \text{transduction rate} \times \text{dilution factor}}{\text{volume}\,(\text{ml})\,\text{of transduction}}$$

For example, if the 1:1000 dilution has a rate of transduction of 10% and that 300 µl of the diluted virus was added to 5000 cells, the titer is as follows:

$$\frac{5000 \times \dfrac{10}{100} \times 1000}{0.3} = 1.6 \times 10^6 \,/\,\text{ml}$$

6. Our titer is on average 6×10^5 IU/ml. While this may appear low, keep in mind that it is determined using cells that are difficult to transduce, contrary to the commonly used 293 or HT1080 cell lines. This gives an MOI of 2 for Schwann cell transduction and of 1.5 for DRG neuron transduction (*see* **Note 33**).

5   Notes

1. "Effective infectious titer" refers to the determination of transduction-competent viral particles. It does not refer to the total amount of viral particles as determined by the quantitation of the capsid protein p24Gag.

2. Chloroquine in solution is translucent to slightly yellowish. It is sensitive to light and stable for about a month at 4 °C. When the color starts to shift orange, discard and use a fresh aliquot.

3. HEPES pH is critical. Variations as little as 0.1 units can compromise the efficacy of calcium phosphate transfection [22]. Transfection kits are commercially available that may help reduce variability associated with transfection efficiency.

4. The cholesterol is available as a powder mixture with methyl-β-cyclodextrin. The powder becomes "gummy" upon contact with PBS. Do not touch with the pipet tip or it will get stuck to it. Do not vortex as the solution will foam. Dissolve by pipetting the PBS up and down, without touching the "gummy" part, for about 10–15 min.

5. Use culture plates sterilized by gamma ray irradiation. Ethylene oxide sterilization leaves residuals to which many cells are sensitive to. The health of the 293 cells will greatly affect virus production and viral titer. To limit variability due to cell health, use cell culture dishes that have been sterilized by gamma ray irradiation.

6. There are many clonal variants of the HEK293 cells (293F, 293FT, 293F, Lenti-X 293T, 293T/17) available from different suppliers specifically for lentiviral production. We have not tested them and have only used the 293T/17 clone. As the main criteria for selection are high levels of protein expressions and fast growth, most if not all of the clone variants should be suitable unless they have been engineered to work with particular lentiviral constructs.

7. Serum quality varies from lot to lot. It is therefore possible that the amount of serum used in the 293-LPM media needs to be adjusted. The Gibco catalog # is provided as a guide to the quality of serum to look for.

8. As mentioned in the introduction, one of the most important criteria that affect the efficient packaging of lentiviral transfer vectors is overall size and size of insert between LTR elements. The pLL3.7 represents a good compromise with an overall size of 7.6 kb and a 5′-LTR to 3′-LTR size of 5 kb. This vector has a U6 promoter for shRNA expression and a GFP reporter gene under a minimal CMV promoter. A cDNA for a protein of interest can easily be cloned in place of the GFP cDNA for protein expression/shRNA studies. Our constructs have so far not exceeded an overall size of 8.3 kb for the packaging of an RNA of 6 kb. This vector does not express a resistance gene for antibiotic selection. We do not find it to be necessary as we routinely transduce close to 100% of the cells.

   Plasmid DNA is prepared using standard midiprep kits, followed by a phenol/chloroform extraction. We routinely obtain plasmid preparations with a 260/280 ratio between 1.85 and 2.0. We do not use DNA preparations with a ratio lower than 1.8 (protein and/or phenol contaminations) or higher than 2 (RNA contaminations). To minimize the amount of TE buffer added when preparing for the calcium phosphate transfection, keep DNA concentration above 1 μg/μl.

9. DMSO is usually present at 5–10% in most freezing media. Even if diluted 10-fold when a 1 ml aliquot is cultured directly in 10 ml of media, the resulting 0.5–1% of DMSO is toxic to cells. It must be washed out.

10. ø100 mm cell culture plates from different manufacturers have different effective growth surfaces. The variation can be as much as 10%. In order to minimize variability in the timing of cell splitting, and later for transfection and viral production, keep cell density as indicated, and calculate the total number of cells to be seeded based on the actual growth surface area of the culture plates you use.

11. We use 12 ml per cultures. If using a smaller volume, the culture media will need to be changed every 2–3 days. This frequent change in media will stimulate 293 cell proliferations, and cells may have to be passaged at an earlier time than 4 days. In any case, cells should look like as shown in Fig. 1a. Note that we are not using the expression "split when 70–90% confluent," a rather loose and variable appreciation depending on who observes the cells. For reproducible results, we recommend that you follow a fixed setup of cell density, volume of media, and time of culture. You may however have to adjust cell density at plating and/or length of culture as the growth rate of cells in culture will be dependent of the lot # and quality of the serum. Although an expensive purchase, we recommend that once you have identified a batch of serum that works well, you purchase a bulk amount so that optimal conditions do not need to be worked out too often.

12. Note: HEK293 cells should always look flat when healthy (*see* Fig. 1a). If they start "bulging up" (*see* Fig. 1b), whether they are still not confluent or because they have been overgrown, discard and start a new batch.

13. The culture dish can be shaken sideways to help cells detach and get into suspension. But do not aspirate the media up and down on cells that are still attached. This will only damage the cells.

14. We use 293T/17 cells up to passage 10 included.

15. Prepare one plate for each lentiviral transfer vector to be used. A typical expression experiment requires two (empty vector and vector with transgene). A knockdown/rescue experiment typically requires four (vector with scrambled shRNA, two vectors each harboring one shRNA targeting the same RNA but at different positions, and a shRNA construct that also expresses a cDNA for rescue; this cDNA must be point mutated to prevent its RNA to be targeted by the shRNA).

16. This media reflects some of the findings by groups that have specifically researched how to improve lentiviral production.

One of the most critical components is the addition of cholesterol, essential to the formation and stability of lipid rafts. Lipid rafts are essential for the formation of the envelope of HIV-1 and HIV-based lentiviruses [10, 13].

17. This truly reflects 24 h, not overnight.

18. Access of the plasmids to the host cell nucleus is further enhanced by nuclear breakdown in proliferating cells.

19. This protocol was developed for the second generation of lentiviral systems. The third generation splits the genes from the packaging plasmid (in this case psPAX2) onto two separate plasmids. If you decide to use the third generation system, we would suggest the following ratios 3/1/1/1 (transfer/packaging 1/packaging 2/VSV-G).

20. The efficiency of the calcium phosphate transfection depends on the size of the precipitate. Large precipitate is also toxic to cells. The usual method to obtain a fine precipitate is to bubble air while adding B to A. We find that a quick and short vortexing between the additions of small amounts of B to A is as efficient.

21. DNA/Ca_2PO_4 precipitate is taken up into the cells by endocytosis. Chloroquine accumulates in lysosomes and enhances transfection most likely by blocking DNAse II function [14, 23].

22. Swirling will collect the DNA/Ca_2PO_4 toward the center of the dish and limit transfection to a smaller amount of cells.

23. As the lentiviral transfer vector expresses GFP, check for its expression in 293 cells. VSV-G is a fusogen and 293 cells should fuse and form syncytia. A successful transfection should result in >90% of cells expressing GFP and the appearance of many syncytia.

24. Always leave at least 500 μl to 1 ml behind to avoid collecting the pelleted debris. Do not have the tip of the serological pipet touch the bottom of the tube. Aspirate slowly to avoid displacing the pellet.

25. The 6-well plates are prepared the day before (PLL coating is done overnight). The number of passage 2 Schwann cell plates to be used will depend on the experimental design. An optimal preparation should have about 4×10^6 cells per ø100 mm plate when at confluence. One plate is enough for an experiment that has only two lentiviral constructs. Two plates allow for the four 6-well plates needed for an experiment with four different lentiviral constructs.

26. Like cells, viruses are negatively charged. It is believed that the increase in viral transduction observed with polycationic agents is due to the neutralization of these negative charges.

Although numerous protocols use polybrene, we have observed significant cell death at concentrations as low as 1 μg/ml. We do not observe such effects with the use of protamine.

27. In the myelinating co-culture system, only a small percentage of cells will myelinate. Therefore in order to detect significant changes in myelination, it is best to have at least a >65% infection rate. For biochemical analysis, transduction efficiency >80% is needed.

28. We have so far achieved knockdown and expression within 5 days. However this will depend on the shRNA efficiency and stability of the protein studied.

29. It is possible that Schwann cell numbers will vary if your target protein has an impact of cell proliferation and/or survival.

30. Transduction efficiency is time dependent and will be reduced considerably as the DRG neurons get older.

31. A volume of at least 600 μl per well of a 24-well plate will ensure an even distribution of the cells, which will facilitate subsequent observations and imaging. If reduced volumes are used, cells will mostly localize to the periphery of the well, making observations difficult.

32. Using data from cultures with a higher rate increases the risk of underestimating the viral titer due to multiple integration events.

33. MOI, or multiplicity of infection, is the ratio of infectious units (IU) over the number of cells to be transduced. For Schwann cells, the MOI is as follows: 6×10^5 IU/ml $\times$ 1 ml per transduction/300,000 cells = 2. For DRG neurons, the MOI is as follows: 3.6×10^6 (virus is concentrated sixfold; *see* Subheading 4.5, **step 2**) $\times$ 10 μl/1000 μl/25000 cells = 1.5.

References

1. Ganem D, Nussbaum AL, Davoli D, Fareed GC (1976) Propagation of a segment of bacteriophage lambda-DNA in monkey cells after covalent linkage to a defective simian virus 40 genome. Cell 7(3):349–359

2. Goff SP, Berg P (1976) Construction of hybrid viruses containing SV40 and lambda phage DNA segments and their propagation in cultured monkey cells. Cell 9(4 PT 2):695–705

3. Nussbaum AL, Davoli D, Ganem D, Fareed GC (1976) Construction and propagation of a defective simian virus 40 genome bearing an operator from bacteriophage lambda. Proc Natl Acad Sci U S A 73(4):1068–1072

4. Mann R, Mulligan RC, Baltimore D (1983) Construction of a retrovirus packaging mutant and its use to produce helper-free defective retrovirus. Cell 33(1):153–159

5. Naldini L, Blomer U, Gallay P, Ory D, Mulligan R, Gage FH, Verma IM, Trono D (1996) In vivo gene delivery and stable transduction of nondividing cells by a lentiviral vector. Science 272(5259):263–267

6. Cronin J, Zhang XY, Reiser J (2005) Altering the tropism of lentiviral vectors through pseudotyping. Curr Gene Ther 5(4):387–398

7. Kaplan AH, Swanstrom R (1991) Human immunodeficiency virus type 1 Gag proteins

are processed in two cellular compartments. Proc Natl Acad Sci U S A 88(10):4528–4532

8. Nie Z, Phenix BN, Lum JJ, Alam A, Lynch DH, Beckett B, Krammer PH, Sekaly RP, Badley AD (2002) HIV-1 protease processes procaspase 8 to cause mitochondrial release of cytochrome c, caspase cleavage and nuclear fragmentation. Cell Death Differ 9(11):1172–1184. https://doi.org/10.1038/sj.cdd.4401094

9. Hoffmann M, Wu YJ, Gerber M, Berger-Rentsch M, Heimrich B, Schwemmle M, Zimmer G (2010) Fusion-active glycoprotein G mediates the cytotoxicity of vesicular stomatitis virus M mutants lacking host shut-off activity. J Gen Virol 91(Pt 11):2782–2793. https://doi.org/10.1099/vir.0.023978-0

10. Mitta B, Rimann M, Fussenegger M (2005) Detailed design and comparative analysis of protocols for optimized production of high-performance HIV-1-derived lentiviral particles. Metab Eng 7(5–6):426–436. https://doi.org/10.1016/j.ymben.2005.06.006

11. al Yacoub N, Romanowska M, Haritonova N, Foerster J (2007) Optimized production and concentration of lentiviral vectors containing large inserts. J Gene Med 9(7):579–584. https://doi.org/10.1002/jgm.1052

12. Kumar M, Keller B, Makalou N, Sutton RE (2001) Systematic determination of the packaging limit of lentiviral vectors. Hum Gene Ther 12(15):1893–1905. https://doi.org/10.1089/104303401753153947

13. Chen Y, Ott CJ, Townsend K, Subbaiah P, Aiyar A, Miller WM (2009) Cholesterol supplementation during production increases the infectivity of retroviral and lentiviral vectors pseudotyped with the vesicular stomatitis virus glycoprotein (VSV-G). Biochem Eng J 44(2–3):199–207. https://doi.org/10.1016/j.bej.2008.12.004

14. Ciftci K, Levy RJ (2001) Enhanced plasmid DNA transfection with lysosomotropic agents in cultured fibroblasts. Int J Pharm 218(1–2):81–92

15. Cribbs AP, Kennedy A, Gregory B, Brennan FM (2013) Simplified production and concen-tration of lentiviral vectors to achieve high transduction in primary human T cells. BMC Biotechnol 13:98. https://doi.org/10.1186/1472-6750-13-98

16. Sakoda T, Kasahara N, Hamamori Y, Kedes L (1999) A high-titer lentiviral production system mediates efficient transduction of differentiated cells including beating cardiac myocytes. J Mol Cell Cardiol 31(11):2037–2047. https://doi.org/10.1006/jmcc.1999.1035

17. Villa-Diaz LG, Garcia-Perez JL, Krebsbach PH (2010) Enhanced transfection efficiency of human embryonic stem cells by the incorporation of DNA liposomes in extracellular matrix. Stem Cells Dev 19(12):1949–1957. https://doi.org/10.1089/scd.2009.0505

18. Coelen RJ, Jose DG, May JT (1983) The effect of hexadimethrine bromide (polybrene) on the infection of the primate retroviruses SSV 1/SSAV 1 and BaEV. Arch Virol 75(4):307–311

19. Manning JS, Hackett AJ, Darby NB Jr (1971) Effect of polycations on sensitivity of BALD-3T3 cells to murine leukemia and sarcoma virus infectivity. Appl Microbiol 22(6):1162–1163

20. Toyoshima K, Vogt PK (1969) Enhancement and inhibition of avian sarcoma viruses by polycations and polyanions. Virology 38(3):414–426

21. Rubinson DA, Dillon CP, Kwiatkowski AV, Sievers C, Yang L, Kopinja J, Rooney DL, Zhang M, Ihrig MM, McManus MT, Gertler FB, Scott ML, Van Parijs L (2003) A lentivirus-based system to functionally silence genes in primary mammalian cells, stem cells and transgenic mice by RNA interference. Nat Genet 33(3):401–406. https://doi.org/10.1038/ng1117.ng1117[pii]

22. Felgner PL (1990) Particulate systems and polymers for in vitro and in vivo delivery of polynucleotides. Adv Drug Deliv Rev 5(3):163–304

23. Howell DP, Krieser RJ, Eastman A, Barry MA (2003) Deoxyribonuclease II is a lysosomal barrier to transfection. Mol Ther 8(6):957–963

Chapter 13

Preservation, Sectioning, and Staining of Schwann Cell Cultures for Transmission Electron Microscopy Analysis

Vania W. Almeida, Margaret L. Bates, and Mary Bartlett Bunge

Abstract

The transmission electron microscope (TEM) enables a unique and valuable examination of cellular and extracellular elements in tissue in situ, in cultured cells, or in pellets derived from suspensions of cells or other materials such as nanoparticles. Here we focus on the preparation of cultured Schwann cells or Schwann cell-containing dorsal root ganglion cultures. To gain as life-like as possible views of the cellular details, it is imperative to achieve excellent preservation of the cellular structure. The steps in the preparation of cultures described in this chapter represent the results of many years of accumulated TEM images to find the best methods of preservation for Schwann cells, myelin, and basal lamina components. All the materials required are listed. The methods for fixing, dehydrating, and embedding a culture are described. Choosing an area in the culture to view, scoring it, cutting it out of the resin-embedded culture, mounting it appropriately for *enface* or cross-sectioning, and performing the semi-thin and thin sectioning are detailed. Explaining the way in which the sections are then stained for TEM completes the Methods section. Preservation of cultured Schwann cells and their myelin sheaths can be outstanding due to the direct and rapid but careful addition of the fixative solution to the culture dish.

Key words Schwann cells, Primary cultures, Embedding, Ultramicrotome sectioning, Transmission electron microscopy

1 Introduction

The transmission electron microscope (TEM) is an invaluable tool for observing the ultrastructure of cellular components by studying high-magnification, high-resolution images. More specifically, it provides an insight into myelination by Schwann cells, how they associate with axons and how they interact with elements of the extracellular matrix. For example, the TEM enabled the discovery that Schwann cells need to form basal lamina on their surface prefatory to forming myelin [1, 2]; the basal lamina can only be resolved in the TEM. In order to obtain high-quality images from which to extract useful data, it is imperative that the tissue be subjected to a series of specialized TEM preparative techniques that have been refined over the years [3]. Our aim here is to describe

Paula V. Monje and Haesun A. Kim (eds.), *Schwann Cells: Methods and Protocols*, Methods in Molecular Biology, vol. 1739, https://doi.org/10.1007/978-1-4939-7649-2_13, © Springer Science+Business Media, LLC 2018

these techniques that have been optimized in our laboratory and to share our expertise to best preserve Schwann cells for TEM analysis. Whereas our focus is on Schwann cell preservation for this chapter, the techniques are suitable for the preparation of a variety of cell types.

2   Materials

The supplies described below are required to prepare the materials used during the fixation, dehydration, embedding, and sectioning of Schwann cell cultures for TEM analysis. In our laboratory, most cultures prepared for TEM analysis are plated on 15 mm glass coverslips and grown in 24-multiwell polystyrene tissue culture dishes. Although we also prepare cultures grown in Permanox™ or Thermanox™ multiwells (2, 4, or 8 wells) or on Aclar™ coverslips, the descriptions provided below are for the preparation of Schwann cell cultures grown on glass coverslips. Additionally, most solutions used are prepared from scratch, but they can also be purchased from laboratory suppliers. Wear gloves and work in a fume hood where appropriate. Handle, collect, and dispose of potentially hazardous fixatives and staining solutions such as Richardson's stain and any leftover solutions of uranyl acetate and lead citrate according to Environmental Health and Safety toxic waste disposal standards.

2.1 Supplies, Materials, and Equipment for the Preparation of Buffers and Fixatives

1. Fume hood.
2. Disposable gloves.
3. 50 ml conical tubes.
4. Pipettor with adaptor for serological pipettes.
5. 100 ml glass bottle with screw cap.
6. 100 ml graduated cylinder.
7. Rotator.
8. 5 cm × 5 cm gauze pads.
9. 1 l wide-mouth glass jar with screw cap (*see* **Note 1**).
10. 100 ml wide-mouth glass jar with screw cap.
11. 1-mm-thick stainless steel wire.
12. Aluminum foil.
13. Forceps.
14. Disposable 5 and 23 cm Pasteur pipettes with bulbs. Do not mix pipettes.
15. Disposable 5 ml serological pipettes.
16. 20 ml scintillation glass vial with a screw cap.

2.2 Buffers and Fixatives

1. 0.2 M PO$_4$ stock buffer: dissolve 3.36 g of anhydrous sodium phosphate monobasic and 10.22 g of anhydrous sodium phosphate dibasic in 500 ml of double-distilled water (DDW).

2. 0.15 M PO$_4$ buffer: mix 75 ml of 0.2 M PO$_4$ stock buffer with 25 ml of DDW.

3. 0.05 M PO$_4$ buffer: mix 10 ml of 0.2 M PO$_4$ stock buffer with 25 ml of DDW.

4. Buffered 2% glutaraldehyde fixative: combine 40 ml of 0.05 M PO$_4$ buffer with 1.36 g of anhydrous sucrose. Pipet out 3.2 ml of this solution, discard, and then add 3.2 ml of 25% glutaraldehyde (*see* **Note 2**).

5. 4% osmium tetroxide (OsO$_4$) solution: dissolve 1 g ampule of OsO$_4$ crystals in 25 ml DDW (*see* **Note 3**).

6. 2% OsO$_4$ in PO$_4$ buffer: mix 5 ml aqueous 4% OsO$_4$ stock solution and 5 ml 0.2 M PO$_4$ stock buffer (*see* **Note 4**).

7. 100% ethanol (EtOH) stored at room temperature (RT).

8. Graded EtOH solutions: prepare 25%, 50%, 70%, and 95% solutions of EtOH in DDW. Make ahead of time and keep in the refrigerator.

2.3 Resin Embedding Mixture

1. Disposable 50 ml Tri-Pour beaker (*see* **Note 5**).

2. Wood applicator stick.

3. 1 ml tuberculin syringe.

4. Disposable transfer pipet.

5. Forceps.

6. Resin mixture: combine 10.5 ml of EMbed-812 with 6.5 ml of dodecenyl succinic anhydride (DDSA), 5.5 ml of nadic methyl anhydride (NMA), and 0.35 ml of 2,4,6-Tris (dimethylaminomethyl)phenol (DMP-30) (*see* **Notes 6** and **7**).

7. Small polyethylene bottle snap caps (15 mm in diameter × 9 mm in depth).

8. Large polyethylene bottle snap caps (20 mm in diameter × 5 mm in depth) (*see* **Note 20**).

9. Polystyrene multiwell tissue culture plate cover.

10. 15 cm diameter plastic tissue culture dish cover to use as a "waste" plate (*see* **step 3** in Subheading 3.1.3 and **Note 19**).

11. Write-on cardboard cap for Tri-Pour beaker.

12. Parafilm™.

13. Sharpie™ permanent marker.

14. Oven set at 64 °C.

2.4 Glass Etching Procedure

1. 12-well polystyrene tissue culture plate.

2. Disposable 250 ml Tri-Pour beaker.

3. Hydrofluoric acid. Use a fume hood, wear gloves, and be extremely careful when handling hydrofluoric acid. Use plastics only because it etches glass.

4. Tap water.

5. Forceps.

6. Single-edge steel razor blade.

2.5 Scoring Cultures Embedded in Small Polyethylene Snap Caps and Immersed in Large Polyethylene Snap Caps

1. Vise. The vise chosen for this purpose should have a large base with a suction mechanism and a wide clamping surface (*see* Fig. 1).

2. Jeweler's saw.

3. Fine jeweler's saw blades.

4. Standard compound light microscope equipped with regular 5× and 10× objectives and another objective modified to contain a scorer.

5. Jeweler's loupe.

6. Glass slide.

7. Time and transparent tape. Use transparent tape for securing the embedded culture to the slide, and time tape for securing the slide to the microscope stage (*see* Fig. 2).

8. Fine tip Sharpie™ permanent marker.

Fig. 1 The vise is secured to a bench by suction. The culture is placed in the vise so that the cell side faces the person sawing it. Rest the jeweler's saw 1–2 mm in back of the cell layer

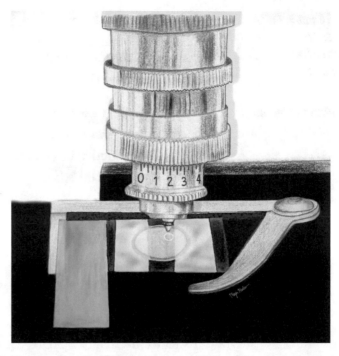

Fig. 2 After securing the disc to the center of the slide with two segments of transparent tape, place the slide on the microscope stage, and tape one end of the slide to the stage with time tape so that it does not move when the score is made on the surface of the disc. The 40× objective is replaced with another objective that has been adapted to contain a scorer. After selecting the area of interest using the 10× objective, move the scorer into position being careful that the tip does not touch the cell surface

2.6 Mounting Cultures

1. Hot plate set at 100 °C (*see* **Note 8**).
2. Double-edge steel razor blade (*see* **Note 9**).
3. Glass slide.
4. Fine tip Sharpie™ permanent marker.
5. Toothpicks.
6. BEEM™ capsules, size 00.
7. BEEM™ capsule holder with 22 cavities.
8. Resin stubs (*see* **Note 10**).
9. Fine forceps.
10. Resin mixture (*see* **Note 11**).
11. Oven set at 64 °C.
12. Pasteboard sliding storage box (5.5 cm × 3 cm × 1.5 cm).

2.7 Trimming Cell Cultures for Ultramicrotome Sectioning

1. Dissecting microscope.
2. Heavy base that accommodates the chuck to hold the stub-carrying block for trimming.
3. Double-edge steel razor blade (*see* **Notes 9** and **12**).

2.8 Ultramicrotome Sectioning

1. Leica EM ultramicrotome UC 7.
2. Histotech diamond knife (6 mm, 45° angle).
3. Ultracut diamond knife (3 mm, 45° angle).
4. Hot plate set at 100 °C.
5. 10 ml syringe.
6. 25 mm syringe filter with 0.2 μm cellulose acetate membrane.
7. DDW.
8. Wood applicator stick.
9. Frosted selected pre-cleaned glass slides (25 mm × 75 mm, 1 mm thick).
10. Slide storage box.
11. Richardson's stain: 1% borax, 1% toluidine blue, and 1% methylene blue in DDW.
12. 500 ml squirt bottle with DDW.
13. Waste container for Richardson's stain.
14. Paper towels.
15. Compressed air duster.
16. Reverse action forceps.
17. 150 mesh Formvar-coated copper grids.
18. 55 mm number 1 filter paper circle.
19. 50 mm × 9 mm petri dish.
20. Pencil.
21. Chloroform.

2.9 Staining Thin Sections on Grids

1. 4% uranyl acetate solution: 8 g of uranyl acetate in 20 ml of 100% methanol (MeOH). The 4% uranyl acetate solution should be made ahead of time and stored in the dark at RT. After the solution is prepared, shake the vial so the uranyl acetate crystals go into solution. The excess will precipitate to the bottom of the vial, so be careful when manipulating it to avoid having contamination on the grids. Use a 0.22 μm acetate syringe filter when staining with the 4% uranyl acetate solution to help eliminate any crystals.
2. 0.25% lead citrate solution: 25 mg of lead citrate in 9 ml of DDW. The 0.25% lead citrate solution should be made ahead of time and stored at RT. Use a 15 ml plastic conical tube and

previously boiled DDW, free of CO_2. Hold your breath while preparing the solution to avoid precipitation of the lead citrate.

3. 10 N sodium hydroxide (NaOH) solution: 4 g NaOH pellets in 9.5 ml of DDW. The 10 N NaOH solution for the preparation of the lead citrate should be made fresh with previously boiled DDW. Use a glass scintillation vial with a screw cap. Be careful as this is an exothermic reaction.

4. 12 multiwell Pyrex™ dish.

5. 8 cm × 8 cm Teflon™ sheet. A clean sheet of Parafilm™ can be used instead of the Teflon™.

6. 1 ml tuberculin syringe.

7. 5 ml syringe.

8. 18 gauge 3.8 cm needles.

9. 0.22 μm acetate syringe filter.

10. Reverse action forceps.

11. 15 cm diameter glass petri dish with opaque cover.

12. Disposable 250 ml Tri-Pour beaker.

13. Fine mesh stainless steel strainer that fits on top of the 250 ml Tri-Pour beaker. The strainer is used as a safety precaution to catch the grid during washes.

14. 100 ml squirt bottle with DDW.

15. 55 mm number 1 filter paper circle.

---

3 Methods

Described below are the methods for (1) fixation and dehydration, (2) resin embedding, (3) scoring the resin-embedded culture and mounting the scored area, (4) semi-thin and thin sectioning, and (5) staining sections.

3.1 Fixation, Dehydration, and Resin Embedding of Schwann Cell Cultures Grown on Glass Coverslips

Good fixation is the first step in the preservation of the cellular ultrastructure, and it is essential for obtaining quality images at the EM level. The glutaraldehyde-osmium tetroxide method used here was developed by Sabatini et al. in 1963 and has since become the standard for most tissue fixation intended for EM analysis [4].

3.1.1 Fixation

1. Working in a fume hood with gloves on, remove the culture medium and replace it with 2% glutaraldehyde fixative. Two percent glutaraldehyde fixative is stored at 4 °C but must be at RT when ready to use. Add enough fixative to cover the culture (*see* **Note 13**). Leave the covered 24-well plate in the hood at RT for at least 30 min before placing it in a 4 °C refrigerator overnight (*see* **Note 14**).

2. Remove the 24-well plate from the refrigerator and work in a hood with gloves on. Rinse each coverslip, one well at a time, by using the perfusion method to avoid drying the culture (*see* **Note 13**). Remove the 2% glutaraldehyde fixative from one side while simultaneously adding the 0.15 M PO_4 buffer to the other side of the well. Carefully lift the edge of the coverslip with fine forceps to rinse underneath it. Repeat the washing step with 0.15 M PO_4 buffer three times, 5–10 min each.

3. During the last wash, prepare the 2% OsO_4 buffered solution in the fume hood. Remove the 0.15 M PO_4 buffer, and flood each well with the 2% OsO_4 solution, lifting the edge of the coverslip to allow the OsO_4 solution to reach underneath it. Add enough OsO_4 solution to cover the culture. Cover the plate with the plastic lid, and place a cardboard box on top of it to protect it from the light. Leave the plate in the hood at RT for 1 h (*see* **Note 15**).

3.1.2 Dehydration

The dehydration step consists of gradually replacing water within the tissue by submitting it through a series of graded alcohols at RT. Do one well at a time, and carefully lift the edge of the coverslip with fine forceps each time to expose the underside to the fluid. Proceed to incubate the coverslips in graded alcohols as follows. Do not let the culture dry out, especially during the 100% EtOH step.

1. Incubate the samples with 25% EtOH (two sequential incubation steps, 5 min each).

2. Immediately transfer them to 50% EtOH (one incubation step, 5 min), 70% EtOH (two sequential incubations, 5 min each) (*see* **Note 16**), 95% EtOH, (one incubation, 5 min), and ultimately 100% EtOH (two incubations, 10 min each) (*see* **Note 17**).

3. Maintain the coverslips in 100% EtOH and proceed to Subheading 3.1.3.

3.1.3 Resin Embedding

1. Prepare the resin in the hood while wearing gloves (*see* **Note 18**).

2. Carefully remove the glass coverslip from the well with fine forceps. Lightly touch the edge of the coverslip to a paper towel to remove the excess EtOH, and using a transfer pipet, quickly drip the resin mixture onto the culture while swirling the resin over the surface to coat the tissue evenly. Do not let any resin drip onto the back of the coverslip.

3. Remove the excess resin by dragging the edge of the coverslip on a plastic "waste" plate. Repeat this step three times (*see* **Note 19**).

4. Place the bottle snap caps filled with resin on a multiwell tissue plate cover that is used as a tray. Embed each culture by carefully inverting the resin-impregnated culture surface onto the

bottle snap cap to contact the resin. Let it float, being careful to prevent the coverslip from sinking (*see* **Note 20**).

5. Gently transfer the tray with floating cultures to the oven for polymerization at 64 °C overnight. Do not leave cultures in the oven longer than 24 h to avoid brittleness.

3.2 Glass Etching Procedure

1. The following day, remove the cultures from the oven and cut away the bottle snap caps.

2. In the hood and wearing gloves, fill the appropriate number of wells in a 12-well polystyrene tissue culture plate with hydrofluoric acid to etch the glass. Do not overfill the wells (*see* **Note 21**).

3. Place the embedded culture on the glass coverslip tissue side down into the well.

4. Leave the 12-well plate in the fume hood for 1–2 h at RT.

5. To check if the glass coverslip has been etched away, remove the embedded culture from the well with forceps, and drop it in a beaker filled with tap water. Go to the sink, and let water run into the beaker for a few minutes.

6. If any glass remains, return the embedded culture to the hydrofluoric acid, and after 1 h repeat the washing step. Continue this process until all the glass has been etched away.

3.3 Sawing the Embedded Culture

1. Put the embedded culture in a vise with the cell side facing the operator, and tighten it well (*see* Fig. 1).

2. Rest the jeweler's saw on the perimeter of the embedded culture, 1–2 mm away from the cell layer. Start sawing down evenly to create a thinner and flat disc. Having a disc with a flat surface is important in order to ensure that an even circle is formed when the culture is scored (*see* Fig. 2).

3. Write the identification number on the outside perimeter of the disc with a fine tip Sharpie™ permanent marker.

3.4 Scoring Areas of Interest

1. Tape the disc to the center of a glass slide (*see* Fig. 2 and **Note 22**).

2. Place the slide in a standard compound light microscope.

3. Remove the 40× objective and replace it with another objective modified to contain a scorer.

4. Look through the 10× objective and select the cell region of interest. Place this region in the center of the viewing area (*see* **Notes 23** and **24**).

5. Attach the slide to the microscope stage with a piece of time tape, so the slide does not move when the culture is being scored, and move the scorer into position (*see* Fig. 2).

6. Using a jeweler's loupe, look close to the slide and focus on the surface of the disc.

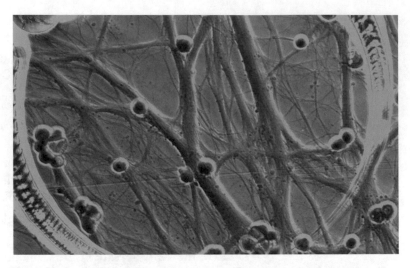

Fig. 3 Close-up of a score on the surface of a resin-embedded cell culture. Numerous Schwann cell axon fascicles can be seen. The dark objects are dorsal root ganglion neuronal somata distributed throughout the scored region (Courtesy of Dr. Kevin Golden, The Miami Project to Cure Paralysis)

7. Slowly bring up the stage (1/4 turn) until the reflection of the scorer tip can be seen on the resin surface, and turn the scorer tip around (*see* **Note 25**).

8. Repeat this step until a complete circle is made (*see* Fig. 3).

9. It is possible to score several regions on one disc, but they should be far enough apart to be able to saw around each one.

3.5 Mounting Embedded Cell Cultures

3.5.1 Mounting Cell Cultures for Enface Sectioning

1. Select the scored areas of the cell culture to be mounted, and circle them with a fine-tip Sharpie™ permanent marker.

2. Place the disc with the culture side up on a hot plate on top of a glass slide. Wait 1 min, and using a double-edge razor blade, make a straight cut across the culture that includes the area of interest (*see* **Note 9**). Then, make two other cuts resulting in an equilateral triangle with the scored area in the middle of it (*see* **Note 26** and Fig. 4).

3. Put a BEEM™ capsule in the holder (*see* **Note 27**). Insert a pre-labeled stub, either already embedded with the appropriate label or marked with a fine tip Sharpie™ permanent marker on the outside, halfway down the BEEM™ capsule. Make sure that the identification number written with a Sharpie™ permanent marker is completely covered by the BEEM™ capsule to protect it from being obscured in case excess resin used to mount the block drips down the side of the stub.

4. Place a very small drop of resin with a toothpick on top of the stub, and carefully mount the triangular piece with the cell

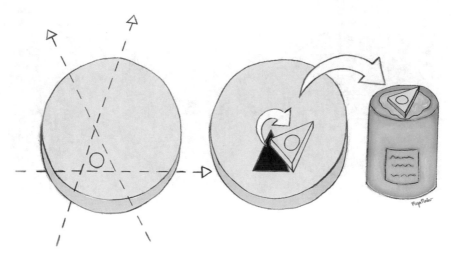

Fig. 4 Mounting for *enface* sectioning. The surface of the equilateral triangle with the scored area is placed facing up on top of a drop of resin mixture to glue the block to the stub

surface facing up. This piece is termed the "block" (*see* **Note 11** and Fig. 4).

5. Put the BEEM™ holder with the stub in a 64 °C oven to polymerize the glue overnight.

3.5.2 Mounting Cell Cultures for Cross-Sectioning

1. With a double-edge razor blade placed perpendicularly to the disc, go straight down, and make a mark on one edge of the scored area to indicate where to start sectioning (*see* **Note 9**). This will save time once the culture has been mounted and is ready to be cut in the ultramicrotome. Then make another parallel mark directly below and outside the scored area to indicate which side will be sawed straight in order to have a flat surface to be glued to the stub (*see* Fig. 5).

2. Using the jeweler's saw, cut a U-shaped arc around the scored area. Change the direction of the saw by turning it 90° and sawing straight across until the completely scored area is isolated. The sawed piece will have the shape of a Quonset hut with a flat bottom (*see* Fig. 5). This piece is termed the "block."

3. Repeat **step 3** in Subheading 3.5.1.

4. Place a very small drop of resin with a toothpick on top of the stub, and carefully mount the block with the flat surface down (*see* **Note 11**). The scored area will be perpendicular to the stub (*see* Fig. 5).

5. Put the BEEM™ holder with the stub in a 64 °C oven to polymerize overnight.

6. The next day, trim away the excess resin with a double-edge razor blade (*see* **Note 9**). Start by placing the edge of the razor

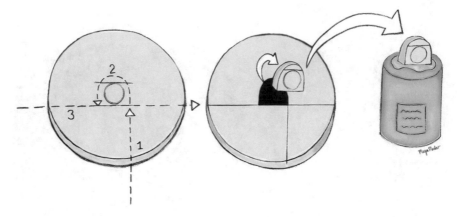

Fig. 5 Mounting for cross-sectioning. The flat surface of the block is placed on top of a drop of resin mixture to glue the block to the stub. The cell layer will be perpendicular to the stub, and the mark on the edge of the scored area will indicate where trimming the excess resin will be started

blade directly on top of the mark made on the scored area the day before. This will serve as the starting point for the trimming. Proceed to make a clean cut, straight across the entire length of the U-shaped surface of the block. This will save time when cutting in the ultramicrotome because the first sections will already contain the area with the cells of interest.

3.6 Sectioning the Cell Culture

3.6.1 Semi-thin Sectioning

1. Trim the face of the block around the score to make a trapezoid (*see* **Note 12**).

2. Put the histotech knife in place in the ultramicrotome. Adjust the cutting thickness to approximately 600 nm, and fill the boat with DDW. Slowly start approaching the knife to the face of the block.

3. Put several drops of DDW on a clean glass slide. As soon as the first sliver of tissue is cut, start collecting the pieces by picking them up with a wood stick and placing them on top of the water on the slide. Make the necessary minor adjustments to align the edge of the knife with the face of the block. Let the ultramicrotome make one cut at a time. The area within the score will start to become complete. Do not continue until the entire circle with the cells has been sectioned. The tissue must be saved for thin sectioning particularly when cutting *enface*. The first semi-thin sections are sufficient to check the orientation of the tissue before proceeding to thin sectioning. The cells are in only a thin layer near the surface of the block, necessitating capture of the tissue in just a few sections when cutting *enface*.

4. Place the slide on a hot place until the water has evaporated and the sections have adhered to the slide. Add enough Richardson's stain to cover the sections and let it stand for

1 min. Rinse with DDW and put the slide back on the hot plate until it is dry. The Richardson's stain should be made ahead of time and can be kept at RT. All the Richardson's stain waste must be collected in a gallon beaker and poured into a container to meet Environmental Health and Safety toxic waste disposal standards.

3.6.2 Thin Sectioning

1. Using an Ultracut diamond knife, adjust the cutting thickness to approximately 100 nm, and fill the boat with DDW. Be careful with the first cut because the block could have expanded due to changes in RT, humidity, etc. and the section could be thicker than desired as a result.

2. Cut only one or two more sections until a change in the texture in the center of the section is observed. This indicates that the cell layer is starting to be cut. Stretch the sections with chloroform vapor, and collect on a Formvar-coated copper grid. Let the grid dry, and place it in a 50 mm × 9 mm petri dish lined with a 55 mm number 1 filter paper circle. This grid can be labeled "Level 1" by writing next to it on the filter paper with a pencil.

3. Resume cutting, and collect a few more sections on a series of grids until the cell layer starts becoming sparse when sectioning *enface*. Label the grids according to the sequence in which they were collected ("Level 2," "Level 3," etc.).

4. If there are leftover sections in the boat, put them on a glass slide, and stain them with Richardson's stain. Label the slide "post thin sectioning." The sections on this slide serve as an indicator of how far the block has been cut.

3.7 Staining Thin Sections

In order for the tissue to be visualized when the electron beam goes through it in the TEM, its contrast must be increased. This is achieved by following a positive staining protocol that uses heavy metal salts as stains. It is important to ensure that the sections are clean, with no contamination from dust particles, so a dirt-free area on a laboratory bench, preferably away from doors and windows, should be dedicated exclusively for staining grids.

1. With reverse action forceps, hold the grid by the edge, and make a notch by bending it slightly upward. This will make it easier to handle the grid during staining.

2. Using a 6-well Pyrex™ dish, stain grids by immersing them with the section side up in a 4% uranyl acetate solution for 12 min under cover. Each well holds a 1 ml volume. Collect any leftover uranyl acetate and lead citrate solutions with a syringe, and discard in a waste container to meet Environmental Health and Safety toxic waste disposal standards.

3. Pick up the grid with forceps and rinse with 50% MeOH followed by rinses with DDW. Use a 250 ml Tri-Pour beaker with a fine mesh stainless steel strainer on top for the washes. Let the grid air-dry.

4. Put one drop of 0.25% lead citrate on a Teflon™ or a Parafilm™ sheet for each grid, and place 3–5 NaOH pellets around them. Slide the grid underneath the lead citrate drop with the section side up, and leave for 3 min under cover (*see* **Note 28**).

5. Pick up the grid, rinse with DDW, and leave to dry. Store in a grid box until ready to be examined in the TEM.

4   Notes

1. Line the inside of the 1 l wide-mouth glass jar with aluminum foil and set aside. Prepare the 4% OsO_4 aqueous solution in a 100 ml wide-mouth glass container, and cover the outside with aluminum foil to protect the contents from light. Then wrap the stainless steel wire tightly around the mouth of the container leaving a 10-cm-long piece at the end that will be used to pick it up once it is placed inside the larger, 1 l wide-mouth glass jar.

2. In a fume hood, carefully break the glass ampules containing the 25% glutaraldehyde solution with a gauze pad. Each ampule with 10 ml provides three batches of 2% glutaraldehyde fixative. Transfer the contents of the ampule to a scintillation vial using a 23 cm Pasteur glass pipet, and keep at 4 °C. Leave empty ampules in the hood to evaporate the glutaraldehyde, and then discard them in a proper glass disposal container the following day.

3. In a fume hood with the lights off, carefully break the OsO_4 glass ampules with a gauze pad. Using forceps, place the top and bottom parts of the ampule containing the OsO_4 crystals inside the 100 ml wide-mouth glass container. Carefully place them upright, resting against the side of the jar. Slowly add the DDW. Using a 23 cm Pasteur pipet, fill the inside of each ampule with the DDW in order to dislodge all of the OsO_4 crystals. This will help the crystals dissolve faster. Put the 100 ml jar inside the 1 l jar, and leave it on the rotator inside the hood until all of the crystals have dissolved and the solution has a light yellow color. If the solution turns gray, discard it. This process could take more than 1 day. If this is the case, keep at 4 °C overnight, and return it to the rotator the next day. All OsO_4 solutions should be protected from light and stored at 4 °C.

4. Prepare the 2% OsO_4 buffered solution in a 20 ml scintillation glass vial with a screw cap. If there is leftover solution, protect it from the light and store at 4 °C.

5. Since the components are viscous and the volume graduations are on the inside of the Tri-Pour beaker, they will disappear once the reagents are added. So, on the outside of the 50 ml beaker, make three marks corresponding to the total volumes for each of the first three components to be added. The fourth reagent, DMP-30, is added last (*see* **Note** 7). Do not use pipettes or graduated cylinders to measure the volumes. For example, to prepare a double batch, make the first mark at the 21 ml level to indicate the volume of the EMbed-812. Make a second mark at the 34 ml level to account for the addition of 13 ml of DDSA and then a third mark at the 45 ml level to account for the addition of 11 ml of NMA.

6. Clean the mouths of the stock bottles with a gauze pad and 100% EtOH after each use. If wearing gloves, check if there is any resin residue on the gloves to prevent transferring stickiness to the bottles. All of the resin reagents should be kept in a desiccator at RT.

7. Since the volume of DMP-30 is small, use a 1 ml tuberculin syringe for measuring.

8. For cultures embedded by immersion, use a jeweler's saw to isolate the scored areas instead of placing the culture on a hot plate and cutting it with a razor blade.

9. Before using the double-edge steel razor blades, leave them wrapped, and carefully bend them in half lengthwise until they break into two parts.

10. Stubs can be formed by adding the resin mixture to size 00 BEEM™ capsules with the identification labels the same day as the cultures are embedded. Close the cap, cut the conical bottom away with a single-edge razor blade, and place the capsule with the closed end down into the BEEM™ holder. Fill each capsule with the resin mixture, leaving room at the top. Insert the label all the way down, and push it against the side of the capsule with a toothpick to remove any air bubbles. Top it off with the resin mixture, but do not overfill it. Put it in a 64 °C oven to polymerize overnight.

11. Use the leftover resin mixture that has been stored in 4 °C as glue for mounting.

12. The face of the block should be trimmed to form a trapezoid shape. The wider base is the first area of the block to touch the edge of the diamond knife. When trimming the excess resin around the score to make a trapezoid shape on the face of the block, make sure that the area of interest is not cut away by mistake.

13. Gently perfuse the fixative across the culture by withdrawing the culture medium with a pipet from one side while adding the fixative to the other side of the well, being careful not to

add the fixative directly on top of the coverslip to avoid disrupting the cells. It is imperative to do one well at a time in order to prevent the cell cultures from drying out and, thus, compromise the quality of the final product.

14. Ideally, cultures should be fixed overnight with 2% glutaraldehyde at 4 °C and the embedding process carried out the next day. However, in case of a delay, it has been our experience that leaving cultures in fixative results in better preservation than keeping them in PO_4 buffer. Cultures should not be left in fixative for more than 5 days. In the event that the cultures are left in fixative in the refrigerator for more than 5 days, check to make sure there is enough liquid covering them, so they do not dry out.

15. In case one cannot proceed with the dehydration step on the same day, rinse off the 2% OsO_4 solution twice, 5 min each time, with 0.15 M PO_4 buffer, fill the well with fresh buffer, seal the plate with Parafilm™, cover it with the lid, and leave at 4 °C overnight. The following day, rinse once with 0.15 M PO_4 buffer for 5 min before continuing with dehydration.

16. If there is a need for a brief interruption in the dehydration process, the cultures can be left at RT in the first wash with 70% EtOH. Change it to fresh 70% EtOH once dehydration is resumed.

17. During the second 100% EtOH change, use the 10 min-interval to prepare the fresh resin mixture.

18. Make 2–3 ml of resin mixture per culture. For ten or more cultures, make a double batch. Also make enough for stubs to be used for mounting (*see* **Note 10**). Prepare one polyethylene bottle cap per coverslip with the appropriate label wrapped around the inside of the cap. Push the label down and against the side with a toothpick. The resin contracts while it is in the oven, so slightly overfill the cap with the resin mixture to form a meniscus, and ensure that the surface of the block will be flat when it is cured. Place caps filled with resin on a plastic multi-well tissue plate cover to serve as a tray. This will help when it is time to transfer the cultures to the oven.

19. A 15 cm diameter plastic tissue culture dish cover can be used for this purpose. Bake the dish in the oven with the cultures and save for reuse.

20. For immersion embedding, fill the large polyethylene bottle snap cap with fresh resin mixture, and wrap the appropriate identification label around the inside of the cap. Push the label down against the side of the cap with a toothpick to remove any air bubbles. Gently slide the coverslip with the cell surface facing up into the cap, making sure it lies flat on the bottom, completely covered by the resin.

21. If the culture was embedded by immersion with the culture side facing up, remove the polyethylene cap, locate the coverslip, and use a marker to draw around it. With a single-edge razor blade, score around the circumference of the coverslip and hatch mark the encircled interior on top of the coverslip and then scrape off all of the resin covering the surface of the glass. Place the culture with the glass side up in a 6-well polystyrene tissue culture plate, and put 2–3 drops of hydrofluoric acid directly on top of the coverslip. Leave it in the hood for 1–2 h before checking if the glass has been completely etched (*see* Subheading 3.2, **step 5**).

22. Take two long pieces of transparent tape, and wrap each one around opposite edges of the disc and continuing on the back of the slide. The disc must be well secured and sit flat on the glass slide (Fig. 2).

23. The diameter of the area that is seen through the 10× objective is a little larger than the circle that the scorer will make.

24. Depending on the orientation chosen for sectioning, either *enface* or cross-section, choose an area with several cells that are close to each other to optimize the amount of cells available per section. For cross-sectioning, choose areas of straight and parallel outgrowth (Fig. 3) so that they can be mounted perpendicularly (Fig. 5) to show optimal circular axons and myelin sheaths rather than obliquely sectioned structures (Fig. 6).

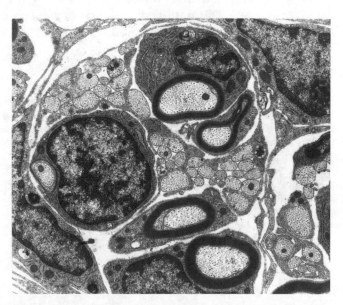

Fig. 6 Schwann cell culture cut in cross-section. Schwann cells wrap around axons to form a myelin sheath. This cross-section orientation is preferred when looking at the condition of the myelin rings and measuring axon diameter, for example. This thin section has been stained with uranyl acetate and lead citrate and photographed in a Philips CM-10 TEM (magnification, 8900×)

25. Use caution to avoid gauging the surface over the cells with the scorer tip. This would damage the area of interest when the score is made.

26. If the culture was embedded by immersion, use the jeweler's saw to cut out the selected area by sawing a straight line across the culture until you reach the score, and then start rotating the disc while moving the saw up and down around the scored circle. Make sure to have a tray underneath to catch the piece if it falls.

27. Take a size 00 BEEM™ capsule, detach the cap, and cut away the conical bottom with a single-edge razor blade to create an open-ended cylinder. Then make a cut with a razor blade on the side of the capsule, all the way down its length. This opening will facilitate inserting the stub into the capsule.

28. Hold your breath while manipulating lead citrate to avoid precipitation. Discard if the solution has a cloudy layer on top.

Acknowledgments

We are grateful for the cartoons prepared by Megan Marlow at the Miami Project to Cure Paralysis. The protocols, developed over many years at the Miami Project, depended primarily upon funding from NINDS (# 09923), the Miami Project to Cure Paralysis, and the Buoniconti Fund. MBB is the Christine E. Lynn Distinguished Professor of Neuroscience. The authors declare no conflicts of interest with the contents of this chapter.

References

1. Bunge RP, Bunge MB, Eldridge CF (1986) Linkage between axonal ensheathment and basal lamina production by Schwann cells. Annu Rev Neurosci 9(1):305–328

2. Clark MB, Bunge MB (1989) Cultured Schwann cells assemble normal-appearing basal lamina only when they ensheathe axons. Dev Biol 133(2):393–404

3. Bates ML, Puzis R, Bunge MB (2011) Preparation of spinal cord injured tissue for light and electron microscopy including preparation for immunostaining. In: Lane E, Dunnett S (eds) Animal models of movement disorder, Neuromethods, vol 62. Springer, New York, pp 381–399

4. Sabatini DD, Bensch K, Barrnett RJ (1963) Cytochemistry and electron microscopy. J Cell Biol 17(1):19–58

Scalable Differentiation and Dedifferentiation Assays Using Neuron-Free Schwann Cell Cultures

Paula V. Monje

Abstract

This chapter describes protocols to establish simplified in vitro assays of Schwann cell (SC) differentiation in the absence of neurons. The assays are based on the capacity of isolated primary SCs to increase or decrease the expression of myelination-associated genes in response to the presence or absence of cell permeable analogs of cyclic adenosine monophosphate (cAMP). No special conditions of media or substrates beyond the administration or removal of cAMP analogs are required to obtain a synchronous response on differentiation and dedifferentiation. The assays are cost-effective and far easier to implement than traditional myelinating SC-neuron cultures. They are scalable to a variety of plate formats suited for downstream experimentation and analysis. These cell-based assays can be used as drug discovery platforms for the evaluation of novel agents controlling the onset, maintenance, and reversal of the differentiated state using any typical adherent SC population.

Key words Primary cultures, cAMP, In vitro assays, Myelin markers, Cell morphology, Krox-20, c-Jun, O1, MAG

1 Introduction

Once isolated from the nerve tissue, Schwann cells (SCs) can undergo successive rounds of expansion in medium containing defined mitogenic factors. The cultures derived from a single nerve biopsy can yield tens of millions of SCs that can be passaged or cryopreserved without loss of biological activity. The unique expansion potential of primary SCs provides an advantage for in vitro and in vivo experimentation as compared to that of cell lines. However, successful in vitro differentiation of cultured SCs can only be achieved when these cells are established in co-culture with dorsal root ganglia (DRG) neurons using complex media formulations supplemented with serum and ascorbic acid [1]. Myelinating co-culture systems of SCs and neurons not only are labor intensive and difficult to carry out but also require at least 2–3 weeks of continued culture under optimal conditions to

Paula V. Monje and Haesun A. Kim (eds.), *Schwann Cells: Methods and Protocols*, Methods in Molecular Biology, vol. 1739, https://doi.org/10.1007/978-1-4939-7649-2_14, © Springer Science+Business Media, LLC 2018

observe the appearance of differentiated, myelin-forming cells. In the absence of neurons, cultured SCs exhibit an immature, proliferative phenotype unless stringent conditions for differentiation are provided.

It has been known for over 30 years that the expression of several key proteins and lipids specific to the myelin sheath, including galactocerebroside (also known as the O1 antigen) and myelin protein zero or MPZ (the most abundant peripheral myelin protein), can be enhanced in the presence of agents that increase the levels of the intracellular second messenger cyclic adenosine monophosphate (cAMP) [2, 3]. cAMP signaling is required at the onset of myelination [4–6]. Without cAMP, SCs fail to express sufficient levels of Krox-20/Egr2 (a transcriptional enhancer of myelination), downregulate the expression of c-Jun/AP1 (a transcriptional inhibitor of myelination) and increase the expression of critical genes necessary for myelin sheath synthesis. In isolated SCs, treatment with high doses of cAMP derivatives is sufficient to drive synchronous cell cycle exit, morphological differentiation, and myelin gene expression within just 3 days [7, 8]. cAMP-treated SCs maintain a differentiated, myelinating SC-like phenotype for as long as they receive a fresh provision of cAMP analogs, but they spontaneously and autonomously dedifferentiate upon the withdrawal of the cAMP stimuli from the culture medium [9] (Fig. 1). Dedifferentiation of adult SCs is a process that typically occurs during nerve repair after injury. Through dedifferentiation, mature SCs regain their capacity to self-renew while losing myelin gene expression and restoring immature SCs features such as high levels of c-Jun/AP1 and GFAP (glial fibrillary acidic protein). In cultured SCs in vitro, cAMP withdrawal allows for an easy and effective way to drive dedifferentiation in a synchronous fashion over a prompt time course of 3 days in the absence of added serum or growth factors [9].

This chapter includes a series of protocols to generate neuron-free systems consisting of differentiated and dedifferentiated SCs generated in response to cAMP addition and removal, respectively. The assays can be set up using defined media and substrate conditions that are nonsupportive of SC proliferation, enabling examination of factors affecting differentiation independently of S-phase progression. Scaling to different plate formats is possible provided the cultures are established and maintained under adherent conditions. These assays are easy to implement and offer an alternative to the use of the more sophisticated myelinating SC-DRG neuron cultures for a variety of applications, including high-throughput analysis. The use of a unique inducing factor, cAMP, in a simplified in vitro setting helps achieve highly reproducible data and consistency among independent experiments. In turn, this facilitates the analysis of results, the interpretation of the data, and the design of follow-up studies.

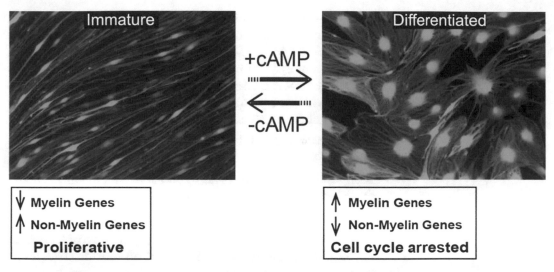

Fig. 1 Reversibility of a differentiated state in cultured SCs subjected to cAMP stimulation and deprivation. SCs experience a dramatic morphological transformation that includes cell flattening and enlargement along with the acquisition of numerous intracellular vacuoles that produce a highly reticulated appearance within the cytoplasm. Morphological changes become apparent within 3–4 days after the addition of cell permeable cAMP derivatives (+ cAMP). Differentiated SCs recover their elongated shape and pattern of cell-to-cell alignment within 3–4 days after cAMP deprivation (− cAMP). These morphological changes are accompanied with a reversion in the pattern of expression of immature versus myelin-related SC genes, as indicated. cAMP-treated SCs are in a cell cycle arrested state that is readily reversed by the elimination of the cAMP stimuli

Throughout this chapter, the terms differentiation and dedifferentiation refer exclusively to the acquisition and loss, respectively, of the differentiated characteristics induced by prolonged exposure of isolated SCs to exogenously provided cAMP-stimulating agents [7, 9]. Subheading 3.1, 3.2 and 3.3 includes a detailed description regarding how to establish, maintain, and analyze differentiated and dedifferentiated SCs in multiwell dishes and large plate formats. Subheading 3.4 contains information on the effect of pharmacological agents with known activity to induce or reduce myelin gene expression for control purposes and assay validation. Information on the most relevant factors affecting the efficiency of differentiation is also provided to help the experimenter set up the assays in the cellular model of choice.

2 Materials

2.1 Cell Material, Media, and Supplies Used for Cell Expansion, Plating, and Starvation

1. Established cultures or cryopreserved stocks of primary SCs (*see* **Note 1**). Refer to Chapter 4 for a detailed protocol on the preparation of a typical primary SC culture suitable for cell-based assays.

2. Laminar airflow hood. Equipped with a pipet aid and a vacuum line system for media filtration and disposal.

3. Disposable sterile plastic ware for culture media preparation, sterilization, storage, and centrifugation. Items include but are not restricted to graduated serological pipettes with filters (volume range, 1–50 mL), 15 and 50 mL graduated conical centrifuge tubes, 1.5 mL microcentrifuge tubes, and assorted bottles with and without bottle-top receptacle containing 0.22 μm filter membranes.

4. Micropipette set and disposable filter tips of 1,000–1 μL volume range.

5. Refrigerated benchtop centrifuge with inserts to adapt 15 and 50 mL conical centrifuge tubes.

6. Temperature-adjustable water or bead bath set at 37 °C.

7. Cell incubator. Humidified standard cell incubator, set to 37 °C and an atmosphere of 8–9% carbon dioxide (CO_2).

8. D10 medium. High-glucose Dulbecco's modified Eagle's medium (DMEM) supplemented with 10% (v/v) heat-inactivated fetal bovine serum (FBS), 200 mM L-glutamine (or GlutaMAX™, Gibco), 25 μg/mL gentamicin, and (optional) 1% penicillin-streptomycin, sterile filtered and kept at 4 °C for up to 1 month. This medium can serve as "dedifferentiation medium" (*see* Subheading 2.2).

9. Complete SC medium. D10 medium containing 10 nM recombinant heregulin-β1 and 2 μM forskolin, sterile filtered and kept at 4 °C for up to 1 month.

10. D1 medium. DMEM supplemented with 1% (v/v) heat-inactivated FBS, 200 mM L-glutamine, 25 μg/mL gentamicin, and 10 mM 4-(2-hydroxyethyl)-1-piperazineethanesulfonic acid (HEPES), sterile filtered and kept at 4 °C for up to 1 month. This medium can serve as "starvation medium" and/or "dedifferentiation medium" (*see* Subheading 2.2).

11. Hanks' Balanced Salt Solution (HBSS) without calcium or magnesium and containing the pH indicator phenol red.

12. 10× trypsin/EDTA solution. 0.5% trypsin and 0.02% ethylenediaminetetraacetic acid (EDTA) prepared in phosphate buffer saline (PBS). Stored in aliquots at −20 °C.

13. 1× trypsin/EDTA solution. Prepared in ice-cold HBSS at the time of use. Used for enzymatic dissociation of non-differentiated (non-treated) or dedifferentiated (cAMP-deprived) SCs.

14. 3× trypsin/EDTA solution. Prepared in ice-cold HBSS at the time of use. Used for detachment of differentiated (cAMP-treated) SCs.

15. (Optional) Repeater pipette or multichannel pipette. Used to dispense an equal volume of cells and perform media changes in multiwell dishes.

16. Coated plates or flasks. Cell culture-treated 100 mm or T75 flasks coated with PLL and/or laminin. Used for routine cell expansion and the generation of large-scale cultures of differentiated SCs. (*See* Chapter 4 for a description of a standard protocol for PLL and laminin coating).

17. Coated multiwell dishes. Cell culture-treated, flat-bottom, multiwell plates in the format of choice coated with PLL and/or laminin. Used to assay variables/agents affecting SC differentiation and dedifferentiation (*see* **Note 2**).

18. Trypan blue solution. 0.4% (w/v) trypan blue prepared in 0.81%(w/v) sodium chloride and 0.06% (w/v) potassium phosphate dibasic solution. (Optional) Propidium iodide (PI) stock solution prepared at 1 mg/mL in distilled water. Used for determination of viability in cells in suspension.

19. Automated cell counter or hemocytometer.

20. Inverted microscope. Phase-contrast microscope equipped with a standard set of objectives (4×–40×) and a digital camera for routine image analysis and documentation. Used to visually monitor and register the density and expansion of the cultures as well as the progression of morphological changes.

2.2 Solutions and Reagents Used in Cell-Based Assays of SC Differentiation and Dedifferentiation

1. DMEM.

2. cAMP stock solution. 5 mM stock solution of 8-(4-chlorophenylthio)adenosine-3′,5′-cyclic monophosphate (CPT-cAMP) or alternatively a 20 mM stock solution of N6 2′-O-dibutyryladenosine-3′,5′-cyclic monophosphate (db-cAMP) both prepared in DMEM. The stock solutions of cAMP are sterilized by filtration and stored in aliquots at −80 °C for up to 1 year. Avoid repeated freezing-thawing of cAMP stocks (*see* **Note 3**).

3. Differentiation medium. D1 or D10 medium supplemented with 250 μM CPT-cAMP or, alternatively, with 1 mM db-cAMP. This medium is used as strong stimuli to induce and maintain high levels of myelin gene expression in SCs.

4. LiCl stock solution. Prepare a 1 M stock of lithium chloride (LiCl) in DMEM. Sterilize by filtration and store in aliquots at −80 °C. LiCl is used as a known inhibitor of cAMP-dependent SC differentiation at a final concentration of 15–30 mM in differentiation medium [10].

5. KT5720 stock solution. Prepare a 2 mM stock of KT5720 (pharmacological inhibitor of protein kinase A, PKA) in dimethyl sulfoxide (DMSO). Store the stock solution protected from light at −20 °C for up to 6 months. KT5720 is used as a known enhancer of cAMP-dependent SC differentiation at a final concentration of 0.05–0.5 μM [8].

6. Dedifferentiation medium. D1 or D10 without CPT-cAMP or db-cAMP or any other cAMP-stimulating agent. This medium is used to deprive SCs of exogenous cAMP and, thus, induce a rapid decay in myelin gene expression.

7. SP600125 stock solution. Prepare a 20 mM stock of SP600125 (anthra[1,9-cd]pyrazol-6(2H)-one) in DMSO. Store the stock solution protected from light at −20 °C for up to 6 months. SP600125 is used as a known inhibitor of SC dedifferentiation at a final concentration of 5–10 µM [9].

8. (Optional) Total homogenate or membrane preparation from DRG neurons. This preparation is used as a known stimulus and positive control for the induction of SC dedifferentiation in medium containing CPT-cAMP or db-cAMP [11].

9. (Optional) Solutions of candidate drugs with potential activity to enhance (or prevent) SC differentiation or dedifferentiation.

3   Methods

The protocols described below assess the reversible changes in the state of differentiation in adherent SC cultures, according to methods established in our lab [6, 7, 9]. The conditions for experimental differentiation and dedifferentiation have been validated using mitogen-expanded, purified adult rat SCs obtained by different methods [12, 13]. Identical differentiation protocols can be used with any other typical SC population, including SCs derived from postnatal nerves [7] and embryonic DRGs [6]. The suggested time for cAMP treatment and removal is 3 days in both cases based on previous data [7, 9]. The efficiency of differentiation can be variable. It is primarily affected by the density of the cells, the type of cAMP-stimulating agent used, and the duration of the stimulus. Other environmental factors, such as the type of media, substrate, and the addition of supplements, serum, or growth factors, can affect the viability of the SC populations and the efficiency of differentiation to various degrees. Still, these factors do not preclude the induction of myelin gene expression in response to cAMP stimulation [7]. In the following sections, a description is provided to aid the experimenter in the expansion and plating of SCs in multi-well dishes, including their starvation in media lacking mitogenic factors and/or serum (Subheading 3.1), and the induction of cell differentiation (Subheading 3.2) with or without subsequent induction of dedifferentiation (Subheading 3.3). The sequence of steps described in this protocol has been represented in the schematic diagram provided in Fig. 2a. The generation of large-scale differentiated SC cultures is described separately (Subheading 3.4) due to particular aspects related to the preparation and collection of cells for replating or analysis.

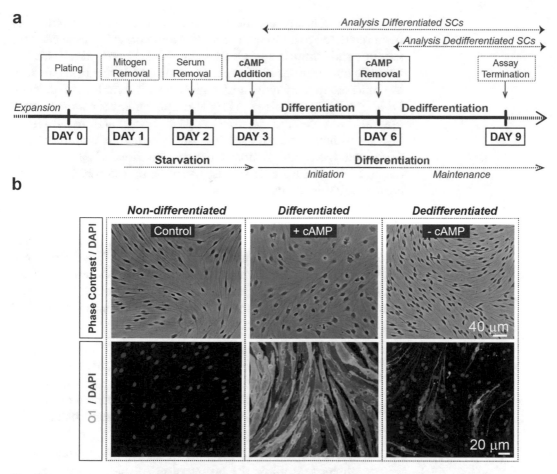

Fig. 2 (**a**) Flowchart depicting the temporal course of experimentation to establish neuron-free assays of SC differentiation and dedifferentiation. The sequence of events in a standard assay has been demarked on top of the time line. The optional steps have been depicted using dashed lines or boxes. The inclusion of mitogen and/ or serum starvation is optional but recommended to reduce the rate of SC proliferation prior to cAMP stimulation. (**b**) Phase-contrast and immunofluorescence microscopy images of non-treated cells (control), cells treated with CPT-cAMP for 3 days (+ cAMP), and cells deprived of cAMP for 3 additional days (− cAMP). These images can be used as a reference to the morphology and levels of O1 expression observable in cultures of non-differentiated, differentiated, and dedifferentiated SCs

3.1 Preparation, Plating, and Starvation of SC Cultures Established in Multiwell Dishes

It is recommended to use primary rat SCs obtained from a fresh or cryopreserved cell suspension subjected to expansion to achieve large enough cell populations for plating into multiwell dishes. Performing assays in a multiwell format is convenient for testing multiple replicate samples and experimental variables simultaneously using single batches of cells. The control of the initial cell density is the determinant factor for successful differentiation if different plate formats are used and results are to be compared. These assays can be carried out in multiwell plates of different formats with nearly identical results. Thus, the description has been restricted to address the cell densities and volumes needed for seeding and stimulating SCs in a standard 24-well dish. This is a versatile and appropriate format for multiple applications, including microscopic analysis, and collection of samples for protein, DNA, or RNA analysis [9]. It is also convenient for most manual operations, including pharmacological manipulations, transfections, and media changes carried out in academic laboratory settings equipped with standard instrumentation. Care should be taken when plating SCs in suspension into multiwell dishes to ensure equivalence in cell loading for a reproducible quantitative assessment of the potency of the differentiating treatment. Performing a medium replacement to eliminate the mitogenic factors and/or the serum in the culture medium prior to cAMP administration (herein referred to as "starvation") is an optional but highly recommended step. Starvation concomitantly reduces the rate of SC proliferation and the levels of myelin gene expression to a minimum in the control condition. If using an established culture, the protocol can commence with **step 4**.

1. Plate the primary SC stocks onto coated 100 mm dishes or, alternatively, T75 flasks at a density of $1–2 \times 10^6$ cells per plate in complete SC medium to allow a fast expansion of the cell population. (*See* Chapter 4 for a detailed protocol on how to establish and expand a SC culture from a frozen stock and the criteria used for passaging into new dishes.)

2. Incubate the cells in a cell incubator up until the cultures reach confluency. Monitor the progression of the cultures daily by phase-contrast microscopy, and perform media changes every 2–3 days with complete SC medium to maintain an optimal rate of growth.

3. Estimate the number of cells needed for experimentation beforehand considering the type and number of dishes needed to run the assays (*see* Subheadings 3.2–3.4). It is possible to subject the cells to a second round of expansion using the conditions described in **steps 1** and **2** if downstream experimentation requires a larger load of cells.

4. Split the cells by trypsinization at the time the cells reach confluency using 8 mL of chilled 1× trypsin/EDTA per 100 mm plate. Stop the action of trypsin by addition of 20 mL of D10 medium.

5. Pass the suspension through a 10 mL serological pipette several times to obtain a homogeneous single-cell suspension.

6. Collect the cells by centrifugation at $200 \times g$ for 8 min at 4 °C. Discard the supernatant and homogeneously resuspend the cell pellet in 1 mL of D10.

7. Count the cells and estimate their viability by using the method of choice. A confluent 100 mm plate of primary rat SCs can render 6–9×10^6 cells/plate. The percentage of viability is usually >90% based on trypan blue staining or PI incorporation.

8. Resuspend the cells in the appropriate volume of complete SC medium making sure that the cells are maintained in suspension at all times. To seed one 24-well plate, prepare a cell suspension containing 1.2×10^6 cells in 12.5 mL of complete SC medium. (*See* **Note 4** on the importance of controlling cell density.)

9. Dispense an equal volume of cells in each well using a repeater, multichannel, or serological pipette. For a 24-well plate, dispense 500 μL of the cell suspension per well. Avoid delays at this step as SCs may precipitate in the tube and/or form aggregates within a few minutes (*see* **Note 5**).

10. Observe the cells at the inverted microscope immediately after seeding to confirm equal loading in each well. Shake the plate 2–3 times using gentle, lateral movements (i.e., side to side and forward-backward) to ensure even distribution of the cells in all wells (*see* **Note 5**).

11. Immediately place the plate in the cell incubator to avoid changes in the pH of the media.

12. Observe the cells under the inverted microscope after overnight incubation to visually appreciate the health of the cultures. By this time, the SCs should have adhered, extended processes, and began to align to one another by forming the expected pattern of growth. (*See* Fig. 2 for a representative phase-contrast image of a SC culture exhibiting optimal density and alignment. Images of typical cultures can also be found in our previous publications [7, 10].)

13. Remove the complete SC medium from each well by gentle aspiration, and rapidly replace it with an equal volume of D10 medium (*see* **Note 6**).

14. Transfer the cells to the cell incubator for overnight growth and adaptation to the lack of mitogenic stimulation.

15. On the next day, replace the culture medium with an equivalent volume of D1 medium if serum withdrawal is needed for experimentation (*see* **Notes 6–8**).

16. Incubate the cells overnight prior to cAMP stimulation.

17. Alternatively, proceed to stimulate the cells directly in D10 medium with or without an additional medium change, as indicated in Subheading 3.2. (*See* **Note 9** on aspects related to the type of media used for the induction of differentiation.)

3.2 Induction of SC Differentiation

SC differentiation is triggered by the addition of membrane permeable cAMP analogs directly to the culture medium (*see* **Note 10**). Experimental variables in the form of pharmacological treatments can be introduced prior to, during, or after cAMP administration. It is recommended to include cultures maintained from the outset in the absence of cAMP-inducing agents to serve as controls of non-differentiated SCs expressing low levels of myelin-related markers (Figs. 1 and 2). Comments on the use of replicate samples, positive and negative controls, and known pharmacological treatments that have been previously tested for their activity to enhance/prevent differentiation are provided in **Notes 11** and **12**. For a reference, Fig. 3 portrays fluorescence microscopy

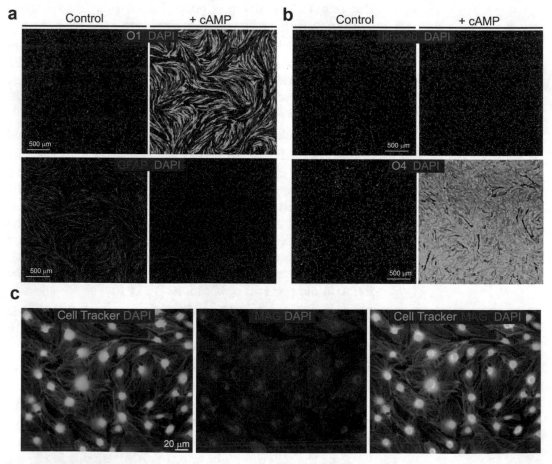

Fig. 3 Typical features of non-differentiated (control) and cAMP-differentiated (+ cAMP) SCs. (**a** and **b**) Panoramic low-magnification images of non-treated (control) and cAMP-treated SCs stimulated with CPT-cAMP (250 μM) for 4 days. Cultures were immunostained using antibodies against O1 and GFAP (**a**) and Krox-20 and O4 (**b**), as indicated. (**c**) High-magnification image of CPT-cAMP-treated SCs labeled with CellTracker™ (green) and MAG-specific antibodies to reveal the internal reticular cytoarchitecture and the myelin-like plasma membrane, respectively. Nuclei were counterstained with DAPI (blue) in all images

images of representative cAMP-treated cultures maintained in culture for 4 days and subsequently analyzed by immunostaining using antibodies against mature (O1, Krox-20, O4, and MAG) and immature (GFAP) SC-specific markers. Images are provided at low and high magnification to provide a reference to the pattern of growth and differentiation of a typical homogeneous SC culture established in a 24-well dish. For assay termination or analysis, one should consider that markers specific to the myelin sheath, such as O1 and myelin-associated glycoprotein (MAG), are detected no earlier than 3–4 days following the onset of stimulation [7]. Cells can be maintained differentiated for extended periods of time with repeated additions of cAMP analogs or other cAMP-inducing treatments [9].

Specific guidelines and information on plate design and analysis are not provided in this chapter given the dependency on the goal(s) of the experiment, the outcome measure(s), and the availability of reagents and resources in each laboratory setting. Examples on how to analyze and interpret the data can be found in published literature. We have used fluorescence microscopy followed by manual or automated image analysis [6], live cell video imaging [10], microplate reading [10], and Western blotting [9, 11] as independent readouts. We have also used combined measurements of proliferation and differentiation in individual wells [7, 11]. Chapter 6 provides a description of our standard immunofluorescence protocol for the analysis of differentiation markers in live and fixed SCs.

1. Initiate the differentiation treatment by adding 500 μL of differentiation medium to the respective wells, according to the designed experimental layout.

2. Transfer the plates immediately after treatment to the cell incubator to avoid changes in the pH of the media.

3. Monitor the cells by phase-contrast microscopy daily for the subsequent 3–4 days to visually appreciate the health of the cultures and morphological changes associated with differentiation (*see* **Note 13** and Fig. 2b).

4. (Optional) If excessive floating debris becomes apparent, replace the medium with freshly prepared differentiation medium, as described in **step 1**. Medium changes can be performed at any time during the 3–4-day incubation period without affecting the efficiency of differentiation.

5. Proceed to analyze the cultures as needed according to the experimental design. (*See* **Note 14** on the criteria used for assay termination and **Notes 15–17** on variables affecting the efficiency of differentiation.)

6. (Optional) Proceed as described in **step 7** to prolong the differentiated state or as indicated in Subheading 3.3 to induce dedifferentiation.

7. To maintain SCs differentiated for an extended period of time, simply carry out a medium replacement using new differentiation medium (as described in **step 1**).

8. Continue to replace this medium on a 3-day schedule bearing in mind that the state of differentiation can be maintained for at least 2 weeks if the survival of the cultures is not compromised [9]. For this reason, it is important to monitor the cultures daily by phase-contrast microscopy to identify early signs of stress and take action on assay management or termination in due time.

3.3 Induction of SC Dedifferentiation

After the SCs have differentiated in response to cAMP treatment, it is possible to induce dedifferentiation simply by removing the cAMP-stimulating agents from the culture medium [9, 11]. **Note 18** provides details on the medium used for dedifferentiation. Experimental variables in the form of pharmacological intervention (or other types of treatments) can be introduced prior to, during, or after the treatment/s inducing dedifferentiation. (*See* **Note 19** for suggested positive and negative controls, replicate samples, and known pharmacological treatments able to enhance or prevent dedifferentiation.) It is highly recommended to use 3- or 4-day-old cultures of cAMP-differentiated SCs to assess their responses on dedifferentiation and the impact of experimental variables on this process. Repeated cycles of addition and removal of cAMP analogs on a 3-day schedule are possible given that the expression of myelin-related markers is reversible under these experimental conditions [9].

1. Observe the cultures of cAMP-differentiated SCs under the inverted microscope to confirm the expected morphological changes prior to applying a potentially dedifferentiating treatment (*see* Fig. 2).

2. To induce dedifferentiation, simply replace the culture medium with an equivalent volume (500 μL) of D1 or D10 media lacking CPT-cAMP or db-cAMP (also referred to as "dedifferentiation medium") (*see* **Note 20**).

3. Immediately transfer the cells to the incubator.

4. Monitor the cultures daily by phase-contrast microscopy for the subsequent 24–72 h to visually appreciate the health of the cultures and morphological changes associated with dedifferentiation.

5. Proceed to collect or analyze the cultures on the third or fourth day after cAMP withdrawal or as needed according to the experimental design. (*See* **Note 21** on the criteria used to determine the time for assay termination and **Note 22** for the visual identification of signs of dedifferentiation.)

6. (Optional) Continue to provide dedifferentiation medium or differentiation medium on a 3-day feeding schedule to maintain a dedifferentiated state or induce a second round of differentiation, respectively.

3.4 Generation and Collection of Large-Scale Cultures of Differentiated SCs

Individual batches of differentiated SCs can be generated in large numbers for downstream applications in vitro or in vivo. With the proper handling, differentiated SCs can be recovered from their dishes by trypsinization while maintaining their viability. These cells can be subsequently collected as a single-cell suspension for direct use, replating [11], purification, or preparation of cryogenic stocks [13]. The description provided below addresses the cell densities and volumes needed for plating, stimulation, and collection of differentiated SCs established in 100 mm dishes. Any plate format can be used if the density of cells per surface area is maintained as recommended.

1. Plate a single-cell suspension of SCs containing 3×10^6 cells in a 100 mm coated plate using 10 mL of complete SC media. Make sure the cells are evenly distributed in the dish.

2. Incubate the cells in the cell incubator overnight. One day after plating, replace the medium with 10 mL of D10 to reduce the rate of proliferation.

3. On the next day, change the culture medium to differentiation medium using a minimum of 8 mL per plate.

4. Incubate the cells in the cell incubator, and monitor them daily by phase-contrast microscopy until changes in cell morphology become apparent, as described in Subheading 3.2.

5. Remove the culture medium, wash the cells with at least 10 mL of HBSS, and add 6 mL of 3× trypsin/EDTA solution per plate.

6. Monitor the progression of the dissociation under the inverted microscope. Tap the dish vigorously multiple times to facilitate cell detachment. Cells may require extensive and continuous tapping up until detachment is observed, as differentiated SCs are very strongly attached to their dishes (*see* **Note 23**).

7. Add 10 mL of D10 to the dish, and pass the cells 3–4 times through the pipette to enable detachment. Collect the cells into a 50 mL tube.

8. Repeat this twice to recover the remainder of the cells using 10 mL of D10 in each collection step. Observe the plates under the phase-contrast microscope to confirm full detachment in all areas of the dish. Ultimately, collect all cells in a minimum volume of 50 mL of D10.

9. Pellet the cells by centrifugation ($200 \times g$ for 8 min at 4 °C).

10. Resuspend the cells in D10 medium, and perform a cell count to determine the total number and the percentage of viable cells, as described in Subheading 3.1, **steps 6–7**. It is expected that 3×10^6 highly viable (>85% viability) differentiated SCs are recovered per plate.

11. Proceed to use the cells in downstream experimentation as needed (*see* **Notes 24–26**).

4 Notes

1. The rat SCs can be derived from the sciatic nerve or the DRGs according to standard protocols. The cultures can be used before or after subjecting them to expansion or purification to eliminate contaminating cells. It is recommended not to expand the cells beyond passage 5–6 to prevent phenotypic departure of the SC cultures from the characteristics of the population of origin. Nevertheless, the effect of passaging needs to be determined empirically considering that excessive subculturing of primary SCs should be avoided regardless of the type of application. It is also advisable to test independent batches of cells for their responses to differentiation before-hand as some variability in the efficiency of differentiation is expected from batch to batch.

2. A laminin substrate is not required but is highly recommended to support adhesion, survival, and differentiation of cultured SCs [7]. The use of plates sequentially coated with PLL and laminin maximizes cell survival and adhesion, in particular if a starvation period is introduced. PLL coating is sufficient to support differentiation in rat SC cultures. One should consider that adhesion to the substrate is a determinant factor for efficient SC differentiation.

3. The preparation, handling, and use of cAMP stock solutions are key to the success of the experiments. Use practices recommended by the manufacturer for the preparation, storage, and handling of cAMP stocks in DMEM. Ensure complete dissolution of cAMP crystals prior to storage and use. Vortexing and heating the cAMP stock solutions in a water bath can help in redissolving precipitated cAMP crystals.

4. The density of the cells at the time of stimulation is one important variable that affects differentiation. If 24-well plates are used, 500 μL media/well containing 40,000–60,000 cells provides an optimal cell density for differentiation. Scale the number of cells up or down according to the available surface area. Avoid using overconfluent cultures as they may be refractory to undergo differentiation [7]. Assays can be performed

in virtually any culture format as long as the cells remain meta-bolically active and adherent throughout the time course of experimentation. If cell aggregation is observed as a reflection of poor substrate adhesion, or stress, avoid performing the cAMP treatment up until the health and adhesion of the cells have been restituted. Only areas containing equivalent cell density are comparable for quantitative estimation of the changes in differentiation (*see* Fig. 3a, b).

5. Clumping of SCs in suspension at the time of plating should be avoided and managed appropriately to guarantee an equal number and even distribution of the cells in each well. Avoid circular movements after plating the cells in multiwell dishes to prevent cell collection in the center of the well.

6. Mitogen starvation by removing neuregulin and forskolin from the culture medium reduces the rate of SC proliferation. Mitogen and serum withdrawal is not required for effective SC differentiation in response to CPT-cAMP, but it may be needed to reduce the potential effect of other experimental variables during the course of differentiation. Progressive removal of mitogens and serum with intermediate periods of incubation is needed to avoid sudden and massive apoptotic cell death due to lack of trophic support. Consider that pro-longed starvation of SCs over 3 days may reduce the efficiency of differentiation in response to cAMP treatment.

7. The use of a non-mitogenic concentration of serum (1%) in the D1 medium is suggested to maintain the health of the cells throughout the course of treatment. Some SC cultures with-stand serum withdrawal, whereas others do not. In the latter case, the cells can be seen to rapidly retract their processes and be lost floating in the culture medium right after the serum is removed. While it is possible to reverse this alarming issue by adding 10% serum back into the wells, doing so may be effec-tive only if serum is applied within 1–2 h after the onset of cell detachment.

8. Performing mitogen and serum reduction/removal prior to assay initiation is optional. Our previous findings have shown that even high concentrations of serum and neuregulin do not reduce the differentiating effect of cAMP [9, 11]. The suscep-tibility of SC batches to serum and mitogen removal needs to be tested in advance. If removal of serum and mitogens is required, this can be done in a progressive manner, first by removing the mitogens neuregulin and forskolin and then by removing the serum to allow for cell adaptation after each media change, as described in **Note 6**.

9. The use of DMEM high glucose is recommended for plating, expansion, and experimentation. The addition of supplements

such as pituitary extract, B27 or N2 should not influence the responses on differentiation if assays are carried out in DMEM-based media [7]. If other supplements are needed, their effect on cAMP-induced SC differentiation should be tested separately as an experimental variable.

10. The type and dosage of the cAMP-stimulating agent used to induce SC differentiation are critical factors that affect the efficiency of differentiation in isolated SCs [7]. Optimal differentiating responses have been achieved using either CPT-cAMP or db-cAMP, which are membrane-permeable, phosphodiesterase-resistant, non-pathway-selective analogs of cAMP. These agents have been chosen on the basis of their demonstrated effectiveness to increase the expression of O1 and MAG as compared to other agents that increase cAMP via different molecular mechanisms [8]. These cAMP derivatives are effective only when provided in the range of 0.1–1 mM.

11. When designing differentiation/dedifferentiation assays, it is useful to leave at least three replicate samples of cells without cAMP treatment to use as negative or reference controls. These controls are needed to estimate the potency of treatment when cAMP is added or removed. The percentage of SCs that naturally express O1 or MAG under basal conditions (i.e., in medium lacking cAMP-inducing activity) is <2% in most SC populations (see Fig. 3). Plan carefully to include a sufficient number of replicate samples so that statistical analysis will be valid using the selected outcome measure. If properly set up, the efficiency of differentiation and dedifferentiation should be highly reproducible in comparing independent wells within single stimulation experiments.

12. If a series of experimental drugs are tested for their activity to prevent or enhance differentiation, known agonists/antagonists can be used as controls and their effects contrasted with the candidate drugs whose activity is unknown. Provision of LiCl in the range of 15–30 mM antagonizes cAMP-induced morphological differentiation, along with Krox-20, O1, and MAG expression [10]. Inhibition of cAMP-dependent O1 expression is also possible by means of using blockers of cAMP signaling such as the EPAC inhibitor ESI-09 and the soluble adenylyl cyclase inhibitor KH7 [6, 8]. By contrast, addition of KT5720 at 0.05–0.5 μM enhances the differentiating effect of CPT-cAMP on the levels of MPZ, Krox-20, O1, and MAG expression [8].

13. The cultured cells should be morphologically differentiated within 3–4 days of cAMP treatment. Typically, the cells lose their spindle-shaped morphology and acquire a large flattened shape with many intracellular vacuoles clearly observable by

phase-contrast microscopy (Figs. 1, 2, and 3). If morphological differentiation is confirmed, the cells can then be fixed and analyzed for the expression of specific differentiation markers (Fig. 3) or treated accordingly for the induction of dedifferentiation. If morphological differentiation is not appreciated by the third or fourth day of treatment, new medium containing fresh CPT-cAMP can be added and additional time allowed for effective differentiation to occur. The induction of myelin-associated markers at the RNA and protein level directly correlates with the extent of morphological change. It has been consistently observed that enhanced O1 or MAG expression is not found in cells that have not experienced morphological differentiation. However, some SCs may be resistant to acquire morphological features even with prolonged and persistent cAMP stimulation. Even though the mechanistic link between changes in cell shape and myelin gene expression is not yet clear, morphological differentiation is a highly effective visual criteria for a straightforward assessment of the effectiveness of treatment in live cell cultures.

14. The time required for effective induction of myelin markers is on average 3–4 days after the provision of CPT-cAMP or db-cAMP [7]. Shorter times of stimulation may be used as a function of the outcome measure and the sensitivity of the detection method. The minimum time needed to detect changes in molecular markers is contingent upon the type of marker used. An increase in the levels of Krox-20 expression (protein) with respect to the control condition (i.e., cells not stimulated with cAMP) is detected in the nucleus of SCs as early as 24 h post-stimulation [6]. MPZ expression (protein) is detected usually within 48 h posttreatment [8]. MAG (protein) and galactocerebroside (i.e., cell surface antigenicity for O1 antibodies) are detected concurrently at 3–4 days post-stimulation but soon reach a plateau where no further increase is detected irrespective of continued stimulation [7]. MBP mRNA rather than protein has been detected under these experimental conditions [6]. Other markers sensitive to CPT-cAMP stimulation are peripheral myelin protein-22 (PMP-22), cyclic nucleotide phosphodiesterase (CNPase), periaxin, and fatty acid 2-hydroxylase (FA2H) [14]. It is recommended that the experimenter confirms the kinetics of expression of the marker of choice at the RNA or protein level prior to initiating systematic experimentation.

15. The efficiency of differentiation is variable among independent SC cultures but consistent within single batch preparations. The full complement of factors that determine the efficiency of SC differentiation, as determined by the percentage of cells that acquire cell surface expression of O1 or MAG

in vitro, is not fully understood. This percentage may vary from 20 to 90% in independent rounds of testing.

16. The degree to which SCs increase the expression of myelin-associated markers may differ considerably according to species-specific variables. The conditions for differentiation need to be adjusted to each particular cell population. It is advisable to test the effect of cell density, preferred time of stimulation, and dosage of the cAMP-stimulating agent in the cell type of choice prior to methodical experimentation of other experimental variables. Some populations of SCs, such as those derived from human donors, are resistant to induce the expression of some critical myelin-associated markers for reasons that are yet ill defined.

17. The percentage of differentiated SCs should be homogeneous across the surface area of the well, and this should be consistent in the experimental replicates (see Fig. 3a, b). If this is not the case, look into local changes in cell density due to uneven distribution of the cells at the time of plating, local detachment, or cell death.

18. To induce SC dedifferentiation, the medium of cAMP-differentiated SCs can be replaced with essentially any type of medium lacking cAMP-stimulating agents.

19. Use the same criteria described in **Note 11** for selecting the number of replicate samples and controls. The JNK inhibitor SP600125 can be used at a final concentration of 5–10 μM to prevent the morphological changes and the loss of O1 expression initiated by cAMP removal [9]. In the presence of cAMP, one can induce O1 loss and proliferation by incubating SCs with a total or membrane preparation obtained from purified DRG neurons, as described in [11].

20. Perform media changes with care while working fast to avoid drying of the cells due to overexposure to the air flow. Avoid alkalization of the media at all times by timely reintroducing the cells in the incubator to stabilize the pH. Alternatively, use D1 media adjusted to pH 6.5 to allow for more time of operation while working outside of the incubator.

21. The average time needed to observe changes associated with SC dedifferentiation, i.e., disappearance of cell surface myelin proteins and lipids, is 3 days. Still, this needs to be determined empirically in each cell type. Some SC batches can be resistant to dedifferentiate fully and may require longer incubation in cAMP-free medium up until observable changes in cellular morphology occur. If morphological change is not observed after 3 days, simply replace the medium with new D1 (or D10), and continue to monitor the changes by phase-contrast microscopy until the cells recover their elongated shape.

22. The most robust criteria of morphological dedifferentiation are the recovery of an elongated shape and the disappearance of intracellular vacuoles (Fig. 2).

23. Fully differentiated SCs experience a dramatic increase in cell size and become flat and expanded while adhering very strongly to the available substrate (Figs. 1, 2, and 3). Detachment of the cells by enzymatic treatment is possible but needs to be carried out with care while monitoring the cells at all times under the phase-contrast microscope to avoid excessive cell damage. Augmenting the concentration of trypsin is recommended to facilitate prompt detachment along with mechanical tapping of the dish. It is also recommended to use the flow of media through a 10 mL pipette to gently detach the cells. Do not use a scraper for mechanical detachment, as this leads to impaired viability. Once detached, differentiated cells can easily be discriminated from non-differentiated SCs by their greater size and presence of intracellular vacuoles.

24. Differentiated SCs should remain highly viable after trypsinization and capable to readily attach, extend processes, and proliferate on a laminin or axonal substrate [11]. Reattachment to the substrate typically occurs within a matter of minutes.

25. If the percentage of cells exhibiting myelin markers needs to be increased, the cultures of differentiated SCs can be subjected to purification by magnetic-activated cell sorting (MACS). Purification can be done via indirect cell labeling using O1 antibodies or direct cell labeling using magnetically labeled Myelin Removal Beads™ [11]. (*See* Chapter 6 for detailed protocols on the use of MACS for SC purification.)

26. With the exception of trypsinization, differentiated SCs can be managed essentially as non-differentiated cells during replating, cryopreservation, and purification. The recovery of viable differentiated SCs should be high if gentle conditions for detachment are used. The generation and collection of differentiated SCs in the absence of other cell types can enable the isolation of myelin-specific proteins, lipids, and RNA for a variety of applications.

Acknowledgments

The author appreciates the guidance provided by Dr. Patrick Wood and the assistance provided by Drs. Jennifer Soto and Ketty Bacallao in the development of these protocols. James Guest, Kristine Ravelo, and Gonzalo Piñero are acknowledged for performing critical review of the manuscript. The work presented in this chapter was generously supported by the NIH-NINDS

(NS084326), The Craig Neilsen Foundation (339576), The Miami Project to Cure Paralysis, and The Buoniconti Fund. The author declares no conflicts of interest with the contents of this article.

References

1. Eldridge CF, Bunge MB, Bunge RP, Wood PM (1987) Differentiation of axon-related Schwann cells in vitro. I. Ascorbic acid regulates basal lamina assembly and myelin formation. J Cell Biol 105(2):1023–1034

2. Sobue G, Pleasure D (1984) Schwann cell galactocerebroside induced by derivatives of adenosine 3′,5′-monophosphate. Science 224(4644):72–74

3. Jessen KR, Mirsky R, Morgan L (1991) Role of cyclic AMP and proliferation controls in Schwann cell differentiation. Ann N Y Acad Sci 633:78–89

4. Glenn TD, Talbot WS (2013) Analysis of Gpr126 function defines distinct mechanisms controlling the initiation and maturation of myelin. Development 140(15):3167–3175

5. Monk KR, Oshima K, Jors S, Heller S, Talbot WS (2011) Gpr126 is essential for peripheral nerve development and myelination in mammals. Development 138(13):2673–2680

6. Bacallao K, Monje PV (2015) Requirement of cAMP signaling for Schwann cell differentiation restricts the onset of myelination. PLoS One 10(2):e0116948. https://doi.org/10.1371/journal.pone.0116948

7. Monje PV, Rendon S, Athauda G, Bates M, Wood PM, Bunge MB (2009) Non-antagonistic relationship between mitogenic factors and cAMP in adult Schwann cell re-differentiation. Glia 57(9):947–961

8. Bacallao K, Monje PV (2013) Opposing roles of PKA and EPAC in the cAMP-dependent regulation of schwann cell proliferation and differentiation [corrected]. PLoS One 8(12):e82354. https://doi.org/10.1371/journal.pone.0082354

9. Monje PV, Soto J, Bacallao K, Wood PM (2010) Schwann cell dedifferentiation is independent of mitogenic signaling and uncoupled to proliferation: role of cAMP and JNK in the maintenance of the differentiated state. J Biol Chem 285(40):31024–31036

10. Pinero G, Berg R, Andersen ND, Setton-Avruj P, Monje PV (2016) Lithium reversibly inhibits Schwann cell proliferation and differentiation without inducing myelin loss. Mol Neurobiol. https://doi.org/10.1007/s12035-016-0262-z

11. Soto J, Monje PV (2017) Axon contact-driven Schwann cell dedifferentiation. Glia. https://doi.org/10.1002/glia.23131

12. Morrissey TK, Kleitman N, Bunge RP (1991) Isolation and functional characterization of Schwann cells derived from adult peripheral nerve. J Neurosci 11(8):2433–2442

13. Andersen ND, Srinivas S, Pinero G, Monje PV (2016) A rapid and versatile method for the isolation, purification and cryogenic storage of Schwann cells from adult rodent nerves. Sci Rep 6:31781. https://doi.org/10.1038/srep31781

14. Maldonado EN, Alderson NL, Monje PV, Wood PM, Hama H (2008) FA2H is responsible for the formation of 2-hydroxy galactolipids in peripheral nervous system myelin. J Lipid Res 49(1):153–161. https://doi.org/10.1194/jlr.M700400-JLR200

Chapter 15

The Pseudopod System for Axon-Glia Interactions: Stimulation and Isolation of Schwann Cell Protrusions that Form in Response to Axonal Membranes

Yannick Poitelon and M. Laura Feltri

Abstract

In the peripheral nervous system, axons dictate the differentiation state of Schwann cells. Most of this axonal influence on Schwann cells is due to juxtacrine interactions between axonal transmembrane molecules (e.g., the neuregulin growth factor) and receptors on the Schwann cell (e.g., the ErbB2/ErbB3 receptor). The fleeting nature of this interaction together with the lack of synchronicity in the development of the Schwann cell population limits our capability to study this phenomenon in vivo. Here we present a simple Boyden Chamber-based method to study this important cell-cell interaction event. We isolate the early protrusions of Schwann cells that are generated in response to juxtacrine stimulation by sensory neuronal membranes. This method is compatible with a large array of current biochemical analyses and provides an effective approach to study biomolecules that are differentially localized in Schwann cell protrusions and cell bodies in response to axonal signals. A similar approach can be extended to different kinds of cell-cell interactions.

Key words Axon-Schwann cell interaction, Boyden chamber, Neuronal membranes, Protrusions, Juxtacrine interaction, Early cell-cell contact, Pseudopod

1 Introduction

The prerequisite to myelination is the recognition of large axons by immature Schwann cells. Schwann cells surround bundles of mixed caliber axons and send cytoplasmic protrusions between them to select and segregate large axons that will be myelinated [1]. The final myelin fate of an axon and the thickness of myelin sheath around it depend on the amount of neuregulin-1-type III present on the axonal membrane, in a strictly contact-dependent signaling event [2, 3]. Upon sensing neurons, Schwann cells probably respond with localized amplification of signaling pathways [2, 4–8] and enrichment of specific proteins [9–13] at the Schwann cell-axon interface. Although localized signaling is critical for the interaction between axons and Schwann cells, the molecular mechanisms

Paula V. Monje and Haesun A. Kim (eds.), *Schwann Cells: Methods and Protocols*, Methods in Molecular Biology, vol. 1739, https://doi.org/10.1007/978-1-4939-7649-2_15, © Springer Science+Business Media, LLC 2018

that mediate this response remain poorly defined. To investigate the mechanisms of Schwann cell-axon recognition and interaction, we adapted the pseudopod system to the study of juxtacrine signals [13]. The original pseudopod system physically separates cell bodies, in the top compartment of a Boyden chamber, from protrusions, extending in the bottom chamber in response to soluble signals [14]. The cell body and protrusion compartments can then be isolated and compared by differential biochemical analyses. Our modification consisted of using a suspension of neuronal membranes as a stimulus [13], postulating that neuronal membranes are a source of active molecules that elicit juxtacrine signals. Here, we provide detailed methods for the effective purification of Schwann cell protrusions using these pseudopod chambers with microporous filters and for identification of the RNAs and proteins that are enriched in protrusions. These methods include the preparation of the neuronal membrane suspension, the isolation of primary Schwann cells, and the induction of Schwann cell protrusions in response to neuronal membranes. This method can be applied in principle to any type of biologically active cell membrane.

2   Materials

2.1   Dissections

1. Horizontal laminar airflow hood.

2. Dissecting microscope.

3. Cell incubator, 37 °C with 5% CO_2.

4. Centrifuge, refrigerated benchtop.

5. Dissection tools, including one set for macrodissection (one blunt forceps and one medium-sized dissecting scissors) and a microdissection set (two blunts forceps and one #5 dissecting forceps) (see **Note 1**).

6. Beaker, 50 ml, sterile, filled with 70% ethanol and with gauze in the bottom (see **Note 2**).

7. Petri dishes, 10 cm diameter.

2.2   Isolation of Primary Schwann Cells

1. Eight Sprague-Dawley rat pups, at postnatal day 3 (P3) (see **Note 3**).

2. Hemocytometer.

3. Conical centrifuge tubes (15 ml).

4. Hank's Balanced Salt Solution (HBSS).

5. Dulbecco's phosphate-buffered saline (D-PBS).

6. 1 g/l collagenase solubilized in H_2O (see **Note 4**).

7. 2.5% trypsin.

8. 10,000 U/ml penicillin/streptomycin.

9. Antibody, anti-Thy1.1 (Sigma, M7898).

10. Complement, rabbit serum, filtered.

11. PLL solution: 0.01 g/l poly-L-lysine solubilized in H_2O.

12. Calf serum media: Dulbecco's modified Eagle medium (DMEM) supplemented with 10% calf serum and 2 mM L-glutamine.

13. Fetal calf serum media: DMEM supplemented with 10% fetal calf serum and 2 mM L-glutamine.

14. ARA-C media: DMEM supplemented with 10% fetal calf serum, 2 mM L-glutamine, and 10 μM cytosine-B-arabinofuranoside hydrochloride.

15. MSC (media for Schwann cell): DMEM supplemented with 10% fetal calf serum, 2 mM L-glutamine, 2 μM forskolin, and 2 ng/ml recombinant human neuregulin.

16. HMEM media: DMEM supplemented with 10% fetal calf serum, 2 mM L-glutamine, 100 U/ml penicillin/streptomycin, and 20 mM HEPES (4-(2-hydroxyethyl)-1-piperazineethanesulfonic acid) at pH = 7.2.

17. Freezing media: DMEM supplemented with 20% fetal calf serum and 10% DMSO.

2.3 Schwann Cell Myelination Assay

1. Epifluorescence microscope.

2. Hemocytometer.

3. Dorsal root ganglia (DRG) neuron culture (*see* **Note 5**).

4. 4-well plates.

5. Glass coverslips, 18 mm, washed in ethanol and sterilized.

6. Glass microscope slides.

7. Nail polish.

8. 100% methanol, store at −20 °C.

9. 4% paraformaldehyde in PBS.

10. Mounting media (i.e., Vectashield).

11. 0.1 μg/ml DAPI (4′,6-diamidino-2-phenylindole) solubilized in D-PBS.

12. CL solution: 10% collagen and 1‰ glacial acetic acid diluted in water, freshly prepared (*see* **Note 6**).

13. C media: Minimum Essential Media (MEM) supplemented with 10% fetal bovine serum, 4 g/l D-glucose, 2 mM L-glutamine, and 50 ng/ml nerve growth factor.

14. C media PLUS: MEM supplemented with 10% fetal bovine serum, 4 g/l D-glucose, 2 mM L-glutamine, 50 ng/ml nerve growth factor, and 50 μg/ml ascorbic acid.

15. MLF blocking: 10% fetal bovine serum, 2% bovine serum albumin, 0.1% Triton-X, and 0.02% sodium azide diluted in PBS.

16. Anti-myelin basic protein antibody (Covance, SMI-99P), anti-neurofilament M (Covance, PCK-593P), anti-mouse Ig 488, and anti-chicken Ig 549.

2.4 Preparation of Dorsal Root Ganglia Neuron Membrane Suspensions

1. One Sprague-Dawley rat female, pregnant at embryonic day E15.5 (*see* **Note 7**).

2. Centrifuge and ultracentrifuge.

3. 6-well plates, coated with PLL solution for 15 min.

4. Microcentrifuge tubes, 1.5 ml.

5. Dounce homogenizer, 1 ml, tight.

6. DMEM.

7. Leibovitz's L-15 medium.

8. 10,000 U/ml penicillin/streptomycin.

9. NB-media: Neurobasal media supplemented with 1× B27 supplement, 4 g/l D-glucose, 2 mM L-glutamine, and 50 ng/ml nerve growth factor.

10. NB+ media: Neurobasal media supplemented with 1× B27 supplement, 4 g/l D-glucose, 2 mM L-glutamine and 50 ng/ml nerve growth factor, 10 mM fluorodeoxyuridine, and 10 mM uridine.

2.5 Neuronal Membrane Suspension Activity Assay

1. 4-well plates, coated with PLL solution for 15 min.

2. Antibody, anti-Akt (Cell Signaling, 9272), anti-Phospho-Akt (Cell Signaling, 9271).

3. Starvation media: DMEM supplemented with 2 mM L-glutamine, 2 μM forskolin, and 100 U/ml penicillin/streptomycin.

4. PN1: 100 mM pH 7.4, 5 mM EDTA, 150 mM NaCl, 1% SDS, 1 mM sodium orthovanadate, protease inhibitors.

2.6 Schwann Cell Protrusions Assay

1. Epifluorescence microscope.

2. Hemocytometer.

3. Microscissors.

4. Parafilm.

5. Forceps.

6. Glass microscope slides.

7. Glass coverslips, 22 × 50 mm.

8. Microcentrifuge tubes, 1.5 ml.

9. Petri dishes, 10 cm diameter.

10. Transwell 24 mm with 3 μm pore polycarbonate membrane (Corning, 3414) (*see* **Note 8**).

11. Cotton swabs.

12. Cell scraper.

13. Pointed cotton swabs.

14. Cell culture tray, containing a humid paper.

15. Nail polish.

16. Vectashield (Vector Labs).

17. D-PBS.

18. Bicinchoninic acid assay (BCA).

19. Trizol.

20. 0.25% trypsin.

21. StarvMed (starvation media): DMEM supplemented with 0.5% fetal calf serum, 2 mM L-glutamine, and 100 U/ml penicillin/streptomycin (*see* **Note 9**).

22. M&A (migration and adhesion media): DMEM supplemented with 0.5% bovine serum albumin, 2 mM L-glutamine, and 100 U/ml penicillin/streptomycin.

23. Lysis buffer: 125 mM Tris pH 7.4, 2 mM EDTA, 1% SDS 1% Triton, protease inhibitors. Store at 4 °C.

24. Quenching solution: 75 mM NH_4Cl, 20 mM glycine diluted in D-PBS.

25. Permeabilizing solution: 0.7% fish skin gelatin, 0.16% saponin diluted in D-PBS. Store at 4 °C.

26. Antibody, anti-His2B (Abcam, ab1790).

27. Rhodamine-coupled phalloidin.

3   Methods

3.1 Isolation of Primary Schwann Cells

The isolation of primary Schwann cells was adapted from Brockes et al. [15].

1. Prior to start the dissection, soak the microdissection set in the beaker filled with 70% ethanol. Also, fill three Petri dishes with D-PBS and 100 U/ml penicillin/streptomycin.

2. Cool animals on ice for 10 min before the procedure to minimize pain and distress. Behead one P3 rat, and place the pup in prone position. Spread out the hind limb laterally (such as the body will become an inverted Y) fixing the paw with needles. Wash the skin with ethanol 70%. The nerves lie beneath an imaginary line going diagonally from the toes up to the spine. A bony protuberance, the sciatic notch, can be felt at

about one-third proximally in this line. Cut skin, then muscle and fascia at about the height of the sciatic notch through this diagonal line. The nerve, identifiable as a whitish string, can be seen underneath the muscle.

3. Using two sterile forceps, expose the sciatic nerves by running parallel to the nerve up to the fourth lumbar vertebra and down to the gastrocnemius muscle. Grab the sciatic nerve near the L4 extremity being careful not to damage any major blood vessels and without poking through the peritoneal cavity, which would contaminate the preparation, and then gently pull away the nerve toward the gastrocnemius muscle. Using the dissection microscope, remove any excess muscle and connective tissue from the sciatic nerves. Sever the sciatic nerve, and place it in the Petri dish with D-PBS and 1× penicillin/streptomycin (*see* **Note 10**).

4. Shred the sciatic nerves into small pieces until there is no structural integrity remaining in the nerve (*see* **Note 10**).

5. Combine and pool the nerves in a 15 ml Falcon with 8 ml of D-PBS. Add 1 ml of collagenase and 1 ml of 2.5% trypsin. Incubate for 30 min at 37 °C in a water bath. Mix by inversion during the incubation.

6. Spin at $930 \times g$-force (g) at room temperature for 10 min to pellet the nerves. Pipette and discard the supernatant.

7. Wash the pellet with 5 ml of calf serum media, and spin at $930 \times g$ at room temperature for 5 min. Repeat the washing step a second time.

8. Suspend the pellet in 2 ml of fetal calf serum media by pipetting at least 20 times up and down.

9. Count the cells with the hemocytometer, and plate 1,000,000 cells per poly-L-lysine-coated Petri dishes. Incubate with fetal calf serum media for 24 h to let the cells adhere in the cell incubator.

10. Wash the cells twice with HBSS, add 5 ml of ARA-C media, and incubate for 48 h in the cell incubator.

11. Wash the cells twice with HBSS, add 5 ml of Fetal calf serum media, and incubate for 48 h in the cell incubator.

12. Remove media, and add 5 ml of MSC and incubate for 72 h in the cell incubator.

13. Wash the cells in HBSS, and then wash in HMEM. Add 2 ml of HMEM containing 40 μl of anti-Thy1.1, and incubate for 15 min at 37 °C. Add 400 μl of complement, and incubate for 40 min at 37 °C (*see* **Note 11**).

14. Wash the cells twice with HBSS, add 5 ml of MSC, and incubate for 72 h in the cell incubator.

15. When cells are confluent, remove MSC, and replace with 1 ml of trypsin. Gently swirl to cover the whole plate with trypsin and incubate at 37 °C for 5 min (*see* **Note 12**).

16. When cells are detached, add 9 ml of fetal calf serum media.

17. Spin at 900 × *g* for 5 min, suspend the pellet in 1 ml of pre-warmed MSC, and count cells with a hemocytometer.

18. Adjust the concentration to 500,000 cells/ml by adding pre-warmed MSC.

19. Transfer 1 ml of cell suspension (500,000 cells) per poly-L-lysine-coated Petri dishes containing 7 ml of pre-warmed MSC.

20. Gently swirl the petri dishes, and incubate the dishes in the cell incubator. When cells reach confluency, repeat **steps 13–20** once.

21. Remove MSC and replace with 1 ml of trypsin and incubate at 37 °C for 5 min.

22. When cells are detached, add 9 ml of fetal calf serum media.

23. Spin at 900 × *g* for 5 min and suspend the pellet in 1 ml of pre-warmed MSC.

24. Discard supernatant, suspend 4,000,000 Schwann cells in 1 ml of freezing media, and store at −80 °C (*see* **Note 13**).

3.2 Schwann Cell Myelination Assay

Isolated Schwann cells can myelinate neurons in vitro. However, after being subcultured more than five times, Schwann cells start to lose this ability [16, 17]. Therefore, to identify relevant proteins for Schwann cell-axon interaction, it is important to ensure first that the batch of Schwann cells used in the protrusion assay retain the capability to myelinate axons.

1. Thaw Schwann cells by resuspending them in 9 ml of DMEM. Spin at 900 × *g* for 5 min.

2. Plate Schwann cells on poly-L-lysine-coated Petri dishes, and incubate with MSC for 6 days in the cell incubator at 37 °C to let the cells reach confluence.

3. Wash cells with D-PBS, and incubate for 5 min at 37 °C with 1 ml of 0.25% trypsin.

4. When cells are detached, inactivate the trypsin by adding 9 ml of fetal calf serum media, and then spin at 900 × *g* for 5 min.

5. Suspend cells in 1 ml of C media.

6. Count the Schwann cells, and adjust the concentration to 500,000 cells/ml.

7. Plate 400 μl of Schwann cells per coverslip of DRG neuron culture (*see* **Note 5**).

8. 24 h later, replace the C media by C media PLUS.

9. Change the media every 2 days for 14 days.

10. Remove C media PLUS, and rinse with D-PBS.

11. Fix co-culture for 15 min in 4% paraformaldehyde, and then wash three times with D-PBS.

12. Permeabilize cells by incubation in cold methanol for 10 min, and then rinse twice with D-PBS.

13. Incubate 1 h in MLF blocking then incubate overnight with anti-myelin basic protein (1:1000) and anti-neurofilament M (1:700) diluted in MLF blocking.

14. Wash three times in D-PBS, and then incubate 1 h with anti-mouse Ig 488 (1:1000) and anti-chicken Ig 549 (1:1000) diluted in MLF Blocking.

15. Wash three times in D-PBS and incubate 10 min in DAPI solution.

16. Wash three times in D-PBS, and then mount the coverslip with Vectashield on a glass slide.

17. Seal the coverslip with nail polish, and analyze myelination under the microscope (*see* **Note 14**).

3.3 Isolation of Dorsal Root Ganglia Neurons

The isolation of primary dorsal root ganglia neurons was adapted from Wood et al. [18].

1. Coat two 6-well plates by adding 500 μl of CL solution in each well, and incubate for 15 min. Then, aspirate the CL solution back, and leave behind a thin film of collagen. Let dry, and then add 1 ml of C media.

2. Prior to start the dissection, soak the microdissection set in the beaker filled with 70% ethanol. Also, fill three Petri dishes with L-15 medium.

3. After euthanasia, place the rat in supine position, and cut the lower abdominal skin and muscles. Gently remove the uterus by dissecting away connective tissues.

4. Rinse the uterus with L-15 to remove as much blood as possible, transfer it to a sterile Petri dish with L-15, and place the dish in the dissection hood.

5. Remove all embryos from the uterus by opening the amniotic sac with blunt forceps.

6. Gently grab the embryos by the head, and transfer them to a fresh Petri dish with L-15.

7. For each embryo, under the dissecting microscope, cut off the head at the cervical flexure and the tail just caudal to the hind limbs using micro-dissecting scissors.

8. With the embryo lying on the side, remove all the ventral portion of the embryo to isolate the dorsal structures containing

the spinal cord. Flip the embryo, and proceed similarly on the other side.

9. Remove all viscera.

10. Place one blade of micro-dissecting scissors between the vertebral column and spinal canal at the rostral end, and very carefully cut through vertebral column proceeding caudally, and then tease apart the right and left halves of vertebral column to expose the spinal cord and DRGs.

11. Gently attempt to lift the spinal cord from dorsal structures by grasping the cord at rostral end and pulling perpendicularly.

12. Once the entire spinal cord with attached DRGs is isolated and can be peeled off from the underlying tissues, transfer it to a clean Petri dish containing L-15.

13. After isolating all spinal cords, pluck off DRGs using the #5 forceps, and transfer them to 6-well plates filled with C media. Place 30 DRGs per well.

14. Once all wells are filled with DRGs, remove the C media for 15 min to let DRGs adhere to the bottom of the well (*see* **Note 15**).

15. Add 1 ml of C media with 100 U/ml penicillin/streptomycin in each well, and then gently transfer 6-well plates to the cell incubator.

16. Twenty-four hours after dissection, replace C media by 1.5 ml of NB+ media per well, and then every 2 days, alternate NB+ and NB- media for three cycles (*see* **Note 16**).

3.4 Preparation of Neuronal Membrane Suspension

The preparation of neuronal membranes that maintain biological activity and can behave for a short time like live neurons was first described in 1980 by the Bunge lab [19]. In this study, authors showed that DRG neurons express on their surfaces pro-mitogenic signals for Schwann cells. Using neuronal membranes, it was then showed that neuregulin 1 type III and the PI3-kinase pathway mediate these signals [2, 4, 20]. The protocol below recapitulates the critical steps for the preparation of neuronal membrane suspension.

1. After eliminating the population of glial cells as described in Subheading 3.3, rinse DRG neurons and their network of axons once with ice-cold D-PBS (*see* **Note 17**).

2. With a blunt forceps, push the network of neurons to the center of the well. All the neurons can be collected in one pinch (*see* **Note 18**).

3. For each well of the 6-well plate, transfer the DRG neurons in a 1 ml Dounce homogenizer, and proceed to 30 tight strokes in 200 µl of ice-cold D-PBS.

4. Transfer the homogenized solution in a 1.5 ml vial on ice, and then rinse the pestle with 200 μl of ice-cold D-PBS at the end.

5. Proceed similarly for the second 6-well plate.

6. Centrifuge at 300 × g for 20 min at 4 °C to remove debris and collagen.

7. Ultracentrifuge at 35,000 × g for 1 h at 4 °C.

8. Resuspend the pellet in 800 μl DMEM, vortex, and store at −80 °C (*see* **Note 19**).

3.5 Neuronal Membrane Suspension Activity Assay

Maurel et al. showed that contact with neuronal membranes was sufficient to activate Akt in Schwann cells [4]. Thus, to determine if new batches of neuronal membrane preparations contain active growth factors, it has to be assessed if isolated neuronal membranes have the ability to elicit phosphorylation of Akt in Schwann cells.

1. Plate 400,000 Schwann cells in a well of a 4-well plate coated with poly-L-lysine.

2. Maintain Schwann cells in MSC for 48 h.

3. 12 h prior to the experiment, starve Schwann cells by changing MSC to Starvation media.

4. Thaw neuronal membrane suspension, and suspend them with a vortex.

5. Put Schwann cells on ice, and keep them always on ice (*see* **Note 20**).

6. Remove media from the well, and then add 33 μl of neuronal membrane suspension and 167 μl of DMEM.

7. Spin the 4-well plate at 200 × g for 10 min at 4 °C, and then incubate at 37 °C for 20 min.

8. When the incubation is over, put the 4-well plate on ice, and wash the Schwann cells with ice-cold D-PBS.

9. Lysate the cells with PN1, boil for 5 min, and spin at 16,000 × g for 8 min at 16 °C.

10. Proceed to sodium dodecyl sulfate polyacrylamide gel electrophoresis (SDS-Page) and detect immunoreactive bands with anti-Akt and anti-phospho-Akt antibodies (*see* **Notes 21** and **22**).

3.6 Schwann Cell Protrusions Assay

The Schwann cell protrusions assay is adapted from the method developed by the Klemke lab [21]. The assay uses a Boyden chamber which is separated into lower and upper chambers by a 3-μm porous filter. First, cells are seeded in the top chamber and attached to the microporous filter. Second, a stimulus is added in the bottom chamber. Third, cells respond to the stimulus by becoming polarized and extending protrusions though the 3 μm pores into the bottom chamber. Because cell bodies cannot pass through the 3 μm pores, they remain on the top. Protrusions passing in the

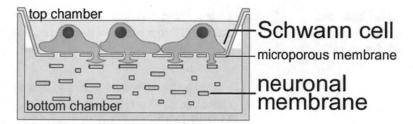

Fig. 1 Schematic illustration of the Boyden chamber used for juxtacrine stimulation of Schwann cells with neuronal membranes

bottom chamber can be separated and their content analyzed. In the original system, cells were stimulated by a chemotactic or haptotactic stimulus on the bottom chamber. This method has been used to isolate novel proteins and RNA in protrusions of migrating cells [22–26]. The following protocol is innovative as it uses cell membranes as stimuli instead of soluble or extracellular matrix molecules (Fig. 1).

3.6.1 Induction of Schwann Cell Protrusions

1. Plate 500,000 Schwann cells in a 10 cm Petri dish coated with poly-L-lysine. Add 7 ml of MSC and incubate at 37 °C for 7 days (*see* **Notes 23** and **24**).

2. Replace the media by StarvMed, 16 h before the protrusion assay (*see* **Note 25**).

3. Coat the transwell with poly-L-lysine for 30 min and rinse once with D-PBS (*see* **Note 26**).

4. Wash Schwann cells twice with D-PBS.

5. Add 1 ml of 2.5% trypsin onto Schwann cells, and incubate for 5 min at 37 °C to detach the cells from the plate.

6. Stop the reaction with 9 ml of MSC. Collect Schwann cells, and spin at $900 \times g$ for 5 min.

7. Suspend the pellet in 1 ml of M&A. Count the cells, and adjust the concentration to 1,500,000 cells/ml.

8. Add 1 ml of the cell suspension in the top chamber of the transwell and 2 ml of DMEM in the bottom chamber of the transwell. Let the Schwann cells attach for 4 h (*see* **Notes 27** and **28**).

9. Thaw the neuronal membrane suspension. Replace the DMEM in the bottom chamber by the neuronal membrane suspension (*see* **Notes 29** and **30**), and let the protrusions grow for 2 h (*see* **Note 31**).

10. Once the protrusions have been induced, wash the top and bottom chambers with D-PBS.

11. Proceed to immunohistochemistry of Schwann cell protrusions (*see* Subheading 3.6.2), RNA extraction (*see* Subheading 3.6.3), or protein extraction (*see* Subheading 3.6.4).

1. To test the extension of Schwann cell protrusion in the bottom chamber, use a cotton swab to remove half of the Schwann cell bodies in the top chamber of the transwell (*see* **Note 32**) (Fig. 2).

2. Place the transwell back in the 6-well. Wash the inside the top chamber of the transwell twice by adding 2 ml and 1 ml of D-PBS in the top and bottom chamber, respectively. Gently swirl the transwell, and remove the D-PBS by pipetting.

3. Add 2 ml and 1 ml of quenching solution in the bottom and top chamber, respectively. Incubate for 10 min at room temperature, and then wash once with D-PBS as in **step 2** (*see* **Note 33**).

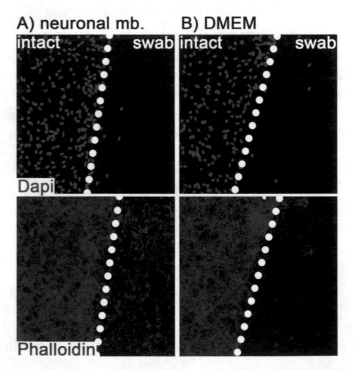

Fig. 2 Immunohistochemistry of Schwann cell protrusions. DAPI and phalloidin staining of Schwann cell protrusions and cell bodies in response to neuronal membrane suspension (**a**, neuronal membrane) or a negative control (**b**, DMEM). In both cases the left side of the membrane is intact, while the cell bodies of the right side on the top chamber were swabbed away (swab). The pseudopods can be seen below the microporous membrane in the swabbed region, because the cell bodies in the upper chamber were removed. The polycarbonate microporous membrane does not block the phalloidin signal in epifluorescent microscopy. The staining with phalloidin can be used to verify that the pseudopods were formed in higher quantity in the membrane-stimulated condition when compared to the DMEM control (no stimulus)

4. Add 2 ml and 1 ml of permeabilizing solution in the bottom and top chamber, respectively. Incubate for 15 min at 37 °C, and then wash once with D-PBS as in **step 2**.

5. Place a piece of Parafilm in a humid cell culture tray. Prepare 150 µl of phalloidin diluted 1:400 in permeabilizing solution. Add a 40 µl drop of the phalloidin solution on the Parafilm. Place the transwell on top of it, and then add the remaining 110 µl in the top chamber. Incubate for 1 h at 37 °C (*see* **Note 34**).

6. Wash twice with D-PBS as in **step 2**.

7. Add 2 ml and 1 ml of permeabilizing solution in the bottom and top chamber, respectively. Incubate for 5 min at room temperature, then twice with D-PBS as in **step 2**.

8. Cut out the membrane of the transwell with micro-dissecting scissors.

9. On a glass microscope slide, put a drop of Vectashield. Grab the membrane with forceps, and place it on the Vectashield drop. Add another drop of Vectashield on the top of the membrane.

10. Cover with a glass coverslip, and seal with nail polish.

11. The production of pseudopods can be evaluated and imaged with an epifluorescence microscope as displayed in Fig. 2 (where half of the top chamber cell bodies were removed). Alternatively, cell bodies and protrusions can be observed in confocal microscopy without swab of the cell bodies from the top chamber, as displayed in Fig. 3.

3.6.3 Extraction of RNA

1. Fix the cell bodies and protrusions with cold methanol for 30 min, by adding 2 ml and 1 ml of methanol in the top and bottom chamber, respectively, to reduce most of the protease and phosphatase activity (*see* **Note 35**).

2. Remove the methanol by pipetting, and wash the inside of the top chamber of the transwell twice by adding 2 ml and 1 ml of D-PBS in the top and bottom chamber, respectively. Gently swirl the transwell. Do not remove the PBS after the second wash.

3. To collect the protrusion fraction, remove the Schwann cell bodies with a cotton swab. Apply a gentle pressure inside the transwell to detach the Schwann cell bodies. Rinse the inside of the transwell with D-PBS, and remove the D-PBS by pipetting. Repeat the procedure until all cell debris floating in the upper chamber are removed.

4. Proceed similarly by using a pointed cotton swab to detach the cell bodies at the edge of the transwell (*see* **Note 36**).

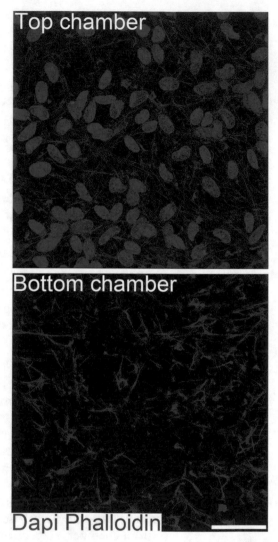

Fig. 3 Confocal imaging of Schwann cell protrusions in response to neuronal membranes. Cells were stained with DAPI (blue) and phalloidin (red). Progressive Z-sections were collected with a step size of 0.15 μm from the whole top or bottom chamber and stacked in Z-projection. No nuclei are seen in the bottom chamber. If using confocal microscopy, there is no need to swap the upper chamber. Bar = 25 μm

5. Alternatively, to collect the cell body fraction, flip over the transwell, and remove the Schwann cell protrusions with a cotton swab.

6. Cut the membrane of the transwell with micro-dissecting scissors (*see* **Note 37**).

7. Place the membrane on a piece of Parafilm with the protrusion side up.

8. Add 100 μl of Trizol on the membrane.

9. Hold the membrane with a forceps, and gently use a cell scraper to free the Schwann cell protrusions (*see* **Notes 38** and **39**).

10. Collect carefully the Trizol in a 1.5 ml vial, and proceed to RNA isolation (*see* **Note 40**).

3.6.4 Extraction of Proteins

1. Perform **steps 1–6** as described above for extraction of RNA (Subheading 3.6.3).

2. Transfer the membrane to the bottom of a 1.5 ml vial with 100 μl of lysis buffer.

3. Chop the membrane with microdissection scissors (*see* **Note 41**).

4. Sonicate for four cycles of 20 s, and place it on ice for 30 s between each cycle.

5. Place the vial on ice for 30 min, and then spin at $16,000 \times g$ for 30 min at 4 °C.

6. Transfer carefully the supernatant to a new 1.5 ml vial, and measure protein concentration by BCA (*see* **Note 42**).

7. Proceed to SDS page analysis for nuclear marker (i.e., His2b) to assess the purity of the Schwann cell protrusion fraction (*see* **Note 43**).

4 Notes

1. Wash all instruments with sterile water, followed by ethanol 70% then autoclave.

2. The gauze at the bottom of the beaker protects the tools from being damaged. Alternatively, a plastic tip can be used as a protecting sleeve for the tools.

3. The average litter size of the Sprague-Dawley rat is ten pups.

4. The collagenase can be stored in 1 ml aliquots at −20 °C for up to 6 months.

5. To prepare coverslips of dorsal root ganglia neurons, proceed as indicated in Subheading 3.3, but instead of plating ganglia in 6-well plates, seed 1 ganglion per 18 mm coverslip in a 4-well plate. As for 6-well plates, coverslips have to be coated in CL solution for 15 min.

6. The CL solution has to be fresh. If the CL solution is too old, it can affect the capability of neurons to adhere to the coverslips. Use a 10% solution for the first month after opening the collagen bottle and a 25% solution between the second and third months. Discard the collagen if it is older than 3 months.

7. To obtain E15.5 embryos, mate a male and female rat on the evening. The next morning, check for a vaginal plug, and if

the female is pregnant, the pups will be at E0.5. Proceed to the dissection exactly 15 days later, in the morning. Dissection can also be performed on the afternoon of the same day (~E16). At E15, the embryo is too soft, which complicates the pulling of the spinal cord from the rest of the tissues. At E16.5, DRGs have started to send axons in the peripheral tissues, and the DRGs will detach from the spinal roots when pulling the spinal cord out.

8. The 3 μm pore of the transwell membrane allows Schwann cells to extend protrusions while blocking their 5 μm nucleus from passing on the lower side of the membrane.

9. StarvMed is different in composition from starvation media in Subheading 2.5.

10. Proceed to these steps as quickly and efficiently as possible. The less time used, the greater the yield of Schwann cells.

11. This procedure selectively removes fibroblasts from the Petri dishes. Isolated fibroblasts, or patches of fibroblasts rounding up and then detaching from the plates, should be observable. Healthy Schwann cells have a slender, elongated, bipolar morphology, exhibit contact-dependent inhibition of proliferation, and, when they are in a confluent monolayer, present swirling patterns.

12. Monitor the detachment of SC from the plate using a light microscope. A gentle tap of the side of the flask can help to determine the proper detachment of the cells from the plate.

13. One litter of rat will give about 60 millions of Schwann cells.

14. An efficient myelination should be reflected by at least 1000 segments of myelin per coverslip. Several factors can affect the efficiency of myelination, i.e., the composition of the fetal calf serum and the presence of mycoplasma. Therefore, it is recommended when trying to myelinate axons with Schwann cells for the first time to test several batch of serum and to test for the presence of mycoplasma.

15. Close the plate with the cap to avoid DRGs to dry out. Drying out DRGs increase cell death.

16. NB+ contains fluorodeoxyuridine, an inhibitor of cellular proliferation that blocks DNA synthesis. Schwann cell and fibroblast proliferation is blocked, and fibroblasts and Schwann cells are eliminated from the culture. Cycling NB+ and NB media to obtain a pure neuronal culture takes 12 days.

17. Membrane suspensions can be prepared from non-neuronal cells (i.e., CHO, COS-7) to be used as a negative control. Non-neuronal membranes do not activate Akt in Schwann cells (*see* Subheading 3.5) [13].

18. The collection of neurons can be done macroscopically. However, it is convenient for the first time to do it under a

microscope to appreciate how the network of neurons is moved by the forceps.

19. Neuronal membrane suspension can be stored up to 6 months at −80 °C, and efficiently activate Akt in Schwann cell (*see* Subheading 3.5). A longer storage has not been tested. It is also recommended to pool several neuronal membrane suspension preparations to reach a volume of 2 ml or above. This ensures to have enough membrane suspension to test its activity (*see* Subheading 3.5) and to perform several protrusion assays (Subheading 3.6). Finally, do not freeze/thaw the neuron membrane suspension more than one time.

20. It is critical to keep Schwann cells on ice, as this limits the activation of Akt.

21. Non-starved Schwann cells and Schwann cell in contact with MSC for 20 min can be used as a positive control for the activation of Akt.

22. As neuronal membranes are not quantified after preparation, it is critical to test that neuronal membranes have the potency to activate Akt.

23. For protein or RNA extraction (*see* Subheadings 3.6.3 and 3.6.4), 12 Petri dishes of Schwann cells are necessary.

24. After 7 days, Schwann cells should have reached confluence, and a Petri dish should contain around 4,000,000 Schwann cells.

25. The starvation of Schwann cells causes 30–40% of cell death.

26. The coating of the transwell with poly-L-lysine is done the day prior to the protrusion assay.

27. As the polycarbonate membrane is opaque, it is not possible to appreciate the health and adhesion of the Schwann cells until the end of the immunochemistry assay (Subheading 3.6.2).

28. For protein or RNA extraction (*see* Subheadings 3.6.3 and 3.6.4), two 6-well plates are necessary (6 wells will be induced with neuronal membranes in the bottom chamber, and 6 wells will be used as control with DMEM in the bottom chamber).

29. The neuronal membranes are the limiting reagent in this experiment. It is possible to reduce the quantity of neuronal membrane suspension to use in the bottom chamber by diluting the neuronal membrane suspension. However, the dilution has to activate efficiently Akt in Schwann cells (*see* Subheading 3.5). Another approach is to reduce the volume of the bottom chamber by lowering the distance between the microporous membrane and the floor of the bottom chamber. This can be done by using a laser cutting machine to reduce the height of the walls supporting the transwell (Fig. 1, vertical structures in gray). However, a minimal distance of 200 μm between the microporous membrane and the floor of the bottom chamber

is necessary to prevent Schwann cells protrusions, 25 μm on average, to contact the bottom of the well.

30. When putting the neuronal membrane suspension in the bottom chamber, remove the transwell, add the suspension, and replace the transwell. Avoid creating air bubble between the transwell and the bottom of the well, as they can disrupt the juxtacrine interaction between neuronal membranes and Schwann cells. Several conditions can be used in parallel to neuronal membrane suspension. DMEM and non-neuronal membranes are good negative controls. MSC and lysophosphatidic acid (100 ng/ml) are efficient positive controls for the induction of Schwann cell protrusions.

31. Two hours of induction is optimal for the production of Schwann cell protrusions [13], similar to other cell types [21, 25, 27]. After this time, pseudopods retract. It is critical to not move the plate nor open the incubator for these 2 h as any movement might disrupt the contact that is forming between Schwann cells and neuronal membranes.

32. Removing only half of the cell bodies allows the visualization of cell bodies on one half (non-swabbed region) and the visualization of protrusions by epifluorescence microscopy on the other half (swabbed region), because the cell bodies in the upper chamber were removed. A similar approach can be adapted for immunolocalization of proteins.

33. The quenching solution is used to reduce autofluorescence.

34. The piece of Parafilm is essential as it will retain the drop of phalloidin solution by surface tension, and avoid any leakage.

35. For RNA and protein extraction, 6 transwells need to be prepared. One will be used for the biochemistry assay on the cell body fraction, four for biochemistry assay on the protrusion fraction, and one for the immunohistochemistry of Schwann cell protrusions.

36. Be very careful. If the integrity of the microporous membrane is affected, it could contaminate the protrusion fraction with the Schwann cell body fraction.

37. Removing the cell bodies properly is critical for the purity of the protrusion fraction. The cell bodies on the edge are often the hardest to remove. When cutting out the membrane, leave a 1 mm edge behind to reduce possible contamination.

38. Because cell bodies are swabbed before cutting the membrane, one transwell cannot be used to isolate both cell body and protrusion fractions. Similarly when cell bodies need to be isolated, pseudopods need to be swabbed before and cannot be isolated.

39. While scraping, avoid the formation of bubbles. Bubbles will reduce the amount of Trizol that can be recovered.

40. From one 24 mm transwell, the RNA concentration will be around 100 ng. It is possible to increase this concentration by scraping several transwell membranes with the same Trizol suspension, which will increase the final yield of RNA.

41. An approach similar to the RNA isolation with a cell scraper can also be done for the proteins. However, as the SDS increases the formation of bubbles, it can limit the amount of Lysis buffer recovered after scrapping. Also the sonication method allows to collect the protrusions proteins inside the transwell membrane (10 μm thick) and therefore to increase the final yield of proteins.

42. From one 24 mm transwell, the protein concentration is around 0.15 g/l. It is possible to increase this concentration by pooling multiple transwells.

43. There is no good loading control for proteins that are equally distributed between the cell body and the pseudopod fraction. Therefore, Amido Black, Coomassie, or Silver staining are recommend to visualize equal loading of proteins. Histones, present at around 15 kDa, are enriched in the cell body fraction, as shown in Fig. 4. The presence of Histones in the protrusion fraction indicates an incomplete removal of the cell bodies in the top chamber.

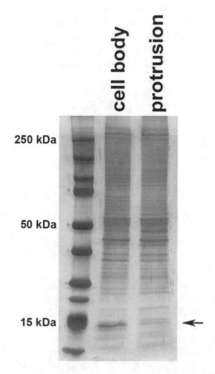

Fig. 4 Silver staining of proteins in the Schwann cell body and protrusion fractions. The silver staining shows lysates from Schwann cell bodies or protrusions separated on SDS-PAGE. Histones are observed at 15 kDa in the cell body fraction (arrow)

Acknowledgments

This work was supported by grants from the National Institute of Health R01NS045630 and R21HD075363 (to M.L.F). We thank current and past members of the Feltri's laboratory and our many collaborators at San Raffaele Scientific Institute, SUNY, at Buffalo, and the University of Salento for help and suggestions.

References

1. Feltri ML, Poitelon Y, Previtali SC (2016) How Schwann cells sort axons: new concepts. Neuroscientist 22(3):252–265. https://doi.org/10.1177/1073858415572361

2. Taveggia C, Zanazzi G, Petrylak A, Yano H, Rosenbluth J, Einheber S, Xu X, Esper RM, Loeb JA, Shrager P, Chao MV, Falls DL, Role L, Salzer JL (2005) Neuregulin-1 type III determines the ensheathment fate of axons. Neuron 47(5):681–694

3. Michailov GV, Sereda MW, Brinkmann BG, Fischer TM, Haug B, Birchmeier C, Role L, Lai C, Schwab MH, Nave KA (2004) Axonal neuregulin-1 regulates myelin sheath thickness. Science 304(5671):700–703. https://doi.org/10.1126/science.1095862

4. Maurel P, Salzer JL (2000) Axonal regulation of Schwann cell proliferation and survival and the initial events of myelination requires PI 3-kinase activity. J Neurosci 20(12): 4635–4645

5. Nishino J, Saunders TL, Sagane K, Morrison SJ (2010) Lgi4 promotes the proliferation and differentiation of glial lineage cells throughout the developing peripheral nervous system. J Neurosci 30(45):15228–15240. https://doi.org/10.1523/JNEUROSCI.2286-10.2010

6. Ozkaynak E, Abello G, Jaegle M, van Berge L, Hamer D, Kegel L, Driegen S, Sagane K, Bermingham JR Jr, Meijer D (2010) Adam22 is a major neuronal receptor for Lgi4-mediated Schwann cell signaling. J Neurosci 30(10):3857–3864. https://doi.org/10.1523/JNEUROSCI.6287-09.2010

7. Trimarco A, Forese MG, Alfieri V, Lucente A, Brambilla P, Dina G, Pieragostino D, Sacchetta P, Urade Y, Boizet-Bonhoure B, Martinelli Boneschi F, Quattrini A, Taveggia C (2014) Prostaglandin D2 synthase/GPR44: a signaling axis in PNS myelination. Nat Neurosci 17(12):1682–1692. https://doi.org/10.1038/nn.3857

8. Woodhoo A, Alonso MB, Droggiti A, Turmaine M, D'Antonio M, Parkinson DB, Wilton DK, Al-Shawi R, Simons P, Shen J, Guillemot F, Radtke F, Meijer D, Feltri ML, Wrabetz L, Mirsky R, Jessen KR (2009) Notch controls embryonic Schwann cell differentiation, postnatal myelination and adult plasticity. Nat Neurosci 12(7):839–847. https://doi.org/10.1038/nn.2323

9. Lewallen KA, Shen YA, De la Torre AR, Ng BK, Meijer D, Chan JR (2011) Assessing the role of the cadherin/catenin complex at the Schwann cell-axon interface and in the initiation of myelination. J Neurosci 31(8):3032–3043. https://doi.org/10.1523/JNEUROSCI.4345-10.2011

10. Maurel P, Einheber S, Galinska J, Thaker P, Lam I, Rubin MB, Scherer SS, Murakami Y, Gutmann DH, Salzer JL (2007) Nectin-like proteins mediate axon Schwann cell interactions along the internode and are essential for myelination. J Cell Biol 178(5):861–874. https://doi.org/10.1083/jcb.200705132

11. Spiegel I, Adamsky K, Eshed Y, Milo R, Sabanay H, Sarig-Nadir O, Horresh I, Scherer SS, Rasband MN, Peles E (2007) A central role for Necl4 (SynCAM4) in Schwann cell-axon interaction and myelination. Nat Neurosci 10(7):861–869. https://doi.org/10.1038/nn1915

12. Wanner IB, Wood PM (2002) N-cadherin mediates axon-aligned process growth and cell-cell interaction in rat Schwann cells. J Neurosci 22(10):4066–4079. doi:20026406

13. Poitelon Y, Bogni S, Matafora V, Della-Flora Nunes G, Hurley E, Ghidinelli M, Katzenellenbogen BS, Taveggia C, Silvestri N, Bachi A, Sannino A, Wrabetz L, Feltri ML (2015) Spatial mapping of juxtacrine axo-glial interactions identifies novel molecules in peripheral myelination. Nat Commun 6:8303. https://doi.org/10.1038/ncomms9303

14. Wang Y, Ding SJ, Wang W, Yang F, Jacobs JM, Camp D II, Smith RD, Klemke RL (2007)

Methods for pseudopodia purification and proteomic analysis. Sci STKE 2007(400):pl4. https://doi.org/10.1126/stke.4002007pl4

15. Brockes JP, Fields KL, Raff MC (1979) Studies on cultured rat Schwann cells. I. Establishment of purified populations from cultures of peripheral nerve. Brain Res 165(1):105–118

16. Feltri ML, Scherer SS, Wrabetz L, Kamholz J, Shy ME (1992) Mitogen-expanded Schwann cells retain the capacity to myelinate regenerating axons after transplantation into rat sciatic nerve. Proc Natl Acad Sci U S A 89(18):8827–8831

17. Porter S, Clark MB, Glaser L, Bunge RP (1986) Schwann cells stimulated to proliferate in the absence of neurons retain full functional capability. J Neurosci 6(10):3070–3078

18. Wood PM, Bunge RP (1975) Evidence that sensory axons are mitogenic for Schwann cells. Nature 256(5519):662–664

19. Salzer JL, Williams AK, Glaser L, Bunge RP (1980) Studies of Schwann cell proliferation. II. Characterization of the stimulation and specificity of the response to a neurite membrane fraction. J Cell Biol 84(3):753–766

20. Ratner N, Bunge RP, Glaser L (1985) A neuronal cell surface heparan sulfate proteoglycan is required for dorsal root ganglion neuron stimulation of Schwann cell proliferation. J Cell Biol 101(3):744–754

21. Cho SY, Klemke RL (2002) Purification of pseudopodia from polarized cells reveals redistribution and activation of Rac through assembly of a CAS/Crk scaffold. J Cell Biol 156(4):725–736. https://doi.org/10.1083/jcb.200111032

22. Lin YH, Park ZY, Lin D, Brahmbhatt AA, Rio MC, Yates JR III, Klemke RL (2004) Regulation of cell migration and survival by focal adhesion targeting of Lasp-1. J Cell Biol 165(3):421–432. https://doi.org/10.1083/jcb.200311045

23. Wang Y, Kelber JA, Tran Cao HS, Cantin GT, Lin R, Wang W, Kaushal S, Bristow JM, Edgington TS, Hoffman RM, Bouvet M, Yates JR III, Klemke RL (2010) Pseudopodium-enriched atypical kinase 1 regulates the cytoskeleton and cancer progression [corrected]. Proc Natl Acad Sci U S A 107(24):10920–10925. https://doi.org/10.1073/pnas.0914776107

24. Wang Y, Klemke RL (2007) Biochemical purification of pseudopodia from migratory cells. Methods Mol Biol 370:55–66. https://doi.org/10.1007/978-1-59745-353-0_5

25. Mili S, Moissoglu K, Macara IG (2008) Genome-wide screen reveals APC-associated RNAs enriched in cell protrusions. Nature 453(7191):115–119. https://doi.org/10.1038/nature06888

26. Howe AK, Baldor LC, Hogan BP (2005) Spatial regulation of the cAMP-dependent protein kinase during chemotactic cell migration. Proc Natl Acad Sci U S A 102(40):14320–14325. https://doi.org/10.1073/pnas.0507072102

27. Thomsen R, Lade Nielsen A (2011) A Boyden chamber-based method for characterization of astrocyte protrusion localized RNA and protein. Glia 59(11):1782–1792. https://doi.org/10.1002/glia.21223

In Vitro Analysis of the Role of Schwann Cells on Axonal Degeneration and Regeneration Using Sensory Neurons from Dorsal Root Ganglia

Rodrigo López-Leal, Paula Diaz, and Felipe A. Court

Abstract

Sensory neurons from dorsal root ganglion efficiently regenerate after peripheral nerve injuries. These neurons are widely used as a model system to study degenerative mechanisms of the soma and axons, as well as regenerative axonal growth in the peripheral nervous system. This chapter describes techniques associated to the study of axonal degeneration and regeneration using explant cultures of dorsal root ganglion sensory neurons in vitro in the presence or absence of Schwann cells. Schwann cells are extremely important due to their involvement in tissue clearance during axonal degeneration as well as their known pro-regenerative effect during regeneration in the peripheral nervous system. We describe methods to induce and study axonal degeneration triggered by axotomy (mechanical separation of the axon from its soma) and treatment with vinblastine (which blocks axonal transport), which constitute clinically relevant mechanical and toxic models of axonal degeneration. In addition, we describe three different methods to evaluate axonal regeneration using quantitative methods. These protocols constitute a valuable tool to analyze in vitro mechanisms associated to axonal degeneration and regeneration of sensory neurons and the role of Schwann cells in these processes.

Key words Dorsal root ganglion, Sensory neurons, Schwann cell, Axonal degeneration, Axonal regeneration, Axonal growth

1 Introduction

Regeneration in the peripheral nervous system (PNS) has been widely used to identify factors associated to axonal regeneration, including regeneration-associated genes and changes in the tissular microenvironment that either inhibit or promote a regenerative response. This knowledge has been used to develop strategies to enhance axonal regeneration in neurons from the central nervous system (CNS) with reduced regenerative capability [1]. In the PNS, neurons have the ability to regenerate axons after damage mostly due to intrinsic neuronal properties as well as the ability of Schwann cells (SCs) to promote a permissive environment for

Paula V. Monje and Haesun A. Kim (eds.), *Schwann Cells: Methods and Protocols*, Methods in Molecular Biology, vol. 1739, https://doi.org/10.1007/978-1-4939-7649-2_16, © Springer Science+Business Media, LLC 2018

axonal growth [2]. These pro-regenerative conditions are initially triggered by axonal degeneration in a tissular program known as Wallerian degeneration [3]. Axonal regeneration is the underlying process for functional recovery after nervous system damage due to mechanical, toxic, or inflammatory processes [4], and involves a programmed death mechanism different from canonical apoptosis and associated to mitochondrial dysfunction [5, 6]. Importantly, axonal degeneration is an early event in several neurodegenerative conditions; therefore it represents a target for neuroprotection.

The dorsal root ganglion (DRG) corresponds to peripheral ganglia that contain sensory neurons. These neurons are pseudo-unipolar and project their axons to the periphery as well as to the CNS. DRGs are located close to the spinal cord as a pair in each spinal cord segment. These neurons represent an excellent model to study axonal regeneration, as their peripheral axons regenerate efficiently, but their central projections have poor regenerative capabilities [7, 8]. In addition, DRGs are easy to dissect from embryos and adult animals, and from them a high amount of neuronal somas with well-characterized populations of different sensory modalities can be isolated. Furthermore, DRG neurons can be cultured in association with SCs or depleted of them. These particularities of the system allow the analysis of signalling mechanisms between neurons and SCs in physiological and pathological conditions.

Here, we present a detailed protocol to obtain and culture DRG neurons along with associated models to study axonal degeneration after two pro-degenerative stimuli in vitro: axotomy and vinblastine treatment. Axotomy of neurons enables the study of intrinsic regenerative capacities of injured axons in a controlled, reproducible, and growth-permissive environment [9–11]. In addition, axonal transport along a peripheral nerve can be disrupted by local application of antineoplastic agents such as vinblastine, leading to axonal degeneration. Vinblastine effectively blocks axonal transport by binding to tubulin and inhibiting microtubule formation [12–14] and represents a well-described model of chemotherapy-induced axonal degeneration. We also describe various protocols to study axonal regeneration, including axonal growth in intact DRG explants, which enables a direct readout of axonal regeneration by immunofluorescence (IF). Regeneration after mechanical axotomy and regeneration after hypotonic lysis of the axons in compartmentalized Campenot chambers [15] also allow the study of axon growth dynamics after damage in mature DRG explants that have been grown for several days. These methods allow the use of genetic or pharmacological modifications such as shRNA or drugs with possible therapeutic applications. We also include qualitative and quantitative methods to evaluate axonal degeneration and regeneration in these model systems.

2   Materials

2.1   Primary DRG Cultures from Day 16 Embryonic Rats

2.1.1   Reagents and Equipment

1. Blunt forceps and rat tooth pickup forceps.

2. One surgical scissors, one curved operating scissors, and two dissecting scissors.

3. Two fine Dumont #5 forceps.

4. Dissecting microscope.

5. Biological safety cabinet certified for level II work.

6. Cell culture incubator (humidified, 37 °C, 5% CO_2).

7. Laminar flow hood.

8. Fluorescence microscope.

9. Sterilized 6- and 10-cm cell culture dishes.

10. Campenot chambers prepared according to the method of Campenot and Lund (2009) [15].

11. One pregnant (E16) Sprague-Dawley rat (~10 embryos).

12. Sterile 24-well cell culture plate.

13. Sterile 13-mm glass coverslips.

14. Disposable biopsy punch, 4.0 mm diameter.

15. Chicken anti-neurofilament medium subunit antibody (Chemicon, Cat# AB5735).

16. Rabbit anti-neurofilament heavy subunit antibody (Sigma, Cat# N4142).

17. Mouse anti-S100β antibody (Sigma, Cat# S2532).

2.1.2   Media and Supplements for Dish Coating

The DRG explants are plated and grown onto coverslips coated with poly-L-lysine (PLL) and collagen. The coated surface is prepared using the materials and protocols described as follows:

1. 10 mg/ml PLL stock solution. Reconstitute 100 mg of PLL in 10 ml of ultrapure water. Sterilize by filtration and store in aliquots at 4 °C for at least 1 year.

2. PLL-coated coverslips. Dilute PLL stock solution in sterile distilled water (1:100). Add 500 μl of this solution onto sterile coverslips placed inside a sterile 24-well plate. Incubate the coverslips with PLL at 37 °C for at least 30 min. After that, wash the coverslips twice with sterile distilled water. Then, continue with collagen treatment (*see* **Note 1**).

3. Collagen I, rat tail solution. Stock concentration 3 mg/ml. Dilute the collagen to 50 μg/ml in 20 mM acetic acid at the final volume needed.

4. Collagen-coated coverslips. Transfer 500 μl of collagen I, rat tail solution onto coverslips placed inside a sterile 24-well cell

culture plate. Leave the lid of the culture dish open inside the hood so as to allow drying of the collagen prior to use. Rinse dishes three times with equal volumes of sterile 1× PBS or media to remove the acid (*see* **Note 2**).

5. 200 mM L-glutamine (100×). Store at −20 °C for up to 24 months.

6. Leibovitz's L-15 medium (L-15 medium).

7. Neurobasal medium.

8. B-27 supplement (50×).

9. Neuronal growth factor 2.5S (NGF). 100 µg/ml stock concentration. 50 ng/ml work concentration. Diluted 2000× in explant medium immediately before use.

10. 1 mM 5-fluoro-2′-deoxyuridine stock solution. Reconstitute 2 mg of 5-fluoro-2′-deoxyuridine in 1 ml ultrapure water. Filter and store in aliquots at −20 °C for at least 1 year.

11. 3 mM aphidicolin stock solution. Reconstitute 1 mg of aphidicolin in 1 ml of dimethyl sulfoxide (DMSO). Filter and store in aliquots at −20 °C for at least 1 year.

12. 100× antibiotic-antimycotic. 10,000 units/ml of penicillin, 10,000 µg/ml of streptomycin, and 25 µg/ml of amphotericin B.

13. 1 mM vinblastine stock solution. Reconstitute 90 µg of vinblastine in 100 µl sterile DMSO. Store in aliquots at −20 °C for a maximum of 1 year.

14. 1× phosphate buffer saline (1× PBS). To prepare 1 l of 10× PBS, dissolve 80 g NaCl, 2 g KCl, 14.4 g Na_2HPO_4, and 2.4 g KH_2PO_4, in 800 ml of distilled water. Adjust the pH to 7.4 with HCl and the volume to 1 l with distilled water. Sterilize by autoclaving.

15. 4% (w/v) paraformaldehyde (PFA). Dissolve 4 g of PFA powder in 90 ml of 1× PBS, and heat to 60 °C while stirring. Add drops of 1 N NaOH until the solution becomes clear. Complete the volume to 100 ml with 1× PBS. Filter and store at −20 °C for 6 months, thaw aliquots immediately before use, and do not refreeze.

16. Blocking solution. Dilute gelatin from cold water fish skin to 5% (v/v) in 1× PBS. Add Triton X-100 to 0.1% (v/v).

17. Antibody solution. Dilute gelatin from cold water fish skin to 1% (v/v) in 1× PBS. Add Triton X-100 to 0.1% (v/v).

18. 5 mg/ml 4′,6-diamidino-2-phenylindole dihydrochloride (DAPI) stock solution. Dissolve 10 mg of DAPI in 2 ml of deionized water. Store in aliquots at −20 °C for at least 2 years protected from light.

19. Explant medium. Combine 0.5 ml of 2 mM L-glutamine, 0.5 ml of antibiotic-antimycotic (100×), 1.0 ml B-27 supplement (50×), and 25 µl of NGF, and bring up to 50 ml with Neurobasal medium. This medium should be used fresh.

20. Explant medium with antimitotic agent. Combine 125 µl 5-fluoro-2′-deoxyuridine and 62.5 µl of aphidicolin in 50 ml of explant medium. This medium should be used fresh. It is used to eliminate non-neuronal cells such as SCs and fibroblasts.

21. Mounting medium.

3 Methods

3.1 Primary DRG Cultures from Day 16 Embryonic Rats

One day before the dissection, place one sterilized coverslip on each 24-well cell culture plate, and coat each coverslip once with PLL and collagen I as described in Subheading 2.1.2.

3.1.1 Dissection and Isolation of DRGs

1. Sacrifice one pregnant rat at gestational day 16 (E16) by CO_2 asphyxiation or according to local regulations. All animal procedures must be carried out in accordance to the institution's animal care committee guidelines.

2. Place the rat on its back on absorbent paper, and thoroughly rinse the abdomen with 70% (v/v) ethanol.

3. Use pickup forceps to grasp and lift the lower abdominal skin at midline.

4. Cut the skin with surgical scissors in an "I" pattern to expose the abdominal muscles.

5. Lift up the muscles and abdominal wall with pick-up forceps, and make a transverse incision with small dissecting scissors taking care not to puncture the viscera.

6. With blunt forceps, grasp the uterus laden with embryos near the cervical insertion, and gently lift it from the peritoneal cavity.

7. Cut away the connective tissue and suspensory ligaments, and place the entire uterus in a sterile 10 cm tissue culture dish.

8. Remove the embryos from the uterus by cutting through the uterus and then through the amniotic sacs. Gently squeeze each embryo through this last incision (*see* **Note 3**).

9. Transfer all embryos into another 10 cm dish containing L-15 medium. Swirl the dish to clean the embryos of blood and debris, and divide them into two 10 cm dishes containing fresh L-15 medium. Next, transfer the individual embryos to a 6 cm dish containing a thin layer of L-15 medium to isolate the DRGs (*see* **Note 4**).

10. Working under a dissecting microscope, lay each embryo on its side. Cut off the head at the cervical flexure and the tail just caudal to the hind limbs using micro-dissecting scissors.

11. Remove the ventral (belly) portion of the embryo to isolate the dorsal (back) structures containing the spinal cord.

12. Position the tissue dorsal side down. And carefully remove any remaining viscera from the posterior wall.

13. Place one blade of micro-dissecting scissors between the vertebral column and the spinal canal at rostral end, and very carefully cut through the vertebral column proceeding caudally. Then, tease apart the right and left halves of the vertebral column to expose the spinal cord and DRGs.

14. Gently lift the spinal cord from dorsal structures by grasping the cord at rostral end while carefully cutting behind and around the DRGs to sever adherent tissues.

15. Transfer all the spinal cords with attached DRGs to another 6 cm dish containing L-15 medium immediately after isolating them from the embryos.

3.1.2 DRG Explant Cultures

1. Add 500 μl of explant medium or explant medium with antimitotic agent per coverslip, and swirl the 24-well cell culture plate to cover the entire surface (*see* **Note 5**).

2. Using dissecting forceps, deposit one DRG onto the center of each coated coverslip. Then gently transfer the plate containing the DRGs into the cell culture incubator. If not done gently, DRGs can move to the side of the well and start floating. DRG will attach permanently to the bottom of the well after approximately 1 h.

3. Grow explant cultures in the cell culture incubator.

4. Change the explant medium (or the explant medium with antimitotic agent) on the fourth day after plating (*see* **Note 6**).

3.2 DRG Explant Immunofluorescence

Immunofluorescence is a common laboratory technique, which is based on the use of antibodies that recognize specific antigens. The antibody-antigen complex is then identified by a secondary antibody chemically conjugated to a fluorescent dye, which can be later detected using a fluorescence microscope. The procedure for IF staining of DRGs includes fixation, permeabilization, blocking of unspecific binding sites, and incubation with primary antibodies followed by secondary antibodies and specimen mounting. The protocol described below has been successfully used to study axonal regeneration and degeneration in vitro [8, 16].

1. Remove the culture medium from wells containing the DRG explants.

2. Wash the cells 2–3 times with 1× PBS every 5 min at room temperature (RT).

3. Fix the cells with 4% PFA for 20 min at RT.

4. Remove the PFA, and wash the cells by adding 1× PBS three times every 5 min at RT.

5. Block and permeabilize the cells for 2 h at RT with blocking solution in a humidified chamber.

6. Incubate the cells with primary antibodies diluted in antibody solution overnight at 4 °C in a humidified chamber.

7. Remove the primary antibody, and wash the cells with 1× PBS three times every 5 min at RT.

8. Incubate the cells with secondary antibodies diluted in antibody solution for 2 h at RT in a dark place in a humidified chamber.

9. Remove the secondary antibodies, and wash the cells with 1× PBS three times every 5 min at RT. If nuclear staining is desired, replace the second 1× PBS washing step with DAPI (stock concentration 5 mg/ml) diluted 1:50,000 and incubate for 5 min at RT. Then, proceed with the third 1× PBS washing step.

10. Mount the coverslips with mounting medium.

11. Samples are ready to be observed by fluorescence microscopy (Fig. 1) (*see* **Note 7**).

3.3 Analysis of Axonal Degeneration

Axonal degeneration in vitro can be evaluated by two methods, mechanical axotomy and vinblastine treatment. The readout is performed by IF (*see* **Note 8**) [9–14].

3.3.1 Axotomy

1. Culture the DRGs in explant medium (with or without antimitotic agent) for 7 days.

2. Axotomy is performed by excising the ganglia using a micropipette tip to eliminate all neuronal somas from the explanted DRG culture. Axonal degeneration is seen after 6 to 24 h after damage.

3. Perform IF staining (*see* Subheading 3.1).

4. Perform study of axonal degeneration as described in [5, 16], (*see* **Note 9**).

3.3.2 Vinblastine

1. Culture the DRGs in explant medium (with or without antimitotic agent) for 7 days.

2. Treat the DRG cultures with 1 μM vinblastine. Axonal degeneration is seen 24 h after treatment.

3. Perform **steps 3** and **4** as described above (*see* Subheading 3.3.1). Results from a typical experiment are provided in Fig. 2.

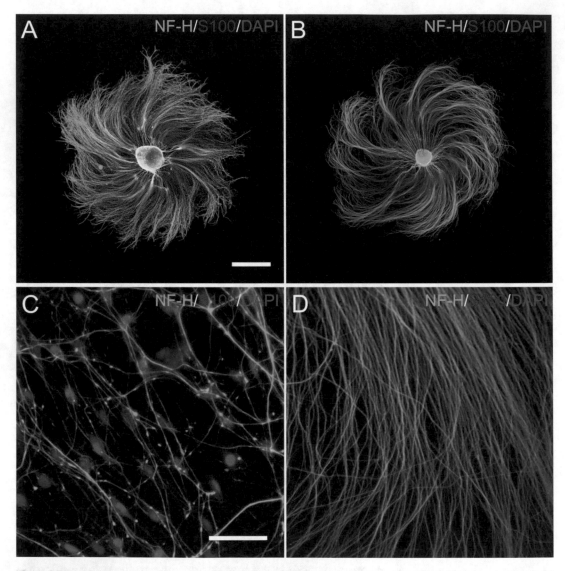

Fig. 1 DRG explant cultures. Representative images of DRGs grown for 4 days in vitro with (**a**) and without (**b**) SCs and immunostained for neurofilament heavy subunit (NF-H, green), the SC marker S100β (red) and the nuclear counterstain DAPI (blue). (**c**, **d**) Higher magnification of distal axons from **a** and **b**, respectively. Scale bars, 400 and 100 μm, for **a**, **b** and **c**, **d**, respectively

3.4 Analysis of Axonal Growth and Regeneration

Axonal regeneration can be evaluated by three methods: (1) assessment of axonal growth after plating intact DRG explants (Subheading 3.4.1), (2) axonal regeneration after allowing DRG axons to grow and then performing an in vitro transection (Subheading 3.4.2), and (3) axonal regeneration in DRG explants plated in compartmentalized chambers (Subheading 3.4.3) (*see* **Note 10**).

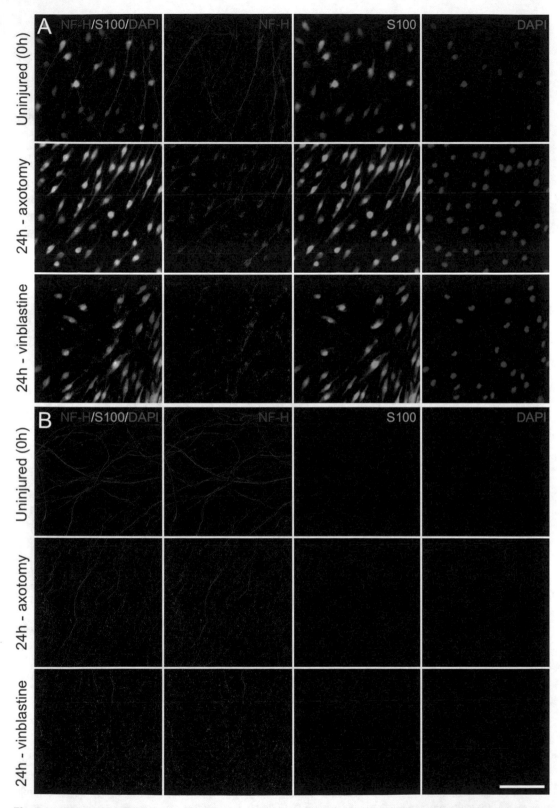

Fig. 2 In vitro model of axon degeneration from DRG neurons with or without Schwann cells. Representative images of distal axons stained for neurofilament heavy subunit (NF-H, red), the SC marker S100β (green) and the nuclear counterstain DAPI (blue). (**a**) DRG culture with SCs and (**b**) DRG culture without SCs. In each panel the upper part corresponds to uninjured cultures (Ctrl 0 h), the middle part to cultures 24 h after axotomy, and the lower part to cultures 24 h after vinblastine treatment. Scale bar, 100 μm

3.4.1 Assessment of Axonal Growth (No Damage In Vitro)

1. Plate intact DRGs, and grow the explants for 4 days in the presence or absence of a specific treatment (*see* **Note 11**).

2. Perform an IF staining (*see* Subheading 3.1).

3. Take pictures, and measure axonal regeneration in mm using the Image J software (*see* **Note 12**) [8].

3.4.2 Axonal Regeneration After In Vitro Transection

1. Plate and grow the DRG explants for 7 days in vitro prior to mechanically transecting the axons. The damage is performed using a sterile biopsy punch of 4.0 mm in diameter to transect away the axons. Locate the biopsy punch over the explant at approximately 1 mm from the somas. Carefully push down, twist, and drag the biopsy punch. Make sure the somas remain attached to the dish. Take a bright field picture immediately after performing the injury.

2. Grow the explants for 4 days in explant medium with or without antimitotic agent.

3. Perform **steps 2** and **3** as described above (*see* Subheading 3.4.1 and Fig. 3).

3.4.3 Axonal Regeneration in Compartmentalized Chambers

1. Plate the DRG explants into the proximal compartment of a Campenot chamber (*see* Subheading 2.1.1).

2. In the Campenot chamber, DRG axons are allowed to grow underneath high-vacuum grease barriers into the left and right distal compartments.

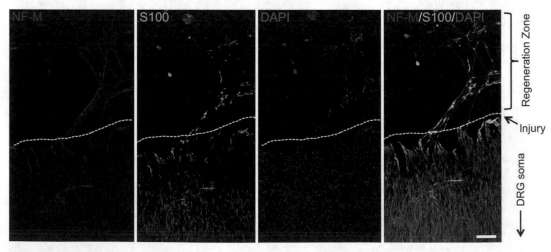

Fig. 3 Analysis of axonal regeneration. DRGs 7 days after seeding in vitro were axotomized by the method described in Subheading 3.4.2. Three days after injury, the DRGs were fixed, and immunofluorescence was performed using antibodies against the axonal marker neurofilament medium subunit (NF-M, green), the SC marker S100β (red), and the nuclear counterstain DAPI (blue). Dashed lines indicate the injury region. Above this region, regenerated axons can be detected and measured. DRG somas are not seen in this picture but are located in the lower part. Scale bar, 100 μm

3. Seven days after plating, remove the axons in the distal compartment by incubating them with distilled water for 10 min at 37 °C.

4. Fill the axonal compartment with explant medium [8].

5. Repeat **steps 2** and **3** as described above (Subheading 3.4.1). *See* **Note 13**.

4 Notes

1. After washing PLL-treated coverslips with sterile water, leave them dry, for better adhesion of collagen.

2. Plates may be used immediately or air dried (stored at 2–8 °C) for future use.

3. **Steps 8–15** should be performed on ice.

4. Microdissection of the DRGs from rat embryos and preparation of media and supplements should be performed in a level II biosafety cabinet or laminar flow hood using aseptic technique and universal safety precautions.

5. For co-culturing DRGs with SCs, use explant medium. To grow DRG neurons depleted of SCs, use explant medium with antimitotic agent (For preparation *see* Subheading 2.1.2 **steps 19–20**.)

6. DRGs become attached to the bottom of the well around 1 h after plating. Neurite extension growing radially from the center of the DRG can be seen using bright-field microscopy. Neurites grow approximately 300 μm/day [8]. If the DRGs are not in good condition, 24 h after plating, they will not attach to the surface of the well. DRGs growing in the presence of SCs can grow faster and cover 20% more distance than those DRGs deprived from SCs.

7. Acquire images of fluorescence microscopy using a 40× objective.

8. For this technique anti-neurofilament (medium or heavy subunit) or anti-acetylated tubulin antibodies should be used as a direct readout of axonal degeneration through IF. After injury, the axonal skeleton disintegrates, and the axonal membrane breaks apart. Usually, the microtubule cytoskeleton fragments before the neurofilament one, and then analysis of these two proteins by IF gives information about early and late stages of axonal degeneration.

9. To quantitatively study axonal degeneration, the degeneration index is based on the ratio of the areas of fragmented axons versus total axonal area immunostained with anti-acetylated tubulin antibody. Degenerated axon fragments are detected

using the particle analyzer algorithm of ImageJ (NIH, USA), and the total fragmented axon area versus total axonal area is used to estimate a degeneration index.

10. To evaluate axonal regeneration, IF with an anti-acetylated tubulin is used. By using this neuronal staining, intact axons (continuous axonal labeling) and degenerated axons (discontinuous or fragmented axonal labeling) can be discriminated even at early times after damage (from 6 h onwards).

11. Cultures of DRGs with or without SCs allow the study of several aspects of neuronal biology and neuron-SC interactions. Through the analysis of the degeneration index, it is possible to evaluate neurotoxic compound in addition to neuroprotective agents. Also, it is possible to evaluate neurite extension and their associated mechanisms after mechanical injuries. Axon-SC interactions can be evaluated by the subsequent addition of modified SCs to DRGs previously depleted of SCs. These added SCs could be previously subjected to gene modification using viral transduction or other suitable technique in order to overexpress or downregulate a gene of interest.

12. Axonal regeneration is obtained by measuring from the cut edge to the tip of the ten longest regenerating axons per DRG. The location of the damage is recognized by overlapping the bright field picture after injury (from **step 1**) with the IF picture.

13. Axonal regeneration is quantified from the grease barrier to the tip of the longest axon in the track. The lengths of the ten longest regenerating axons per chamber are measured in the axonal compartment.

Acknowledgments

We wish to thank all the members of the Court Lab for their contributions to this protocol. This work was supported by Center for Integrative Biology, Universidad Mayor, FONDECYT-1150766, Geroscience Center for Brain Health and Metabolism (FONDAP-15150012), Ring Initiative ACT1109, and Canada-Israel Health Research initiative, jointly Funded by the Canadian Institutes of Health Research; the Israel Science Foundation; the International Development Research Centre, Canada; and the Azrieli Foundation, Canada.

References

1. Scheib J, Höke A (2013) Advances in peripheral nerve regeneration. Nat Rev Neurol 9:668–676. https://doi.org/10.1038/nrneurol.2013.227

2. Ferguson TA, Son Y-J (2011) Extrinsic and intrinsic determinants of nerve regeneration. J Tissue Eng 2:2041731411418392. https://doi.org/10.1177/2041731411418392

3. Waller A (1850) Experiment on the section of the glossopharyngeal and hypoglossal nerves of the frog, and observations of the alterations produced thereby in the structure of their primitive fibres. Phil Trans R Soc London 140:423–429

4. Allodi I, Udina E, Navarro X (2012) Specificity of peripheral nerve regeneration: interactions at the axon level. Prog Neurobiol 98:16–37. https://doi.org/10.1016/j.pneurobio.2012.05.005

5. Barrientos SA, Martinez NW, Yoo S et al (2011) Axonal degeneration is mediated by the mitochondrial permeability transition pore. J Neurosci 31:966–978. https://doi.org/10.1523/JNEUROSCI.4065-10.2011

6. Court FA, Coleman MP (2012) Mitochondria as a central sensor for axonal degenerative stimuli. Trends Neurosci 35:364–372. https://doi.org/10.1016/j.tins.2012.04.001

7. Dodd J, Solter D, Jessell TM (1984) Monoclonal antibodies against carbohydrate differentiation antigens identify subsets of primary sensory neurones. Nature 311:469–472

8. Lopez-Verrilli MA, Picou F, Court FA (2013) Schwann cell-derived exosomes enhance axonal regeneration in the peripheral nervous system. Glia 61:1795–1806. https://doi.org/10.1002/glia.22558

9. Cho Y, Cavalli V (2012) HDAC5 is a novel injury-regulated tubulin deacetylase controlling axon regeneration. EMBO J 31:3063–3078. https://doi.org/10.1038/emboj.2012.160

10. Oblinger MM, Lasek RJ (1988) Axotomy-induced alterations in the synthesis and transport of neurofilaments and microtubules in dorsal root ganglion cells. J Neurosci 8:1747–1758

11. de Waegh SM, Lee VM, Brady ST (1992) Local modulation of neurofilament phosphorylation, axonal caliber, and slow axonal transport by myelinating Schwann cells. Cell 68:451–463. https://doi.org/10.1016/0092-8674(92)90183-D

12. Kashiba H, Senba E, Kawai Y et al (1992) Axonal blockade induces the expression of vasoactive intestinal polypeptide and galanin in rat dorsal root ganglion neurons. Brain Res 577:19–28

13. Zhuo H, Lewin AC, Phillips ET et al (1995) Inhibition of axoplasmic transport in the rat vagus nerve alters the numbers of neuropeptide and tyrosine hydroxylase messenger RNA-containing and immunoreactive visceral afferent neurons of the nodose ganglion. Neuroscience 66:175–187

14. Dilley A, Richards N, Pulman KG, Bove GM (2013) Disruption of fast axonal transport in the rat induces behavioral changes consistent with neuropathic pain. J Pain 14:1437–1449. https://doi.org/10.1016/j.jpain.2013.07.005

15. Campenot RB, Lund K, Mok S-A (2009) Production of compartmented cultures of rat sympathetic neurons. Nat Protoc 4:1869–1887. https://doi.org/10.1038/nprot.2009.210

16. Villegas R, Martinez NW, Lillo J et al (2014) Calcium release from intra-axonal endoplasmic reticulum leads to axon degeneration through mitochondrial dysfunction. J Neurosci 34:7179–7189. https://doi.org/10.1523/JNEUROSCI.4784-13.2014

Chapter 17

A Culture Model to Study Neuron-Schwann Cell-Astrocyte Interactions

Susana R. Cerqueira, Yee-Shuan Lee, and Mary Bartlett Bunge

Abstract

In vitro models using Schwann cell and astrocyte co-cultures have been used to understand the mechanisms underlying the formation of boundaries between these cells in vivo. Schwann cell/astrocyte co-cultures also mimic the in vivo scenario of a transplant in a spinal cord injury site, thereby allowing testing of therapeutic approaches. In this chapter, we describe a triple cell culture system with Schwann cells, astrocytes, and neurons that replicates axon growth from a Schwann cell graft into an astrocyte-rich region. In vitro studies using this model can accelerate the discovery of more effective therapeutic combinations to be used along with Schwann cell transplantation after spinal cord injuries.

Key words Neurons, Schwann cells, Astrocytes, Co-cultures, Axon growth, Glial barrier

1 Introduction

Co-culture systems are versatile and valuable tools to uniquely discover cell-cell interactions that are difficult to discern in animal models [1]. They also allow high-throughput analysis for rapid and large data collection [2]. Combination cultures of neurons and glia provide test systems that enable studies of their interactions in normal conditions or when modulated by many types of factors. Numerous investigations of factors influencing myelin formation in these cultures have been published [3–5]. It was the combination of sensory neurons and purified populations of Schwann cells that enabled the discovery that molecules on the axon surface cause Schwann cells to divide [6]. Changes in cell shape, spatial distribution, migration, proliferation, neurite extension, secreted factors, and molecular pathways all can be profitably investigated.

For the fields of neural injury and repair, combination cultures are valuable tools as well. Astrocytes can be cultured along with Schwann cells to represent the borders created by Schwann cell implantation after spinal cord injury [7, 8]. Whereas these populations generally do not mix, manipulations can be made to see what

Paula V. Monje and Haesun A. Kim (eds.), *Schwann Cells: Methods and Protocols*, Methods in Molecular Biology, vol. 1739, https://doi.org/10.1007/978-1-4939-7649-2_17, © Springer Science+Business Media, LLC 2018

will cause them to instead mix in the culture dish [9], providing possibilities for testing in the animal. These co-cultures have been used to investigate mechanisms underlying Schwann cell migration into astrocyte areas and the role of inhibitory molecules secreted by astrocytes [10]. Transplanted Schwann cells are typically surrounded by reactive astrocytes that contribute to an inhibitory boundary for both Schwann cell migration and axon crossing [11–13]. Schwann cell transplantation is one of the most comprehensively investigated therapies for spinal cord injury. Results from spinal cord injured animal models have spurred initiation of clinical trials testing the safety of transplanting human Schwann cells into subacute and chronic lesions [14, 15].

Neurons can be added to Schwann cell/astrocyte co-cultures [16] to detect ways in which neurite growth can be coaxed across the astrocyte-Schwann cell interface [17], again suggesting paradigms to test in vivo. Studies in spinal cord injured animal models have demonstrated that with transplantation of Schwann cells alone, regenerating axons enter but do not substantially exit the transplant [18]. Effective ways to elicit axon growth from the transplant into the spinal cord need to be found. Thus the combination of neurons, Schwann cells, and astrocytes in culture can be utilized to accelerate the discovery of more effective therapies for spinal cord injury.

In this chapter we describe the preparation of a spatially defined combination culture of neurons, Schwann cells, and astrocytes to mimic the environment of a Schwann cell graft within a lesion site in the spinal cord. The spatial distribution of the cells can be accomplished by introducing cell dividers to create individual compartments. Instead of randomly seeding cells into the culture dish, a physical separation can be temporarily created, resulting in a design that resembles more closely the organization of the in vivo tissue architecture [19]. Once Schwann cell and astrocyte monolayers are established in separate compartments, neurons are added to the Schwann cell monolayer. After the cell divider is removed, a Schwann cell/astrocyte border is formed, and Schwann cell migration and neurite growth can be measured across the boundary. It is hoped that this combination culture system may allow faster discovery of effective therapeutic interventions for neural repair after injury.

2   Materials

All solutions, reagents, and supplies used in cell culture should be sterile and tissue culture grade.

2.1   Basic Equipment

1. Laminar-flow hood equipped with an aspiration pump (peristaltic or vacuum).

2. Humidified 6% CO_2 incubator.

3. Water bath (set to 37 °C).

4. Bench-top centrifuge.

5. Refrigerator (4 °C) and freezer (−20 °C).

6. Cell counter (automated cell counter, with capability to discriminate live/dead cells based on trypan blue staining, or hemacytometer).

7. Pipettes (5 and 15 mL), pipette tips (0.02, 0.2, and 1 mL), and pipettors (P20, P200, P1000).

8. Polypropylene centrifuge tubes (10 and 50 mL) and conical bottom tubes (1.5 mL).

9. Inverted phase-contrast microscope (with 10× and 20× objectives).

10. Inverted fluorescence microscope.

11. Autoclave.

12. 70% ethanol.

13. Paraffin film.

14. Orbital shaker.

2.2 Coating and Inserting 2-Well Silicone Dividers into the Wells

1. 24-well culture dishes.

2. 2-well silicone cell dividers (*see* **Note 1**).

3. 200 µg/mL poly-L-Lysine (PLL) solution prepared in distilled water.

4. Distilled water.

5. Forceps.

6. Glass coverslips (optional).

2.3 Seeding Single-Cell Suspensions of Schwann Cells and Astrocytes

1. Schwann cell medium: Dulbecco's modified Eagle's medium (DMEM) supplemented with 10% fetal bovine serum (FBS), 1% penicillin-streptomycin (PenStrep), and 3 mitogens (3 M, 2 µM forskolin, 20 µg/mL bovine pituitary extract, and 2.5 nM heregulin).

2. Complete medium: DMEM supplemented with 10% FBS and 1% PenStrep.

3. Primary cultures of adult rat sciatic nerve Schwann cells. In our lab, Schwann cells are obtained from sciatic nerves of adult female Fischer rats, as described previously [20]. Briefly, nerves are cut into small pieces and placed in culture dishes with Schwann cell medium without mitogens. After 2 weeks, the pieces are enzymatically dissociated and replated in Schwann cell medium supplemented with three mitogens. The purity of the resulting cultures is 95–98%.

4. Primary cultures of rat postnatal cortical astrocytes. Astrocyte cultures were obtained from postnatal (P4–P5) rat pups, as described elsewhere [21]. In short, neonatal rat cortices are mechanically and enzymatically dissociated, plated in culture flask, and maintained for a week. Next, the culture flasks are agitated in an orbital shaker overnight. The medium containing detached cells (microglia and oligodendrocytes) is then discarded, and the remaining cells are trypsinized and plated. The purity of the resulting astrocyte cultures is >90%.

5. Hanks balanced salt solution (HBSS).

6. 0.25% trypsin-EDTA solution.

7. 0.4% trypan blue solution.

2.4 Seeding a Dissociated Neuron Suspension on the Schwann Cell Monolayer

1. Primary cultures of dissociated rat embryonic dorsal root ganglion (DRG) neurons. Briefly, DRGs are dissected from rat embryos (E15) and incubated in trypsin. The trypsin reaction is stopped and followed by mechanical dissociation. Neurons are then placed in laminin-coated culture dishes and treated with fluorodeoxyuridine, to eliminate non-neuronal cells. After this treatment, DRG neurons are kept in supplemented neuronal media.

2. Neuronal medium: Neurobasal-A medium supplemented with 1% L-glutamine, 2% B-27, and 1% PenStrep.

3. HBSS.

4. 0.25% trypsin-EDTA solution.

5. 0.4% trypan blue solution.

2.5 Immunostaining

1. Phosphate-buffered saline (PBS).

2. Permeabilization buffer: PBS containing 0.1% (v/v) Triton X-100.

3. Fixative: 4% paraformaldehyde (PFA) in PBS.

4. Blocking solution: PBS containing 10% FBS.

5. Primary antibody solution: PBS containing 5% FBS.

6. Primary antibodies (commercially available): rabbit polyclonal p75 antibody (dilution 1:500), for labeling Schwann cells; chicken monoclonal GFAP antibody (dilution 1:1000), for identifying astrocytes; and mouse monoclonal neurofilament antibody (dilution 1:500), for neuron labeling. These are suggested primary antibodies that were useful in our laboratory, but other combinations may be used. All primary antibodies are diluted in primary antibody solution. All secondary antibodies are diluted 1:1000 in secondary antibody solution.

7. Secondary antibody solution: PBS containing 0.5% FBS.

8. Fluorescent secondary antibodies (commercially available): anti-rabbit 488, anti-chicken 568, and anti-mouse 660. These are suggestions but different fluorophores also may be used successfully.

9. Anti-fading mounting medium.

3 Methods

All procedures should be performed inside a laminar flow hood and using appropriate aseptic technique as well as safe laboratory practices. Most solutions are purchased from laboratory suppliers. Handling, collection, and disposal of potentially hazardous solutions, such as PFA, should follow environmental health and safety waste disposal standards.

3.1 Coating and Inserting Cell Dividers into the Wells

1. Add PLL solution to each well of a 24-well plate for 30 min at room temperature (RT) (*see* **Note 2**).

2. Aspirate the PLL from the wells, and wash the wells three times with distilled water. Let them dry completely at RT inside the laminar flow hood.

3. Carefully hold the silicone cell divider inserts using forceps, and place one insert in the center of each well of the 24-well plate (Fig. 1).

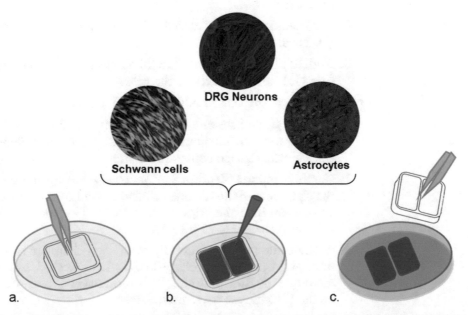

DRG Neurons

Schwann cells

Astrocytes

a.

b.

c.

Fig. 1 Representative illustration of the insertion of cell dividers into the wells and cell seeding to obtain a triple culture of primary DRG neurons, Schwann cells, and astrocytes: (**a**) cell divider insertion into the well, (**b**) seeding of cells into each chamber, and (**c**) removal of the cell divider and filling the well with culture medium

4. Press gently the edges of the insert with the forceps, and verify that it adheres to the bottom of the well, creating a seal (*see* **Notes 3** and **4**).

3.2 Seeding SCs and Astrocytes

The following steps require fresh single-cell suspensions of Schwann cells and astrocytes, obtained from previously prepared purified cultures. If the cells need to be collected from Schwann cell and astrocyte monolayers in culture flasks or dishes, perform the following steps in order to obtain single-cell suspensions. The volumes suggested below are for use with cells cultured in T-75 flasks.

1. Remove the respective culture medium and rinse the flask twice with HBSS.

2. Add 5 mL of trypsin-EDTA to each flask, and wait 2–5 min until the cells start to detach from the substrate. Check the cells under an inverted phase-contrast microscope to verify that they are detaching from the flasks (*see* **Note 5**).

3. When most of the cells are detached from the substrate, add 5 mL of complete medium to inactivate the trypsin. Collect the cell suspension from the flask and put it into a conical tube.

4. Rinse the dish twice with 5 mL complete medium, and add it to the conical tube containing the previously collected cells.

5. Centrifuge the cells at $300 \times g$ for 10 min, at 4 °C.

6. Slowly aspirate the supernatant and discard it. A cell pellet should be seen at the bottom of the conical tube with the naked eye.

7. Resuspend the cell pellet in 3 mL of the respective cell medium (*see* **Note 6**).

8. Add 20 μL of trypan blue to a 20 μL aliquot of the cell suspension (1:1 ratio), and count the cells.

9. Centrifuge the cells at $300 \times g$ for 10 min at 4 °C.

10. Resuspend the Schwann cells at a density of 12×10^5 cells/mL in Schwann cell medium and the astrocytes at a density of 4×10^5 cells/mL in complete medium.

11. Carefully pipette 50 μL of Schwann cell suspension into one of the chambers of the silicone cell divider and 50 μL of astrocyte suspension into the adjacent chamber. Place the cells in the incubator for 2 h (*see* **Note 7**).

12. Aspirate the medium carefully from each chamber to remove unbound cells and debris. Add 70 μL of fresh Schwann cell medium to the Schwann cell monolayers, and complete medium to the astrocyte monolayers.

13. Place the culture plates in the humidified incubator, with 6% CO_2, for 24 h.

3.3 Seeding DRG Neurons onto the Schwann Cell Monolayer

In order to assess axon extension from a SC region into an astrocyte-rich area, DRG-dissociated neurons are seeded on top of the previously established SC monolayer.

1. Proceed as described before, if the DRG neurons need to be trypsinized from the flasks/dishes into a single-cell suspension (*see* Subheading 3.2, **steps 1–8**).

2. Resuspend the DRG neurons in order to have a density of 4×10^4 cells/mL, in a freshly prepared mixture of Schwann cell medium and neuronal culture medium (1:1).

3. Carefully aspirate the medium from the chamber with the Schwann cell monolayer.

4. Add 50 µL of DRG neuron suspension on top of the Schwann cell monolayer.

5. Place the cells in the humidified incubator, with 6% CO_2, overnight.

6. Aspirate the medium from both sides of the cell divider, and, with forceps holding the middle separator, gently remove the silicone insert from the well.

7. Add 400 µL of freshly prepared mixture of complete medium and neuron culture medium (1:1). Confirm the formation of a distinct cell separation boundary under an inverted phase-contrast microscope (Fig. 2).

8. Refresh the cultures with medium every other day until fixation, typically 5 days after seeding the DRG neurons.

9. Follow the progression of the Schwann cell/astrocyte border by monitoring the cultures daily by phase-contrast microscopy.

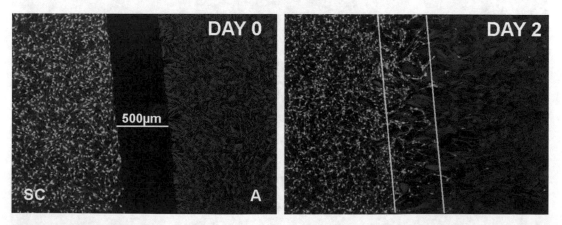

Fig. 2 Fluorescence microscopy images of an early established co-culture of Schwann cells and astrocytes: *Day 0*, co-culture immediately after the removal of the cell divider, showing a gap between Schwann cell and astrocyte areas; *Day 2*, 2 days after removal of the cell divider, the cells have migrated to appose each other. *SC* Schwann cells (in green), *A* astrocytes (in red)

3.4 Immunostaining In order to visualize and easily identify each cell type, we suggest immunostaining with specific antibodies (primary and secondary) followed by observation under an inverted fluorescence microscope (Fig. 3). The fixation step should be performed inside a chemical fume hood.

1. Aspirate slowly the medium from the wells, and add 500 μL of fixative to each well for 20 min to ensure adequate fixation of the cells.

2. Wash the cells three times with PBS.

3. Add 500 μL of permeabilization buffer for 5 min.

4. Wash the cells with PBS and add 500 μL of blocking solution for 1 h.

5. Aspirate the blocking solution, and incubate for 2 h at RT with 400 μL of the respective primary antibody solutions (*see* **Note 8**).

6. Wash the cells three times with PBS, 10 min each.

7. Remove the PBS, and add the secondary antibodies diluted in secondary antibody solution, for 1 h at RT.

8. Wash the cells three times with PBS, 10 min each wash.

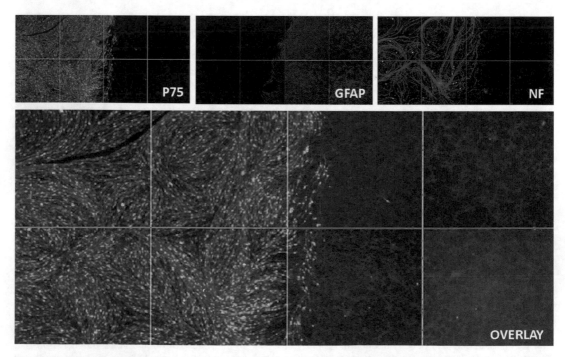

Fig. 3 Fluorescence microscopy images following immunostaining of a triple culture 5 days after insert removal. Single-channel images are shown in the top row. These images are merged below. *P75* Schwann cells (in green), *GFAP* astrocytes (in red), *NF* DRG neurons (in blue). Pseudo coloring was performed for the neurons to appear in blue

9. Add some drops of anti-fading mounting medium until the bottom of the well is totally covered. Avoid bubble formation.

10. Cover the plates with paraffin film, and keep them in the fridge until microscopic observation.

4 Notes

1. We recommend the use of silicone cell dividers with a defined central separator of 500 μm. The maximum volume of each chamber is 70 μL. The underside of the dividers should easily stick to the surface of the wells. Before use, sterilize the cell dividers in the autoclave. We have successfully reused cell dividers 3–4 times.

2. For imaging purposes and simplicity, we seed the cells directly onto the culture plate well. It is possible, however, to perform this technique on glass coverslips placed inside a 24-well plate, following the same steps. Glass coverslips are very fragile, so gentle and careful handling is necessary to perform this assay using them.

3. To check if a seal is formed between the cell divider and the well, two different approaches may be used: (a) flip the plate upside down, and verify that the inserts remain in place and attached to the wells, and (b) carefully hold the insert with forceps, and lift it to check if you can lift the plate while holding the insert. As this step is crucial for the success of the experiment, make sure that all the inserts are in place and sealed to the bottom of the wells. If plates are not used immediately, they can be covered with paraffin film and kept at 4 °C until use.

4. Creating a seal is important if the goal of the study is to measure the extent of SC migration or axon growth into the astrocyte region. We advise to carefully draw a line underneath the culture plate to mark the initial borders of the SC and astrocyte regions. This will make the measurements more accurate.

5. The volume of trypsin-EDTA to add to the cultures varies according to the size of the flasks or dishes used. In our lab, as T-75 culture flasks are routinely used, the suggested volumes are for these. To facilitate cell detachment, tapping the flasks vigorously for a short period of time is effective.

6. The volume of the cell suspension should be adjusted in order to be in the range of 1×10^5–5×10^6 cells/mL.

7. An optimal Schwann cell to astrocyte seeding density ratio is 3:1, due to the smaller dimensions of Schwann cells compared to astrocytes.

8. Dilution of the primary antibodies should follow the recommendations from the supplier. Optimal concentrations will vary between different products. Incubation times may also vary; a 2-h incubation is usually adequate to stain cultured cells.

Acknowledgments

We would like to thank Yelena Pressman for preparing the Schwann cell and DRG neuron cultures and Yan Shi for assistance with fluorescence imaging.

References

1. Goers L, Freemont P, Polizzi KM (2014) Co-culture systems and technologies: taking synthetic biology to the next level. J R Soc Interface 11(96):20140065. https://doi.org/10.1098/rsif.2014.0065

2. Al-Ali H, Beckerman SR, Bixby JL, Lemmon VP (2017) In vitro models of axon regeneration. Exp Neurol 287(Part 3):423–434. https://doi.org/10.1016/j.expneurol.2016.01.020

3. Eldridge CF, Bunge MB, Bunge RP, Wood PM (1987) Differentiation of axon-related Schwann cells in vitro. I. Ascorbic acid regulates basal lamina assembly and myelin formation. J Cell Biol 105(2):1023–1034

4. Wood PM, Schachner M, Bunge RP (1990) Inhibition of Schwann cell myelination in vitro by antibody to the L1 adhesion molecule. J Neurosci 10(11):3635–3645

5. Ogata T, Iijima S, Hoshikawa S, Miura T, S-i Y, Oda H, Nakamura K, Tanaka S (2004) Opposing extracellular signal-regulated kinase and Akt pathways control Schwann cell myelination. J Neurosci 24(30):6724–6732

6. Wood PM, Bunge RP (1975) Evidence that sensory axons are mitogenic for Schwann cells. Nature 256(5519):662–664

7. Afshari FT, Fawcett JW (2012) Astrocyte–Schwann-cell coculture systems. Methods Mol Biol 814:381–391

8. Afshari FT, Kwok JC, Fawcett JW (2011) Analysis of Schwann-astrocyte interactions using in vitro assays. J Vis Exp 47:e2214

9. Afshari FT, Kwok JC, Fawcett JW (2010) Astrocyte-produced ephrins inhibit Schwann cell migration via VAV2 signaling. J Neurosci 30(12):4246–4255

10. Afshari FT, Kwok JC, White L, Fawcett JW (2010) Schwann cell migration is integrin-dependent and inhibited by astrocyte-produced aggrecan. Glia 58(7):857–869

11. Fortun J, Hill CE, Bunge MB (2009) Combinatorial strategies with Schwann cell transplantation to improve repair of the injured spinal cord. Neurosci Lett 456(3):124

12. Oudega M, Xu X-M (2006) Schwann cell transplantation for repair of the adult spinal cord. J Neurotrauma 23(3–4):453–467

13. Tetzlaff W, Okon EB, Karimi-Abdolrezaee S, Hill CE, Sparling JS, Plemel JR, Plunet WT, Tsai EC, Baptiste D, Smithson LJ (2011) A systematic review of cellular transplantation therapies for spinal cord injury. J Neurotrauma 28(8):1611–1682

14. Anderson KD, Guest JD, Dietrich WD, Bunge MB, Curiel R, Dididze M, Green BA, Khan A, Pearse DD, Saraf-Lavi E, Widerstrom-Noga E, Wood P, Levi AD (2017) Safety of autologous human schwann cell transplantation in subacute thoracic spinal cord injury. J Neurotrauma. https://doi.org/10.1089/neu.2016.4895

15. Bunge MB, Monje PV, Khan A, Wood PM From transplanting Schwann cells in experimental rat spinal cord injury to their transplantation into human injured spinal cord in clinical trials. Progr Brain Res. https://doi.org/10.1016/bs.pbr.2016.12.012

16. Guenard V, Gwynn L, Wood P (1994) Astrocytes inhibit Schwann cell proliferation and myelination of dorsal root ganglion neurons in vitro. J Neurosci 14(5):2980–2992

17. Adcock KH, Brown DJ, Shearer MC, Shewan D, Schachner M, Smith GM, Geller HM,

Fawcett JW (2004) Axon behaviour at Schwann cell-astrocyte boundaries: manipulation of axon signalling pathways and the neural adhesion molecule L1 can enable axons to cross. Eur J Neurosci 20(6):1425–1435

18. Bunge MB (2016) Efficacy of Schwann cell (SC) transplantation for spinal cord repair is improved with combinatorial strategies. J Physiol 594(13):3533–3538

19. Bogdanowicz DR, Lu HH (2013) Studying cell-cell communication in co-culture.

Biotechnol J 8(4):395–396. https://doi.org/10.1002/biot.201300054

20. Morrissey TK, Kleitman N, Bunge RP (1991) Isolation and functional characterization of Schwann cells derived from adult peripheral nerve. J Neurosci 11(8):2433–2442

21. Chen Y, Balasubramaniyan V, Peng J, Hurlock EC, Tallquist M, Li J, Lu QR (2007) Isolation and culture of rat and mouse oligodendrocyte precursor cells. Nat Protoc 2(5):1044–1051. https://doi.org/10.1038/nprot.2007.149

Chapter 18

Preparation of Matrices of Variable Stiffness for the Study of Mechanotransduction in Schwann Cell Development

Mateusz M. Urbanski and Carmen V. Melendez-Vasquez

Abstract

Extracellular matrix (ECM) elasticity may direct cellular differentiation and can be modeled in vitro using synthetic ECM-like substrates with defined elastic properties. However, the effectiveness of such approaches depends on the selection of a range of elasticity and ECM ligands that accurately model the relevant tissue. Here, we present a cell culture system than can be used to study Schwann cell differentiation on substrates which model the changes in mechanical ECM properties that occur during sciatic nerve development.

Key words Schwann cells, Cell culture, Differentiation, Mechanotransduction, Extracellular matrix, Sciatic nerve, Elastic modulus, Atomic force microscopy

1 Introduction

The field of mechanotransduction, the study of cellular responses to mechanical forces, has undergone a great deal of growth during recent years. Seminal work in the field has demonstrated that extracellular matrix (ECM) stiffness has powerful effects on cell fate choice and differentiation [1], and it has since been repeatedly shown that ECM-derived mechanical forces and tissue elasticity regulate cell behavior in development and disease (reviewed in [2–4]).

Since modification of ECM composition and its mechanical properties in a developing organism presents significant technical difficulties, mechanotransduction research is highly reliant on in vitro models which use synthetic substrates that mimic the properties of natural ECM. Polyacrylamide hydrogels coated with covalently bound ECM proteins are a commonly used system [1, 5–8] since their elastic properties can be adjusted in a reproducible manner by varying the relative concentrations of acrylamide and bisacrylamide cross-linker (*see* Table 1) and because their optical properties make them highly compatible with most types of fluorescent microscopy. To maximize physiological relevance, these

Paula V. Monje and Haesun A. Kim (eds.), *Schwann Cells: Methods and Protocols*, Methods in Molecular Biology, vol. 1739,
https://doi.org/10.1007/978-1-4939-7649-2_18, © Springer Science+Business Media, LLC 2018

Table 1
AFM measurements of the elastic modulus of acrylamide/bis-acrylamide hydrogels

Acrylamide %	Bis-acrylamide %	$E \pm$ st. dev. (kPa)
7.5	0.050	1.35 ± 0.09
5.0	0.150	2.52 ± 0.29
5.0	0.300	6.67 ± 0.28
10.0	0.225	27.41 ± 1.17
12.0	0.280	33.15 ± 2.87
12.0	0.450	65.27 ± 3.47

The values presented here are the Young's moduli of polymerized PA gels attached to glass coverslips. The gel mixtures were prepared in PBS with the indicated acrylamide/bis-acrylamide concentrations. The range includes elasticities both within and outside the physiological range relevant for Schwann cell development (*see* Table 2) (At least three independent substrates were analyzed by AFM nano-indentation per condition, with at least 100 individual indentations per substrate)

substrates must accurately model the elastic properties of specific animal tissues (which can range in elastic modulus from 0.1 to over 100 kPa [9]) and incorporate sufficient concentrations of relevant ECM ligands.

In this protocol, we present an in vitro method designed to examine the effect of matrix stiffness on the differentiation of primary rat Schwann cells (SC). Glass coverslips are coated with an ~50 μm thick layer of polyacrylamide gel of defined stiffness, following which the polyacrylamide surface is functionalized for cell attachment by treatment with a UV-activated cross-linking agent (sulfo-SANPAH) and the subsequent covalent binding of ECM proteins (*see* Fig. 1). The range of acrylamide gel elasticity suggested here is based on atomic force microscopy (AFM) measurements of rat sciatic nerves performed in our laboratory, with the elastic modulus values of 1.5 kPa and 30.0 kPa chosen to reflect the increase in stiffness observed during sciatic nerve development and the concurrent SC differentiation and myelination (*see* Table 2). Mouse laminin at 0.02 mg/ml is the essential ECM ligand, as work from our laboratory has established that increased matrix stiffness promotes the expression of SC promyelinating factors only in combination with a high density of laminin [7], and it has also been shown that in SC, the key effectors of mechanotransduction YAP/TAZ [10] are only responsive to changes in ECM stiffness in the presence of laminin [11]. Collagen Type 1 is also included in the ECM protein mixture at 0.1 mg/ml to enhance cell attachment and viability.

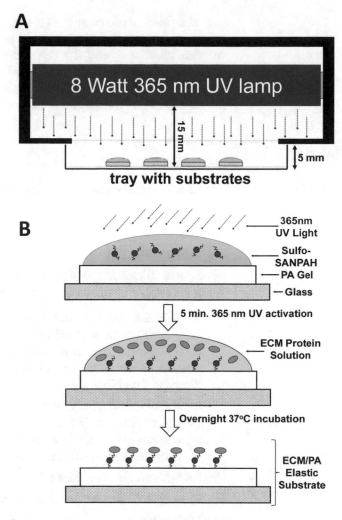

Fig. 1 Overview of the substrate activation and protein coating protocol

Table 2
AFM measurements of the elastic modulus of rat sciatic nerves

Age	$E \pm$ st. dev. (kPa)
P0	6.04 ± 2.79
P2	13.77 ± 7.47
P5	24.17 ± 14.04
Adult	49.4 ± 19.0
Adult (single myelinated fiber, teased)	5.3 ± 0.9

AFM measurements of fresh rat sciatic nerves harvested during the first week of postnatal development and from adult animals, demonstrating a consistent increase in tissue stiffness during sciatic nerve development and myelination. (3–4 nerves were analyzed per time point, with a least 1000 individual measurements per nerve.) (Modified from Urbanski et al. 2016 [7], as permitted under the Creative Commons Attribution 4.0 International Public License, http://creativecommons.org/licenses/by/4.0/)

The functionalized substrates are entirely compatible with cell culture protocols designed for cells grown on tissue culture plastic or glass substrates, and the resulting cultures can be readily processed for immunocytochemistry. We have successfully used them to examine the effect of ECM stiffness on SC morphology and on SC expression of transcription factors regulating myelination [7].

2   Materials

2.1   Variable Stiffness Acrylamide Gel Substrates

1. 12 mm circular glass coverslips, #0.

2. 25 mm × 75 mm glass slides.

3. Two ceramic 12-slot coverslip racks.

4. Stainless steel 24-slot slide holder.

5. Source of running deionized H_2O (diH_2O).

6. 0.1 M NaOH: Prepared in diH_2O, non-sterile.

7. 0.5% 3-aminopropyltrimethoxysilane (APTMS) solution: 1 ml of 3-aminopropyltrimethoxysilane (APTMS) in 199 ml of diH_2O, prepared immediately before use. Keep in a covered glass container large enough to completely immerse the coverslip racks (*see* **Note 1**).

8. 0.5% glutaraldehyde solution: 4 ml of 25% glutaraldehyde in 196 ml of phosphate-buffered saline (PBS), prepared immediately before use. Keep in a covered glass container large enough to completely immerse the coverslip racks (*see* **Note 2**).

9. Rain-X solution (*see* **Note 3**).

10. Kimwipes or similar delicate task wipes.

11. 100% ethanol.

12. 1.5 kPa (7.5% acrylamide, 0.05% bis-acrylamide) mix: Combine 1.875 ml of 40% acrylamide solution, 0.25 ml of 2% bis-acrylamide solution, and 7.875 ml of tissue culture-grade PBS. Store at 4 °C in a container covered with aluminum foil (*see* **Note 4**).

13. 30.0 kPa (12.0% acrylamide, 0.28% bis-acrylamide) mix: Combine 3.0 ml of 40% acrylamide solution, 1.4 ml of 2% bis-acrylamide solution, and 5.6 ml tissue culture-grade PBS. Store at 4 °C in a container covered with aluminum foil (*see* **Note 4**).

14. 10% (w/v) ammonium persulfate (APS) solution in reagent grade water (*see* **Note 5**).

15. N,N,N',N'-tetramethylethane-1,2-diamine (TEMED).

16. 1.5 ml Eppendorf tubes.

17. 10 cm petri dish with 30–40 ml of tissue culture-grade PBS.

18. 24 well tissue culture plate with 1 ml of tissue culture-grade PBS per well.

19. Scalpel blade or fine needle.

20. Fine-point forceps.

2.2 ECM Protein Coating

1. 50 mM HEPES buffer pH 8.5, sterile: Add 450 ml of deionized H_2O and 25 ml of 1 M HEPES to a beaker. Adjust pH to 8.5 with concentrated NaOH, add water to bring the volume up to 500 ml. Filter sterilize, and store at 4 °C.

2. Sulfo-SANPAH stock solution: 20 µl aliquots of 25 mg/ml sulfo-SANPAH solution in anhydrous DMSO, light protected and stored at −80 °C (*see* **Note 6**).

3. Reagent grade H_2O.

4. Tissue culture-grade phosphate-buffered saline (PBS) on ice.

5. Sodium bicarbonate: 7.5% aqueous solution, tissue culture-grade.

6. Rat tail collagen Type 1 (5 mg/ml), on ice.

7. Natural mouse laminin (1 mg/ml), on ice (*see* **Note 7**).

8. ECM protein mix: 0.1 mg/ml collagen and 20 µg/ml laminin. Place a 15 ml Falcon tube on ice. Add 5 ml of ice-cold PBS, 100 µl of collagen Type 1, and 100 µl of natural mouse laminin. Vortex briefly or mix thoroughly with a 5 ml serological pipette. Add 25 µl of 7.5% sodium bicarbonate and mix again. Prepare directly before beginning the protein coating protocol and keep on ice (*see* **Note 8**).

9. Fine-point forceps: Store in 50 ml tube of 70% ethanol when not in use to keep sterile.

10. Six 10 cm petri dishes lined with parafilm (*see* **Note 9**).

11. Flat tray with ~5 mm high lip, lined with parafilm (*see* **Note 10**).

12. 8 W 365 nm ultraviolet lamp (*see* **Note 10**).

13. 24-well tissue culture plate.

2.3 Schwann Cell Culture

1. 10 mM forskolin stock: Add 10 mg forskolin to 2.4 ml DMSO. Aliquot and store at −20 °C.

2. 10 µg/ml rhNRG1β1-EGF domain stock: Reconstitute 50 µg of the cytokine in 5 ml of sterile PBS with 0.1% bovine serum albumin. Aliquot and store at −20 °C.

3. 200 µM dibutyryl-cAMP stock (db-cAMP): Add 1000 µl of reagent grade H_2O to a 100 mg bottle of db-cAMP. Aliquot and store at −20 °C.

4. Hank's balanced salt solution (HBSS).

5. 0.25% trypsin-EDTA.

6. SC growth medium: Combine 44 ml of modified eagle's medium (MEM), 5 ml of fetal bovine serum (FBS), 0.5 ml of 200 mM L-glutamine, and 0.5 ml of 10,000 units/ml penicillin and 10,000 μg/ml streptomycin mixture (P/S). Filter sterilize through a 0.22 μm membrane. After filtration, supplement with 10 μl of forskolin stock (final concentration 2 μM) and 12.5 μl of rhNRG1β1-EGF domain stock (final concentration 2.5 ng/ml). Store at 4 °C for up to 1 month (*see* **Note 11**).

7. SC low serum medium: Combine 48.5 ml of MEM, 0.5 ml of FBS, 0.5 ml of 200 mM L-glutamine, and 0.5 ml of P/S. Filter sterilize through a 0.22 μm membrane. Store at 4 °C for up to 1 month.

8. SC differentiation medium: Combine 48.5 ml of MEM, 0.5 ml of FBS, 0.5 ml of 200 mM L-glutamine, 0.5 ml of P/S, and 250 μl of db-cAMP stock (1 mM db-cAMP final). Filter sterilize through a 0.22 μm membrane. Store at 4 °C for up to 1 month.

9. Schwann cells (SC): ~1 × 10^6 rat SC in growth medium (*see* **Note 12**).

10. 37 °C/5% CO_2 tissue culture incubator.

11. 24-well tissue culture plate.

3 Methods

3.1 Preparation of Variable Stiffness Acrylamide/ECM Protein Substrates

Due to the extensive washing required, preparation of hydrophilic coverslips and hydrophobic slides cannot be readily performed under sterile conditions but should nevertheless be done as cleanly as possible to protect the coverslips and slides from fouling and dust. Standard protective equipment (gloves, lab coat) should be worn to prevent exposure to hazardous chemicals. The cross-link activation and protein coating steps are best performed in a tissue culture cabinet – the UV exposure during the cross-linking step should be sufficient to sterilize the substrates, and they must be treated using standard sterile technique from that point on.

3.1.1 Preparation of Hydrophilic Coverslips

1. Place 24 coverslips in ceramic racks, and immerse them overnight in 0.1 M NaOH in a covered glass container.

2. Transfer the racks to a new container, and rinse for 2 h under running diH_2O, not allowing the stream of water to flow directly on to the coverslips to prevent damage or loss of coverslips.

3. Place the coverslip racks in a tissue culture cabinet, remove excess moisture using vacuum suction, and allow them to dry completely.

4. Immerse the coverslips racks in 200 ml of 0.5% APTMS for 10 min.

5. Discard the APTMS solution in a chemical waste container, rinse the coverslips by slowly immersing the racks in diH_2O 4–5 times (fresh diH_2O each time), and then place in a container of diH_2O for 10 min (*see* **Note 13**).

6. Transfer the racks to a container with 200 ml of 0.5% glutaraldehyde for 30 min.

7. Discard the glutaraldehyde solution in a chemical waste container, rinse the coverslip racks under running diH_2O for 30 min, not allowing the stream of water to flow directly on to the coverslips.

8. Place the coverslip racks in a tissue culture cabinet, remove excess liquid using vacuum suction, and allow them to dry completely (*see* **Note 14**).

3.1.2 Preparation of Hydrophobic Slides

1. Soak a Kimwipe with Rain-X, and use it to spread the solution over the entire surface of a 25 × 75 mm slide. Prepare one slide per every six coverslips, and place coated slides in holder. Wait until the Rain-X is visibly dry (2–3 min).

2. Soak a Kimwipe in 100% ethanol and wipe off visible Rain-X residue from the slides. Allow the slides to dry. If there is still visible residue, repeat the ethanol wipe.

3. Place the slides in racks under running diH_2O for 10 min.

4. Briefly immerse slides in 100% EtOH (*see* **Note 15**), then transfer to a tissue culture cabinet, and allow them to completely dry (*see* **Note 16**).

3.1.3 Preparation of Gel-Coated Substrates

1. In a tissue culture cabinet, lay out the hydrophobic slides and hydrophilic coverslips. Prepare the APS and TEMED solutions, as well as the 1.5 and 30.0 kPa polyacrylamide mixtures and the 24 well plate with PBS. Remove the fine-point forceps from the 70% ethanol, and allow them to air dry inside the cabinet.

2. Add 0.5 ml of the desired polyacrylamide mixture to a 1.5 ml Eppendorf tube (*see* **Note 17**). Initiate the polymerization reaction by adding 3 μl of 10% APS and 2 μl of TEMED, inverting the tube several times to mix after each addition. Proceed to **step 3** immediately.

3. Pipette up to six 6 μl droplets of acrylamide solution onto the surface of a hydrophobic slide, and then use forceps to gently lower the hydrophilic coverslips onto the droplets. Take care not to let the coverslips fall onto the droplets to avoid introducing air bubbles (*see* **Note 18**).

4. Allow the gels to polymerize for 5–10 min; monitor the mix remaining in the tube to verify that polymerization has occurred (*see* **Note 19**).

5. Transfer the entire slide with coverslips to a 10 cm petri dish filled with tissue culture-grade PBS. Use a scalpel blade or the tip of a fine needle to slightly pry up the edge of each coverslip, allowing the liquid to flow in and separate it from the slide. The coverslips should now be coated with a uniform layer of polyacrylamide, with no visible traces remaining on the slide (*see* **Note 20**).

6. Using forceps, transfer the gel-coated coverslips (gel side up) to the 24 well plate (*see* **Note 21**). Discard the slide.

7. Repeat **steps 1–6** as needed to produce 12 1.5 kPa gel-coated coverslips and 12 30.0 kPa ones (*see* Fig. 2 and **Note 22**).

8. Wash the gel-coated coverslips five times with 1 ml of PBS per well. Leave the final wash in the wells, and store the plate at 4 °C until the protein coating step (*see* **Note 23**).

3.2 Substrate Cross-Link Activation and ECM Protein Coating

1. In a tissue culture cabinet, prepare the UV lamp, parafilm-lined tray, 10 cm parafilm-lined dishes, well plate containing the gel-coated coverslips in PBS, sterile forceps, ECM protein mixture on ice, 50 mM HEPES pH 8.5, 4 sulfo-SANPAH aliquots on dry ice, and 5–10 ml of reagent grade H_2O. *See* Fig. 1 for overview of the subsequent steps.

2. Lay out four gel-coated coverslips gel side up on the parafilm-lined tray. Tilt the tray slightly, and use light suction to remove excess PBS, but do not allow the gel to dry out (*see* **Note 24**).

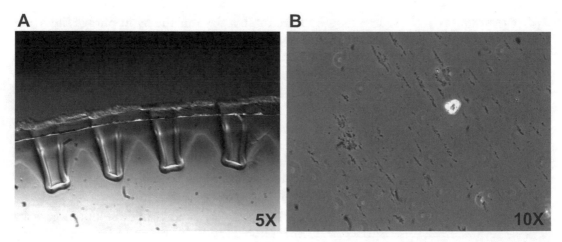

Fig. 2 Phase-contrast images of common artifacts which occur during gel formation and protein cross-linking, showing a 30.0 kPa PA gel substrate treated with a mixture of 0.1 mg/ml collagen Type 1 and 20 µg/ml laminin (magnification indicated in the lower right). (**a**) "Ruffles" along the edges of the hydrogel are relatively common and do not affect the elastic properties of the remainder of the substrate. (**b**) Image of the surface of a substrate unsuitable for cell culture because of the presence of protein aggregates. These distort the elastic properties of the substrate and cause cells to preferentially cluster in areas of high ECM protein density. Properly coated substrates should appear to be optically clear in phase-contrast images

3. Working as quickly as possible, add 480 μl of reagent grade H$_2$O to an aliquot of sulfo-SANPAH, pipette up and down several times to completely dissolve it, and add 75 μl of the solution to each coverslip, making sure the entire surface is covered (*see* **Note 25**).

4. Place the tray with the coverslips under a 365 nm UV source for 5 min. The sulfo-SANPAH solution should uniformly turn from a bright orange color to rusty brown (*see* **Note 10**).

5. Using a 5 ml serological pipette, rinse the coverslips several times with 50 mM HEPES pH 8.5 until there is no visible brownish residue.

6. Transfer the coverslips to a 10 cm dish lined with parafilm. Tilt the dish and carefully aspirate any liquid visibly pooling on the surface of the coverslips, but do not allow the coverslips to become completely dry.

7. Add 150 μl of the ECM protein solution to each coverslip, close the dish, and place it in a tissue culture incubator at 37 °C overnight.

8. Repeat **steps 2–7** to activate and protein coat all the remaining coverslips of both stiffness conditions.

9. On the following day, prepare a new 24 well tissue culture plate, and add 1 ml of ice-cold PBS to all the wells. Transfer the coated coverslips from the 10 cm dishes to the 24 well plate, and then rinse 4–5 times with 1 ml of ice-cold PBS to remove the protein solution. Leave the last PBS wash in the wells. The substrates are now ready for seeding but can be stored in PBS at 4 °C for 24 h (*see* Fig. 2 and **Note 26**).

10. Although the substrates are now ready for cell culture, additional validation steps are recommended to verify both the elastic properties of the substrates and the uniformity of ligand coating (*see* **Note 27** and Fig. 4).

3.3 Schwann Cell Culture

1. Place the SC growth medium and 0.25% trypsin-EDTA in a 37 °C water bath.

2. Aspirate the PBS from the well plate holding the protein-coated substrates, add 300 μl of SC growth medium per well, place the plate in the 37 °C incubator, and wait for 30–45 min.

3. Wash a plate of confluent Schwann cells in SC growth medium three times with 5 ml of room temperature HBSS. After removing the final wash, immediately add 5 ml of 0.25% trypsin-EDTA, and incubate at 37 °C for 3–5 min.

4. Add 5 ml of SC growth medium to the plate to stop the trypsin, rinse several times using a 5 ml serological pipette to fully detach all cells, and transfer the suspension to a waiting 15 ml centrifuge tube. Centrifuge at 1000 rpm/200 × *g* for 5 min at room temperature to pellet the cells.

5. Remove the supernatant, tap the tube repeatedly to break up the pellet, add 5 ml of SC growth medium, and pipette up and down until the cells form a uniform suspension.

6. Count the cells, and dilute them to a final concentration of 6×10^4 cells/ml, using 0.5 ml of SC growth medium per substrate (*see* **Note 28**).

7. Remove the well plate with the substrates from the incubator, aspirate the medium, and add 0.5 ml of cell suspension to each of the wells. This should result in a density of $\sim 2 \times 10^4$ cells per substrate (*see* **Note 29**).

8. Incubate the cells overnight to allow the cells to attach (*see* Fig. 3a), and then replace the medium with 0.3 ml per well of SC low serum medium. Culture the cells for another 24 h (*see* **Note 30**).

9. Replace the medium with 0.3 ml of SC differentiation medium, and culture for 48 h while monitoring the cells for morphological signs of differentiation (*see* Fig. 3b and **Note 31**).

10. Fix and perform immunocytochemistry using standard protocols (*see* **Note 32**).

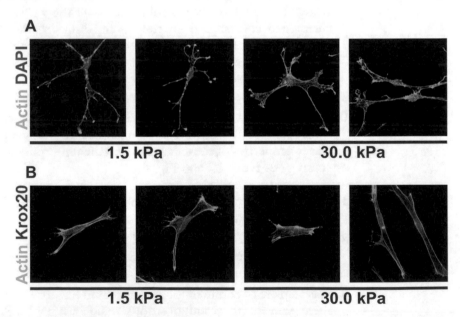

Fig. 3 Effects of matrix stiffness on SC morphology and Krox20 expression, representative cells shown at 63× magnification. (**a**) SC cultured for 24 h in proliferation medium. Although the cells appear highly arborized in both stiffness conditions, those grown on the soft matrix display thinner and more elongated cells processes, as well as reduced membrane spreading. (**b**) SC serum starved for 24 h and treated with 1 mM db-cAMP for 48 h. The induction of differentiation by cAMP abolishes most of the morphological differences due to matrix elasticity, but stiff matrix causes a significant increase in the number of cells expressing high levels of Krox20, a key regulator of SC differentiation and myelination. *See* Urbanski 2016 et al. [7] for detailed discussion of the quantitation of SC morphology and Krox20 expression in this tissue culture system

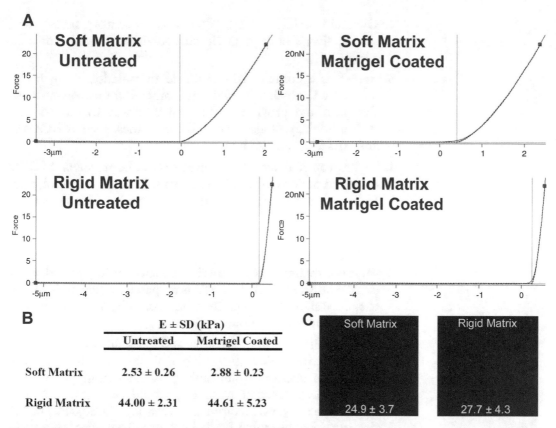

Fig. 4 AFM and immunostaining-based validation of PA/ECM substrates comparable to those described in this protocol. (**a**) Representative AFM force indentation curves of either uncoated substrates (left panels) or matrigel-coated substrates (right panels). The vertical dashed lines indicate the initial contact point between probe and substrate; protein coating has no readily observable effect on the indentation profile. (**b**) Quantitation of the effects of matrigel coating on soft and rigid substrate elasticity (six substrates tested per condition, ~64 measurements per substrate). Treatment with cross-linker and protein coating appear to cause a slight increase in elastic modulus but do not significantly affect the elastic properties of the substrates. (**c**) Representative images of soft and rigid matrigel-coated substrates immunostained for laminin, average fluorescence intensity per condition shown. The difference in laminin density is not statistically significant ($p = 0.22$, arbitrary units ± SD, three substrates stained per condition, four random fields imaged per substrate) (Modified from Urbanski et al. 2016 [7], as permitted under the Creative Commons Attribution 4.0 International Public License, http://creativecommons.org/licenses/by/4.0/)

4   Notes

1. APTMS is mildly corrosive and will react with some laboratory plastics. Covered glass containers should be used to hold the APTMS solution and any plastic pipettes tested for reactivity before use. The reaction is not violent, but results in the production of a precipitate that may attach to the surface of the coverslips being treated.

2. Glutaraldehyde is an eye, skin, and respiratory irritant and should always be kept in closed containers in well-ventilated areas.

3. Rain-X is a commercially available formulation of hydroxy-terminated polydimethylsiloxane marketed for use on automotive glass and provides a safe and easy way to increase the hydrophobicity of glass slides. Other silanes, such as DCDMS and APTMS, can also be used.

4. Light-protected acrylamide mixtures can be stored at 4 °C for extended periods of time, but to maintain consistent polymerization, we prepare fresh stocks monthly. Unpolymerized acrylamide is a neurotoxin; gloves should be worn to prevent skin contact.

5. Although it is possible to prepare frozen stocks of 10% APS solution, we find that polymerization speed is higher and much more consistent when 10% APS is prepared immediately before use. Care should be taken that the stock of solid APS does not absorb atmospheric moisture.

6. Sulfo-SANPAH is light-sensitive and has a short aqueous half-life of ~5 min. Since preparing a fresh batch before every experiment and discarding any unused solution (as per the manufacturer's recommendation) would be impractical and extremely costly, we recommend preparing aliquots in DMSO that can safely be stored for 3–4 months. To prepare the stock, dissolve 50 mg of sulfo-SANPAH in 2 ml of anhydrous DMSO, divide into 20 µl aliquots while minimizing light exposure, and freeze at −80 °C. (Although darkroom conditions are ideal from the point of view of light sensitivity, they are rarely compatible with sterile technique—it is enough to work in a tissue culture cabinet without any direct light sources and the minimal amount of ambient light that allows for accurate pipetting.) During the activation protocol, necessary aliquots should be light-protected and kept on dry ice until needed.

7. Concentrated laminin should be allowed to thaw on ice and kept refrigerated or on ice at all times to prevent aggregate formation. In addition, 50–100 µl working aliquots should be prepared and stored at −20 °C to avoid multiple freeze-thaw cycles.

8. Collagen precipitates easily in neutral or basic solution. While it is possible to dilute collagen by adding it to a basic solution (such as 50 mM HEPES pH 8.5) with extensive vortexing, we recommend adding the collagen to a neutral medium first, then increasing the pH with sodium bicarbonate.

9. To sterilize the parafilm-lined dishes, wipe down with 70% ethanol, dry, and irradiate for 10 min using the standard UV lamp of a tissue culture cabinet.

10. Various UV sources can be used to initiate the cross-linking step of this protocol. In our laboratory, we use a portable 8 W 365 nm UV lamp, which can easily be placed inside a tissue culture cabinet (approximate dimensions $400 \times 75 \times 120$ mm). The lamp is placed directly on top of a plastic tray with a 5 mm raised lip, resulting in a distance of ~15 mm between the light tube and the substrates placed within the tray. UV wavelengths from 320 to 365 nm are ideal for cross-link activation. While 254 nm can be used as well, it requires longer exposure times and causes photo-destruction of protein. This can be an important consideration in cases when a single cross-linking step does not allow for uniform protein binding or when the protein-coated substrates need to be re-sterilized after protein coating.

11. Prepare sterile stocks of both rhNRG1β1-EGF domain and forskolin, and add them after the filtration step to avoid losses resulting from binding to the filter membrane.

12. We normally use rat Schwann cells purified from P2 sciatic nerves [12]. The cells are stored cryogenically at passage 3 or 4 and propagated on poly-L-lysine (0.01 mg/ml)-coated dishes in SC growth medium until confluent and split as necessary. Typically, cells of passage 5 are used for experiments, and cells are not propagated past passage 6.

13. It is important to rinse away all the APTMS, since it will react with the glutaraldehyde used in the next step and form an orange residue on the coverslip surface that may interfere with imaging.

14. The hydrophilic coverslips can be briefly stored in a dry place before use (24–48 h) or up to a week in a sealed container with desiccant.

15. This step can be optional, but it reduces water spotting, which can be helpful in keeping the surface of the acrylamide gel as clean and uniform as possible.

16. The hydrophobic slides can be stored for 1–2 weeks with no special precautions.

17. Although protocols involving acrylamide polymerization frequently recommend degassing of the mixture prior to the addition of APS and TEMED, we do not find it to be necessary.

18. While it is technically possible to prepare larger numbers of substrates at a time, using one reaction tube to prepare too many substrates results in inconsistent polymerization and

greater variability in the elastic properties of the polyacrylamide coating and is not recommended.

19. The exact rate of polymerization will vary depending on the quality of the reagents used, and the optimal time must be determined by the user. However, polymerization times longer than 10 min most likely indicate either a low quality of reagents (in particular APS and TEMED) or errors in the preparation of the reaction mix. A small amount of drying and deformation might be observed at the edges of the gel, but this does not affect the elastic properties of the rest of the gel surface.

20. Do not attempt to separate the coated coverslip from the slide by forcibly pulling or prying them apart, since this will most likely break the coverslip and tear the gel. A very small amount of force applied to rise the edge of the coverslip should be sufficient to let it break free when submerged in PBS. We recommend using a disposable scalpel blade or a fine (25 G or thinner) needle, since most forceps are either not fine enough or would be damaged in the attempt.

21. Try to grasp the coverslips no deeper than 0.5–1.0 mm from the edge (this area will often already have some imperfections resulting from the polymerization process and should not be imaged) to minimize damage to the gel surface. When transferring the coverslips to a well containing PBS, insert them into the liquid at a 45° angle to minimize their tendency to float or turn over.

22. We recommend preparing 50% more substrates than called for in the experimental design, since they are very delicate and easy to damage during manufacture. Substrates with tears in the gel coating (aside from minor ones on the periphery) or gaps in gel coating resulting from air bubbles should be discarded. Any excess but undamaged substrates may be useful for the validation steps described below.

23. The gel-coated coverslips can be stored for extended periods of time (up to a month) provided they're hydrated and kept at 4 °C.

24. Turn down the vacuum as low as possible, and do not bring the tip into direct contact with the edge of the coverslip to avoid ripping the gel substrates.

25. Since sulfo-SANPAH is light-sensitive and has a short half-life (~5 min) in aqueous solution, attempting to activate more than four coverslips at a time is not recommended, as it will most likely result in large inconsistencies in cross-link density and protein coating.

26. Verify the absence of visible protein aggregates, gelled protein solution, or physical damage to the gel surface by microscopic observation (5× or 10× objective) of the substrates. While a small amount of fouling (occasional dust particles, fibers) is unavoidable, large amounts of either gelled protein or visible deposits on the gel surface can be an indication that the protein concentration is higher than indicated in the protocol, or that the pH of the protein solution was higher than ideal.

27. The protocol described here should result in elasticity values close to those presented in Table 1 as well as consistent ligand coating. However, the quality of the reagents used to prepare the PA hydrogels and human error can result in defects that will not be apparent on visual inspection. Therefore, independent validation (at least on a small scale) is highly recommended. In our laboratory, we have used AFM nano-indentation to collect force curves (fitted using the Hertz model) and verify the elastic properties of the gel substrates, both before and after ECM protein coating (*see* Fig. 4a, b). A probe with a 6.1 μm polystyrene bead was used for the AFM measurements, with a trigger point of 100 nN and an indentation velocity of 50 μm/s, with the relatively high indentation velocity selected to allow for the collection of clean force curves despite the gels' high rate of creep. The coated substrates were also immunostained for one of the ECM components used and imaged confocally to verify uniform and consistent protein deposition (*see* Fig. 4c) and to ensure that the protein coating was limited to a thin layer on the surface of the hydrogel. Addition of a small (10% of total protein) quantity of fluorescently labeled ECM protein to the master mix can also be used to verify ligand density in lieu of immunostaining.

28. The number of cells recommended here assumes that some will settle on the plastic of the well, leaving approximately 20×10^4 cells on the substrate itself. The relatively high volume of medium used per well is meant to minimize meniscus formation and promote even cell dispersal. If it is necessary to seed cells directly onto the substrates (e.g., due to a limited number of transgenic cells), the substrates can be placed in dishes lined with sterile parafilm, seeded with ~150 μl of cell suspension, incubated overnight to allow for cell attachment, and returned to a 24 well plate for the remainder of the experiment.

29. SC responses to mechanical stimuli are significantly affected by culture density. We have observed that the upregulation of Krox20 by increased matrix stiffness described here [7] is abolished at high cell densities (unpublished data), and it has also

been reported that SC density affects the localization and activity of YAP/TAZ, one of the key mediators of mechano-transduction [11].

30. The serum starvation is necessary for the efficient induction of differentiation. Optionally, the cells can be rinsed once with HBSS to make sure all the residual high-serum SC growth medium has been removed.

31. As SC differentiate, there should be a visible reduction in the number of cytoskeletal stress fibers, significant membrane expansion, shortening of cell processes, and a development of a more polygonal morphology (*see* Fig. 3b).

32. The SC cultures described here can be treated similarly to ones grown on glass or plastic in most respects, although they are more delicate—extra care should be taken to make all washing and pipetting steps as gentle as possible to minimize cell loss through detachment. The cultures can be fixed for 10 min in 4% paraformaldehyde in PBS, followed by 2–3 PBS washes, and stored at 4 °C for a period of several weeks. For immunocytochemistry, the cells are permeabilized for 5 min with 0.5% Triton X-100, while both primary and secondary antibody solutions are prepared in 5% BSA, 1% donkey serum, and 0.2% Triton X-100 in PBS, which also serves as the blocking solution.

References

1. Engler AJ, Sen S, Sweeney HL, Discher DE (2006) Matrix elasticity directs stem cell lineage specification. Cell 126(4):677–689. https://doi.org/10.1016/j.cell.2006.06.044

2. Wozniak MA, Chen CS (2009) Mechanotransduction in development: a growing role for contractility. Nat Rev Mol Cell Biol 10(1):34–43. https://doi.org/10.1038/nrm2592

3. Janmey PA, Miller RT (2011) Mechanisms of mechanical signaling in development and disease. J Cell Sci 124(Pt 1):9–18. https://doi.org/10.1242/jcs.071001

4. Handorf AM, Zhou Y, Halanski MA, Li WJ (2015) Tissue stiffness dictates development, homeostasis, and disease progression. Organogenesis 11(1):1–15. https://doi.org/10.1080/15476278.2015.1019687

5. Tse JR, Engler AJ (2010) Preparation of hydrogel substrates with tunable mechanical properties. Curr Protoc Cell Biol Chapter 10:Unit 10.16. https://doi.org/10.1002/0471143030.cb1016s47

6. Fischer RS, Gardel M, Ma X, Adelstein RS, Waterman CM (2009) Local cortical tension by myosin II guides 3D endothelial cell branching. Curr Biol 19(3):260–265. https://doi.org/10.1016/j.cub.2008.12.045

7. Urbanski MM, Kingsbury L, Moussouros D, Kassim I, Mehjabeen S, Paknejad N, Melendez-Vasquez CV (2016) Myelinating glia differentiation is regulated by extracellular matrix elasticity. Sci Rep 6:33751. https://doi.org/10.1038/srep33751

8. Georges PC, Miller WJ, Meaney DF, Sawyer ES, Janmey PA (2006) Matrices with compliance comparable to that of brain tissue select neuronal over glial growth in mixed cortical cultures. Biophys J 90(8):3012–3018. https://doi.org/10.1529/biophysj.105.073114

9. Levental I, Georges PC, Janmey PA (2007) Soft biological materials and their impact on cell function. Soft Matter 3(3):299–306. https://doi.org/10.1039/B610522J

10. Dupont S, Morsut L, Aragona M, Enzo E, Giulitti S, Cordenonsi M, Zanconato F, Le

Digabel J, Forcato M, Bicciato S, Elvassore N, Piccolo S (2011) Role of YAP/TAZ in mechanotransduction. Nature 474(7350):179–183. https://doi.org/10.1038/nature10137

11. Poitelon Y, Lopez-Anido C, Catignas K, Berti C, Palmisano M, Williamson C, Ameroso D, Abiko K, Hwang Y, Gregorieff A, Wrana JL, Asmani M, Zhao R, Sim FJ, Wrabetz L, Svaren J, Feltri ML

(2016) YAP and TAZ control peripheral myelination and the expression of laminin receptors in Schwann cells. Nat Neurosci 19(7):879–887. https://doi.org/10.1038/nn.4316

12. Porter S, Clark MB, Glaser L, Bunge RP (1986) Schwann cells stimulated to proliferate in the absence of neurons retain full functional capability. J Neurosci 6(10):3070–3078

Purification of Exosomes from Primary Schwann Cells, RNA Extraction, and Next-Generation Sequencing of Exosomal RNAs

Cristian De Gregorio, Paula Díaz, Rodrigo López-Leal, Patricio Manque, and Felipe A. Court

Abstract

Exosomes are small (30–150 nm) vesicles of endosomal origin secreted by most cell types. Exosomes contain proteins, lipids, and RNA species including microRNA, mRNA, rRNA, and long noncoding RNAs. The mechanisms associated with exosome synthesis and cargo loading are still poorly understood. A role for exosomes in intercellular communication has been reported in physiological and pathological conditions both in vitro and in vivo. Previous studies have suggested that Schwann cell-derived exosomes regulate neuronal functions, but the mechanisms are still unclear. Here, we describe protocols to establish rat neonatal Schwann cell cultures and to isolate exosomes from the conditioned medium of these cultures by differential ultracentrifugation. To analyze the RNA content of Schwann cell-derived exosomes, we detail protocols for RNA extraction and next-generation sequencing using miRNA and mRNA libraries. The protocol also includes RNA sequencing of Schwann cells, which allows the comparison between RNA content from cells and the secreted exosomes. Identification of RNAs present in Schwann cell-derived exosomes is a valuable tool to understand novel roles of Schwann cells in neuronal function in health and disease.

Key words Exosomes, Microvesicles, Transcriptomics, mRNA, miRNA, Schwann cell, Neuron, Intercellular communication

1 Introduction

Exosomes are a subclass of extracellular vesicles (EVs) secreted by most types of cells and found in multiple body fluids. Exosomes are the smallest described EVs, ranging from 30 to 150 nm, and originate by the fusion of multivesicular bodies (MVBs) with the plasma membrane. Exosomes can act locally or distally by travelling via different extracellular fluids and mediating intercellular communication with certain specificity. Exosomes have a wide range of functions, including immunogenic modulation [1], spreading of infectious agents [2], secretion of proteins as well as lipids for

Paula V. Monje and Haesun A. Kim (eds.), *Schwann Cells: Methods and Protocols*, Methods in Molecular Biology, vol. 1739, https://doi.org/10.1007/978-1-4939-7649-2_19, © Springer Science+Business Media, LLC 2018

disposal [3], and promotion of metastasis [4]. In addition, bio-chemical analyses have shown that exosomes also contain specific lipids and proteins. The exosomal membrane is composed mainly of sphingomyelin, cholesterol, and ceramide-rich lipid rafts [5]. Most exosomes contain proteins like endosome-associated pro-teins, tetraspanins (e.g., CD63, CD81, CD9), GTPases, chaper-ones and cytoskeletal proteins, and other proteins specific to the cell from which they were originated [5].

Since the discovery that exosomes are highly enriched in RNA molecules, the interest in these nanovesicles has expanded. Valadi et al. [6] demonstrated that exosomes derived from cells in culture were loaded with about 1300 different mRNA species, suggesting for the first time a role for exosomes in the regulation of protein synthesis in the recipient cell. Furthermore, exosomes are also loaded with over 100 different miRNA species. Given that a single miRNA molecule can modulate the translation of over 200 different mRNAs [7], exosomes constitute a powerful cellular mechanism for the reg-ulation of gene expression in the recipient cell [6, 8].

Many recent studies support a role for exosomes in a variety of pathological conditions, including the spreading of several types of cancer, as well as in metabolic and degenerative diseases [9, 10]. Skog et al. [11] proposed that exosomes could be used as diagnos-tic markers, as they showed that exosomes derived from patients with glioblastomas were loaded with specific mRNAs, which were absent in healthy individuals. To date, several studies have identi-fied specific molecular signatures associated with exosomes obtained from pathological conditions, which can be eventually used as noninvasive diagnostic tools for clinical purposes. Furthermore, exosomes have been proposed to act as immune sys-tem modulators and specific delivery vehicles for drugs and genetic material [12].

Despite the explosive growth in EVs research, there is still insufficient information related to their molecular mechanisms of action, including the mechanisms for cargo of molecules into MVs, specific cell recognition, and the modulation of genetic programs in target cells. Exosomal omics (transcriptomics and proteomics) analysis has been important to identify their molecular content, which is a first step to understand the role of these nanovesicles in health and disease [6, 13, 14].

We have previously reported that axons internalize exosomes secreted by Schwann cells (SCs) and that these exosomes enhance axonal regeneration in vitro and in vivo [15]. However, the mole-cules present in exosomes responsible for this pro-regenerative effect remain to be identified. To this end, we have used next-generation sequencing (NGS) technology as a highly sensitive method for the identification of RNA molecules contained in SC exosomes. Here, we present a protocol to generate primary rat neonatal SC cultures and isolate exosomes from the SC conditioned medium by differential ultracentrifugation, in order to analyze their RNA content by NGS using miRNA and mRNA libraries.

2 Materials

2.1 Rat Schwann Cell Primary Cultures

1. Biological safety cabinet certified for level II use.
2. Cell culture incubator (humidified, 37 °C, 5% CO_2).
3. Laminar flow hood provided with a Bunsen burner.
4. Dissecting microscope.
5. 10 cm glass Petri dishes and Pasteur pipettes, sterilized by autoclaving.
6. T25 and T75 cell culture flasks.
7. 15 ml conical centrifugation tubes.
8. Postnatal day 3 (P3) Sprague-Dawley rat pups.
9. Sterilized surgical instruments. Three straight scissors (16, 12, and 8 cm) and three forceps (one 12 cm forceps and two fine Dumont #5 forceps).
10. Ethanol 70% in distilled water.
11. Hank's balanced salt solution (HBSS).
12. Leibovitz's L-15 medium (L-15 medium).
13. 0.25% trypsin/1% collagenase solution. Add 1 ml of 2.5% trypsin solution (10× stock) and 1 ml of 1% (w/v) collagenase type 1 to 8 ml of L-15 medium.
14. 1× phosphate buffer saline (1× PBS). To prepare 1 l of 10× PBS, dissolve 80 g NaCl, 2 g KCl, 14.4 g Na_2HPO_4, and 2.4 g KH_2PO_4, in 800 ml of distilled water. Adjust the pH to 7.4 with HCl and the volume to 1 l with distilled water. Sterilize by autoclaving.
15. Bovine pituitary extract (BPE) stock solution. Dissolve lyophilized BPE extract in 1× PBS to reach a concentration of 20 mg/ml. Store in 50 μl aliquots at −20 °C for up to 3 months.
16. 1 mM cytosine β-arabinofuranoside hydrochloride (AraC) stock solution. Dissolve AraC in distilled water and sterilize by filtration. Store in 1 ml aliquots at −80 °C for up to 6 months.
17. 10 mM forskolin stock solution. Dissolve 10 mg of forskolin in 2.43 ml of sterile dimethyl sulfoxide (DMSO). Store at −20 °C for up to 4 months.
18. 100× antibiotic-antimycotic solution. 10,000 units/ml penicillin, 10,000 μg/ml streptomycin, and 25 μg/ml amphotericin B in 0.9% sterile-filtered sodium chloride solution.

19. DMEM-10% FBS. DMEM supplemented with 10% FBS and 1× antibiotic-antimycotic solution.

20. SC growth medium. DMEM supplemented with 10% exosome-free FBS (*see* Subheading 2.4, **item 1**), 2 μM forskolin, 20 μg/ml BPE, and 1× antibiotic-antimycotic solution.

21. DMEM-AraC. DMEM supplemented with 10% FBS, 10 μM AraC, and 1× antibiotic-antimycotic solution.

22. Laminin solution. Dilute mouse purified laminin to 400 ng/ml in sterile 1× PBS just before use.

23. Freezing medium. DMEM containing 10% FBS and 10% DMSO. To prepare 8 ml of freezing medium, mix 7.2 ml of DMEM-10% exosome-free FBS (*see* Subheading 2.4, **item 1**) with 800 μl of sterile DMSO. Use 1 ml of freezing medium per each cryovial. Prepare it just before use.

24. 0.25% trypsin solution. Add 1 ml of 2.5% trypsin solution (10× stock) to 9 ml of 1× PBS.

25. 0.4% trypan blue solution.

26. Hemocytometer.

27. Hand tally counter.

2.2 Complement-Mediated Cell Lysis

1. Mouse CD90 monoclonal antibody (Thy1.1, 1 mg/ml), cloneT11D7e (AbD Serotec/Biorad).

2. Rabbit complement stock solution. Reconstitute rabbit complement (lyophilized powder, Sigma Cat#S7764) with 5 ml of sterile ice-cold deionized water or according to the manufacturer's protocol. Store in 400 μl aliquots at −80 °C for up to 12 months.

2.3 Immuno-fluorescence Microscopy

1. 4% (w/v) paraformaldehyde (PFA). Dissolve 4 g of PFA powder in 90 ml of 1× PBS solution and heat to 65 °C while stirring. If needed, add drops of 1 N NaOH until the solution becomes clear. Bring to 100 ml with 1× PBS solution. Cool and filter. Store in 1 ml aliquots at −20 °C for up to 6 months.

2. Gelatin from cold water fish skin (Sigma, Cat#G7765).

3. Primary antibodies against SC-specific markers. The use of S100β (Sigma, Cat#S2532) and p75[NTR] (Millipore, Cat#07-476) primary antibodies are recommended to specifically analyze the purity of SC cultures (*see* **Note 1**).

4. Secondary antibodies. 488 and 546 Alexa Fluor-conjugated anti-rabbit and anti-mouse antibodies (Thermo Fisher).

5. Blocking solution. Dilute gelatin from cold water fish skin to 5% (v/v) in 1× PBS. Add 0.1% (v/v) final of Triton X-100 to permeabilize the cells.

6. Antibody solution. Dilute gelatin from cold water fish skin to 1% (v/v) in 1× PBS. Add 0.1% (v/v) final of Triton X-100.

7. 5 mg/ml 4′,6-diamidino-2-phenylindole dihydrochloride (DAPI) stock solution. Dissolve 10 mg of DAPI in 2 ml of deionized water. Store in aliquots at −20 °C for at least 2 years protected from light.

8. Mounting medium. As a reference, we use Mowiol 4-88.

9. Microscope slides and glass coverslips.

10. Fluorescence microscope.

2.4 Exosome Purification and Characterization

1. DMEM-10% exosome-free FBS. To prepare 1 l of DMEM-10% exosome-free FBS, ultracentrifuge 100 ml of FBS at 100,000 × *g* for 2 h at 4 °C, and collect the supernatant. Then, add 100 ml of this supernatant to 900 ml of DMEM medium and 10 ml of 100× antibiotic-antimycotic solution. Sterilize this medium by filtering (0.2 μm pore). Store at 4 °C for up to 3 weeks.

2. 50 ml conical tubes.

3. Ultracentrifuge and rotor for 100,000 × *g*. A benchtop ultracentrifuge (Optima™ MAX-XP Beckman, USA) provided with a fixed angle MLA-55 rotor supporting eight tubes can be used.

4. Ultracentrifuge tubes sterilized by UV irradiation. 13.5 ml tubes suitable for ultracentrifugation.

5. Exosomes storage solution: 1× PBS containing 10% glycerol.

2.5 Characterization of Exosomes by Transmission Electron Microscopy

1. 2% (w/v) PFA in 1× PBS. Prepare it from 4% PFA solution (*see* Subheading 2.3, **item 1**).

2. 0.1 M cacodylate buffer. Dissolve 21.4 g of sodium cacodylate in 350 ml of distilled water. Adjust the pH to 7.4 with HCl and the volume to 500 ml.

3. 2% (w/v) uranyl acetate solution. Dissolve 1 g of uranyl acetate in 50 ml of distilled water. Uranyl acetate crystals are difficult to dissolve, and it may be required to use a rotating wheel for mixing. Store for up to 4 months at 4 °C in a 20 ml plastic syringe protected from light. Filter the amount needed of uranyl solution with a 0.2 μm pore filter just before use. Observe the Institutional Radiation Safety Office guidelines concerning proper handling and disposal of this solution.

4. 2% (w/v) methyl cellulose. Dissolve 4 g of methyl cellulose in 196 ml of distilled water previously heated to 90 °C. Cool on ice with stirring until the solution reaches 10 °C, and continue to stir overnight at 4 °C. This solution should be kept for 3 days at 4 °C. Ultracentrifuge the solution at 100,000 × *g* at 4 °C and collect the supernatant. Store at 4 °C for up to 3 months.

5. Formvar-carbon-coated electron microscopy grids.

6. Forceps (fine Dumont #5).

7. Whatman #1 filter paper.

8. 80–120 kV transmission electron microscope (TEM). As a reference, we use a Tecnai 12 electron microscope (Philips) operated at 80 kV.

2.6 Exosomal and SC RNA Extraction

1. RNA extraction kit. As a reference, we use the Quick-RNA™ MiniPrep Kit (Zymo Research, Cat# R1054).

2. Bioanalyzer. As a reference, we use a 2100 Bioanalyzer (Agilent Technologies, USA).

3. RNA 6000 Pico Assay Kit (Agilent Technologies, USA).

4. Refrigerated centrifuge suitable for 1.5 ml tubes.

3   Methods

3.1 Primary Cultures of SCs with Contaminating Fibroblasts

1. Sterilize surgical instruments by autoclaving.

2. Treat a T25 flask with 2–3 ml of 400 ng/ml laminin solution for 0.5–2 h at 37 °C in a cell culture incubator. Then, discard the laminin and wash the flask with HBSS 3 times.

3. Sacrifice the rat pups by decapitation or according to local regulations. All animal procedures must be carried out in accordance to the institutional Animal Care Committee guidelines.

4. Spray the body of the animals with ethanol 70% and cut through the rat skin around the abdominal region. Remove the skin above the gluteus maximus muscles and position the rat pup ventral side down while stretching out lower limbs, i.e. splaying the hind legs into an inverted V shape.

5. With the 12 cm forceps and scissors, bilaterally dissect the sciatic nerve by separating the muscle tissue from the upper dorsal thigh. Cut the nerve just above this position at one end and immediately above the knee at the other end to obtain nerves of a length of approximately 1–1.5 cm.

6. Dissect both sciatic nerves from each rat pup and place them into 10 cm glass Petri dish containing cold L-15 medium on ice. Perform the dissection step as quickly as possible.

7. Using the dissecting microscope and two sterile forceps (fine Dumont #5), remove connective tissue from the sciatic nerve (epineurium and perineurium) and transfer the nerve to a second glass Petri dish containing cold L-15 medium on ice. Remove the connective tissue under wet conditions without letting the nerves dry.

8. Carefully shred the nerve fascicles using the fine Dumont #5 forceps while visualizing the procedure in a dissecting microscope (*see* **Note 2**).

9. In a 15 ml conical tube, combine the nerves and add 10 ml of the trypsin and collagenase solution (*see* Subheading 2.1, **item 13**). Incubate for 30 min at 37 °C in a water bath. Shake the tubes manually every 10 min.

10. Centrifuge at 400 × *g* for 5 min at room temperature (RT) to pellet the digested cellular suspension.

11. Remove the supernatant and add 5 ml of DMEM-10% FBS to the pellet. Disaggregate the nerve fragments by pipetting through a blunt-ended Pasteur pipette 10 times. To blunt the end of the pipette, flame it in a Bunsen burner.

12. Plate the cells in a T25 laminin-pretreated flask in DMEM-10% FBS. Let the cells adhere overnight in the cell culture incubator.

13. After 24 h, remove the medium and refill the flask with 5 ml of DMEM-10 μM AraC (*see* **Note 3**).

14. After 48 h, change the medium to remove residual AraC and allow the cells to recover in 5 ml of DMEM-10% FBS.

15. After 48 h completely change the medium and add 5 ml of SC growth medium (Fig. 1).

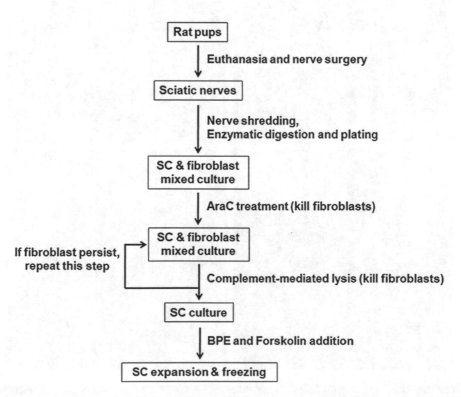

Fig. 1 Schwann cell culture protocol. Flowchart showing the main steps for establishing a primary SC culture from rat neonatal pups. This method includes three main steps: (1) sciatic nerve dissection and dissociation, (2) purification of SCs from a mixed culture (with fibroblasts), and (3) expansion and freezing of purified SCs

3.2 Complement-Mediated Cell Lysis

This technique eliminates contaminant fibroblasts by complement-mediated cell lysis using IgM-class mouse CD90 monoclonal antibody and rabbit complement proteins. CD90 is an antigen located in the fibroblast membrane, which allows for the selective elimination of the remaining contaminating fibroblasts by complement-mediated cytolysis. After this procedure, the percentage of Schwann cells in the culture can be measured by immunofluorescence using antibodies against S100 and p75NTR together with DAPI (Fig. 2). The following protocol is intended for a T25 flask.

1. Wash the plate with 3 ml of HBSS.
2. Add 2 ml of DMEM-10% FBS containing 10 μl of CD90 antibody and incubate for 15 min in the cell culture incubator.

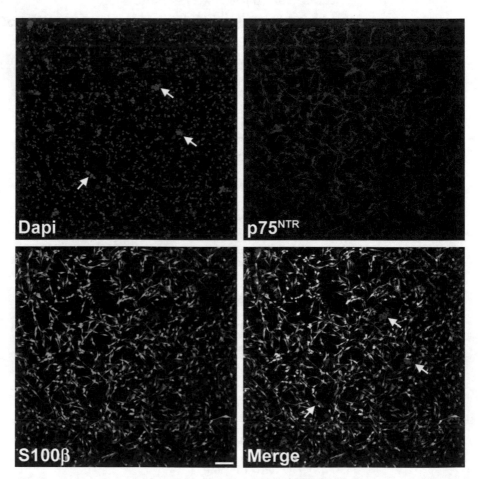

Fig. 2 Establishment of primary rat Schwann cell cultures. Representative images of immunofluorescence of primary SC cultures (passage 2) stained with the SC markers p75 (red) and S100β (green) and the nuclei marker DAPI (blue). White arrows show some contaminant cell clusters (probably fibroblasts, *see* **Notes 6** and **7**). Scale bar, 50 μm. The protocol for immunofluorescence is detailed in Subheading 3.6

3. Add 400 μl of rabbit complement solution to the culture medium and mix by gently swirling the dish. Incubate for 30–40 min in the cell culture incubator. Observe the progress of killing under the microscope every 10 min making sure that SCs remain attached to the dish (*see* **Note 4**).

4. Remove the medium and wash twice with 3 ml of HBSS.

5. Add 5 ml of SC growth medium.

6. After 3 days, repeat the procedure described in Subheading 3.2 if needed (*see* **Note 5**).

7. Culture the cells for 7 days or until they become confluent.

3.3 Schwann Cell Expansion

When cells are confluent in the T25 flask, the cells should be sub-cultured to passage 2. Essentially, the SC suspension derived from one T25 flask is split into four T75 flasks. The following steps are recommended to passage the cells (Fig. 1).

1. Remove and discard the culture medium.

2. Rinse the cells with 5 ml of HBSS.

3. Detach and dissociate the cells by adding 5 ml of 0.25% trypsin solution.

4. Place the flask in the cell culture incubator for 5 min. After the incubation, add 5 ml of DMEM-10% FBS.

5. Transfer the detached cell suspension to a sterile 15 ml conical tube.

6. Centrifuge at $700 \times g$ for 5 min at RT.

7. Resuspend the cell pellet with 5 ml of DMEM-10% FBS.

8. Centrifuge again at $700 \times g$ for 5 min at RT.

9. Resuspend the cell pellet in 1 ml of DMEM-10% FBS.

10. Count the cells in the suspension (*see* **Note 6**).

11. Plate 3×10^5 cells per laminin-coated T75 flask.

12. Add 10 ml of SC growth medium to the flask.

13. Change the medium every 2–3 days. Culture the cells for 7 days or until they reach confluency.

3.4 Freezing of SCs at the Second Passage

Proceed to freeze the cells after 7 days of in vitro culture or when reaching confluency.

1. Trypsinize the cells as described in Subheading 3.3.

2. Count the cells in the suspension as described in Subheading 3.3.

3. Resuspend 7×10^5 cells in 1 ml of freezing media (*see* Subheading 2.1, **item 23**) in a sterile cryovial.

4. Place the vials in the −80 °C freezer and then transfer them to liquid nitrogen.

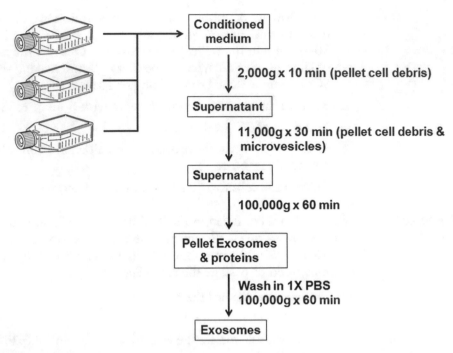

Fig. 3 Exosome purification protocol. Flowchart showing the main steps for exosome purification from SC conditioned medium. This method is based on serial centrifugation and ultracentrifugation, which eliminates cell debris, microvesicles, and contaminating proteins. Further analysis may be necessary to determine the purity of the final exosome-containing pellet (*see* **Note 13**)

3.5 Exosome Purification

For a single RNA extraction, use 120 ml of SC conditioned medium derived from 12 confluents (90–100%) T75 plates. All the exosome purification steps described here are carried out using cells collected at passage 4 (*see* Fig. 3, **Note 8**).

1. Thaw a SC vial containing 7×10^5 cells and plate them in a laminin pretreated T75 flask containing 10 ml of SC growth medium.

2. Allow the cells to proliferate for 7 days or until the cells become confluent. Passage the cells as described in Subheading 3.3 and expand them for 7 days or until they reach confluency (90–100%). Remove the cell medium and add 10 ml of SC growth medium. After 48 h, transfer the conditioned medium (10 ml per flask) to 50 ml conical centrifuge tubes. Add 10 ml of SC growth medium to the T75 flasks (*see* **Note 9**).

3. Centrifuge the tubes at $2000 \times g$ for 10 min at 4 °C, and collect the supernatant (*see* **Note 10**).

4. Transfer the supernatant to ultracentrifuge tubes.

5. Centrifuge at $11,000 \times g$ for 30 min at 4 °C, collect the supernatant, and transfer to a new clean ultracentrifuge tube.

6. Label one side of each ultracentrifuge tube with a waterproof marker. Orient the tube in the rotor with the label facing up and centrifuge at $100,000 \times g$ for 60 min at 4 °C. Recover the pellet by removing the supernatant. Use the mark as a reference for the localization of the pellet at the end of the ultracentrifugation.

7. Resuspend the pellet of each tube in 1 ml of sterile 1× PBS pipetting up and down several times. Combine the resuspended pellets in a single ultracentrifuge tube. Then, add 1× PBS to fill the tube completely.

8. Centrifuge for 1 h at $100,000 \times g$ at 4 °C (*see* **Note 11**).

9. Remove the supernatant. Keep the tube upside down, and aspirate the remaining liquid on the sides and the mouth of the tube with a micropipette.

10. Add a small volume (50–100 μl) of cold 1× PBS to resuspend the pellet containing exosomes. Pipette up and down several times to facilitate the resuspension of the exosomal pellet.

11. Estimate the amount of exosomes secreted by SCs in each preparation. The total protein obtained in the exosome suspension may be quantified by the method of preference (e.g. Bradford or BCA assay). The amount of collected exosomes is usually expressed as μg of exosomal protein per 1×10^6 cells. This number ranges between 0.1 and 0.5 μg of protein/1×10^6 cells, which is in the same range obtained by other authors [16–18] (*see* **Note 12**).

12. Store exosomes at 4 °C for up to 3 weeks or freeze them at −80 °C in exosome storage solution for up to 1 year. Avoid repeated freezing and thawing.

3.6 Characterization of SC Primary Cultures by Immunofluorescence

To assess the purity of the primary SC culture preparation, an immunofluorescence assay should be performed using specific SC markers (*see* **Note 1**). The procedure is performed as follows.

1. Place sterilized coverslips (13 mm) into a multi-well plate (24 wells) and add 500 μl of 400 ng/ml laminin solution. After 2 h remove the laminin and wash 3 times with HBSS.

2. Seed the cells obtained from Subheading 3.5, **step 2**, at a density of 5000–10,000 cells/well.

3. Grow the cells with SC growth medium until they reach the desired confluence.

4. Discard the growth medium from the wells and wash the cells three times with 1× PBS every 5 min at RT.

5. Fix the cells with 4% PFA for 20 min at RT.

6. Remove the PFA and wash the cells by adding 1× PBS. Repeat this step 3 times every 5 min at RT.

7. Treat cells for 2 h at RT with blocking solution.

8. Incubate the cells with primary antibody diluted in antibody solution overnight at 4 °C.

9. Remove primary antibody and wash the cells with 1× PBS 3 times every 5 min at RT.

10. Incubate the cells with secondary antibody diluted in antibody solution for 2 h at RT in a dark and humidified chamber.

11. Remove the secondary antibody solution and wash the samples with 1× PBS 3 times every 5 min at RT. If nuclear staining is desired, replace the second 1× PBS washing step with DAPI solution and incubate for 10 min at RT before proceeding with the third 1× PBS washing step.

12. Mount the coverslips with 10 μl mounting medium. For liquid mounting media, apply clear nail polish along the edges of the coverslip to seal.

13. Samples are ready to be observed by fluorescence microscopy (Fig. 2).

3.7 Exosome Characterization by TEM

1. Deposit a drop containing 10% of a single exosomal preparation volume (5–10 μl aliquot of the exosomal ultracentrifuged pellet) on Formvar-carbon-coated EM grids. Cover and let the membranes adsorb for 20 min in a dry environment.

2. Fix the exosomes with 2% PFA for 10 min and wash the grids with 0.1 M cacodylate buffer.

3. Contrast the grids with 2% uranyl acetate solution for 15 min.

4. Transfer the grids with the membrane side down to a 50 μl drop of methyl cellulose using clean forceps. Incubate for 10 min on ice (*see* **Note 14**).

5. Remove the grids and drain excess fluid using a Whatman filter paper. Air dry the grid for 5–10 min.

6. Examine the exosomes under the electron microscope at 80–120 kV (*see* Fig. 4, **Note 13**).

3.8 RNA Extraction from SCs and Exosomes

Exosomal RNAs are isolated from the exosomal-enriched fractions obtained by ultracentrifugation. RNA extractions from SC cultures should be performed in parallel for the bioinformatics analysis of exosome-secreting cells. To perform RNA extractions, a column-based kit is used in order to obtain a salt-free, protein-free, highly concentrated RNA sample. It is important to use an extraction method compatible with downstream applications. In this case, the Quick-RNA™ MiniPrep Kit efficiently isolates RNAs of a wide range of sizes, including small RNAs ≥17 nucleotides, necessary for miRNA sequencing. With this protocol, highly concentrated RNA suitable for RT-PCR and RNA sequencing is obtained. All buffers and purification columns are supplied by the Kit.

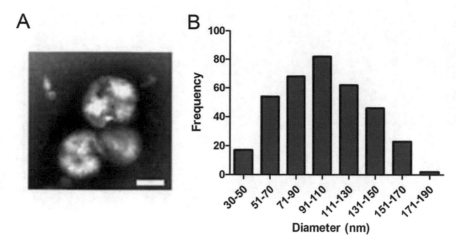

Fig. 4 Characterization of purified exosomes by TEM. (**a**) Representative image of exosomal fractions obtained by negative-stained preparations. Particles of approximately 100 nm are detected. Scale bar, 50 nm. (**b**) Histogram showing the exosome diameter distribution from our preparations. Figure adapted from reference [15] (*see* **Note 15**)

1. Resuspend the exosome pellet in 500 μl of chilled RNA lysis buffer and homogenize the pellet using a vortex. Place the samples on ice.

2. In parallel, resuspend the SC cultures from a plate containing ~2 × 10⁶ cells with 500 μl of chilled RNA lysis buffer. Collect the cell lysate with the aid of a scraper and transfer it to a 1.5 ml tube. Vortex to enhance cell rupture.

3. Clear the lysates from **steps 1** and **2** by centrifugation at 14,000 × *g* for 1 min at 4 °C.

4. To remove genomic DNA from the SC extracts, transfer the aqueous phase from the previous step to a Spin-Away™ filter in a collection tube. Proceed to centrifuge the samples at 14,000 × *g* for 1 min at 4 °C. Collect the flow-through and proceed to RNA purification. From this step on, both samples (SCs and exosomes) are subjected to the same protocol.

5. Mix the lysate vigorously by vortexing with 1 volume of ethanol 100% (500 μl) for 10 s, and then transfer this lysate to a Zymo-Spin™ column. Centrifuge the lysate at 14,000 × *g* for 1 min at 4 °C and discard the flow-through.

6. Wash the column with 400 μl of RNA wash buffer and centrifuge it at 14,000 × *g* for 30 s. Discard the flow-through.

7. Treat the samples in the column matrix with DNase I prepared in DNase I digestion buffer (provided with the kit) for 15 min at RT to remove remaining DNA. Then, centrifuge the column at 14,000 × *g* for 30 s.

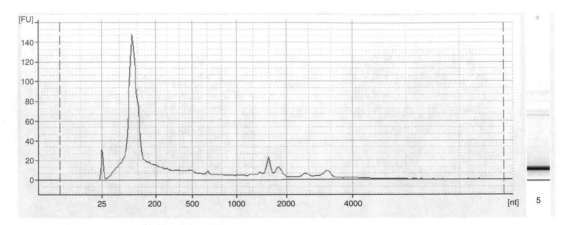

Fig. 5 Characterization of exosome RNA using Bioanalyzer. Representative image displaying the total RNA content from a sample of SC-secreted exosomes obtained by RNA PicoChip Kit (Agilent Technologies, USA). Bioanalyzer data show that SC-derived exosomes contain a broad range of RNA sizes including a high amount of small RNAs (25–200 nucleotides). This profile also shows a low level of intact rRNA, according to previous reports [14]

8. Wash the column twice with 400 µl of RNA wash buffer and centrifuge it at $14,000 \times g$ for 2 min each time.

9. Transfer the column to an RNase-free tube and add 10–15 µl of RNase-free water to the column matrix (*see* **Note 16**). Centrifuge the column for 1 min at $14,000 \times g$ at 4 °C, and store the eluted RNA immediately at −80 °C.

10. Evaluate RNA integrity using a Bioanalyzer (*see* Fig. 5 and **Note 17**). RNA samples should show similar profiles to the ones found in the literature [14] to be considered of enough quality for downstream applications.

11. Perform the sequencing analysis according to the goals of the experiment and the manufacturer's instructions (*see* **Note 18**).

4   Notes

1. In vitro SC cultures share similar expression profiles to dedifferentiated SCs after nerve injury. S100β, p75[NTR], and Sox10 are some appropriate markers for dedifferentiated SCs [19].

2. The shorter the time used, the better the yield. An optimal time is 1.5 h for dissection to then start the enzymatic digestion of the nerves.

3. SCs do not proliferate efficiently in medium containing serum but fibroblasts do. To avoid fibroblast proliferation, it is recommended to use antimitotic agents such as AraC, which impairs DNA synthesis and kills rapidly dividing cells while leaving SCs intact.

4. SCs have a bipolar, spindle-shaped morphology, clearly distinguishable from the flattened and spread-out morphology of fibroblasts.

5. If contaminating fibroblasts remain in the plate, repeat the complement-mediated lysis step. We obtain a final SC purity of ~95% with 2–3 steps of complement-mediated cell lysis.

6. To count the cells, mix 10 μl of cellular suspension with 10 μl of trypan blue. Place 10 μl of this mixture in a hemocytometer and count the cells using a hand tally counter. Do not count dark, dead cells.

7. This protocol produces a ~95% pure SCs culture (*see* Fig. 2). It is important to consider this fact if using highly sensitive techniques such as RT-PCR or qPCR that may detect the expression of non-SC transcripts.

8. Our observations indicate that SC cultures are not stable through time, as SCs may change their phenotype after passage 6. Thus, we strongly recommend defining a specific passage for your studies and characterize the state of differentiation of the cultures if you are interested in using cells expanded over passage 6.

9. Purification of exosomes from the SC conditioned medium can be performed every 48 h until SCs are completely confluent. Nevertheless, we have found that the maximum amount of exosomes in the conditioned medium is reached at around 12 h after changing the medium. We usually perform up to three extractions from the same culture plate. In order to decide the amount of time after extracting exosomes from the SC conditioned medium, the amount of exosomes secreted by SCs should be assessed. To this end, extract exosomes from conditioned medium collected at 12 h and 1, 2, and 3 days and measure protein content from the exosomal sample after ultracentrifugation.

10. From this point on, work on ice or at 4 °C to better preserve exosome structure and content.

11. The pellet at this point is very loose, and you should be very careful in removing the supernatant (use a micropipette).

12. Exosome secretion seems to be a general mechanism, but the amount of secreted exosomes is quite variable depending on the cell type and proliferative rate.

13. Analysis of ultracentrifuged exosomal pellet by TEM is essential to evaluate possible contamination with non-exosomal vesicles, membranes, and cell debris that may be present in the conditioned medium as a consequence of cell lysis and centrifugation steps.

14. To keep the grids cold, use a prechilled glass dish covered with Parafilm and perform all the procedures on ice.

15. By negative staining and TEM exosomes are 50–150 nm vesicles with rounded or mushroom-shaped morphology (Fig. 4). Smaller (<30 nm) or bigger (>200 nm) particles cannot be considered exosomes and may constitute contaminants such as cell debris or microvesicles.

16. NGS analysis requires a highly concentrated RNA sample (ideally >300 ng/μl). Thus, we recommend using small volumes of nuclease-free water to elute the RNA from the purification columns during the RNA extraction procedure (10–15 μl). Manufacturer's protocols normally suggest an elution volume of 30–50 μl. To obtain most of the RNA from the column, pass the same eluted volume more than once through the column to increase the concentration of RNA.

17. RNA yield, concentration, and integrity from the preparations of SCs and exosomes must be evaluated using a Bioanalyzer. The RNA profiles obtained by this method allow an accurate assessment of RNA integrity.

18. To proceed with NGS, we recommend a sample input ≥1 μg of total RNA for each sample. For RNA-Seq library preparation, exosomal and SC-derived RNA should be treated with a kit to remove rRNA.

Acknowledgments

We wish to thank all the members of the Court Lab for their contributions to this protocol. This work was supported by the Center for Integrative Biology, Universidad Mayor, FONDECYT-1150766, Geroscience Center for Brain Health and Metabolism (FONDAP-15150012), Ring Initiative ACT1109, and Canada-Israel Health Research Initiative, jointly funded by the Canadian Institutes of Health Research; the Israel Science Foundation; the International Development Research Centre, Canada; and the Azrieli Foundation, Canada.

References

1. Montecalvo A, Shufesky WJ, Stolz DB et al (2008) Exosomes as a short-range mechanism to spread alloantigen between dendritic cells during T cell allorecognition. J Immunol 180:3081–3090

2. Vella LJ, Sharples RA, Lawson VA et al (2007) Packaging of prions into exosomes is associated with a novel pathway of PrP processing. J Pathol 211:582–590. https://doi.org/10.1002/path.2145

3. Johnstone RM, Adam M, Hammond JR et al (1987) Vesicle formation during reticulocyte maturation. Association of plasma membrane activities with released vesicles (exosomes). J Biol Chem 262:9412–9420

4. Peinado H, Alečković M, Lavotshkin S et al (2012) Melanoma exosomes educate bone marrow progenitor cells toward a pro-metastatic phenotype through MET. Nat Med 18:883–891. https://doi.org/10.1038/nm.2753

5. Mathivanan S, Ji H, Simpson RJ (2010) Exosomes: extracellular organelles important in intercellular communication. J Proteome 73:1907–1920. https://doi.org/10.1016/j.jprot.2010.06.006

6. Valadi H, Ekström K, Bossios A et al (2007) Exosome-mediated transfer of mRNAs and microRNAs is a novel mechanism of genetic exchange between cells. Nat Cell Biol 9:654–659. https://doi.org/10.1038/ncb1596

7. Krek A, Grün D, Poy MN et al (2005) Combinatorial microRNA target predictions. Nat Genet 37:495–500. https://doi.org/10.1038/ng1536

8. Mittelbrunn M, Gutiérrez-Vázquez C, Villarroya-Beltri C et al (2011) Unidirectional transfer of microRNA-loaded exosomes from T cells to antigen-presenting cells. Nat Commun 2:282. https://doi.org/10.1038/ncomms1285

9. Lin J, Li J, Huang B et al (2015) Exosomes: novel biomarkers for clinical diagnosis. Sci World J 2015:1–8. https://doi.org/10.1155/2015/657086

10. Zhang J, Li S, Li L et al (2015) Exosome and exosomal microRNA: trafficking, sorting, and function. Genomics Proteomics Bioinformatics 13:17–24. https://doi.org/10.1016/j.gpb.2015.02.001

11. Skog J, Würdinger T, van Rijn S et al (2008) Glioblastoma microvesicles transport RNA and proteins that promote tumour growth and provide diagnostic biomarkers. Nat Cell Biol 10:1470–1476. https://doi.org/10.1038/ncb1800

12. Ludwig A-K, Giebel B (2012) Exosomes: small vesicles participating in intercellular communication. Int J Biochem Cell Biol 44:11–15. https://doi.org/10.1016/j.biocel.2011.10.005

13. Schageman J, Zeringer E, Li M et al (2013) The complete exosome workflow solution: from isolation to characterization of RNA cargo. Biomed Res Int 2013:253957–253915. https://doi.org/10.1155/2013/253957

14. Jenjaroenpun P, Kremenska Y, Nair VM et al (2013) Characterization of RNA in exosomes secreted by human breast cancer cell lines using next-generation sequencing. PeerJ 1:e201. https://doi.org/10.7717/peerj.201

15. Lopez-Verrilli MA, Picou F, Court FA (2013) Schwann cell-derived exosomes enhance axonal regeneration in the peripheral nervous system. Glia 61:1795–1806. https://doi.org/10.1002/glia.22558

16. Segura E, Nicco C, Lombard B et al (2005) ICAM-1 on exosomes from mature dendritic cells is critical for efficient naive T-cell priming. Blood 106:216–223. https://doi.org/10.1182/blood-2005-01-0220

17. van Niel G, Mallegol J, Bevilacqua C et al (2003) Intestinal epithelial exosomes carry MHC class II/peptides able to inform the immune system in mice. Gut 52:1690–1697

18. Thery C, Amigorena S, Raposo G, Clayton A (2006) Isolation and characterization of exosomes from cell culture supernatants and biological fluids. Curr Protoc Cell Biol Chapter 3:Unit 3.22–Un3.22.29. https://doi.org/10.1002/0471143030.cb0322s30

19. Liu Z, Jin Y-Q, Chen L et al (2015) Specific marker expression and cell state of Schwann cells during culture in vitro. PLoS One 10:e0123278. https://doi.org/10.1371/journal.pone.0123278

Chapter 20

3D Cancer Migration Assay with Schwann Cells

Laura Fangmann, Steffen Teller, Pavel Stupakov, Helmut Friess, Güralp O. Ceyhan, and Ihsan Ekin Demir

Abstract

In pancreatic cancer, neural invasion is one of the most common paths of cancer dissemination. Classically, cancer cells actively invade nerves and cause local recurrence and pain. Three-dimensional (3D) neural migration assay has become a standard tool for scientists to study neural invasion by confronting the involved cell types. This protocol introduces Schwann cells, i.e., the most prevalent cell type in peripheral nerves, in a novel heterotypic, glia-cancer-neuron, 3D migration assay for assessing their relevance in the early pathogenesis of neural invasion. Particularly, this assay allows the monitoring of the early Schwann cell migratory activity.

Key words 3D migration assay, Schwann cells, Neural invasion, Pancreatic cancer, Dorsal root ganglia/DRG, Schwann cell carcinotropism

1   Introduction

The 3D neural migration assay is a special co-culture system aimed at studying neural invasion in pancreatic cancer [1]. Neural invasion can be defined as the continuous growth of tumor cells along nerves and is typically encountered around the epi- and perineurium of nerves. The cell types that are thought to play a major role in neural invasion (NI) are confronted simultaneously and photo-documented for 72 hours (h) in order to observe their migratory behavior, cell-to-cell interactions, and morphological changes.

In the original migration assay setup, these cell types were (a) pancreatic cancer (PCa) cell lines and (b) murine dorsal root ganglia (DRG), which were used to represent the neural component (Fig. 1). A characteristic feature of the present assay is that Schwann cells, as the most prevalent cell type in peripheral nerves, are added to the procedure as a third cell type, representing the glial component in the process of neural invasion. This addition revealed that, before cancer cells start to invade the DRG, Schwann cells migrate toward cancer cells due to their carcinotropic features [2].

Paula V. Monje and Haesun A. Kim (eds.), *Schwann Cells: Methods and Protocols*, Methods in Molecular Biology, vol. 1739, https://doi.org/10.1007/978-1-4939-7649-2_20, © Springer Science+Business Media, LLC 2018

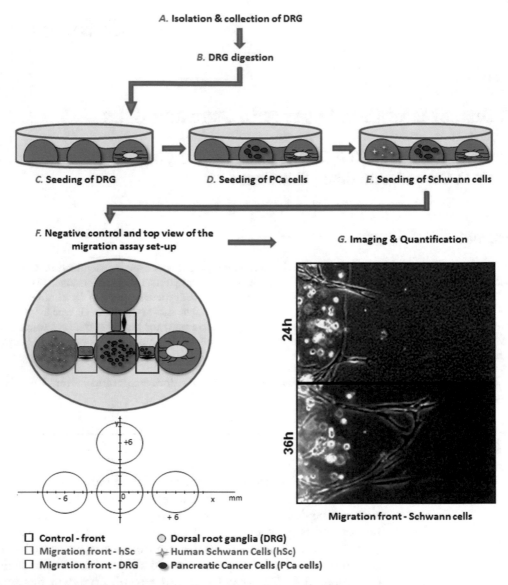

Fig. 1 (**a–e**) Overview of the pipetting steps for the 3D migration assay including pancreatic cancer cells (PCa cells), murine or rat DRGs, and human (or murine) Schwann cells (hSC). (**f**) Millimeter-scaled template designed as a pipetting scheme for the migration assay. First, place an ECM gel drop containing the DRG cells at millimeter +6 of a 35 mm culture dish. Next, place an additional ECM gel drop containing PCa cells at millimeter 0. Then, place the ECM cell drop containing the Schwann cell suspension at millimeter −6. Lastly, build an additional cell-free ECM gel drop at millimeter +6 of the y-axis serving as a negative control. (**g**) After 24 h, the migration of Schwann cells can be tracked via digital time-lapse microscopy and quantified for the extent of targeted migration

Therefore, the 3D migration assay with Schwann cells is suitable for simulating the mutual tropism between (a) PCa cells and DRG neurons, as well as (b) between glial cells, e.g., Schwann cells, and PCa cells [2]. Hence, in this triple-culture system, two sites of confrontation are present, i.e., first between PCa cells and DRG and second between PCa cells and Schwann cells (Fig. 1). The cancer cell population additionally has a back front (Fig. 1) that is directed toward an empty gel suspension. All three cell populations are suspended in an extracellular matrix (ECM) gel, giving the co-culture system its 3D structure and placed in a culture dish at an exact distance of 1 mm by means of a mini-ruler beneath the culture dish. To allow migration, all three cell types are connected by an ECM gel bridge for establishing a chemoattractive gradient. After applying a defined neuronal medium, the assay is incubated for 48 h at 37 °C in a CO_2 incubator. At 24 h, the assay is taken out of the incubator and placed into a digital time-lapse microscope and photo-documented periodically for the time course of the experiment. The digital time-lapse microscopy allows exact tracking and quantification of the migratory activity. At 72 h after initial seeding of the assay, ECM gels containing each cell type are excised from the culture dish, dissociated, and lysed in radioimmunoprecipitation buffer for protein extraction. The extracted protein can be used for immunoblotting against, e.g., neurotrophic factor receptors to track their expression in migrating versus nonmigrating cancer or Schwann cells.

In summary, this assay shows that Schwann cells are attracted to cancer cells even before these start to migrate toward neurons. This observation suggests that Schwann cells serve as the first access for cancer cells during the process of NI, which ultimately causes tumor progression. Collectively, this chemoattraction of Schwann cells is termed Schwann cell carcinotropism [2]. Advantages of this system are the following: (1) the usage of a 3D cell culture model as opposed to a 2D culture; (2) the possibility to quantitatively assess cell migration by determining, e.g., the forward migration index (FMI), the migration velocity, and the linear (Euclidean) distance migrated by individual cells; and (3) the capacity to compare cancer cell migration toward different cell types within the same experimental setting.

2   Materials

2.1   Primary Cell Cultures and Established Cell Lines

1. Human PCa cell line SU86.86. This cell line can be purchased from American Type Culture Collection (ATCC) and should be cultured according to supplier's recommendations.

2. Human PCa cell line T3M4. This cell line is a kind gift by Dr. Metzgar (Durham, North Carolina) [1, 2].

3. Human Schwann cells. Primary human Schwann cells can be purchased from ScienCell Research Laboratories (Carlsbad, CA). Cells are derived from human spinal nerves. Cells can be cultured up to a maximum of ten passages when grown in complete human Schwann cell medium [1]. Alternatively,

murine (i.e., mouse or rat) Schwann cells provided by the same company can also be used in this assay.

4. Primary DRG neurons. Neurons are freshly isolated from newborn Wistar rats or C57BL/6J mice, as described below (*see* Subheading 3.1).

2.2 Preparation of Culture Media and Buffers

1. Human Schwann cell medium (ScienCell, Carlsbad, CA). The complete medium contains 5% fetal calf serum (FCS) and human Schwann cell growth supplements, as suggested by the manufacturer [1].

2. Neurobasal medium for DRG neurons. Neurobasal medium supplied with 100 U/ml penicillin, 100 μg/ml streptomycin, 2% B-27 supplement, and 0.5 mM L-glutamine.

3. Minimal essential medium (MEM, Sigma-Aldrich, Taufkirchen, Germany) supplied with 0.04 mg/ml gentamicin and 0.05 mg/ml metronidazole.

4. Hank's balanced salt solution (HBSS).

5. Gentamicin stock solution (10 mg/ml).

6. Metronidazole stock solution (5 mg/ml).

7. Collagenase type II. Dissolve 100 mg collagenase type II (Worthington Biochemical, Lakewood, NJ) in 10 ml HBSS to obtain a concentration of 10 mg/ml (10× stock solution). For digestion (*see* Subheading 3.1, **step 4**), add 100 μl of collagenase type II 10× stock solution and 900 μl of HBSS for a final collagenase concentration of 1 mg/ml (1× working solution). Store on ice.

8. ECM gel. It is recommended to use ECM from Engelbreth-Holm-Swarm (EHS) mouse sarcoma (E1270, Sigma-Aldrich, Munich, Germany). Thaw ECM gel on ice 1 h prior to experimentation. Keep ECM gel strictly on ice to avoid unwanted polymerization.

9. 1 M HEPES buffer solution prepared in water.

10. Radioimmunoprecipitation buffer (RIPA). 150 mM sodium chloride, 1.0% NP-40, 0.5% sodium deoxycholate, 0.1% sodium dodecyl sulfate, 50 mM Tris–HCl, 2 mM EDTA, 50 mM NaF. Adjust to pH 8.0. Add and dissolve one tablet of the protease inhibitor Complete (Roche, Penzberg, Germany) in the RIPA buffer for optimal protease inhibition.

2.3 Supplies, Equipment, and Software

1. 3.5 cm Petri dishes.

2. Millimeter-scaled template.

3. Micro-scissors.

4. Micro-forceps.

5. Hemocytometer.

6. CO_2 incubator (Heracell™). Set at a constant temperature of 37 °C and a humid atmosphere saturated with 5% CO_2.

7. Time-lapse microscope. Inverted light microscope (Zeiss Axio Observer D1) equipped with Zeiss CCD camera.

8. ImageJ software version 1.44 (NIH, Wayne Rasband) with manual tracking plug-in.

9. Chemotaxis and migration tool (Ibidi).

3   Methods

3.1   Isolation of DRG Neurons

1. Primary DRG neurons should be freshly isolated from three newborn Wistar rats or C57BL/6J mice between postnatal day (P) 2 and 12 after decapitation, anterior laminectomy, and stereomicroscopic dissection from the cervical to lumbar region (Fig. 2) [2] (*see* **Notes 1** and **2**).

2. Remove peripheral and central projections (nerve roots) of the DRGs by means of micro-scissors (Fig. 2) (*see* **Note 3**).

3. Collect DRGs in iced-cold MEM (Fig. 2).

4. Aspirate medium and resuspend the DRGs in HBSS supplied with collagenase type II and incubate them for 20–30 min, at 37 °C.

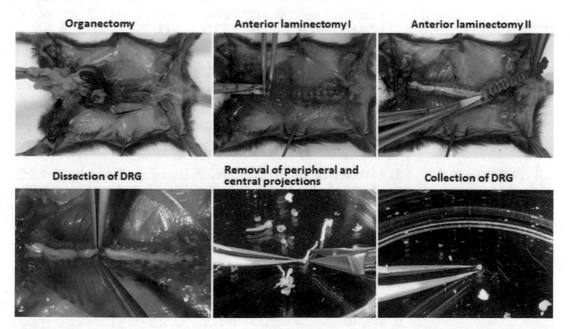

Fig. 2 Photographic depiction of DRG isolation from C57BL/6J mice. After removal of internal organs, an anterior laminectomy is performed for exposing the spinal cord. The collected DRGs are released from their projections to ease the subsequent trituration step

5. Triturate the DRGs through syringes of decreasing diameter (*see* **Notes 4** and **5**).

6. Centrifuge the DRG suspension at $93.9 \times g$ for 5 min at 4 °C, aspirate old medium, and directly resuspend the cells in ECM gel. Use 25 µl of ECM gel per drop (*see* **Notes 6** and 7).

7. Use a millimeter-scaled template for placing the ECM gel drop containing the DRG suspension in a 3.5 cm culture dish (*see* **Note 8**).

8. Incubate for 10 min at 37 °C, in a CO_2 incubator to allow polymerization of the ECM gel.

3.2 Preparation of Schwann Cells and PCa Cells

1. Count Schwann cells and PCa cells by means of a hemocytometer. The use of 10^5 cells of each cell type is recommended for one assay. Cells should be harvested for migration assay after reaching 100% confluence. The dissociation can be achieved by trypsinization in the culture flask.

2. Directly resuspend cells in ECM gel. Use 25 µl per drop (*see* **Notes 6** and 7).

3. Use a millimeter-scaled template for placing the ECM gel drop containing PCa cell suspension in the culture dish, at an exact 1 mm distance to the left-hand side of the ECM gel drop containing the DRG suspension (Fig. 1) (*see* **Note 9**).

4. Repeat the procedure described above with Schwann cells. Place an ECM gel drop containing Schwann cell in the suspension at an exact 1 mm distance to the left-hand side of the ECM gel drop containing PCa cells (*see* **Note 10**).

5. In order to exclude unspecific cellular interaction, build an additional cell-free ECM gel drop and place it at an exact 1 mm distance to the upper front of the ECM drop containing PCa cells (*see* **Note 11**).

6. Incubate the dish for 10 min at 37 °C in a CO_2 incubator to allow polymerization of the ECM gel.

3.3 Setting Up the 3D Neural Migration Assay

1. Connect the ECM drops containing each cell type by building an ECM gel bridge in between adjacent ECM gel drops. Use 3 µl of ECM gel per bridge (*see* **Note 12**).

2. Incubate the dish for 15 min at 37 °C in a CO_2 incubator to allow polymerization of the ECM gel.

3. Cautiously add neurobasal medium for DRG neurons to submerge the ECM gel structures in a liquid phase of 3 ml of medium.

4. Incubate the dish for 48 h at 37 °C in a CO_2 incubator.

3.4 Time-Lapse Microscopy

1. Forty-eight hours after seeding the assay, add HEPES buffer (25 µl of HEPES buffer per ml of culture medium) for ionic stabilization purposes. Transfer the assay into the incubator of the time-lapse microscope and start the periodic photo-documentation every 15 min for up to 48 h.

2. Convert the obtained image sequences to an avi-formatted file compatible with the ImageJ software.

3. Using the manual tracking plug-in of the ImageJ, click on the cell to be tracked on every image of the image sequence. For this purpose, randomly select 30 cells at each migration front and track each cell with the manual tracking plug-in. The x/y calibration of the tracking plug-in should match that of the time-lapse microscope (e.g., 0.645 µm).

4. Save the file containing the results of the tracking as an .xls file and open it with the Ibidi's chemotaxis and migration tool for ImageJ.

5. Click on "Import data," enter the number of images in each tracked video, and select the statistic feature to obtain the summary of results. This analysis yields the following parameters: (1) the accumulated distance covered by migrating cells, (2) the Euclidean distance, (3) the velocity of migration, and (4) the FMI.

3.5 Protein Lysis for Immunoblotting

1. At 72 h after the initial seeding, excise the ECM gels containing each cell type from the culture dish and incubate them in HBSS containing collagenase type II (1 mg/ml) for 15 min to allow dissociation of the cells from the ECM gel.

2. Centrifuge the cells obtained after collagenase dissociation at $93.9 \times g$ for 5 min. Aspirate the supernatant and add the RIPA buffer containing protease inhibitor for cell lysis.

3. Cell lysates may be subjected to immunoblot analysis together with samples of native Schwann cells or PCa cancer cells not subjected to co-culture. Analysis can be performed to determine the amount of neurotrophic factor receptors such as TrkA or p75NTR [2]. Analysis may show dynamic differences in the expression of such receptors upon co-culture of these cell types [2].

4 Notes

1. To achieve the best possible neuronal growth, it is recommended to avoid the use of mitotic inhibitors or additional neurotrophic factors in the DRG cultures. The above-described protocol confronts cancer cells with pure cultures of human Schwann cells on one side and with mixed neuronal-glial cul-

tures (murine or rat) on the other side. Mitosis inhibitors can also be applied to the ECM gel suspension containing the DRG cells to suppress glial cell division on the DRG side.

2. In order to have sufficient DRG neurons, collect all cervical to lumbar DRGs of each newborn rat or mouse (equaling 52 DRGs per rat/mouse).

3. Leaving the projections in place impedes trituration and increases the contamination risk of culture by fibroblasts.

4. For trituration, use 20 gauge/G needles followed by 23 G/gauge needles. However, extensive trituration can negatively affect the viability of the DRG neurons more than the viability of glial cells. Typically, five to seven times of trituration with a 20 G needle and two times of trituration with a 23 G needle enable sufficient dissociation of the cells in the DRGs. Their viability can be confirmed with trypan blue staining.

5. ECM gel should be thawed on ice and kept on ice during the entire experiment to avoid early polymerization.

6. Strictly avoid air bubbles during resuspension of the cells in the ECM gel. Air bubbles decrease the quality of the migration assay.

7. It is recommended to use 35 mm culture dishes and place the ECM gel drop containing the DRG at millimeter 6 on the x-axis on the scaled template (Fig. 1f). The total diameter of one ECM gel plug should be 5 mm.

8. Carefully place the ECM gel drop containing the PCa cell suspension at millimeter 0 of the scaled template.

9. Carefully place the ECM gel drop containing the Schwann cell suspension at millimeter—6 of the scaled template.

10. Carefully place the cell-free ECM drop at millimeter 6 on the y-axis of the scaled template.

11. First, build a bridge for connecting DRG and PCa cells by placing 3 μl of ECM gel exactly in between these two drops. Build a second bridge for connecting PCa cells and Schwann cells in the same manner and finally a third bridge for connecting PCa cells with cell-free ECM drop.

12. The accumulated distance corresponds to the total path of cell migration. The Euclidean distance corresponds to the linear distance (net displacement) covered by migrating cells. The FMI is an index value equaling the proportion of Euclidean to accumulated distances and expresses the extent of targeted migration.

References

1. Ceyhan GO, Demir IE, Altintas B (2008) Neural invasion in pancreatic cancer: a mutual tropism between neurons and cancer cells. Biochem Biophys Res Commun 374:442–447

2. Demir IE, Boldis A, Pfitzinger PL (2014) Investigation of Schwann cells at neoplastic cell sites before the onset of cancer invasion. J Natl Cancer Inst 106:dju184

Part III

Protocols for Studying Schwann Cell Development and Myelination In Vivo

Teased Fiber Preparation of Myelinated Nerve Fibers from Peripheral Nerves for Vital Dye Staining and Immunofluorescence Analysis

Alejandra Catenaccio and Felipe A. Court

Abstract

Glial cells regulate a wide variety of neuronal functions during physiological and pathological conditions. Therefore, the study of glial cells and their association with axons is of paramount importance in order to understand the physiology of the nervous system. This chapter describes a detailed protocol to prepare and stain teased nerve fibers from peripheral nerves using fluorescent indirect immunolabeling and staining with vital dyes. For immunofluorescence analysis, we describe techniques to study the axonal compartment and the expression of cytoplasmic and plasma membrane proteins in Schwann cells.

Key words Peripheral nervous system, Schwann cell, Nerve fiber, Immunofluorescence, Teased fibers

1 Introduction

In the peripheral nervous system (PNS), myelin-forming Schwann cells (SCs) associate with a single axonal segment. In this organization, a nerve fiber composed by an axon and the associated SCs can be individually teased apart and analyzed by different techniques. Teased fibers can be centimeters long and have several SCs associated to a single axon. Teased fibers have been used from the times of Santiago Ramon y Cajal to generate a detailed characterization of the SC cytoarchitecture at the light microscopy level, including the organization of the myelin sheath, the nodal, paranodal, and juxtaparanodal regions, as well as the external cytoplasm arranged in Cajal bands [1–4]. Many published studies have used immunofluorescence techniques as well as vital dyes in order to stain different SC compartments. In addition, the relationship of SCs with the axon can be examined using the teased fiber technique.

Here, we present a detailed protocol to prepare and stain teased fibers using fluorescent indirect immunolabeling. This protocol also allows the short-term analysis of live teased fibers stained

Paula V. Monje and Haesun A. Kim (eds.), *Schwann Cells: Methods and Protocols*, Methods in Molecular Biology, vol. 1739, https://doi.org/10.1007/978-1-4939-7649-2_21, © Springer Science+Business Media, LLC 2018

with vital dyes or genetically engineered to express fluorescent protein markers in SCs. For immunofluorescence analysis, two independent protocols of double immunolabeling are presented: one to study the axon and its associated SC and another to study cytoplasmic and plasma membrane proteins in SCs. These protocols are extremely useful to study the biology of SCs and associated axons, the process of axonal myelination, and pathological conditions in which the neuron-glial unit is compromised and result in demyelination and axonal degeneration.

2   Materials

Prepare all solutions using distilled water. Prepare and store all reagents at room temperature (RT) unless otherwise indicated.

2.1   Nerve Dissection and Fixation

1. Vanna microdissecting scissors.
2. 1× phosphate-buffered saline (1× PBS). To prepare 1 L of 10× PBS, dissolve 80 g NaCl, 2 g KCl, 14.4 g Na_2HPO_4, and 2.4 g KH_2PO_4 in 800 mL of distilled water. Adjust the pH to 7.4 with HCl and the volume to 1 L with distilled water. Sterilize by autoclaving.
3. 4% (w/v) paraformaldehyde (PFA). Dissolve 4 g of PFA powder in 90 mL of 1× PBS and heat to 60 °C while stirring. Add drops of 1 N NaOH until the solution becomes clear. Complete the volume to 100 mL with 1× PBS. Filter and store at −20 °C for 6 months, thaw aliquots immediately before use and do not refreeze (*see* **Note 1**).
4. Whatman filter paper N° 4.
5. 1.5 mL microcentrifuge conical tubes.

2.2   Preparation of TESPA-Coated Slides

1. Superfrost slides.
2. 100% acetone.
3. 4% (v/v) (3-Aminopropyl)triethoxysilane (TESPA) diluted in acetone.
4. Coplin glass jars.

2.3   Nerve Teasing

1. Dissecting microscope with transmitted light.
2. 35 mm cell culture dishes.
3. 1× PBS.
4. Two ultra-thin Dumont #5 forceps.
5. Sterile acupuncture needles (size 32 G × 1″ 0.25 × 0.25 mm).
6. TESPA-coated slides.
7. Graphite pencil.

2.4 **Immunostaining**	1. 1× PBS.

1. 1× PBS.

2. Chilled (−20 °C) acetone.

3. Hydrophobic Pap pen.

4. Blocking solution. Dilute gelatin from cold water fish skin to 5% (v/v) in 1× PBS. Add Triton X-100 to a final concentration of 0.1% (v/v).

5. Primary antibodies against SC-specific markers, such as S100, myelin-associated proteins, and cytoskeletal elements.

6. Primary antibodies against axon-specific markers, such as neurofilament high chain (anti-neurofilament 200, Sigma, N° N4142), tubulin (anti-tubulin, acetylated, Sigma, N° T6793), and myosin (anti-myosin Va, Cell Signaling Technology, N°3402S).

7. Secondary antibodies conjugated with the fluorophore of choice, such as AlexaFluor™ or Cyanine dyes (Cy2 ™, Cy3™, or Cy5™).

8. Mounting medium. Fluoromount ™, Mowoil ™, or similar aqueous mounting medium.

9. Cover slides.

10. 24- or 96-well cell culture plate for floating immunostaining.

11. Glass Pasteur pipette for floating immunostaining.

12. Washing culture medium: 200 mM L-glutamine, 1× B-27 supplement, neuronal growth factor (NGF), and 1× antibiotic antimycotic prepared in Neurobasal medium. This medium is used for incubation of live teased fibers (*see* Subheading 3.6). Prepare NGF aliquots at a concentration of 100 μg/mL and store them at −20 °C.

13. Fluorescent vital dyes. Commercially available probes for membranes, such as FM 1-43™ and RH 414 ™ and/or organelles, such as MitoTracker™, LysoTracker™, or ERTracker™.

3 Methods

Here we describe different techniques to perform immunostaining in teased fiber preparations. Depending on the cellular compartment being analyzed, conventional or floating fiber immunostaining can be used. The conventional method (Subheading 3.4) uses acetone in order to fully permeabilize the preparation, allowing efficient myelin and axonal staining. The floating method (Subheading 3.5) is preferred in order to perform staining of cytoplasmic and membrane proteins in SCs, as the structure of the cells is better preserved by a milder permeabilization step. In addition, we describe a method to study live preparations of teased fibers using vital dyes for membranes and organelles. Carry out all procedures at RT unless otherwise indicated.

3.1 Nerve Extraction and Fixation

1. Expose a 1.5 cm section of peripheral nerve and separate it from the surrounding muscular tissue. The nerve tissue is usually the sciatic nerve, but any type of sensory or mixed nerve from any species (including mouse, rat, or human) can be used.

2. Hold the nerve with one ultra-thin Dumont #5 forceps from one end and cut it in a segment of approximately 1 cm long using a Vanna microdissecting scissors.

3. Place the nerve segment onto a 1.5 × 0.5 cm Whatman filter paper so that it is well stretched and quickly immerse the paper in a 1.5 mL microcentrifuge conical tube containing 1 mL of 4% PFA. Incubate the tissue for 1 h at RT. Do not overstretch the nerve while placing it on the filter paper or the fibers may be damaged. Avoid drying the nerve by performing this step as fast as possible.

4. After fixation separate the nerve from the filter paper while still in the tube. Notice that the nerve is hard and separates easily from the paper simply by gentle touching. Wash 3 times for 10 minutes (min) each in 1× PBS (*see* **Note 2**). The nerve can also be detached from the filter after the first washing step.

3.2 TESPA-Coated Slides

1. Immerse the superfrost slides in a Coplin glass jar containing 100% acetone for 5 min.

2. Transfer slides to a Coplin glass jar containing 4% TESPA and incubate for 2 min. Use this Coplin jar only for the TESPA solution.

3. Transfer the slides to a Coplin glass jar containing fresh 100% acetone and incubate them for 30 seconds (s). Repeat this step one more time with fresh acetone in a new clean Coplin jar.

4. Let the slides air-dry at RT and store them in a light-protected box at RT (*see* **Note 3**).

3.3 Teasing of the Nerve Fibers

1. Put the fixed and washed nerve section in a 35 mm culture dish containing 1× PBS.

2. Under the dissecting microscope, remove the epineurium and perineurium using fine Dumont #5 forceps (*see* **Note 4**).

3. After removing the epineurium and perineurium, cut the portion of the nerve into two 0.5 cm segments. Nerve fibers without epineurium and perineurium are sticky; perform this and the following steps carefully to prevent damage and loss of tissue (*see* **Note 5**).

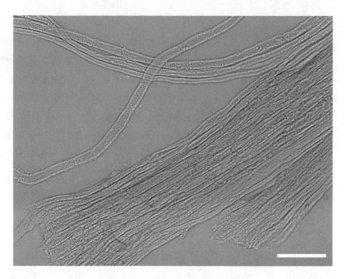

Fig. 1 Bundle of fibers observed under a dissecting microscope. This is the first step in the procedure to generate teased fibers after removing the epineurium and perineurium. The fibers are arranged in bundles similar to their organization in the complete nerve. This step is followed by the teasing of individual fibers. Scale bar, 20 μm

4. Separate small bundles consisting of 30–50 fibers each by using a pair of acupuncture needles (Fig. 1). For floating immuno-fluorescence, proceed to Subheading 3.5.

5. Using a single acupuncture needle, place a small bundle of fibers in a drop of 1× PBS previously placed on top of a TESPA-coated slide.

6. Using the acupuncture needle, stick the nerve bundle to a TESPA-coated slide by touching the bundle in different places. Bundles have a tendency to stick easily to TESPA-coated surfaces.

7. Once the bundle is stuck to the slide, tease the fibers by touching individual fibers with the acupuncture needle and moving them along the bottom of the slide carefully. The individual fibers should remain attached to the slide during this procedure (*see* **Note 6**).

8. Carefully take out the excess of 1× PBS using an absorbent paper without touching the nerve fibers.

9. Leave the slides with teased fibers at RT inside a slide box until the 1× PBS dries out (usually within 10–20 min). Label the slide with a graphite pencil and store at −20 °C in a slide box.

3.4 Conventional Immunostaining

1. Take slides with teased fibers from the −20 °C freezer directly into prechilled (−20 °C) acetone and leave them in acetone for 20 min at −20 °C.

2. Take out the slides from the −20 °C acetone and allow them to air dry for 30 s. Circle the teased fibers with the Pap pen to circumscribe the area to be stained.

3. Block/permeabilize teased fibers with blocking solution added to the circumscribed area for 1 h at RT.

4. Incubate teased fibers with primary antibodies in blocking solution overnight at 4 °C in a humidified chamber.

5. Wash 3 times for 10 min in 1× PBS.

6. Incubate teased fibers with secondary antibodies in blocking solution for 2 h at RT or overnight at 4 °C in a humidified dark chamber.

7. Wash 3 times for 10 min in 1× PBS.

8. Mount the stained samples in mounting medium (Fig. 2).

3.5 Floating Immunostaining

1. Put a small bundle of fibers in a well of a 96-well dish containing 1× PBS.

2. Very carefully take out the 1× PBS using a glass Pasteur pipet and add blocking solution until the fibers are completely

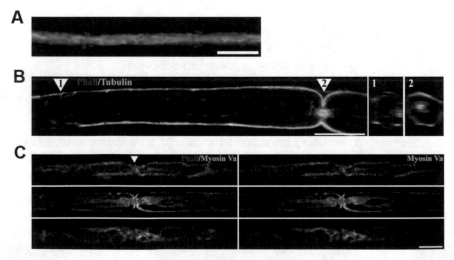

Fig. 2 (a) Confocal microscopy of fixed teased nerve fibers. (**a**) Morphology of the axon-Schwann cell unit. Teased fibers from mice stained for actin with rhodamine-conjugated phalloidin (*red*) which heavily label Schmidt-Lanterman incisures (SLI) and immunostained against neurofilament heavy chain to visualize the axon (*green*). Scale bar, 20 μm. (**b**) Distribution of actin and microtubule networks in Schwann cells. Teased fiber immunostained for acetylated tubulin and actin filaments stained with rhodamine-conjugated phalloidin (*red*). *Left panel*, projection through the middle of the fiber, a SLI (*arrowhead 1*) and the nodal region (*arrowhead 2*) can be seen. Right panels, Z-projections at the levels indicated by arrowheads. Notice the enrichment of microtubules in the longitudinal Schwann cell axis, while in the radial axis, actin filament predominates. Scale bar, 10 μm. (**c**) Distribution of myosin Va in nerve fibers. Teased fiber immunostained for myosin Va and actin filaments stained with rhodamine-conjugated phalloidin. Left panels, merged signals. *Arrowhead*, node of Ranvier. Right panels, myosin Va channel. Center panels pass through the midplane where the node of Ranvier is more apparent. Notice that the paranodal loops are rich in both actin and myosin Va. *Upper* and *lower planes* correspond to planes away from the midplane, which show the Cajal bands of the Schwann cell. Scale bar, 10 μm

immersed in the solution (50–100 μL). Incubate the fibers in blocking solution for 1 h (*see* **Note 7**).

3. Add primary antibodies in blocking solution overnight at 4 °C. Use the dish lid or parafilm to avoid evaporation of the solution.

4. Wash the fibers 3 times for 10 min in 1× PBS.

5. Add secondary antibodies in blocking solution for 2 h at RT in the dark.

6. Wash 3 times for 10 min in 1× PBS using a glass Pasteur pipette adding and removing the liquid carefully.

7. Tease the individual fibers by attaching the bundle of fibers to a TESPA-coated slide following the steps described in Subheading 3.3.

8. Using the acupuncture needle, stick the nerve bundle to the slide and proceed to tease the individual fibers. Note that the nerve bundles easily stick to the TESPA-coated slide.

9. Mount the nerve bundles in mounting medium (Fig. 3).

3.6 Short-Term Incubation and Analysis of Sciatic Nerve Fibers with Vital Dyes

1. Dissect the nerve tissue as described in Subheadings 3.1, **steps 1** and **2** and take out the epineurium. As the tissue is not fixed be careful to not damage nerve fibers during this and following steps.

2. Tease the fibers as described in Subheading 3.3 (*see* **Note 8**).

3. Incubate the teased fibers with the fluorescent dye of choice for the required 3 time in a dark box.

4. Wash the slides in culture medium 3 times for 10 min each.

5. Proceed to analyze the nerve fibers by live cell fluorescence microscopy. Nerve fibers labeled with fluorescent dyes can be visualized without fixation by time-lapse confocal microscopy in a special chamber that holds the nerve fiber-containing coverslip [5]. *See* **Note 9** for a description of other techniques used for in vivo labeling of SCs and neurons.

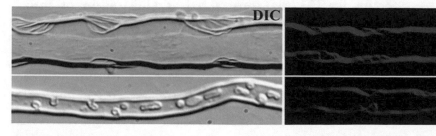

Fig. 3 Staining of membranes with vital dyes in teased nerve fibers. *Right panels*, fluorescence microscopy of fibers labeled with FM 1-43 and RH 414 applied to previously teased nerve fibers. *Left panels*, fibers seen by differential interference contrast (DIC) microscopy. Notice that the dye remains in the plane of the membrane. In the lower left panel, the Schwann cell protrudes into the axoplasm; in the axoplasmic space, vesicles or protrusion is still connected to the Schwann cell in another plane. Scale bar, 10 μm

4   Notes

1. Dissolving the PFA powder takes a long time. This process can be accelerated by adding 3–6 drops of 1 N NaOH. The fixative stock can be stored at −20 °C for up to 6 months.

2. The nerves can be stored in 1× PBS at 4 °C for up to 2 days before proceeding with the teasing step.

3. The slides can be stored for up to 1 month in a dark box at RT.

4. To dissect out the epineurium and perineurium, simply hold the fibers with one set of forceps and remove the epineurium and perineurium by gently pulling it back with another set of forceps. This step has to be done under a dissecting microscope. Be careful to not stretch nerve fibers. After removing the epineurium and perineurium, the nerve fibers become exposed. At this point they are extremely fragile and sticky.

5. Fiber bundles can be stored for a couple of days in 1× PBS at 4 °C before starting or finishing the teasing procedure.

6. Adding too many fibers or moving them around repeatedly causes the TESPA slide to get covered by extracellular proteins derived from the fibers. As a result, the fibers do not attach efficiently, and teasing of individual fibers becomes difficult.

7. The fibers tend to stick to the wall of the dish and the plastic tips if they are not in solution. Thus, do not leave them without liquid. Use a glass pipet to add and remove the liquid. Always try to not touch the fiber bundle.

8. Perform this step as fast as possible. Do not try to carefully separate the individual fibers as this takes time and introduces a risk to damage the unfixed preparation [6, 7].

9. For local administration of labeling agents in vivo, it is possible to inject intraperineurially microliter boluses of drugs or viral particles and allow them to infiltrate a few millimeters inside the nerve [8]. It is also possible to locally infect sensory neurons or Schwann cells with viruses encoding fluorescent proteins [9]. For protracted administration of drugs, silicone rubber cuffs loaded with the drug can be applied to short nerve segments, or alternatively, by means of miniosmotic pumps [10, 11].

Acknowledgments

We wish to thank all the members of the Court Lab for their contributions to this protocol. This work was supported by the Center for Integrative Biology, Universidad Mayor, FONDECYT-1150766, Geroscience Center for Brain Health and Metabolism (FONDAP-15150012), Ring Initiative ACT1109, and Canada-Israel Health Research initiative, jointly funded by the Canadian Institutes of Health Research; the Israel Science Foundation; the International Development Research Centre, Canada; and the Azrieli Foundation, Canada.

References

1. Court FA, Sherman DL, Pratt T et al (2004) Restricted growth of Schwann cells lacking Cajal bands slows conduction in myelinated nerves. Nature 431:191–195. https://doi.org/10.1038/nature02841

2. Yoshimura T, Rasband MN (2014) Axon initial segments: diverse and dynamic neuronal compartments. Curr Opin Neurobiol 27:96–102. https://doi.org/10.1016/j.conb.2014.03.004

3. Zhang A, Desmazieres A, Zonta B et al (2015) Neurofascin 140 is an embryonic neuronal neurofascin isoform that promotes the assembly of the node of Ranvier. J Neurosci 35:2246–2254. https://doi.org/10.1523/JNEUROSCI.3552-14.2015

4. Wu LMN, Williams A, Delaney A et al (2012) Increasing internodal distance in myelinated nerves accelerates nerve conduction to a flat maximum. Curr Biol 22:1957–1961. https://doi.org/10.1016/j.cub.2012.08.025

5. Calixto A, Jara JS, Court FA (2012) Diapause formation and downregulation of insulin-like signaling via DAF-16/FOXO delays axonal degeneration and neuronal loss. PLoS Genet 8:e1003141–e1003115. https://doi.org/10.1371/journal.pgen.1003141

6. Court FA, Hewitt JE, Davies K et al (2009) A laminin-2, dystroglycan, utrophin axis is required for compartmentalization and elongation of myelin segments. J Neurosci 29:3908–3919. https://doi.org/10.1523/JNEUROSCI.5672-08.2009

7. Court FA, Zambroni D, Pavoni E et al (2011) MMP2-9 cleavage of dystroglycan alters the size and molecular composition of Schwann cell domains. J Neurosci 31:12208–12217. https://doi.org/10.1523/JNEUROSCI.0141-11.2011

8. Lopez-Verrilli MA, Picou F, Court FA (2013) Schwann cell-derived exosomes enhance axonal regeneration in the peripheral nervous system. Glia 61:1795–1806. https://doi.org/10.1002/glia.22558

9. Oñate M, Catenaccio A, Martínez G et al (2016) Activation of the unfolded protein response promotes axonal regeneration after peripheral nerve injury. Sci Rep 6:21709. https://doi.org/10.1038/srep21709

10. Mohammadi R, Sanaei N, Ahsan S et al (2015) Stromal vascular fraction combined with silicone rubber chamber improves sciatic nerve regeneration in diabetes. Chin J Traumatol 18:212–218

11. Abbasipour-Dalivand S, Mohammadi R, Mohammadi V (2015) Effects of local administration of platelet rich plasma on functional recovery after bridging sciatic nerve defect using silicone rubber chamber; an experimental study. Bull Emerg Trauma 3:1–7

Whole Mount Immunostaining on Mouse Sciatic Nerves to Visualize Events of Peripheral Nerve Regeneration

Xin-Peng Dun and David B. Parkinson

Abstract

Injury to the peripheral nervous system triggers a series of well-defined events within both neurons and the Schwann cells to allow efficient axonal regeneration, remyelination, and functional repair. The study of these events has previously been done using sections of nerve material to analyze axonal regrowth, cell migration, and immune cell infiltration following injury. This approach, however, has the obvious disadvantage that it is not possible to follow, for instance, the path of regenerating axons in three dimensions within the nerve trunk or the nerve bridge. In order to provide a fuller picture of such events, we have developed a whole mount staining procedure to visualize blood vessel regeneration, Schwann cell migration, axonal regrowth, and remyelination in models of nerve injury.

Key words Whole mount immunostaining, Peripheral nerve injury, Axon regeneration, Schwann cell migration, Blood vessel regeneration

1 Introduction

Whole mount in situ hybridization analysis or antibody labeling has been used in a number of developmental systems to study gene and protein expression in a variety of model organisms [1]. Immunolabeling of whole embryos or tissue followed by a simple tissue-clearing protocol and confocal microscopy allows cell migration, protein expression, and tissue morphology to be visualized in three dimensions.

The events of Wallerian degeneration, axonal regeneration, and functional repair following injury require interactions between many cell types, such as the neurons of the peripheral nervous system (PNS), Schwann cells, immune cells, endothelial cells, and nerve fibroblasts [2–6]. These interactions lead to a directional migration of Schwann cells and regrowth of axons toward the distal part of the nerve, but it is often impossible to follow the paths of individual or groups of axons as they may well leave the plane of a thin cryostat or paraffin tissue section. To solve this problem, we

Paula V. Monje and Haesun A. Kim (eds.), *Schwann Cells: Methods and Protocols*, Methods in Molecular Biology, vol. 1739,
https://doi.org/10.1007/978-1-4939-7649-2_22, © Springer Science+Business Media, LLC 2018

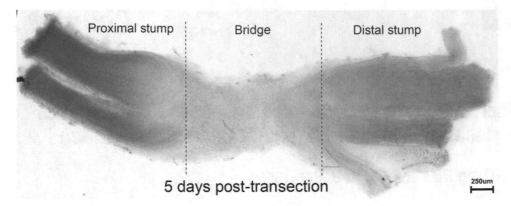

Fig. 1 Phase contrast image of a stained and mounted mouse sciatic nerve. The adult mouse (C57BL/6) nerve sample was dissected at 5 days post-transection injury. The presence of the newly formed nerve bridge between the proximal and distal stumps can easily be visualized at this time point

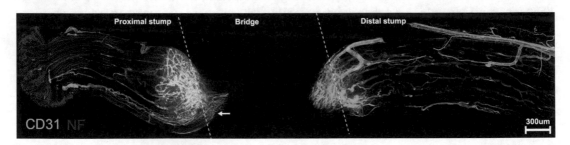

Fig. 2 Visualization of blood vessel and axon regeneration in the mouse sciatic nerve bridge at 5 days following a transection injury. Whole mount immunostaining of mouse sciatic nerve at 5 days post-transection injury, new blood vessel formation within the nerve bridge was labeled with endothelial cell marker CD31; existing blood vessels within the proximal and distal nerve stump also have been labeled with CD31. At this time point, regenerating axons from the proximal stump could be visualized with neurofilament (NF) staining (*white arrow*), and axonal breakdown in the distal stump is complete

have developed a modified whole mount staining protocol for sciatic nerves to analyze peripheral nerve repair following both nerve transection and crush models of injury [7]. This protocol has been used for our own research as well as successfully by other researchers in the field of peripheral nerve development and repair [8, 9]. In contrast to the "two-dimensional" staining of thin cryostat or paraffin sections of peripheral nerves, the method is ideal for visualizing progression of blood vessel regeneration, Schwann cell migration, axonal regrowth, and remyelination during the process of PNS regeneration following injury.

Figure 1 in this chapter shows a stained and mounted mouse sciatic nerve 5 days after transection injury. The proximal and distal stumps of the nerve together with the nerve bridge are labeled. Figure 2 shows the use of the whole mount immunostaining to

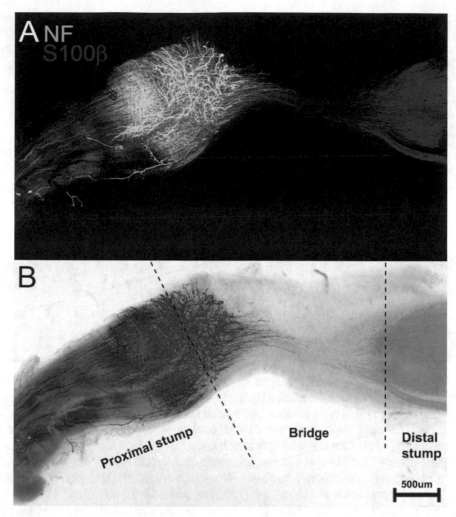

Fig. 3 Visualization of Schwann cell migration and axonal regeneration in the mouse sciatic nerve bridge 7 days after transection injury. (**A**) Labeling of regenerating axons using neurofilament (NF) shows axons crossing the bridge into the distal stump at this time point. Labeling with S100β shows Schwann cells migrating into the nerve bridge from both proximal and distal stumps. (**B**) Overlay of phase contrast image and fluorescence shows the nerve structure. The positions of the nerve bridge and the proximal and distal nerve stumps are shown

visualize blood vessels (marked with the endothelial cell marker CD31) and axon regeneration (marked with neurofilament, NF) 5 days after sciatic nerve transection injury. Figure 3 shows axonal regeneration (NF) and Schwann cell (S100β) staining in the nerve bridge 7 days after transection injury. Finally, Fig. 4 shows the whole mount immunostaining using a transgenic mouse line, with a Schwann cell-specific expression of green fluorescent protein (PLP-EGFP mice) [10], together with a neurofilament stain to visualize both axonal regeneration and Schwann cell migration.

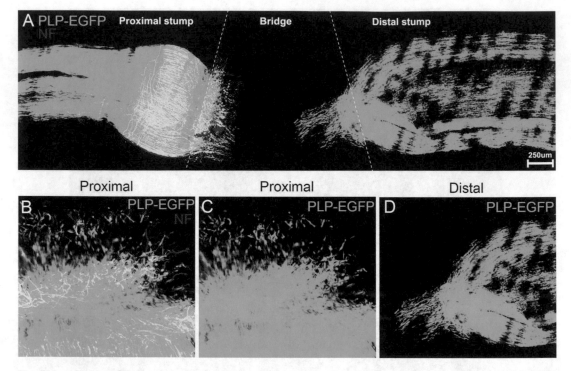

Fig. 4 Visualization of Schwann cell migration and axonal regeneration in the sciatic nerve bridge of PLP-EGFP transgenic mice at 6 days following transection injury. PLP-EGFP mice express an enhanced green fluorescent protein (EGFP) under the control of the Schwann cell-specific proteolipid protein (PLP) promoter [10]. (**A**) Whole mount immunostaining with the neurofilament (NF) antibody shows regenerating axons from the proximal nerve stump localized in the front of migrating Schwann cells at 6 days after sciatic nerve transection injury. Migrating EGFP-positive Schwann cells from the distal nerve stump could also be clearly observed. Positions of the nerve bridge and the proximal and the distal nerve stumps are shown. (**B**) Higher magnification image shows regenerating axons from the proximal nerve stump are localized in the front of migrating Schwann cells at 6 days after sciatic nerve transection injury. (**C**, **D**) Higher magnification images show Schwann cell migration toward the nerve bridge from both the proximal (**C**) and the distal (**D**) nerve stumps

2   Materials

2.1   Mouse Sciatic Nerve Injury, Nerve Dissection, and Fixation

1. Isoflurane anesthetic setup for small animal surgery.

2. Thermostatically heated surgical pad.

3. Small animal hair clipper.

4. Forceps (#5 size, e.g., Fine Science Tools (FST), Cat No. 11251-10; two pairs).

5. Scissors for skin cut and separation of muscles (12 cm, e.g. FST Cat No. 14002-12).

6. 0.4 mm tip angled, 9 cm delicate forceps (e.g., FST, Cat No. 11063-07).

7. Needle holder for use with silk sutures (e.g., World Precision Instruments, Cat No. 500223).

8. 8.0 size silk sutures.

9. 0.025% bupivacaine solution: prepare 1/10 dilution of 0.25% stock (e.g., Marcaine) in sterile PBS.

10. Autoclip surgical staple applier (e.g., FST, Cat No. 12020-09).

11. 70% alcohol in a spray bottle.

12. Corkboard.

13. Aluminum foil.

14. 7 ml screw top bijou containers.

15. Dissecting microscope.

16. 4 °C fridge.

17. Phosphate-buffered saline (PBS): 8 g NaCl, 0.2 g KCl, 1.44 g Na_2HPO_4, and 0.24 g KH_2PO_4 dissolve the reagents in 800 ml of water. Adjust the pH to 7.4 with HCl and then add water to 1 L.

18. Fixing solution: 4% (w/v) paraformaldehyde (PFA) prepared in PBS, pH 7.4. Paraformaldehyde powder is dissolved in warmed (50 °C) PBS on a magnetic stirrer, and a few sodium hydroxide pearls or drops of 1 M sodium hydroxide solution is added to help the powder dissolve. Allow the solution to cool down to room temperature, and then readjust the pH if necessary to pH 7.4 with HCl. The solution can be stored at 4 °C for 2–3 months or aliquoted and frozen at −20 °C for long-term storage.

2.2 Whole Mount Staining of Mouse Sciatic Nerve Tissue

1. PBS (*see* Subheading 2.1, **step 17**).

2. Permeabilization/blocking solution: 1% Triton X-100 detergent (v/v) and 10% (v/v) fetal bovine serum prepared in PBS. Make up and store at 4 °C.

3. Staining solution: 0.1% Triton X-100 detergent (v/v) and 10% (v/v) fetal bovine serum prepared in PBS. Make up and store at 4 °C.

4. Primary antibodies.

5. Appropriate fluorescently labeled species-specific secondary antibodies.

6. Clearing solutions: 25%, 50%, and 75% glycerol (v/v) prepared in PBS.

7. Glass slides.

8. Vacuum grease.

9. CitiFluor anti-fade mounting medium.

10. 20 mm × 40 mm glass coverslips.

11. Nail varnish.

12. Epifluorescence and confocal microscope.

3 Methods

3.1 Mouse Sciatic Nerve Transection or Crush Injury

Sciatic nerve crush or cut procedures are performed on adult wild-type or transgenic mice, usually between 2 and 3 months of age (weight greater than 20 g).

1. Prepare clean cages and labels for mice undergoing surgery.

2. Prepare a sterile surgical area with drapes, the anesthesia machine, sterilized surgical tools, a diluted bupivacaine solution (1/10 dilution of 0.25% stock in sterile PBS to make 0.025% preparation for use), and heating pad in the designated surgery room.

3. Anesthetize the animals with isoflurane in the anesthesia box.

4. Transfer the mouse to the mask and remove the hair around the incision area with a hair clipper.

5. Use sticky tape to gently secure the rear legs on heating pad and test the depth of anesthesia by checking the toe reflex to ensure the animal is fully anesthetized.

6. Identify the position for incision on the mid-thigh of the animal and perform a skin cut (approximately 1 cm in length) using scissors.

7. Carefully separate the gluteal and the hamstring muscles with scissors and forceps to expose the sciatic nerve.

8. Perform a sciatic nerve crush or cut 3 mm proximal to the trifurcation site of sciatic nerve (*see* **Note 1**). For nerve crush experiments, use a pair of delicate forceps (Fine Science Instruments; 0.4 mm tip angled, Cat No. 11063-07), and crush the sciatic nerve once for 30 s, and again for 30 s at the same site but orthogonal to the initial crush. For a nerve cut, the nerve is transected using fine scissors at the same point as for crush.

9. Close the gluteal and the hamstring muscles using an 8.0 suture and topically apply one drop of 0.025% bupivacaine solution above the muscle for local analgesia.

10. Close the skin with a surgical clip using an Autoclip applier.

11. Weigh the mouse and put the mouse in the heated recovery box.

12. Wait until the mouse fully wakes up and then return back to the cage.

13. Repeat **steps 3–12** for the next mouse undergoing surgery.

14. Check the mice that have undergone surgery each day. The weight/condition of the animal postsurgery should be checked and recorded daily until they are euthanized for analysis.

3.2 Dissection and Fixation of Mouse Sciatic Nerve Tissue

1. Clean the dissecting area with 70% (v/v) alcohol/water.

2. Cover a corkboard with tin foil and then wash with 70% alcohol.

3. Fill the 7 ml screw top bijou containers with 5 ml 4% PFA.

4. Kill mice by using the appropriate approved procedure such as using carbon dioxide inhalation and/or cervical dislocation.

5. Pin the mouse carcass out onto a corkboard and spray the mouse body with 70% alcohol to keep the fur wet and to not interfere with the dissection.

6. Use forceps and scissors and remove the skin from lower part of the mouse's body.

7. Dissect out the sciatic nerves (*see* **Notes 2** and **3**) and fix them in 4% PFA for a minimum of 5 hours (h) up to 24 h at 4 °C.

8. Clean the dissecting area and dissecting tools.

3.3 Permeabilization, Blocking, and Staining of Samples

1. Following fixation, nerve samples are washed three times in PBS, each wash lasting 10 min, at room temperature in a 7 ml screw top bijou container (*see* **Notes 4** and **5** for tissue preparation and multiple nerve staining). Nerve tissue samples are then permeabilized and blocked with permeabilization/blocking solution (*see* Subheading 2.2 solutions) overnight at 4 °C to prevent non-specific binding of primary antibody.

2. Desired primary antibodies (*see* **Note 6**) are then diluted into staining solution and incubated with the nerve sample for 72 h at 4 °C to allow primary antibodies to fully penetrate into the nerves.

3. Following incubation with primary antibody, nerve samples are then washed three times in PBS at room temperature, each wash lasts for 15 min. After these three 15 min washes, the nerves are then washed using a further six changes of PBS, 1 h each wash, to ensure complete removal of unbound primary antibody.

4. Nerve samples are then incubated for 48 h at 4 °C with fluorescently labeled species-specific secondary antibodies diluted into staining solution. Hoechst dye to stain nuclear DNA can also be added to the secondary antibody solution at this point (*see* **Note 6** for concentration of secondary antibody and Hoechst dye and **Note 7** for the monitoring of secondary antibody binding).

5. Excess secondary antibody is then removed by washing in PBS for 6 h at room temperature, changing the PBS each hour.

6. Nerve samples are then cleared for imaging with increasing concentrations of glycerol at 25%, 50%, and 75% (v/v) glycerol in PBS. The stained nerve is placed in 25%, 50%, and 75%

glycerol solutions sequentially, incubated 24 h at 4 °C for each increasing glycerol concentration. The cleared nerve tissue is then ready to be mounted onto a glass slide for imaging (*see* **Notes 8** and **9**).

3.4 Mounting and Imaging of Whole Mount Stained Nerve

Following completion of the whole mount immunolabeling, the nerve tissue is now ready to be mounted onto a glass slide for imaging.

1. Place the stained nerve tissue in the center of the glass slide and gently pull both ends of nerve tissue with forceps to keep the nerve tissue straight (*see* **Note 10**).

2. Apply vacuum grease on the glass slide around the four corners where the glass coverslip will be placed (we would usually use 22 mm × 40 mm glass coverslips).

3. Place a coverslip on the top of the vacuum grease and apply a gentle pressure to allow the coverslip to just contact the nerve on the glass slide.

4. Apply CitiFluor anti-fade mounting medium on one side of the coverslip to allow penetration through the whole area covered under the glass coverslip.

5. Seal the coverslip around using clear nail varnish and allow the varnish to fully dry (*see* **Note 11** for storage).

6. Stained and cleared nerves are imaged using a confocal microscope and several Z-series stacks are captured covering the entire field of interest. Individual series may then be flattened into a single image for each location and then combined using image analysis software such as Adobe Photoshop (Figs. 2–4). 3D imaging and reconstructions may also be generated using appropriate 3D software such as Fiji and Imaris software.

4 Notes

1. In order to visualize blood vessel regeneration, Schwann cell migration and axonal regeneration in the nerve gap, no re-anastamosis of the severed nerve, either by suture or by fibrin glue, is performed in the nerve transection procedures. Without re-anastamosis, a nerve bridge will be formed between the proximal and distal stumps of the nerve 4 days following transection (Fig. 1) [11]. The nerve bridge length may vary between 1.5 and 2.5 mm in the mouse [7].

2. For analysis of repair following nerve crush injury, the entire sciatic nerve proximal and distal to the injury, including tibial, peroneal, and sural branches, are carefully dissected as distally as possible. To ensure good antibody penetration into the

nerve, then postfixation, using fine forceps, the epineurium is carefully removed from the entire length of nerve tissue.

3. During dissection, great care must be taken to not damage the delicate bridge tissue formed between the proximal and distal nerve stumps following nerve transection. In order to preserve the nerve bridge tissue, the sciatic nerves (3–5 mm proximal and distal to the injury site) together with the muscles underlying the nerve bridge are dissected out for fixation. Muscle tissue still attached to the nerve will be removed after the paraformaldehyde fixation.

4. The epineurium in the distal nerve stump prevents antibody penetration, and it needs to be removed in order to visualize axonal regeneration and remyelination in the distal nerve stump. However, the blood vessel regeneration, Schwann cell migration, and axonal regeneration can be visualized in the nerve bridge without removing the epineurium.

5. For samples of transected nerves, several nerve preparations can be stained in the same 7 ml bijou container. For crush injury, we normally dissect out a segment of nerve longer than 2 cm, and it is better to process the nerve crush samples separately, otherwise the nerves often tangle with each other.

6. For whole mount staining, both primary and secondary antibodies are normally used at twice the concentration as used for immunohistochemistry on cryostat sections or cell labeling on coverslips. This increased concentration allows full penetration of the antibodies into the nerve tissue and bridge. Hoechst dye for staining of nuclear DNA is used at the same concentration as for standard immunohistochemistry or cells on coverslips.

7. The binding of secondary antibody can be checked under the epifluorescence microscope daily.

8. Following whole mount staining, surrounding tissue around the nerve bridge can be further removed under the epifluorescence microscope after the 75% glycerol clearing step. This may improve the quality of imaging for the sample.

9. For samples that have undergone crush injury, the crush site is recognizable on the slide by eye after the 75% glycerol clearing step; the crush site is always more transparent than the rest of the tissue. We use this method to measure from the crush site to the furthest growing axon and study the speed of axon regeneration in the mouse. The longest time point that we have used for this method is 7 days following crush injury as leading axons are still visible in the tibial nerve in C57BL/6 mice. At later time points, axons have regenerated fully from the crush site through the entire distal nerve, and the relative

axonal regeneration rates between different mouse lines cannot be measured.

10. The diameter of mouse sciatic nerve varies between mouse lines and even between male and female mice. For 2-month-old control C57BL/6 mice, the sciatic nerve is normally between 0.6 and 0.8 mm in diameter.

11. Mounted slides of whole mount stained nerves can be stored at 4 °C. For staining with antibodies such as neurofilament, the fluorescence signal is visible even up to 1 year after staining. However, background signal and auto-fluorescence do appear to increase over time with storage.

Acknowledgments

This work was supported by a Wellcome Trust grant (WT088228) and Medical Research Council grant (MR/J012785/1) to D.B.P. We are grateful to Prof. Wendy Macklin (University of Colorado Denver School of Medicine, USA) for the gift of the PLP-EGFP transgenic mouse line.

References

1. Klymkowsky MW, Hanken J (1991) Whole-mount staining of Xenopus and other vertebrates. Methods Cell Biol 36:419–441

2. Kim HA, Mindos T, Parkinson DB (2013) Plastic fantastic: Schwann cells and repair of the peripheral nervous system. Stem Cells Transl Med 2:553–557

3. Parrinello S, Napoli I, Ribeiro S et al (2010) EphB signaling directs peripheral nerve regeneration through Sox2-dependent Schwann cell sorting. Cell 143:145–155

4. Cattin AL, Lloyd AC (2016) The multicellular complexity of peripheral nerve regeneration. Curr Opin Neurobiol 39:38–46

5. Chen YY, McDonald D, Cheng C et al (2005) Axon and Schwann cell partnership during nerve regrowth. J Neuropathol Exp Neurol 64:613–622

6. McDonald D, Cheng C, Chen Y et al (2006) Early events of peripheral nerve regeneration. Neuron Glia Biol 2:139–147

7. Dun XP, Parkinson DB (2015) Visualizing peripheral nerve regeneration by whole mount staining. PLoS One 10:e0119168

8. Painter MW, Brosius Lutz A, Cheng YC et al (2014) Diminished Schwann cell repair responses underlie age-associated impaired axonal regeneration. Neuron 83:331–343

9. Tian F, Yang W, Mordes DA et al (2016) Monitoring peripheral nerve degeneration in ALS by label-free stimulated Raman scattering imaging. Nat Commun 7:13283

10. Mallon BS, Shick HE, Kidd GJ et al (2002) Proteolipid promoter activity distinguishes two populations of NG2-positive cells throughout neonatal cortical development. J Neurosci 22:876–885

11. Cattin AL, Burden JJ, Van Emmenis L et al (2015) Macrophage-induced blood vessels guide Schwann cell-mediated regeneration of peripheral nerves. Cell 162:1127–1139

Chapter 23

The Use of Low Vacuum Scanning Electron Microscopy (LVSEM) to Analyze Peripheral Nerve Samples

Peter Bond and David B. Parkinson

Abstract

The use of electron microscopy allows analysis of the microarchitecture of peripheral nerves both during development and in the processes of nerve regeneration and repair.

We describe a novel method for the rapid analysis and quantification of myelin in peripheral nerve using a low vacuum scanning electron microscopy protocol. For this methodology, excised nerves are prepared for traditional transmission electron microscopy (TEM) imaging, but at the stage where semi-thin sections would be taken, the resin block is instead imaged at low vacuum in the scanning electron microscope (SEM) using the backscattered electron signal. Any features in the tissue which have incorporated high concentrations of osmium from the fixation process (e.g., myelin) appear as bright regions in the image.

Myelin therefore is easily identifiable in the images, and since there is a high contrast difference between it and the surrounding tissue, automated measurements for myelin thickness (e.g., G ratio) using standard and freely available image analysis software (e.g., ImageJ) are easily achievable, consistent, and repeatable. This method therefore greatly speeds up the analysis of nerve samples and, in many cases, will obviate the need for ultrathin sections and TEM for analyzing nerve morphology.

Low vacuum (LV) SEM imaging has benefits compared with light microscopy; magnification is continuous, resolution is higher, and contrast and brightness are controllable. Since the resin block does not need a metal coating for imaging in LV mode, once the images have been collected, the block is still ready to section for both light microscopy (LM) and TEM.

Key words Myelin, Schwann cell, G ratio, Low vacuum, Scanning electron microscopy

1 Introduction

Transmission electron microscopy (TEM) is an important tool for studying nerve ultrastructure, calculating myelin thickness (G ratio) and assessing Wallerian degeneration and repair [1–4]. However, in many instances it is difficult or impossible to image a whole nerve in cross section to estimate myelin/axon distribution and density. To overcome this problem, semi-thin sections for light microscopy are usually cut from the TEM block prior to sectioning specifically for TEM. The semi-thin sections are generally stained

Paula V. Monje and Haesun A. Kim (eds.), *Schwann Cells: Methods and Protocols*, Methods in Molecular Biology, vol. 1739,
https://doi.org/10.1007/978-1-4939-7649-2_23, © Springer Science+Business Media, LLC 2018

with either methylene or toluidine blue to enhance contrast and make the myelin visible for counting and measuring.

Recent advances in scanning electron microscope (SEM) design have made low vacuum scanning electron microscopes (LV-SEM) widely available. These instruments make it possible to produce high resolution images of non-conductive samples in a low vacuum environment. Traditionally the SEM operates at high vacuum and non-conductive samples are coated with a thin heavy metal or carbon layer to prevent electron charge build up on the specimen surface, which causes distortion. In the LV-SEM, residual gas in the specimen chamber is ionized to form a cloud of positive charge around the sample, which dissipates the electron beam energy without the need to physically coat the sample.

Image formation in the LV-SEM is normally from backscattered electrons, which are generated more strongly from high atomic number material (e.g., heavy metal stained myelin) and therefore areas of the sample with high atomic number appear as bright regions compared with lower atomic number areas. Biological tissue prepared for TEM is normally fixed with osmium tetroxide and stained with uranyl acetate. Osmium has a strong affinity for lipids, and consequently myelin is heavily osmicated. Myelin thus produces a strong backscatter electron signal in the LV-SEM and is brighter than the surrounding cellular matrix making axons very easy to define with respect to the rest of the tissue section.

We describe here a simple procedure for LV-SEM in peripheral nerve samples, which allow the rapid assessment of nerve morphology, measurement of G ratio, and identification of other cell types and morphological changes in mouse transgenic nerves. This technique in many cases greatly speeds up the analysis of transgenic mouse nerves before and following peripheral nerve injury by removing the need to cut and analyze ultrathin sections by traditional TEM.

Figure 1 in this chapter shows a low magnification image of an adult mouse sciatic nerve imaged using the LV-SEM procedure. Figure 2 shows an adult mouse sciatic nerve imaged using LV-SEM. The high contrast of the myelin makes it simple to perform measurements of myelin thickness on such images (e.g., G ratio). Figure 3 shows an adult mouse tibial nerve sample at 21 days post-crush injury. The LV-SEM allows the visualization of both the re-myelinating axons and features such as the lipid-filled macrophages that have been phagocytosing myelin during the processes of PNS repair. Also clearly visible in this image is a small blood vessel within the regenerating nerve.

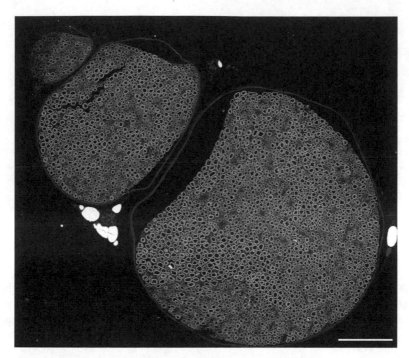

Fig. 1 Low-magnification LV-SEM image of adult mouse sciatic nerve. Scale bar 100 μm

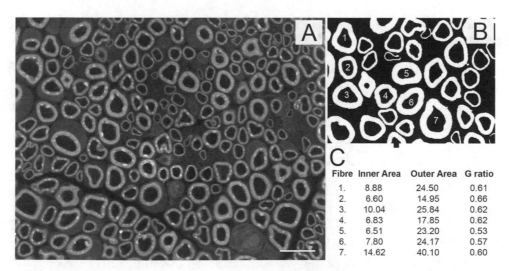

Fibre	Inner Area	Outer Area	G ratio
1.	8.88	24.50	0.61
2.	6.60	14.95	0.66
3.	10.04	25.84	0.62
4.	6.83	17.85	0.62
5.	6.51	23.20	0.53
6.	7.80	24.17	0.57
7.	14.62	40.10	0.60

Fig. 2 LV-SEM image of intact mouse sciatic nerve with *G* ratio measurements. (**a**) LV-SEM image of adult mouse sciatic nerve. Scale bar 10 μm. (**b**) Higher magnification thresholded image of sciatic nerve. Inner area of axon and outer area of fiber (axon plus myelin) are calculated using ImageJ. Myelinated fibers used for calculations in (**c**) are numbered. Values for inner area (axon) and outer area (axon plus myelin) are shown, together with calculated *G* ratio for myelinated fibers shown in B

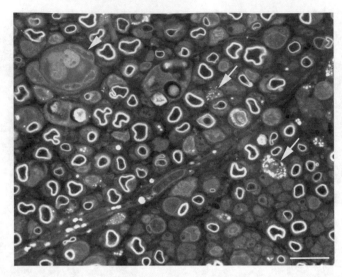

Fig. 3 LV-SEM image of tibial nerve 21 days after sciatic nerve crush injury. Image shows the presence of regenerated axons with ongoing Schwann cell re-myelination. Features such as macrophages containing lipid droplets are still easily visible at this time point (*arrows*), together with small blood vessels (*arrowhead*) within the repairing nerve. Scale bar 10 μm

2   Materials and Equipment

2.1   Fixation and Sample Preparation

All aqueous solutions are to be prepared with distilled or deionized water.

1. Sodium phosphate buffer (2×): 0.2 M Na_2HPO_4, 0.2 M NaH_2PO_4, pH 7.2 (*see* **Notes 1–3**).

2. Glutaraldehyde fixation solution: 2.5% glutaraldehyde in 0.1 M phosphate buffer at pH 7.2 (*see* **Notes 4** and **5**).

3. Filter paper, cut into rectangular strips to fit into 1.5 ml Eppendorf tubes.

4. Eppendorf tubes.

5. 10 ml sealable glass vials.

6. 1.0% osmium tetroxide in 0.1 M phosphate buffer at pH 7.2 (*see* **Notes 6** and **7**).

7. Agar low viscosity resin kit—medium hard (*see* **Note 8**).

8. 100% Analar ethanol.

9. 30%, 50%, and 90% (v/v) ethanol in distilled water.

10. 2% (w/v) uranyl acetate in 70% (v/v) ethanol.

11. BEEM "bottleneck" plastic embedding capsules.

12. Embedding oven set at 60 °C.

2.2 Microtomy	1. Glass knives (*see* **Note 9**).
	2. Ultramicrotome, such as Leica EM UC7 ultramicrotome (*see* **Note 10**).
2.3 Low Vacuum SEM	1. Low vacuum scanning electron microscope, such as JEOL 6610 LVSEM (*see* **Note 11**).

---

3 Methods

3.1 Fixation and Preparation of Samples

Fixation, washing, and dehydration steps for nerve samples can be performed at room temperature. However, if there is a need to store tissue in buffer for an extended period after fixation, then they should be kept in a 4 °C fridge.

1. Dissect peripheral nerves and place individual nerve on to a small, rectangular filter paper strip to keep them straight.

2. Fix the nerves by immersing in 2.5% glutaraldehyde in 0.1 M phosphate buffer at pH 7.2 for 2 h in 1.5 ml Eppendorf tubes.

3. Transfer the nerves to 10 ml glass vials and rinse twice for 15 min in 0.1 M phosphate buffer.

4. Post-fix the nerves for 1 h in 1.0% osmium tetroxide in 0.1 M phosphate buffer.

5. Rinse the nerves twice for 15 min in 0.1 M phosphate buffer.

6. Dehydrate the nerves by immersing sequentially in 30% and 50% ethanol, 15 min at each ethanol concentration.

7. Immerse the nerves in 2% (w/v) uranyl acetate in 70% (v/v) ethanol for 30 min (*see* **Note 12**).

8. Immerse the nerves in 100% ethanol twice, 15 min each.

9. Replace the alcohol with 30% (v/v) resin in ethanol and place on a slow rotator for 12 h. Repeat this process with 50%, 70%, and 90% (v/v) resin in ethanol and two changes of 100% resin, each step for 12 h.

10. After resin infiltration, transfer the nerves to bottleneck plastic-embedding capsules and orient individual nerves as near vertically as possible. Fill the capsules with resin and polymerize in an embedding oven at 60 °C for 24 h (*see* **Note 13**).

3.2 Microtomy

1. Trim excess resin from around the nerve on the ultramicrotome with a glass knife to expose the relevant region of the nerve. To optimize the block face for LV-SEM, it is important to remove as much excess resin as possible from around the nerve tissue.

2. Cut and discard several sections from the block at a thickness of between 200–300 nm before finally "polishing" the block

face by taking several 100 nm sections. This can all be done with a dry glass knife and letting the sections simply pile up on the knife edge since they are not required for SEM observation (*see* **Note 14**).

3. Mount the trimmed and polished block with the exposed nerve facing uppermost in a standard specimen holder for the SEM (*see* **Note 15**).

3.3 Low Vacuum SEM

1. Switch the SEM to low vacuum mode.

2. Insert the specimen (or specimens if a multiple holder is available).

3. Evacuate the SEM and set to 70 Pa pressure. This can be adjusted later if necessary.

4. If the SEM has a navigation feature, locate the resin block and position it under the electron beam.

5. Adjust the working distance to 10 mm.

6. Select the backscatter detector in compositional mode.

7. Switch on the electron beam and adjust focus, contrast, and brightness (*see* **Note 16**).

3.4 Image Analysis and G Ratio Data Collection

Images obtained by LV-SEM can be analyzed for G ratio measurement to assess myelin thickness. This analysis is performed in our laboratory using ImageJ software (available from https://imagej.nih.gov/ij/) and can be performed directly by measuring the areas of both the axon and myelinated fiber (axon plus myelin) to calculate the average areas to determine the G ratio.

1. Launch ImageJ and open the LV-SEM image.

2. To calibrate, select the Line Tool from the Toolbar and draw a horizontal line along the scale bar on the image.

3. From the Analyze menu, select Set Scale and enter Known Distance (scale bar length) and units (nm, μm, mm, etc.)

4. If all your images are taken at the same magnification, click on Global so calibration will be fixed for images presently open and any new ones until the calibration data is changed. You must recalibrate for any new image if the magnification is different.

5. From the Image menu, select Adjust and then Threshold. Using the Auto feature will often be perfectly adequate, but adjust the threshold sliders if necessary to optimize separation of myelin from the surrounding cellular matrix.

6. Click Apply and close the threshold window.

7. From the Analyze menu, select Set Measurements and then Area, Display Label, Add to Overlay, and click OK.

8. Select the Magic Wand Tool from the Toolbar and click inside an axon to select the inner axon area. From the Analyze menu, select Measure to add data to the results table which will appear.

9. Using the Magic Wand tool, click to the left of the outside of the myelin sheath of the same axon to select the total fiber (axon + myelin) area. From the Analyze menu, select Measure to add data to the results table.

10. Labels 1 and 2 will be written to the image, and the results table will have a pair of area measurements for that axon from which the G ratio is calculated (*see* Fig. 2).

11. Repeat the selection and measurement process for an appropriate number of axons from each image (*see* **Notes 17–20**).

4   Notes

1. Make 0.2 M Na_2HPO_4 and 0.2 M NaH_2PO_4 solution separately using distilled water. To make 50 ml, mix 36 ml of 0.2 M Na_2HPO_4 and 14 ml of 0.2 M NaH_2PO_4. The pH should be at 7.2. To make 0.1 M sodium phosphate buffer, mix equal parts of 0.2 M sodium phosphate buffer and distilled water to the desired volume.

2. A 0.2 M phosphate buffer solution is required to prepare fixatives in buffer with a final concentration of 0.1 M when using concentrated fixative stock solutions (e.g., 25% glutaraldehyde or 2% osmium tetroxide). Phosphate buffer should be made fresh for each batch of tissue but can be frozen and stored in small aliquots (10 ml) for convenience to extend its shelf life and thawed when required.

3. A 0.1 M phosphate buffer solution is used for washing and storing tissue and preparing buffered fixatives from solids (e.g., osmium tetroxide crystals).

4. Glutaraldehyde solutions must be prepared and used in the fume cupboard.

5. Prepare 2.5% glutaraldehyde in 0.1 M phosphate buffer by adding 10 ml of 25% EM grade glutaraldehyde to 50 ml of 0.2 M phosphate buffer and make the total volume up to 100 ml by adding 40 ml distilled water.

6. Osmium tetroxide is highly toxic, so to avoid skin contact and inhalation, use only in a fume cupboard. Always use protective gloves. We routinely use ampoules of pre-prepared 2% osmium tetroxide solution to reduce potential hazards when working with this chemical.

7. Osmium tetroxide should be made fresh and used immediately.

To prepare 10 ml of 1% OsO_4 in 0.1 M phosphate buffer at pH 7.2, add 5 ml of 2% osmium solution to 5 ml of 0.2 M phosphate buffer, pH 7.2.

8. Resin can be prepared as per manufacturer's instructions with the product and stored in 10 ml aliquots at -20 °C and thawed when required.

9. Glass knives are adequate for preparing the resin block face for LV-SEM. We make our own knives using a Leica Knife Maker.

10. The ultramicrotome must be well maintained and capable of cutting sections at 100 nm thickness.

11. Any modern variable pressure SEM is capable of producing similar images using this methodology.

12. Treatment of nerve samples with 2% uranyl acetate (in 70% ethanol) at this stage enhances contrast, in case TEM analysis may be performed at a later time on the sample.

13. Using long nosed capsules keeps the nerve orientated conveniently for cutting transverse sections for LV-SEM.

14. At this stage, transverse sections can be cut for light microscopy, if required.

15. Conveniently, the TEM resin blocks fit standard JEOL SEM specimen holders, so the resin block is mounted with the cut block face uppermost. Specimen holders for a range of sample types can be obtained from EM suppliers to fit most makes of SEM.

16. At low magnification, it is difficult to optimize contrast and brightness because the polished nerve section produces a strong backscatter signal with respect to the other areas of the resin block and the space around it. Center the nerve and increase magnification to fill the frame with the nerve tissue to avoid the high contrast between it and the surrounding empty space. Next, adjust the contrast and brightness to optimize the myelin as bright rings in a dark background and focus the image. Adjust the magnification to obtain an image displaying appropriate features, and at this point, contrast and brightness can be readjusted to suit the required outcome. This may be high contrast for delineating axons, or with a greater range of grey scales to show more general tissue and cell features, for instance the presence of immune cells such as macrophages.

17. We would normally measure a minimum of 200 myelinated axons from each nerve sample to generate both average G ratio values and to make scatter plots of axon diameter versus G ratio [5, 6].

18. Data can be copied and exported to a spreadsheet application for later G ratio calculations.

19. Thresholding will not separate axons touching, even by only one pixel. G ratio calculations must be taken only from individual axons.

20. The G ratio is normally approximately 0.60–0.65 for an adult mouse sciatic nerve. For hypomyelinated nerves, the G ratio may increase toward a value of 1, whereas for mouse mutants with increased myelination, there will be a corresponding reduction in G ratio values [7]. Figure 2b and c shows an example of a thresholded image and calculation of G ratios for myelinated axons from an adult mouse nerve using ImageJ to measure axon inner area and myelinated fiber (outer area). G ratio is calculated from the square root of inner area/outer area which provides a value that is shape independent (*see* also http://cifweb.unil.ch/index.php?option=com_docman&task=doc_details&gid=31&Itemid=57).

References

1. Arthur-Farraj PJ, Latouche M, Wilton DK, Quintes S, Chabrol E, Banerjee A, Woodhoo A, Jenkins B, Rahman M, Turmaine M, Wicher GK, Mitter R, Greensmith L, Behrens A, Raivich G, Mirsky R, Jessen KR (2012) C-Jun reprograms Schwann cells of injured nerves to generate a repair cell essential for regeneration. Neuron 75(4):633–647

2. Napoli I, Noon LA, Ribeiro S, Kerai AP, Parrinello S, Rosenberg LH, Collins MJ, Harrisingh MC, White IJ, Woodhoo A, Lloyd AC (2012) A central role for the ERK-signaling pathway in controlling Schwann cell plasticity and peripheral nerve regeneration in vivo. Neuron 73(4):729–742

3. Goebbels S, Oltrogge JH, Wolfer S, Wieser GL, Nientiedt T, Pieper A, Ruhwedel T, Groszer M, Sereda MW, Nave KA (2012) Genetic disruption of Pten in a novel mouse model of tomaculous neuropathy. EMBO Mol Med 4(6):486–499

4. Atanasoski S, Scherer SS, Sirkowski E, Leone D, Garratt AN, Birchmeier C, Suter U (2006) ErbB2 signaling in Schwann cells is mostly dispensable for maintenance of myelinated peripheral nerves and proliferation of adult Schwann cells after injury. J Neurosci 26(7):2124–2131

5. Mindos T, Dun XP, North K, Doddrell RD, Schulz A, Edwards P, Russell J, Gray B, Roberts SL, Shivane A, Mortimer G, Pirie M, Zhang N, Pan D, Morrison H, Parkinson DB (2017) Merlin controls the repair capacity of Schwann cells after injury by regulating Hippo/YAP activity. J Cell Biol 216(2):495–510

6. Roberts SL, Dun XP, Dee G, Gray B, Mindos T, Parkinson DB (2017) The role of p38alpha in Schwann cells in regulating peripheral nerve myelination and repair. J Neurochem 141(1):37–47

7. Michailov GV, Sereda MW, Brinkmann BG, Fischer TM, Haug B, Birchmeier C, Role L, Lai C, Schwab MH, Nave KA (2004) Axonal neuregulin-1 regulates myelin sheath thickness. Science 304(5671):700–703

Chapter 24

Analysis of Myelinating Schwann Cells in Human Skin Biopsies

Mario A. Saporta and Renata de Moraes Maciel

Abstract

The human skin is richly innervated by nerve fibers of different calibers and functions, including thickly myelinated large fibers that act as afferents for mechanoreceptors in the dermal papillae. Skin biopsies offer minimally invasive access to these myelinated fibers, in which each internode represents an individual myelinating Schwann cell. Using this approach, human myelinated nerve fibers can be analyzed by several methods, including immunostaining, morphometric and ultrastructural analysis, and molecular biology techniques. This analysis can reveal important aspects of human Schwann cell biology in health and disease, such as in the case of demyelinating neuropathies. This technique has revealed Schwann cell phenotypes in Charcot-Marie-Tooth disease type 1 and acquired inflammatory neuropathies.

Key words Skin biopsy, Immunohistochemistry, Myelinated nerve fibers, Schwann cells, Charcot-Marie-Tooth disease, Neuropathy

1 Introduction

Unmyelinated (C) or thinly myelinated small (Aδ) nerve fibers terminate at the dermis or epidermis as free endings that convey pain and temperature perception. Myelinated dermal nerve fibers are associated with mechanoreceptors, including Merkel and Meissner corpuscles, conveying tactile sensations of vibration and light touch, and prolonged sensations of touch and pressure, respectively (Fig. 1). The use of skin biopsies to study unmyelinated epidermal fiber density in conditions where small-diameter fibers are preferentially affected (small fiber neuropathies), such as HIV neuropathy [1], diabetes [2], and idiopathic small fiber neuropathies [3], is now well established [4]. More recently, techniques to study dermal myelinated nerve endings have been developed [5, 6] and applied to conditions where inherited or acquired demyelination is a feature of the disease process. These include the inherited neuropathies (collectively known as Charcot-Marie-Tooth disease) [7–10] and the inflammatory demyelinating

Paula V. Monje and Haesun A. Kim (eds.), *Schwann Cells: Methods and Protocols*, Methods in Molecular Biology, vol. 1739, https://doi.org/10.1007/978-1-4939-7649-2_24, © Springer Science+Business Media, LLC 2018

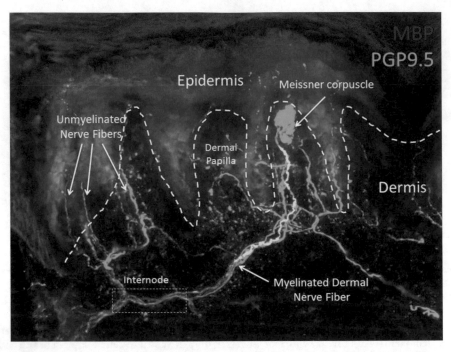

Fig. 1 Typical appearance of a skin biopsy section stained with PGP9.5 (axons) and MBP (myelinating Schwann cells). The skin is organized in three main layers. The more external layer, the epidermis, is composed of stratified squamous epithelium, and its outermost layer, the stratum corneum, is visualized in red in this figure due to autofluorescence. The stratum corneum, or corneal layer, is composed mainly of anucleated corneocytes and keratin and is not innervated. The epidermis, on the other hand, receives innervation from nonmyelinated (intraepidermal) nerve fibers. The dermis, the layer immediately below the epidermis, is the connective tissue layer of the skin, composed mainly of collagen fibers. The dermis extends up into the epidermis in structures called dermal papillae. The dermal papillae are richly innervated by myelinated dermal nerve fibers that end in skin receptors, such as Meissner corpuscles, which are mechanoreceptors that convey vibratory sensation. The hypodermis, the deepest skin layer consisting mainly of adipocytes, nerves, and blood vessels, are usually not sampled by a punch skin biopsy

neuropathies, such as Guillain-Barré syndrome and chronic inflammatory demyelinating polyradiculoneuropathy (CIDP) [9, 11]. This technique offers an alternative to invasive open nerve biopsies as a source of human tissue for peripheral nervous system research.

Myelinated nerve fibers travel the dermis in a tridimensional fashion. Therefore, skin biopsies should be sectioned in thick (>50 μm) slices to allow for the isolation of long nerve fiber segments for imaging analysis. Immunostaining of thick sections require free-floating techniques to efficiently expose tissue to antibody solutions. Immunostaining of skin biopsies can be used to study the molecular architecture of myelinated fibers and to perform morphometric analysis of the internodal, paranodal, and nodal regions [9, 11] providing insights into mechanisms of demyelination in human diseases (Fig. 2). Morphometric analysis of myelinated dermal nerve fibers from patients with Charcot-Marie-Tooth disease type 1A, caused by duplication of peripheral

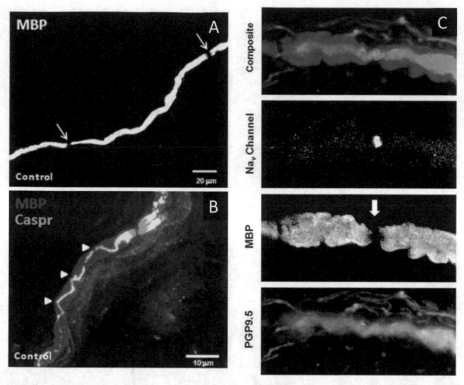

Fig. 2 Molecular architecture analysis of dermal myelinated nerve fibers by immunostaining. Z-stack reconstruction of a skin biopsy section demonstrates the typical myelinated nerve fiber domains described in human and rodent nerve biopsy samples: (**a**) the internode, a segment between two nodes of Ranvier, corresponding to the full length of one Schwann cell, immunoreactive to the compact myelin protein Myelin Basic Protein (MBP); (**b**) the paranode, as determined by contactin-associated protein (CASPR) immunostaining; and (**c**) the node of Ranvier, as determined by the clustering of sodium channels. Adapted from Saporta et al., Brain 2009 [9]

myelin protein 22 (PMP22), has revealed reduced internodal lengths [9], confirming findings from sural nerve biopsy studies [12]. A skin biopsy study of a patient homozygous for a PMP22 deletion has revealed impairment of myelination of dermal fibers, with Schwann cells contacting axons but failing to form compact myelin [13] (Fig. 3).

Skin biopsies can also be used to study the ultrastructure of myelinated fibers by electron microscopy, allowing the visualization of organelles such as mitochondria [9] and the analysis of protein expression by immune electron microscopy [10]. Gene expression analysis by real-time polymerase chain reaction (rt-PCR) has also been successfully conducted using skin biopsy samples from patients with inherited [7, 8, 10] and acquired demyelinating neuropathies [14]. Here we describe the punch skin biopsy procedure and the processing of samples for free-floating immunostaining for the study of dermal myelinated nerve fibers by confocal microscopy.

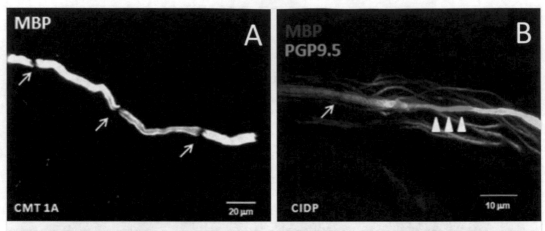

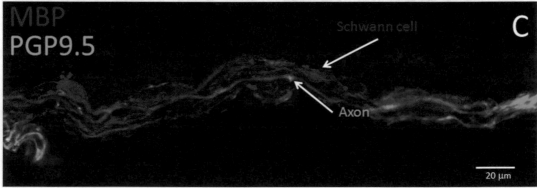

Fig. 3 Abnormal findings in dermal myelinated nerve fibers from a patient with demyelinating peripheral neuropathies. (**a**) Internodes are shorter in patients with Charcot-Marie-Tooth disease type 1A (compare with normal internodal length shown in Fig. 2a); (**b**) paranodal demyelination can be demonstrated as regions of the axon (PGP9.5 immunoreactive) deprived of myelin (MBP immunoreactivity, arrowheads) in patients with chronic inflammatory demyelinating polyradiculoneuropathy (CIDP); (**c**) a patient homozygous for PMP22 deletion demonstrates impaired myelination of dermal nerve fibers characterized by Schwann cells contacting axons but failing to form compact myelin or organized internodes. Adapted from Saporta et al., Brain 2009 [9] and Saporta et al., Arch Neurol 2011 [13]

2 Materials

2.1 Punch Skin Biopsy Procedure

1. 2 and 3 mm sterile, disposable skin biopsy punches.

2. 2% lidocaine hydrochloride with epinephrine 1:100,000.

3. Two TB tuberculin 1 mL syringes.

4. Alcohol swabs.

5. Packages of sterile gauze.

6. One suture removal kit containing forceps and dissecting scissors.

7. Disposable scalpel.

8. Sterile gloves.

9. Self-adherent wound dressing wrap.

10. Adhesive bandages.

11. 2 mL microtubes.

2.2 Supplies and Equipment

1. Sterile pipette tips for 1000, 300, 100, and 10 μL micropipettes.

2. Glass slides.

3. Coverslips 22 × 22 mm and 22 × 50 mm.

4. 96-Well plate with lid.

5. 12 × 12 mm embedding molds.

6. Styrofoam container.

7. Freezer storage bags.

8. Small paintbrush.

9. Cryostat.

10. Perfect loop (EMS, Catalogue number 70944).

11. Clear nail polish.

12. Confocal microscope.

2.3 Reagents and Solutions

1. 10× phosphate-buffered saline (PBS).

2. OCT medium.

3. Mounting medium with 4′,6-diamidino-2-phenylindole, dihydrochloride (DAPI).

4. Primary antibodies: anti-protein gene product 9.5 (PGP9.5); anti-myelin basic protein (MBP); anti-pan voltage-gated sodium channel (NaV); anti-contactin-associated protein (CASPR).

5. Fluorescent secondary antibodies.

6. 1× phosphate-buffered saline (1× PBS). Store at room temperature.

7. PBST. 0.3% Triton X-100 in 1× PBS. Store at room temperature.

8. 25% sucrose in 1× PBS. Store at 4 °C.

9. 4% paraformaldehyde in 1× PBS. Store at 4 °C and replace stock solution every 5–6 months.

10. Phosphate buffer (for antifreeze solution): 0.312 M sodium phosphate monobasic; 0.214 M sodium hydroxide. Prepare the buffer in double-distilled water. Mix thoroughly and store at room temperature.

11. Glycerol antifreeze solution: 30% glycerol; 30% ethylene glycol; 10% phosphate buffer (see above). Prepare the solution in double-distilled water. Leave on a shaker until well mixed and store at room temperature.

12. Blocking solution: 5% fish gelatin; 0.02% sodium azide; 0.1% Triton X-100. Prepare the solution in 1× PBS, mix well, and aliquot. Fish gelatin will melt from a gel to a thick liquid when left at room temperature for several hours. Store the blocking solution at 4 °C when in use. Freeze extra aliquots at −20 °C and thaw when needed.

3   Methods

3.1   Skin Biopsy

The punch skin biopsy is a simple, minimally invasive, outpatient procedure. It can be performed under local anesthesia and can usually be completed in 10–15 min. A 2 or 3 mm punch in used to remove a conical sample encompassing the epidermis and dermis. Patients can be sent home the same day with an adhesive bandage covering the surgical lesion, which should heal completely within 2 weeks.

1. Sterilize biopsy area(s) with 70% alcohol (*see* **Note 1**).
2. Inject 0.3–0.5 mL lidocaine at biopsy site to make a weal (i.e., a transient elevation on the site of injection).
3. Position biopsy punch in the center of the weal and apply pressure while rotating the punch in one direction only until the depth reaches approximately 4 mm or about half the length of the punch.
4. Tilt the punch to one side and remove it slowly so that the biopsy is exposed on the surface of the skin (*see* **Note 2**).
5. Cut any connective tissue to completely excise the skin biopsy by using either a scalpel or autoclaved fine point scissors.
6. Immediately immerse tissue in 4% paraformaldehyde solution in a 2 mL microtube. Record the time (*see* **Note 3**).

3.2   Tissue Freezing and Sectioning

3.2.1   Cryoprotection

1. After the appropriate fixation time, remove the fixative solution and dispose of properly in a formaldehyde waste container.
2. Wash the biopsies three times in 1× PBS solution, 5 min per wash.
3. After the third and final wash, add 25% sucrose solution and store the biopsies at 4 °C at least overnight. Initially, the tissue will float to the top of the sucrose solution. Once it sinks to the bottom of the Eppendorf tube, it is sufficiently cryoprotected and ready for embedding.

3.2.2   Embedding

1. Place a flat piece of dry ice in a Styrofoam container.
2. Label the 12 mm embedding molds with the patient's code, date, and biopsy orientation (e.g., dermal side and epidermal side).

3. Fill the molds with OCT, taking care to squeeze with a constant pressure to limit the number of bubbles (*see* **Note 4**).

4. Pipette out the sucrose solution from the Eppendorf tube containing the biopsy. Dip the forceps in OCT prior to using them to lift out the tissue from the tube without applying any pressure. Submerge biopsy in OCT and push it to near the bottom of the mold with the forceps. Orient the biopsy specimen so that the labels on the mold indicate the dermal and epidermal layers.

5. Set the mold on slab of dry ice and allow it to harden for 5 min.

6. Store the biopsies in labeled freezer storage bags at −20 or −80 °C. Tissues prepared as indicated may be stored indefinitely.

3.2.3 Cryosectioning of Skin Biopsy Samples

1. Pipette 200 µL of glycerol antifreeze solution in each well of a 96-well plate.

2. Set cryostat chamber temperature to −19 °C and allow it to cool if necessary. If a temperature setting for the specimen stage is available, set it between −17 and −19 °C.

3. Put a new blade in, or move to a new area of a used blade.

4. Apply OCT to the specimen disk, cut biopsy mold with a razor blade, and invert the block so that the top will be frozen to the disk. Take care to position the block with the epidermis toward the orientation notch.

5. After allowing the specimen to freeze to the disk for a few minutes in the chamber, lift the lever on the specimen holder and insert the disk. For glabrous biopsies, cut from the epidermis to the dermis.

6. Adjust desired section thickness. For glabrous skin biopsies, sections should be 60 µm thick.

7. Approach the knife with the specimen by using the controls.

8. Adjust the angle of the specimen holder with the levers if the blade is cutting the sample disproportionally. For example, angle the specimen down if after several cuts, only the very bottom edge is being cut.

9. Start cutting and periodically collect a section on a slide to check whether the tissue has been reached.

10. After reaching the tissue, trim the outside of the block down with a razor blade. Throw away the next section in case the pressure from trimming has altered the section thickness with this sample.

11. Dip a small paintbrush in glycerol antifreeze solution and begin collecting sections by touching the brush to a section and placing it in a well of the 96-well plate.

12. After cutting all sections, remove remaining OCT frozen to the specimen disk by briefly warming the disk in your hand and popping off the block with a razor blade. Either vacuum up any excess OCT in the chamber or wipe it down with a paper towel and ethanol.

3.3 Immunostaining of Skin Biopsy Sections for Visualization of Dermal Myelinated Nerve Fibers

A combination of antibodies can be used to visualize dermal myelinated fibers and study their molecular architecture. Dermal axons can be visualized using the pan-axonal marker PGP9.5. Peripheral myelin has a strong immunoreactivity to MBP antibodies, and these can be used to visualize the internodes. The paranodes, the regions immediately adjacent to the node of Ranvier, can be visualized using CASPR antibodies. The size of a paranode is proportional to the internodal length of the corresponding Schwann cell. Therefore, asymmetries between two adjacent paranodes are suggestive of a segmental demyelinating/remyelinating process, where the remyelinated internode is smaller in length compared to the original internodal length. This is seen in acquired inflammatory neuropathies [9]. Lastly, nodes of Ranvier can be visualized using a Nav antibody. Dispersion of the nodal Nav channel expression into the paranodes and internodes is a hallmark of demyelinated axons.

1. Select the sections to stain. If they are in glycerol solution, pipette out the solution and add 200 μL of 1× PBS to the wells (*see* **Note 5**).

2. Transfer the sections to a new plate with a perfect loop.

3. Wash the sections twice with 1× PBS, 5 min per wash.

4. Add 200 μL of chilled 100% acetone to the wells for 5 min (*see* **Note 6**).

5. Wash the sections three times with 1× PBS, 5 min per wash.

6. Wash the sections three times with PBST, 5 min per wash.

7. Add 200 μL of blocking solution to each well and block for 1 h at room temperature.

8. Dilute the primary antibodies against PGP9.5, MBP, CASPR, and Nav in blocking solution at the optimal concentrations for a final volume of 100 μL per well. A combination of primary antibodies can be used according to the goal of the experiment (*see* **Notes 7** and **8**).

9. Dispense 100 μL of primary antibody solution per well. Incubate at 4 °C overnight.

10. After overnight primary antibody incubation, transfer the sections to a new row of wells filled with PBST. After transferring, wash twice with PBST, 5 min per wash.

11. Dilute the fluorescent secondary antibodies in blocking solution at 1:200 concentration and add 100 μL per well.

12. Cover the plate with aluminum foil to limit bleaching from overhead lights. Incubate either 2 h at room temperature, or overnight at 4 °C.

13. Following incubation, transfer the sections to a new row filled with 1× PBS and wash for 5 min.

14. Wash two more times with 1× PBS, 5 min per wash.

15. Fill a weighing boat or similar container with 1× PBS. Submerge at least three quarters of a slide in the container.

16. Transfer a section to the weighing boat and move it toward the surface where the PBS meets the slide.

17. Slowly pull the slide out of the weighing boat and lay the section flat. Mount sections in the same orientation and position them close together to minimize the amount of mounting medium used.

18. After mounting the last section, allow the slide to dry slightly for 2 h or even overnight to avoid sections to move when adding the coverslip. Protect the slides with aluminum foil to prevent contact with dust.

19. Choose a coverslip size that will adequately cover all sections. It is suggested to use 22 mm coverslips for three or four sections, and 50 mm coverslips for more than four sections.

20. Add 1–3 drops of mounting medium with DAPI depending on the size of the coverslip. Make sure all sections are covered and the areas to be measured have no bubbles. If there are bubbles, try to push them out with forceps or reposition coverslip by soaking the slide in 1× PBS and removing it.

21. Remove any excess of mounting medium from the edges of the coverslip with filter paper and seal with clear nail polish.

22. Store the slides at −20 °C to preserve the intensity of fluorescence up until observation by fluorescence microscopy.

23. Use confocal microscopy imaging to visualize and take images of the stained samples for the morphometric analysis of dermal myelinated fibers. Typical confocal microscopy images of normal and abnormal findings in dermal myelinated nerve fibers are provided in Figs. 1–3 (*see* **Note 9**).

4   Notes

1. For dermal myelinated nerve fiber analysis, glabrous skin from the lateral aspect of the fingers or from the fingertips will yield the highest density of fibers and may be necessary especially in patients with a length-dependent peripheral neuropathy.

2. Removing the punch quickly or more perpendicular to the skin causes the tissue to stay within the biopsy site so that it is

difficult to remove without using forceps and potentially creating excision artifacts.

3. Fixation time can vary according to the specific antigen of interest. A short fixation time of 30 min at room temperature is adequate for most antigens. Overnight fixation can be done if necessary, but should not exceed 24 h.

4. Inverting the bottle of OCT medium for a minute allows bubbles to rise and prevents them from getting into the mold.

5. PBS is much less viscous, and sections are less likely to tear when lifted out of PBS than glycerol.

6. Acetone will dissolve the plastic in the wells to some extent, but this will not affect the staining if it is only left in for 5 min. It will, however, make the wells cloudy, which makes transferring somewhat more difficult. Alternatively, a glass EM plate can be use instead.

7. The following combinations of primary antibodies are suggested. For morphometric studies of myelinated fibers (e.g., measurement of internodal length), combine PGP9.5 and MBP antibodies. For analysis of paranodal length and symmetry, combine CASPR and MBP antibodies. For the study of sodium channel clustering, combine Nav, PGP9.5 and MBP antibodies. The primary antibodies should be used as follows: anti-PGP9.5 (1:1000), anti-MBP (1:800), anti-CASPR (1:100), and anti-Nav (1:500).

8. The following primary antibodies have been optimized at the above-cited concentrations for use in free-floating staining of skin biopsy sections: (a) polyclonal rabbit anti-PGP9.5 (Catalogue number: 13C4, Ultraclone, Isle of Wight, UK), (b) monoclonal mouse anti-MBP (Catalogue number 2C3, Ultraclone, Isle of Wight, UK), (c) polyclonal rabbit anti-Caspr (Catalogue number ab34151, Abcam, Cambridge, MA, USA), and (d) monoclonal mouse anti-pan sodium channels (Catalogue number S8809, Sigma-Aldrich, St. Louis, MO, USA).

9. Skin biopsy sections should be examined using a laser confocal microscope in order to obtain Z-stack images for 3D reconstruction. Sections should be imaged with a 20× objective for internodal length measurement. Confocal images should be obtained at 2-μm increments. Measurements should be performed using image software to trace the course of myelinated nerve fibers through the Z-stack images. Internodal length can be measured in nerve fibers located either in the dermal papillae or the subepidermal plexus. This measurement can only be done in nerve fibers that allow single internodes to be individualized from other nerve fibers or internodes (Fig. 3).

Acknowledgements

The authors would like to thank Dr. Jun Li and Michael Shy for mentoring and training in the performing of skin biopsy and dermal myelinated nerve fiber analysis.

References

1. Polydefkis M, Yiannoutsos CT, Cohen BA et al (2002) Reduced intraepidermal nerve fiber density in HIV-associated sensory neuropathy. Neurology 58:115–119

2. Polydefkis M, Hauer P, Griffin JW, McArthur JC (2001) Skin biopsy as a tool to assess distal small fiber innervation in diabetic neuropathy. Diabetes Technol Ther 3:23–28

3. Holland NR, TO C, Hauer P et al (1998) Small-fiber sensory neuropathies: clinical course and neuropathology of idiopathic cases. Ann Neurol 44:47–59. https://doi.org/10.1002/ana.410440111

4. England JD, Gronseth GS, Franklin G et al (2009) Practice parameter: evaluation of distal symmetric polyneuropathy: role of laboratory and genetic testing (an evidence-based review). Report of the American Academy of Neurology, American Association of Neuromuscular and Electrodiagnostic Medicine, and American Academy of Physical Medicine and Rehabilitation. Neurology 72:185–192. https://doi.org/10.1212/01.wnl.0000336370.51010.a1

5. Nolano M, Provitera V, Crisci C et al (2003) Quantification of myelinated endings and mechanoreceptors in human digital skin. Ann Neurol 54:197–205. https://doi.org/10.1002/ana.10615

6. Provitera V, Nolano M, Pagano A et al (2007) Myelinated nerve endings in human skin. Muscle Nerve 35:767–775. https://doi.org/10.1002/mus.20771

7. Li J, Bai Y, Ghandour K et al (2005) Skin biopsies in myelin-related neuropathies: bringing molecular pathology to the bedside. Brain 128:1168–1177. https://doi.org/10.1093/brain/awh483

8. Sabet A, Li J, Ghandour K et al (2006) Skin biopsies demonstrate MPZ splicing abnormalities in Charcot-Marie-Tooth neuropathy 1B. Neurology 67:1141–1146. https://doi.org/10.1212/01.wnl.0000238499.37764.b1

9. Saporta MA, Katona I, Lewis RA et al (2009) Shortened internodal length of dermal myelinated nerve fibres in Charcot-Marie-Tooth disease type 1A. Brain 132:3263–3273. https://doi.org/10.1093/brain/awp274

10. Katona I, Wu X, Feely SME et al (2009) PMP22 expression in dermal nerve myelin from patients with CMT1A. Brain 132:1734–1740. https://doi.org/10.1093/brain/awp113

11. Doppler K, Werner C, Sommer C (2013) Disruption of nodal architecture in skin biopsies of patients with demyelinating neuropathies. J Peripher Nerv Syst 18:168–176. https://doi.org/10.1111/jns5.12023

12. Dyck PJ, Gutrecht JA, Bastron JA et al (1968) Histologic and teased-fiber measurements of sural nerve in disorders of lower motor and primary sensory neurons. Mayo Clin Proc 43:81–123

13. Saporta MA, Katona I, Zhang X et al (2011) Neuropathy in a human without the PMP22 gene. Arch Neurol 68:814–821. https://doi.org/10.1001/archneurol.2011.110

14. Lee G, Xiang Z, Brannagan TH et al (2010) Differential gene expression in chronic inflammatory demyelinating polyneuropathy (CIDP) skin biopsies. J Neurol Sci 290:115–122. https://doi.org/10.1016/j.jns.2009.10.006

Chapter 25

Whole Mount In Situ Hybridization and Immunohistochemistry for Zebrafish Larvae

Rebecca L. Cunningham and Kelly R. Monk

Abstract

In situ hybridization enables visualization of mRNA localization, and immunohistochemistry enables visualization of protein localization within a tissue or organism. Both techniques have been extensively utilized in zebrafish (Thisse et al., Development 119:1203–1215, 1993; Dutton et al., Development 128:4113–4125, 2001; Gilmour et al., Neuron 34:577–588, 2002; Lyons et al., Curr Biol 15:513–524, 2005) including for visualization of mRNA localization in Schwann cells (Lyons et al., Curr Biol 15:513–524, 2005; Monk et al., Science 325:1402–1405, 2009). For in situ hybridization, here, we outline how to generate RNA probes, conduct whole mount in situ hybridization for larvae, and list RNA probes that label different stages of Schwann cell development in zebrafish. For immunohistochemistry, the protocol we outline can be used to mark Schwann cells of sensory and motor nerves to examine properties such as developmental stage, morphology, proliferation, and apoptosis.

Key words In situ hybridization, RNA, Immunohistochemistry, Antibody, Fluorescence

1 Introduction

1.1 Whole Mount In Situ Hybridization in Zebrafish

Whole mount in situ hybridization (WISH) in zebrafish has been employed for decades (e.g., [1]) to examine mRNA expression patterns. The fact that zebrafish are optically clear as larvae enables easy visualization of Schwann cell RNA transcripts during development. WISH enables analyses of Schwann cell development because distinct genetic markers can delineate specific developmental stages (Table 1). For example, when studying zebrafish with mutations affecting Schwann cell development, WISH enables a researcher to assess the earliest stages of Schwann cell development that is affected based upon changes in mRNA expression levels on lateral line nerves (prominent sensory nerves in zebrafish) and/or motor nerves (Fig. 1). Additionally, WISH for *myelin basic protein* (*mbp*) has been used to identify novel regulators of Schwann cell myelination in forward genetic screens [13, 14]. In brief, the WISH process entails generation of a digoxigenin-labeled antisense

Paula V. Monje and Haesun A. Kim (eds.), *Schwann Cells: Methods and Protocols*, Methods in Molecular Biology, vol. 1739, https://doi.org/10.1007/978-1-4939-7649-2_25, © Springer Science+Business Media, LLC 2018

RNA probe that anneals to a native mRNA of interest. This labeled probe can then be detected by an antibody and developed with alkaline phosphatase treatment. The protocol takes 3 days to complete and has been modified from Thisse et al. [15]. Any number of probes can be generated to analyze mRNA localization in Schwann cells; probes that have been shown to label Schwann cells in zebrafish larvae are listed in Table 1.

1.2 Immunohisto chemistry in Zebrafish

Immunohistochemistry (IHC) visualizes protein localization within Schwann cells on the lateral line and motor nerves of zebrafish larvae (Fig. 1). Briefly, IHC utilizes a primary antibody

Table 1
Genetic markers that can be used to delineate specific Schwann cell developmental stages

Schwann cell stage	WISH probes
Neural crest ~12.5 hpf	*sox10* [2], *foxd3* [3], *erbb3* [4], *crestin* [5]
Schwann cell precursor ~18–48 hpf	*sox10, erbb3, foxd3, gpr126* [6]
Immature Schwann cell beginning ~48 hpf	*sox10, erbb3, foxd3, gpr126*
Promyelinating Schwann cell beginning ~48 hpf	*oct6/pouf3f1* [6, 7], *krox20/egr2b* [8], *sox10, erbb3, foxd3, gpr126*
Mature Schwann cell beginning ~60 hpf	*mbp* [4, 9], *mid1ip1b* [10], *ctnnd2* [10], *cldnk* [11], *zwillinga & b* [12], *krox20/egr2b, sox10, foxd3, erbb3, gpr126*

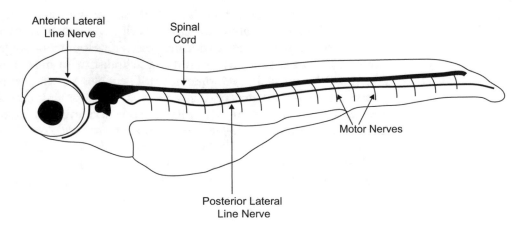

Fig. 1 Schematic of a zebrafish larva. Schwann cells are located on the lateral line and motor nerves. Note that the anterior lateral line nerve actually extends around the entire eye

that binds to a native protein of interest; a secondary antibody conjugated to a fluorophore amplifies the signal to enable visualization. Usually, it is helpful to visualize Schwann cells with an antibody for a protein such as myelin basic protein (Mbp) and an antibody to visualize axons, such as acetylated tubulin [4, 6]. IHC to label Schwann cells in zebrafish can be conducted on whole zebrafish larvae or on sections [4, 16]. Transgenic fluorescence within larvae can also be preserved while performing IHC protocol for another protein of interest. The methodology for IHC can vary depending upon the antibody of interest [17]. If IHC for an antibody of interest does not produce a result, we identify several steps to troubleshoot in Subheading 4. This whole mount IHC protocol presented here has been successfully used for the Mbp antibody [4, 6].

2 Materials

2.1 Materials for WISH

Conduct all steps at room temperature (RT) unless otherwise noted. Follow appropriate disposal of hazardous materials in accordance with environmental health and safety standards at your institution.

2.1.1 Generation of Digoxigenin-RNA Probe

1. 2–5 µg DNA of interest (e.g., isolated by a miniprep) (*see* **Note 1**).

2. Appropriate enzyme to linearize plasmid and enzyme buffer.

3. PCR purification kit (e.g., Qiagen PCR Purification Kit).

4. 1.5% agarose gel made with TAE buffer.

5. 10× transcription buffer associated with polymerase.

6. DIG RNA labeling mix, 10× concentration.

7. RNase I inhibitor.

8. Transcription enzyme (SP6, T7, or T3).

9. DEPC water.

10. DNase I (RNase-free).

11. Microcolumn for RNA purification (e.g., GE Illustra ProbeQuant G-50 Micro Columns).

12. Hybridization buffer [Hyb(+)]: 50% formamide, 5× SSCTw, 0.1% Tween, 9 mM citric acid, 500 µg/mL *Torula* yeast RNA, in ultrapure water (*see* **Note 2**).

2.1.2 Whole Mount In Situ Hybridization

1. 0.003% phenylthiourea (PTU) in egg water (*see* **Note 3**).

2. 4% paraformaldehyde (PFA) prepared in PBS (*see* **Note 4**).

3. 1.5 mL microcentrifuge tubes.

4. Nutator.

5. 100% methanol (MeOH).

6. Hyb(+)—*see* **item 12** in Subheading 2.1.1.

7. PBSTween (PBSTw): 1× PBS, 0.2% Tween (*see* **Note 5**).

8. Proteinase K (ProtK) 20 mg/mL.

9. 2× SSCTween (SSCTw): 2× SSC, 0.2% Tween (*see* **Note 6**).

10. 0.2× SSCTween (SSCTw): 0.2× SSC, 0.2% Tween (*see* **Note 7**).

11. Hyb(+) at 65 °C (*see* **Note 8**).

12. 75% Hyb(+): 25% 2× SSCTw at 65 °C (*see* **Note 8**).

13. 50% Hyb(+):50% 2× SSCTw at 65 °C (*see* **Note 8**).

14. 25% Hyb(+): 75% 2× SSCTw at 65 °C (*see* **Note 8**).

15. Maleic acid buffer (MAB): 0.1 M maleic acid, 150 mM NaCl, pH 7.5 (*see* **Note 9**).

16. MABTriton (MABTr): 0.1–0.2% TritonX-100 in MAB. Store at RT.

17. Blocking solution: 2% blocking reagent for nucleic acid hybridization and detection from Roche in MAB (*see* **Note 10**), 0.2% TritonX-100, 10% sheep serum.

18. Anti-digoxigenin AP Fab fragments.

19. Alkaline phosphatase (AP/NTMT) buffer: 0.1 M Tris–HCl, 0.4 M MgCl$_2$, 0.1 M NaCl, 0.1% Tween in ultrapure water (*see* **Note 11**).

20. Developing solution: Per mL AP/NTMT buffer, 2.2 μL nitroblue tetrazolium (NBT), 1.6 μL 5-bromo-4-chloro-3′-indolyphosphate (BCIP) (*see* **Note 12**).

21. 24-Well tissue culture plate.

22. 30% glycerol in ultrapure water.

23. 50% glycerol in ultrapure water.

24. 70% glycerol in ultrapure water.

25. 100% ethanol.

26. Aluminum foil.

27. Nutator.

2.1.3 Mounting for Imaging

1. 70% glycerol in distilled water.

2. 35 mm dish filled with silicone, such as prepared from Sylgard.

3. Dumont #6 forceps.

4. Glass slides.

5. Vacuum grease.

6. 22 × 22 mm coverslip.

2.2 Materials for IHC	1. Microcentrifuge tubes.

2.2 Materials for IHC

2.2.1 Staining

1. Microcentrifuge tubes.

2. Fixative: 2% paraformaldehyde (PFA)/1% trichloroacetic acid (TCA) (*see* **Note 13**).

3. PBSTr: 1% TritonX-100 in PBS.

4. Distilled water.

5. Acetone (*see* **Note 14**).

6. Blocking solution: PBSTr supplemented with 10% normal goat serum (NGS) (*see* **Note 15**).

7. Primary antibody.

8. Secondary antibody.

9. 50% glycerol in PBS.

2.2.2 Mounting

1. 1× PBS.

2. Glass slide.

3. Vacuum flask with small pipette tip attached.

4. Dissecting pin.

5. Anti-fade mounting medium.

6. Coverslips.

7. Nail polish.

3 Methods for WISH

3.1 Generation of Digoxigenin-Labeled RNA Probe

1. Linearize 2–5 μg of plasmid using appropriate unit of restriction enzyme (*see* **Note 16**).

2. Purify linearized plasmid using PCR purification kit according to the manufacturer's instructions.

3. Run 2 μL of linearized and purified DNA from the total volume eluted in **step 2** on a 1.5% agarose gel to ensure integrity of linearized DNA.

4. Transcribe RNA using 1–2 μg linearized and purified DNA, 2 μL 10× transcription buffer, 2 μL DIG RNA labeling mix, 1 μL RNase inhibitor, and 1 μL RNA polymerase up to 20 μL with DEPC water. Incubate for 2 h at 37 °C.

5. Remove the DNA template by adding 1 μL DNase I (RNase free) to the transcription reaction and incubate for 20 min at 37 °C.

6. Add 10 μL DEPC water.

7. Purify RNA on column using an RNA purification kit according to the manufacturer's instructions.

8. Test 2 μL of RNA from the total volume eluted in **step 7** on a 1.5% agarose gel to ensure there is no degradation (*see* **Note 17**).

9. Add 100 μL Hyb(+) to RNA prepared in **step 7** and store at −80 °C (*see* **Note 18**).

3.1.1 Fixation of Larvae for WISH

To conduct in situ hybridization on larvae older than 24 h postfertilization (hpf), add phenylthiourea (PTU) to embryos anytime between 70% epiboly and 24 hpf to inhibit pigmentation. Keep embryos in PTU-supplemented egg water until fixation. Additionally, if the larvae have not hatched out of their chorions, they must be dechorionated prior to fixation. Conduct all steps at room temperature (RT), unless otherwise noted. Solutions can be removed and added with disposable transfer pipettes.

1. Fix zebrafish embryos in 4% PFA in 1.5 mL microcentrifuge tubes overnight at 4 °C or at RT for 3 h on a nutator (*see* **Note 19**).

2. Remove PFA and replace with 100% MeOH. Wash with MeOH five times with each wash lasting 5 min at RT on nutator.

3. Store embryos in MeOH at −20 °C for at least 2 h (*see* **Note 20**).

3.1.2 Day 1 WISH

1. Remove half of the MeOH and replace with PBSTw. Wash once for 5 min on nutator.

2. Remove half of the liquid and replace with PBSTw. Wash once for 5 min on nutator.

3. Remove all of the liquid and replace with PBSTw. Wash once for 5 min on nutator.

4. Wash with PBSTw four times with each wash lasting 5 min on nutator (*see* **Note 21**).

5. Add Proteinase K (20 mg/mL) diluted in PBSTw and incubate with microcentrifuge tube lying on its side. Do not nutate. Concentration of Proteinase K and length of incubation depend on the age of embryo (Table 2) (*see* **Note 22**).

Table 2
Proteinase K treatment on different ages of zebrafish embryos

	Time
Proteinase K 1:2000	
24 hpf	2 min
36 hpf	21 min
48 hpf	27 min
54 hpf	30 min
60 hpf	45 min
Proteinase K 1:900	
3 and 4 dpf	45 min
5 dpf and older	1 h

6. Remove Proteinase K solution and wash with PBSTw two times, quickly.

7. Remove PBSTw and add 4% PFA. Postfix for 20 min on nutator.

8. Remove PFA and wash with PBSTw five times with each wash lasting 5 min on nutator (*see* **Note 23**).

9. Remove PBSTw and add Hyb(+). Incubate for 1–2 h on side at 65 °C (*see* **Note 24**).

10. Remove Hyb(+) and add 400 μL of probe diluted 1:100 in Hyb(+) that was made in Subheading 3.1.1, **step 9** and incubate at 65 °C overnight. Tubes should be lying on sides (*see* **Notes 18** and **25**).

3.1.3   Day Two WISH

1. Make hot wash solutions and heat to 65 °C in a water bath (*see* **Note 26**).

2. Remove probe from tubes (*see* **Note 27**).

3. Wash with 100% Hyb(+) for 5 min at 65 °C.

4. Remove previous Hyb(+) solution and wash with 75% Hyb(+): 25% 2× SSCTw for 5 min at 65 °C.

5. Remove previous Hyb(+) solution and wash with 50% Hyb(+): 50% 2× SSCTw for 5 min at 65 °C.

6. Remove previous Hyb(+) solution and wash with 25% Hyb(+): 75% 2× SSCTw for 5 min at 65 °C.

7. Remove previous Hyb(+) solution and wash with 2× SSCTw two times with each wash lasting 30 min at 65 °C (*see* **Note 28**).

8. Remove 2× SCCTw and wash with 0.2× SSCTw two times with each wash lasting 30 min at 65 °C (*see* **Note 29**).

9. Remove 0.2× SSCTw and wash with MABTr for 10 min at RT on nutator.

10. Remove MABTr and add blocking solution. Incubate at RT for 1–2 h on nutator or overnight at 4 °C.

11. Remove blocking solution and add anti-DIG AP Fab fragments (1:2000) in blocking solution (*see* **Note 30**). Incubate overnight at 4 °C on nutator.

3.1.4   Day Three WISH

1. Remove antibody in blocking solution and wash with MABTr six times with each wash lasting 30 min on nutator (*see* **Note 31**).

2. Remove antibody in blocking solution and wash with AP/NTMT buffer for 10 min on nutator.

3. Transfer all the larvae from the microcentrifuge tube to a well in a 24-well plate (*see* **Note 32**).

4. Remove AP/NTMT buffer and add developing solution. Agitate plate(s) during development (*see* **Note 33**). After developing, move to **step 5**. If larvae have significant background staining, *see* Subheading 3.1.5 Cleaning Background.

5. Remove developing solution and wash with PBSTw three times quickly, to stop the reaction.

6. Postfix with 4% PFA for 30 min at RT or overnight at 4 °C.

7. Pass larvae through 30% glycerol. Remove 30% glycerol when larvae have sunk to the bottom of the well. Repeat sequentially with 50% glycerol and then 70% glycerol (*see* **Note 34**).

8. Store larvae at 4 °C in 70% glycerol (*see* **Note 35**).

3.1.5 Clearing Background

If there is a significant amount of nonspecific background stain, these steps will aid in clearing the background but will also reduce the staining of the probe overall.

1. Remove developing solution and wash with PBSTw for 5 min.

2. Remove PBSTw and wash with 100% ethanol two times with each wash lasting 5 min (*see* **Note 36**).

3. Remove ethanol and wash with PBSTw two times with each wash lasting 5 min.

4. Continue to **steps 6–8** in Subheading 3.1.4 Day Three WISH.

3.1.6 Mounting for Imaging

Below we outline how to mount larvae following WISH for imaging on a dissecting or compound microscope.

1. If desired, de-yolk larvae before imaging using pins or Dumont #6 forceps. Place a larva in a drop of 70% glycerol on a 35 mm dish filled with silicone and slowly remove the yolk. Try to remove the yolk in one piece so that yolk granules do not remain attached to the body of the larva (*see* **Note 37**).

2. Place larva in a drop of 70% glycerol in the center of the glass slide and position in the orientation desired for imaging.

3. Make four small columns of vacuum grease in corners around, but not touching, the glycerol drop containing the larva. Make sure the columns will fit within the space of the 22 × 22 mm coverslip.

4. Slowly press 22 × 22 mm coverslip on top of vacuum grease columns and over the larva until the coverslip touches the glycerol drop and the larva is stably positioned. Do not flatten the larva.

5. Flood the coverslip with 70% glycerol (*see* **Note 38**).

6. Image.

3.2 Methods for IHC

3.2.1 Staining

Conduct all steps at RT unless otherwise noted. Nutate tubes for each step.

1. Fix larvae for 10 min in microcentrifuge tubes (*see* **Note 39**).

2. Remove fixative and wash with PBSTr three times, quickly.

3. Remove PBSTr and wash with distilled water once, quickly.

4. Wash once for 5 min with distilled water.

5. Remove distilled water and wash once for 7 min with acetone at -20 °C (*see* **Note 40**).

6. Remove acetone and wash with distilled water once, quickly.

7. Wash once for 5 min with distilled water.

8. Remove distilled water and wash with PBSTr two times, quickly.

9. Remove PBSTr and block for 1 h at RT (*see* **Note 41**).

10. Add primary antibody at the appropriate dilution in block and incubate overnight at 4 °C (*see* **Notes 42** and **43**).

11. Remove primary antibody and perform three quick washes with PBSTr (*see* **Note 44**).

12. Wash with PBSTr three times with each wash lasting 30 min.

13. Wash once for 1 h with PBSTr.

14. Remove PBSTr and incubate secondary antibody in block for 3 h at RT or overnight at 4 °C (*see* **Notes 45** and **46**).

15. Remove secondary antibody and wash with PBSTr three times with each wash lasting 1 h.

16. Remove PBSTr and store in 50% glycerol/PBS.

3.2.2 Mounting

1. Replace 50% glycerol/PBS with 1× PBS.

2. Pipette larvae onto glass slide.

3. With a small pipette tip attached to a vacuum glass flask, carefully pipette up any excess liquid. Leave some liquid so that embryos are not completely dry.

4. Using a pin or needle, position larvae in rows along the glass slide on side to image the lateral line nerve and motor nerves.

5. With the small pipette tip attached to the vacuum flask, remove any excess liquid so that larvae are nearly dry (*see* **Note 47**).

6. Place a drop of anti-fade mounting medium over the larvae.

7. Cover with a coverslip, taking care that no bubbles form (*see* **Note 48**).

8. Seal the coverslip with nail polish.

9. Image embryos on compound or confocal fluorescent microscope (*see* **Note 49**).

4   Notes

1. Refer to Table 1 for a list of commonly employed probes to visualize Schwann cells in zebrafish.

2. To make 1 L of Hyb(+): 500 mL formamide, 250 mL 20× SSC, 1 mL 100% Tween, 9 mL 1 M citric acid, 0.5 g *Torula* yeast RNA, 240 mL ultrapure water. Stir together with stir bar at RT. Add *Torula* yeast RNA last while the solution is stirring. Store at −20 °C.

3. To make a 50× PTU stock, dissolve 0.75 g PTU in 500 mL distilled water and heat to 60 °C for 1–2 h while stirring in a fume hood. Store at RT. Add 1 mL of 50× PTU to 50 mL of egg water to make 0.003%.

4. For 100 mL of 4% PFA, heat 50 mL distilled water to 60 °C while stirring. Do not boil water. Measure 4 g of PFA in fume hood and add to water. Increase pH until the solution becomes clear, which usually requires 2–3 drops of concentrated NaOH. Filter after the PFA has dissolved and the solution has cooled. Add 10 mL 10× PBS. Add distilled water until a total volume of 100 mL is reached. Recheck pH and adjust as needed to reach 7.4. Store at −20 °C. Fresh 4% PFA can be aliquoted and frozen at −20 °C. Once thawed, store at 4 °C and use within 3 days.

5. To make 1 L of PBSTw: 2 mL 100% Tween, 100 mL 10× PBS, 898 mL ultrapure water. Store at RT.

6. To make 1 L of 2× SSCTw: 100 mL 20× SSC, 2 mL 100% Tween, 898 mL ultrapure water.

7. To make 1 L of 0.2× SSCTw: 10 mL 20× SSC, 2 mL 100% Tween, 988 mL ultrapure water.

8. Make and heat up hot wash solutions at the beginning of day 2.

9. To make 500 mL of MAB: 5.8 g maleic acid, 4.38 g NaCl, 400 mL ultrapure water. Adjust pH to 7.5 and then bring to final volume of 500 mL with ultrapure water. Filter-sterilize and store at RT. The pH of MAB will initially be low (~1.3 pH). Add concentrated NaOH until a pH of 5 is reached and then carefully add NaOH droplets until pH 7.5 is reached.

10. To make 2% blocking reagent in MAB, dissolve blocking reagent in MAB at 65 °C while stirring with stir bar. Aliquot and store at −20 °C. When ready for use, add TritonX-100 and sheep serum.

11. AP/NTMT buffer must be made fresh each day as needed because bacteria can quickly grow in the buffer. If the buffer is cloudy, discard. To make 50 mL: mix 5 mL 1 M Tris–HCl

pH 9.5, 2 mL 1 M MgCl$_2$, 1 mL 5 M NaCl, 0.5 mL 10% Tween, 41.5 mL ultrapure water.

12. NBT/BCIP is light sensitive. Keep developing solution covered in foil.

13. Fresh PFA can be stored at −20 °C but, once thawed, should be used within 3 days. The fixative can vary depending on the antibody [17]. Fluorescence of a transgene can be preserved by supplementing PFA with 0.1 M pipes, 1.0 mM MgSO$_4$, and 2 mM EGTA in 1× PBS [18].

14. Acetone permeablizes the larvae so that the antibody can penetrate through the skin. Other permeabilizing reagents, such as Proteinase K or trypsin, can be used and vary depending upon the antibody of interest. The skin of the larvae can also be removed (by peeling it back with Dumont #6 forceps) to aid in permeabilization.

15. Blocking reagents can also vary depending upon the animal source of the antibody.

16. If the probe has not been previously tested, also generate a sense RNA probe as a control.

17. Clean gel rig before running. Mix 2 μL RNA with 0.5 μL RNase inhibitor to slow down RNA degradation. Let the gel run for only a few minutes and image before RNA degradation. A second photo can be taken later after the gel has run further.

18. Before use during the WISH process, probes should be diluted 1:100 in Hyb(+). Diluted probe can be stored in large volumes at −20 °C.

19. We typically fix no more than 30 larvae (at any stage) per microcentrifuge tube.

20. Larvae must be stored in MeOH for at least 2 h at −20 °C before conducting WISH. Larvae can be stored for months in MeOH. We typically store larvae overnight at −20 °C before beginning WISH.

21. PBSTw washes can go longer than the noted time, but washes should not last on the order of hours.

22. Make sure tubes are lying on side but not nutating while treated with ProtK; nutating could cause the larvae to fall apart. This is a critical step for the WISH process. Too little ProtK treatment will result in poor WISH quality due too poor probe penetration, whereas too much ProtK treatment will cause larvae to fall apart.

23. Washes can go longer than 5 min.

24. It is important that tubes are lying on their sides so that the larvae are evenly heated.

25. If a probe is not working well, try heating the probe to 98 °C before using.

26. Do not let the solutions or larvae in tubes cool below 65 °C as this can increase nonspecific background stain after developing.

27. The probe can be saved and reused three to four more times, depending on signal strength. Store used probe at −20 °C.

28. Washes can go longer than 30 min.

29. Washes can go longer than 30 min.

30. If staining looks weak, the concentration of the antibody can be increased.

31. Washes can go longer than 30 min.

32. We typically develop up to 30 larvae in 24-well tissue culture plates.

33. The developing reaction is light sensitive. Keep the plate covered in aluminum foil while developing. Check every 30–60 min until fully developed. If the larvae have not developed sufficiently by the end of the day, remove developer and replace with AP/NTMT buffer, and store at 4 °C overnight, replacing with fresh developer solution the next day. If a probe is weak, larvae can be left in fresh developer solution overnight at 4 °C, only if the probe has been tested before.

34. Let embryos sink to the bottom of the well before moving to the next glycerol pass. The higher the concentration of glycerol, the longer it takes for the embryos to sink to the bottom of the well.

35. Larvae can be stored for at least 2 years at 4 °C.

36. Ethanol clears all staining, so be sure to carefully monitor these washes so that only background stain is cleared.

37. Imaging of the lateral line does not always require removal of the yolk, depending on the orientation of the larva for imaging.

38. If the larva moves to an undesired orientation after flooding with glycerol, the coverslip can be manipulated gently from side to side in order to change the position of the larva or removed entirely to reposition the larva directly.

39. Fixation times can also be varied depending upon the strength of the staining.

40. This is an important step; this step must occur at −20 °C.

41. Blocking can go overnight at 4 °C.

42. If the antibody has not been tested before, a no primary antibody control should also be performed to control for nonspecific binding.

43. Depending upon the strength of the staining, incubation with the primary antibody can extend for another day at 4 °C.

44. The primary antibody can be reused. Store at 4 °C with 0.02% NaN₃.

45. Incubation time of the secondary antibody can also vary depending upon the antibody.

46. Make sure tubes are covered with aluminum foil to prevent light from quenching the fluorescence once the secondary antibody is added.

47. Carefully position the pipette tip near the head of the larvae to pipette up any liquid underneath the larvae.

48. The heads of the larvae will be compressed. If the fluorescence intensity of the antibody staining or transgene is bright in the head and could affect imaging of the posterior lateral line nerve, then the head can be removed prior to mounting.

49. To genotype embryos after imaging, the nail polish seal can be removed with acetone, and the larvae can be recovered for DNA extraction.

References

1. Thisse C, Thisse B, Schilling TF et al (1993) Structure of the zebrafish snail1 gene and its expression in wild-type, spadetail and no tail mutant embryos. Development 119:1203–1215

2. Dutton KA, Pauliny A, Lopes S et al (2001) Zebrafish colourless encodes sox10 and specifies non-ectomesenchymal neural crest fates. Development 128:4113–4125

3. Gilmour DT, Maischein HM, Nüsslein-Volhard C (2002) Migration and function of a glial subtype in the vertebrate peripheral nervous system. Neuron 34:577–588

4. Lyons DA, Pogoda HM, Voas MG et al (2005) erbb3 and erbb2 are essential for Schwann cell migration and myelination in zebrafish. Curr Biol 15:513–524

5. Rubinstein AL, Lee D, Luo R et al (2000) Genes dependent on zebrafish cyclops function identified by AFLP differential gene expression screen. Genesis 26:86–97

6. Monk KR, Naylor SG, Glenn TD et al (2009) AG protein–coupled receptor is essential for Schwann cells to initiate myelination. Science 325:1402–1405

7. Levavasseur F, Mandemakers W, Visser P et al (1998) Comparison of sequence and function of the Oct-6 genes in zebrafish, chicken and mouse. Mech Dev 74:89–98

8. Oxtoby E, Jowett T (1993) Cloning of the zebrafish krox-20 gene (krx-20) and its expression during hindbrain development. Nucleic Acids Res 21:1087–1095

9. Brösamle C, Halpern M (2002) Characterization of myelination in the developing zebrafish. Glia 39:47–57

10. Takada N, Appel B (2010) Identification of genes expressed by zebrafish oligodendrocytes using a differential microarray screen. Dev Dyn 239:2041–2047

11. Takada N, Kucenas S, Appel B (2010) Sox10 is necessary for oligodendrocyte survival following axon wrapping. Glia 58:996–1006

12. Schaefer K, Brösamle C (2009) Zwilling-A and -B, two related myelin proteins of teleosts, which originate from a single bicistronic transcript. Mol Biol Evol 26:495–499

13. Pogoda HM, Sternheim N, Lyons DA et al (2006) A genetic screen identifies genes essential for development of myelinated axons in zebrafish. Dev Biol 298:118–131

14. Kazakova N, Li H, Mora A et al (2006) A screen for mutations in zebrafish that affect myelin gene expression in Schwann cells and oligodendrocytes. Dev Biol 297:1–13

15. Thisse C, Thisse B (2008) High resolution in situ hybridization to whole-mount zebrafish embryos. Nat Protoc 3:59–69

16. Langworthy MM, Appel B (2012) Schwann cell myelination requires Dynein function. Neural Dev 7:1

17. Macdonald R (1999) Zebrafish immunohistochemistry. In: Guielle M (ed) Molecular methods in developmental biology: Xenopus and Zebrafish, vol 147. Humana Press, Totowa, pp 77–88

18. Ng AN, de Jong-Curtain TA, Mawdsley DJ et al (2005) Formation of the digestive system in zebrafish: III. Intestinal epithelium morphogenesis. Dev Biol 286:114–135

Chapter 26

Transmission Electron Microscopy for Zebrafish Larvae and Adult Lateral Line Nerve

Rebecca L. Cunningham and Kelly R. Monk

Abstract

Transmission electron microscopy (TEM) enables visualization of the ultrastructure of the myelin sheath. Schwann cells on the posterior lateral line nerves and motor nerves can be imaged by TEM. Here, we detail the multiday processing of larval trunks and dissected posterior lateral line for TEM, as well as how to trim embedded samples, section, and stain grids for imaging.

Key words Transmission electron microscopy, Myelin ultrastructure, Lateral line nerve, Motor nerves, Adult zebrafish lateral line nerve

1 Introduction

Transmission electron microscopy (TEM) allows for analysis of the myelin sheath ultrastructure. Schwann cell stages and the electron dense myelin sheath are identifiable by morphology. Immature, promyelinating, and myelinating Schwann cells can be visualized by TEM in zebrafish. These developmental stages of Schwann cells in zebrafish are similar in mice (Fig. 1); however, myelin in zebrafish larvae is not as compact compared to mouse myelin due, at least in part, to differences in molecular composition [1], resulting in a myelin sheath wavy in appearance. In zebrafish, Schwann cells on the lateral line nerves and motor nerves are the most commonly imaged by TEM (e.g., [2–4]) (Fig. 2). In addition to general nerve integrity, other factors including the percent myelinated axons, number of axons, number of Schwann cell nuclei, and number of Schwann cell processes can be quantified from TEM images. The processing of specimens for TEM is laborious, but the results yielded are valuable. The fixation process is critical, as poor fixation will prevent attainment of clear images for quantification. We have used the methodology presented here for 2 days post fertilization (dpf) to 21 dpf zebrafish, and it has been largely adapted from Czopka et al. [5] but includes more detailed notes.

Paula V. Monje and Haesun A. Kim (eds.), *Schwann Cells: Methods and Protocols*, Methods in Molecular Biology, vol. 1739, https://doi.org/10.1007/978-1-4939-7649-2_26, © Springer Science+Business Media, LLC 2018

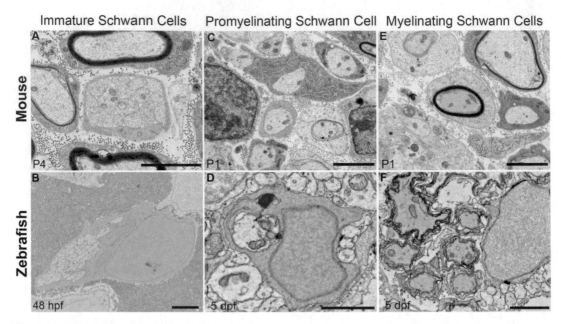

Fig. 1 A comparison of mouse and zebrafish Schwann cell developmental stages. (**a, b**) Immature Schwann cells, pseudocolored blue, surround bundles of axons. (**c, d**) Promyelinating Schwann cells, pseudocolored blue, have performed radial sorting and are associated with a single axon. (**e, f**) Schwann cells myelinate axons in a one-to-one ratio. Myelin in zebrafish is wavy in appearance compared to mouse myelin at larval stages with chemical fixation. Myelinated axons are pseudocolored in blue. Scale bars, 2 μm in mouse images and 1 μm in zebrafish images

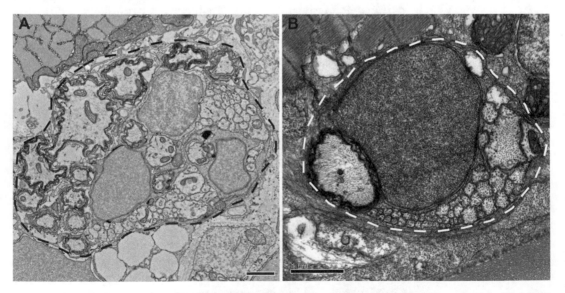

Fig. 2 (**a**) A TEM image of a cross section of the posterior lateral line nerve at 5 dpf in zebrafish. (**b**) A TEM image of a wild-type motor nerve at 7 dpf. The dashed line demarcates nerves. Scale bars, 1 μm

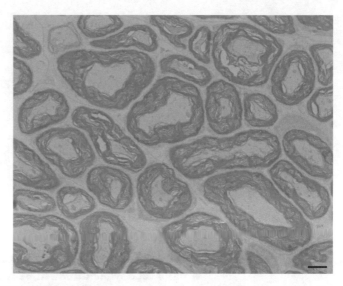

Fig. 3 TEM image showing Schwann cell myelin ultrastructure from a 2-year-old adult zebrafish posterior lateral line nerve. Scale bar, 2 μm

In adult zebrafish, the most commonly studied Schwann cells are located on the posterior lateral line nerve (Fig. 3), a prominent sensory nerve located on either side of the body of the fish [3]. Schwann cells within the barbels of zebrafish have also been analyzed [6]. TEM for the adult posterior lateral line nerve is similar to that described for larvae, although the chemicals used and the fixation process do vary slightly. Protocols for embedding, trimming, sectioning, and staining methodologies do not vary between larval and adult samples. The protocol presented here for fixation of adult zebrafish posterior lateral line nerve has been adapted from previous reports [3, 7].

2 Materials

It is critical that all reagents used throughout this methodology are EM grade. Conduct all steps at room temperature (RT) unless otherwise noted.

2.1 Materials for Larval TEM

2.1.1 Primary Fixation

1. Zebrafish larvae (any stage).

2. Modified Karnovsky's fixative: 2% glutaraldehyde, 4% paraformaldehyde, in 0.1 M sodium cacodylate buffer pH 7.4 (*see* **Note 1**).

3. PCR strip tubes.

4. 4 mg/mL tricaine prepared in 21 mM Tris ultrapure water.

5. 35 mm dish filled with silicone, such as Sylgard.

6. Disposable stab knife (e.g., Accutome Sharpoint 3.0 mm stab knives).

7. Glass pipettes.

8. Power stable microwave (e.g., Ted Pella brand Pelco BioWave Pro).

9. Water bath set to 15 °C linked to microwave or ice bath with thermometer.

10. 0.1 M sodium cacodylate pH 7.4.

2.1.2 Secondary Fixation and En Bloc Stain

1. Glass pipettes.

2. 1.5 mL microcentrifuge tubes.

3. 0.1 M sodium cacodylate buffer pH 7.4.

4. Secondary fixative: 2% osmium tetroxide prepared in 0.1 M sodium cacodylate buffer supplemented with 0.1 M imidazole, pH 7.5 (*see* **Note 2**).

5. Fine tip plastic pipette.

6. Ultrapure water.

7. Saturated uranyl acetate: 8% w/v in ultrapure water (*see* **Note 3**).

2.1.3 Dehydration

1. 30% EM-grade ethanol prepared in ultrapure water.

2. 50% EM-grade ethanol prepared in ultrapure water.

3. 70% EM-grade ethanol prepared in ultrapure water.

4. 85% EM-grade ethanol prepared in ultrapure water.

5. 90% EM-grade ethanol prepared in ultrapure water.

6. 100% EM-grade ethanol.

7. 100% EM-grade acetone.

2.1.4 Embedding and Preparing Specimen Blocks for EM

1. Embedding media (e.g., Araldite 502/Embed 812): make according to the manufacturer's instructions (*see* **Note 4**).

2. Aluminum foil.

3. EM-grade acetone.

4. Standard disposable transfer pipettes.

5. Flat polyethylene-embedding molds.

6. Toothpicks.

7. Oven heated to 65 °C.

8. Butler block trimmer.

9. Dissecting microscope.

10. Razor blades.

11. Gelatin capsules (e.g., EMS 70110).

2.1.5 Sectioning and Staining for EM

1. Ultramicrotome.

2. Toluidine blue stain (optional): 1% (w/v) toluidine blue O and 1% (w/v) sodium borate in ultrapure water (*see* **Note 5**).

3. Heat pen.

4. Loop.

5. Copper grids (e.g., Formvar carbon film on 100-mesh copper grids).

6. Saturated uranyl acetate: 8% w/v in distilled water (*see* **Note 6**).

7. Syringe with 45 μm filter unit.

8. Glass slides.

9. Parafilm.

10. Glass Petri dishes with lids.

11. Fine plastic transfer pipette.

12. Self-closing forceps.

13. 50% ethanol in ultrapure water in wash bottle.

14. Clean, autoclaved beaker.

15. Whatman filter paper.

16. Sato's lead stain (*see* **Note 7**).

17. NaOH pellets.

18. Surgical mask.

2.2 Materials for Adult Lateral Line Nerve TEM

1. Adult zebrafish.

2. Ice water bath.

3. Razor blade.

4. 1.5 mL microcentrifuge tube.

5. Modified Karnovsky's fixative: 2% glutaraldehyde and 4% paraformaldehyde prepared in 0.1 M sodium cacodylate buffer pH 7.4.

6. Power stable microwave (e.g., Ted Pella brand Pelco BioWave Pro).

7. Water bath set to 15 °C linked to microwave or ice bath with thermometer.

8. Dissecting microscope.

9. Petri dish filled with silicone, such as Sylgard.

10. Hypodermic needles, 22 g × 1″ L.

11. Watchmaker forceps or Dumont #7 forceps (*see* **Note 8**).

12. Student Vannas Spring Scissors.

13. 0.1 M sodium cacodylate.

14. Secondary fixative: 2% osmium tetroxide, 0.1 M sodium cacodylate.

15. Ultrapure water.

16. 25% EM-grade ethanol in ultrapure water.

17. 50% EM-grade ethanol in ultrapure water.

18. 75% EM-grade ethanol in ultrapure water.

19. 95% EM-grade ethanol in ultrapure water.

20. 100% EM-grade ethanol in ultrapure water.

21. Propylene oxide (*see* **Note 9**).

22. Embedding media as described in larval TEM section.

3   Methods

3.1   Methods for Larval TEM

This protocol should be conducted in a fume hood unless otherwise stated. TEM of zebrafish larvae also requires the use of a power stable microwave to enable efficient penetration of the chemicals into the specimens. A water bath of 15–18 °C within the microwave is also required so that the specimens are not overheated. Any steps requiring the microwave should have the water bath on and set to 15 °C. An ice bath with a thermometer can also be used. All chemicals should be used with extreme caution and disposed of properly. We recommend wearing a lab coat.

3.1.1   Primary Fixation

1. Turn on water bath set to 15 °C.

2. Pipette 200 μL of modified Karnovsky's fixative into PCR strip tubes and keep cold on wet ice.

3. Anesthetize larvae with tricaine and pipette onto 35 mm dish filled with silicone.

4. With a stab knife, cut embryos between body segments 5 and 6. Save the head for DNA extraction and genotyping. With a glass pipette, pipette the trunk into modified Karnovsky's fixative in strip tube (*see* **Note 10**). Repeat until eight strip tubes are filled with one trunk in 200 μL fixative in each tube.

5. Microwave at 100 W for 1 min and off for 1 min. Repeat this step one more time.

6. Microwave 450 W for 20 s and off for 20 s. Repeat this step five more times.

7. Continue primary fixation by leaving larvae in fixative-containing strip tubes for 2 h at RT or overnight at 4 °C (*see* **Note 11**).

8. Repeat **steps 2–7** until the desired number of larvae is fixed in modified Karnovsky's fixative.

9. Remove Karnovsky's fixative with a glass pipette and wash three times with 0.1 M sodium cacodylate buffer at RT, with each wash lasting 10 min (*see* **Note 12**).

3.1.2 Secondary Fixation and En Bloc Stain

1. After larvae have been genotyped, group together like genotypes with a glass pipette in 1.5 mL microcentrifuge tubes containing 0.1 M sodium cacodylate.

2. Make secondary fixative (*see* **Note 13**).

3. Remove 0.1 M sodium cacodylate with a fine tip plastic transfer pipette and add secondary fixative to 1.5 mL tubes (*see* **Note 14**).

4. Microwave at 100 W for 1 min, then off for 1 min. Repeat this step one more time.

5. Microwave 450 W for 20 s and off for 20 s. Repeat this step five more times.

6. Continue secondary fixation by leaving larvae in fixative-containing tubes for 1 h at RT.

7. Remove secondary fixative with fine tip plastic pipette and wash with ultrapure water three times, with each wash lasting 10 min (*see* **Note 15**).

8. Centrifuge uranyl acetate for 10 min at maximum speed so that any particulates accumulate at the bottom of the tube.

9. Replace water (**step 7**) with uranyl acetate and microwave at 450 W for 1 min, off for 1 min, and 450 W for 1 min (*see* **Note 16**).

3.1.3 Dehydration

1. Remove uranyl acetate with a fine tip plastic transfer pipette. Replace with 30% ethanol and microwave 250 W for 45 s followed by incubation at RT for 10 min (*see* **Note 17**).

2. Repeat **step 1** with 50%, 70%, 85%, and 90% ethanol washes (*see* **Note 18**).

3. Replace with 100% ethanol and microwave 250 W for 1 min, off for 1 min, 250 W for 1 min followed by incubation at RT for 10 min. Repeat this step once.

4. Remove the ethanol and add 100% acetone and microwave 250 W for 1 min, off for 1 min, 250 W for 1 min followed by incubation at RT for 10 min (*see* **Note 19**). Repeat three more times.

3.1.4 Embedding and Preparing Specimen Blocks

1. Thaw embedding media from freezer for ~20 min until it becomes fluid at RT before using (*see* **Note 20**).

2. Mix the embedding media with acetone at 1:1 ratio.

3. Remove acetone from the larvae and fill the tube with the embedding media/acetone mix to the rim of the tubes. Rotate overnight at RT (*see* **Note 21**).

4. Thaw embedding media as described in **step 1**.

5. Replace embedding media/acetone mix with fresh embedding media. Open the tube lids and let any acetone evaporate out for at least 5 h.

6. Thaw fresh embedding media (*see* **Note 22**).

7. Make labels for each larval trunk and place in flat polyethylene-embedding molds. A font size of 4 pt. works well (*see* **Note 23**).

8. Under a dissecting microscope, with a standard disposable transfer pipette, pipette embedding media into each mold. Push down labels and remove any air bubbles (*see* **Note 24**). Make sure that the embedding media is neither too convex nor concave in the mold.

9. With a toothpick, carefully remove a trunk from the microcentrifuge tube taking care not to crush it against the side of the tube (*see* **Note 25**).

10. If the lateral line nerve is to be imaged in cross section, place the cut side (anterior) near the face of the block (Fig. 4). If imaging motor nerves, orient the trunk so that the dorsal side will be closest to the face of the block.

Fig. 4 A schematic of a trimmed TEM block prepared for posterior lateral line nerve cross-section imaging. Larva is embedded with the cut side (anterior) at the face of the block. As much resin as possible is removed from around the trunk

11. Once all trunks have been embedded, bake at 65 °C for 2 days.

12. Once blocks have solidified, remove from molds.

13. Place a block, anterior up, in an apparatus, such as a Butler block trimmer, that can firmly hold the block in place.

14. Under a dissecting microscope, using a razor blade, first shave off the face of the block so that the resin is clear and you can see the trunk. Then slowly trim away resin from the sides of the trunk until the trunk is surrounded only by a small amount of resin (Fig. 4) (*see* **Note 26**).

15. Store trimmed blocks individually in plastic capsules to prevent the trimmed area containing the larva from breaking.

3.1.5 Sectioning

1. If imaging a cross section of the posterior lateral line nerve, section 100 μm into the trunk (~one body segment) by making 1000 nm sections with an ultramicrotome so that any debris from the initial dissection is removed (*see* **Note 27**).

2. Make 70 nm sections using ultramicrotome. The sections should be silver in color.

3. Flatten desired sections by hovering a heat pen slightly over the section.

4. Place sections on a copper grid either using a loop or touching the grid to the sections floating on the water.

5. If imaging motor nerves, make 1000 nm sections through the spinal cord. Once the ventral spinal cord is reached, stain the 1000 nm sections with toluidine blue solution (*see* **Note 27**) to see if motor nerves are in the section. Continue to section until motor nerves are visible by toluidine blue staining. Then continue to **steps 2–4**.

3.1.6 Staining Grids

1. Filter uranyl acetate into microcentrifuge tube and centrifuge at maximum speed for 10 min.

2. While uranyl acetate is centrifuging, label glass slides with genotypes of grids and then cover slides with Parafilm.

3. Place the Parafilm-covered slide in a glass Petri dish.

4. With a fine plastic pipette, make small droplets of uranyl acetate on the glass slides covered with Parafilm. Make sure the drops are large enough so that a grid can sit on top. Place grids on top of droplets using self-closing forceps, and cover the Petri dish with the lid. Let the grids incubate at RT for 5 min (*see* **Note 28**).

5. Holding a grid with self-closing forceps, gently rinse the grid suspended over a clean, autoclaved beaker for 1 min using 50% ethanol in a wash bottle. The grid face should be perpendicular to the bench top (Fig. 5) (*see* **Note 29**).

Fig. 5 A schematic of a grid being washed after staining with uranyl acetate or lead stain

6. Place rinsed grid on labeled Whatman filter paper in a glass Petri dish. Let the grids completely dry before moving to the lead stain (*see* **Note 30**).

7. Once grids are dry, using a syringe filter unit, filter lead stain with a 0.45 μm pore filter into microcentrifuge tube and centrifuge at maximum speed for 10 min (*see* **Note 31**).

8. Replace the Parafilm on the glass slides and put the glass slides in Petri dishes. Place about 20 NaOH pellets on either side of the glass slide inside the Petri dish (*see* **Note 32**).

9. While wearing a mask, pipette droplets of lead stain onto Parafilm-covered glass slide.

10. Place grid on lead stain droplet and incubate for 5 min (*see* **Note 33**).

11. Rinse as in **step 5**.

12. Place rinsed grid on filter paper and let dry. The samples are now ready to be imaged (*see* **Note 34**).

3.2 Preparing Adult Lateral Line Nerve TEM

1. Prepare dissection tools and dissection area.

2. Euthanize adult fish in accordance with your institution's animal protocol. Wait until the gills have stopped moving.

3. With a razor blade, sever the hindbrain of the zebrafish just before the gills (Fig. 7).

4. Place zebrafish in a 1.5 mL microcentrifuge tube filled with modified Karnovsky's fixative (*see* **Note 35**).

5. Microwave at 250 W for 1 min, off for 1 min, 250 W for 1 min.

6. Microwave at 450 W for 1 min, off for 1 min, 450 W for 1 min.

7. Under a dissecting microscope, position zebrafish on its side on the Petri dish filled with silicone. Place one hypodermic needle at the most posterior part of the body, just before the caudal fin begins. Place another hypodermic needle through the gills to stabilize the fish during the dissection.

8. Using forceps, begin to peel away the skin slowly, beginning at the gills and working from anterior to posterior (*see* **Note 36**).

9. In wild-type fish, the posterior lateral line nerve should be white and run along the entire body of the fish (Fig. 6). Carefully scrape away any tissue that may be on top of the lateral line nerve. Cut the nerve at an anterior position and a posterior position. The length of the dissected nerve can vary. Gently remove the nerve with forceps so as to not cause disruption to the morphology of the axons and myelin sheath.

10. Place the nerve in a 1.5 mL microcentrifuge tube filled with modified Karnovsky's fixative.

11. Flip the fish to the other side and repeat **steps 7–10** to dissect the other lateral line nerve.

12. Let the nerves incubate in fixative at RT for 1–2 h or overnight at 4 °C.

13. Remove the fixative and wash three times with 0.1 M sodium cacodylate with each wash lasting 15 min (*see* **Notes 37** and **38**).

    Steps beyond this point should be conducted in a fume hood.

    Conduct all steps in a fume hood.

14. Remove sodium cacodylate with a fine tip plastic transfer pipette and add secondary fixative to nerves and incubate at RT for 1 h.

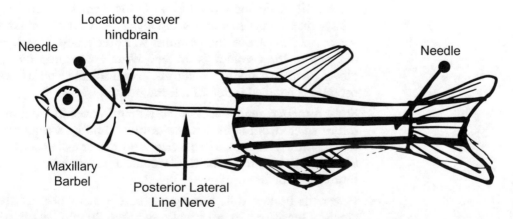

Fig. 6 Schematic of adult zebrafish for posterior lateral line nerve dissection. Dissection pins are placed in the caudal fin and through the gills to stabilize the fish. The skin is carefully peeled away until the posterior lateral line nerve is visible

15. Remove secondary fixative and wash with distilled water three times with each wash lasting 15 min at RT (*see* **Note 39**).

16. Remove distilled water and wash with 25% ethanol and incubate 20 min at RT (*see* **Note 40**).

17. Repeat **step 15** with 50%, 75%, and 95% ethanol (*see* **Note 41**).

18. Wash with 100% ethanol two times with each wash lasting 20 min.

19. Incubate nerves in propylene oxide for 15 min at RT (*see* **Note 42**).

20. Thaw embedding media for ~20 min at RT until it becomes fluid (embedding media can begin to thaw during the 100% ethanol washes).

21. Mix a 1:1 solution of propylene oxide/embedding media and incubate nerves while rotating overnight at RT (*see* **Notes 43** and **44**).

22. Replace 1:1 solution with fresh embedding media. Leave lids of tubes open for 4 h at RT to let any residual propylene oxide evaporate.

23. Embed (*see* **Note 45**), trim, section (*see* **Note 46**), and stain samples as described in the larval TEM section.

4   Notes

1. We purchase 10 mL glass ampules of 16% PFA and 8% glutaraldehyde; pipette the reagents into a 50 mL conical. Then add 20 mL of 0.2 M sodium cacodylate buffer to bring the total volume of the modified Karnovsky's fixative to 40 mL.

2. A 50 mL stock of 0.1 M sodium cacodylate and 0.1 M imidazole can be made and stored at 4 °C for several months. It is critical that the 0.1 M sodium cacodylate and 0.1 M imidazole are at pH 7.5 or else the osmium will precipitate. Check the pH before the osmium is added. Wear eye protection and gloves while handling osmium. Additionally, make sure osmium is handled inside of a fume hood.

3. Store saturated uranyl acetate at RT and cover in aluminum foil to shield from light. The solution should be a bright yellow. If the solution looks too clear and the uranyl acetate has settled at the bottom, shake the conical. Centrifuge at maximum speed before use.

4. Cover the beaker while mixing the components because they are light sensitive. Do not mix longer than 20 min because the mix will begin to harden. We aliquot ~25 mL embedding media per conical tube and store aliquots at −20 °C. If any

disposable transfer pipettes were used to make the embedding media, pipette some embedding media into the pipette and then expel it back into the beaker. This ensures that the waste inside the pipettes will harden in the oven, which then permits them to be thrown away. Place all tubes, beaker, and stir bar in 65 °C oven to harden.

5. Dissolve sodium borate in water and then add the toluidine blue O. Filter with a 0.20 μm pore filter.

6. Filter desired volume with a syringe 0.45 μm pore filter before use. Centrifuge at maximum speed before use.

7. Boil filtered ultrapure water and then cool to RT. Cap as soon as possible to prevent carbon dioxide buildup, which can lead to lead precipitation. With the filtered, boiled water, make a 5 M stock of NaOH. Add three drops of 5 M NaOH to 100 mL distilled water and rinse a stir bar, beaker, and six 10 mL syringes with the NaOH-treated water. Then, rinse beaker, stir bar, and syringes with water. In a glass Petri dish in a fume hood, heat 0.2 g lead citrate powder for 5–10 min until it is a pale tan color. Pour 48.2 mL of the filtered, boiled water in the beaker with stir bar stirring. Add 0.2 g lead citrate, 0.15 g lead nitrate, 0.15 g lead acetate, and 1 g sodium citrate. Continue stirring and add 1.8 mL of 5 M NaOH made with boiled and filtered water. The solution should be clear. Quickly filter solution with filter top bottle. Quickly fill 10 mL capped syringes one at a time with 8 mL of lead stain. Insert plunger, invert, uncap, and extrude all air from the syringe. Cap and wrap tightly with Parafilm to prevent air from accessing lead stain.

8. The type of forceps used for dissection depends upon personal preference.

9. Wear nitrile gloves while handling propylene oxide.

10. A plastic transfer pipette cannot be used because the trunks will stick to the inside of the pipette.

11. Samples can be kept at 4 °C for several days.

12. Samples can be kept at 4 °C for several days.

13. Make secondary fixative fresh before use.

14. Make sure that the lids of the tubes are closed tightly or else the secondary fixative will precipitate during the microwaving process.

15. The trunks should be a deep brown/black color.

16. Once microwaved with uranyl acetate, the specimens can be kept at 4 °C overnight or can proceed to Subheading 3.1.3.

17. Incubation at RT must be at least 10 min but can go longer. We also recommend laying tubes on their sides or rotating tubes while incubating at RT.

18. Slow dehydration is important to prevent poor fixation. Additionally, samples can be kept at 4 °C overnight at the 70% ethanol step (but only at this step).

19. Incubation at RT must be at least 10 min but can go longer. We also recommend laying tubes on their sides while incubating.

20. For this step, embedding media can have been previously thawed. Do not use embedding media if it is too viscous (this indicates that polymerization has occurred). Additionally, we recommend covering your work space with aluminum foil while working with embedding media in case it comes into contact with the fume hood counter. Liquid embedding media is both toxic and extremely difficult to remove.

21. Any embedding media waste or materials that come in contact with embedding media should be baked at 65 °C until the embedding media has become a solid polymer. This usually takes about 2 days. Once the embedding media is solid, it is no longer toxic.

22. Cover work space and knobs of dissecting microscope with aluminum foil to prevent contact with embedding media.

23. Before embedding, ensure that the printer's ink will not be dissolved in embedding media. If needed, pencil, but not pen, can be used to make handwritten labels.

24. It is important that there are no air bubbles because they can cause problems with sectioning.

25. The trunks should be hard and brittle.

26. In our experience, blocks are easier to trim directly from baking at 65 °C.

27. 1000 nm sections, or thick sections, can be stained with toluidine blue solution to check the fixation of the larvae before continuing to cut sections for imaging. Place sections in a small drop of water on a glass slide. Heat the glass slide on a hot plate until the water has evaporated. Place several droplets of toluidine blue solution (enough to cover the sections) and heat until a green ring forms around the outer edges of the droplet (this should take less than a minute). Rinse off the stain with water and then visualize the stained section under a compound microscope.

28. Grids can incubate from 5 min to an hour.

29. Do not stream 50% ethanol onto the grid; rather, slowly drip the water at a speed of about one droplet per second. If the grid face is parallel to the bench, the force of the 50% ethanol rinsing the grid could ruin the grid.

30. Once the grids are stained with uranyl acetate, the lead stain can proceed the same day or at another convenient time.

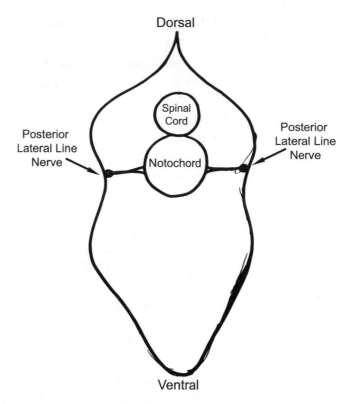

Fig. 7 A schematic of a zebrafish larva in cross section. The posterior lateral line nerves can be found on either side of the notochord

31. The lead stain will precipitate upon exposure to carbon dioxide. Wear a surgical mask while working with the lead stain and try not to breathe on it.

32. The lead stain is essential to clearly see myelinated axons; however, it can also ruin sample if lead crystals form on the grid. NaOH pellets should help absorb some carbon dioxide.

33. If you have numerous grids to stain (e.g., 20 or more grids), we recommend not lead staining all the grids at one time as they will be exposed to the stain for longer than 5 min and thus run the risk of precipitate forming on the grids.

34. The posterior lateral line nerve can be located under the TEM by finding the notochord and following a seam along muscles out to either lateral side of the larva (Fig. 7).

35. Fixation of the whole fish at this step results in better fixation of the posterior lateral line nerve for TEM.

36. Be careful to work slowly so that the lateral line nerve does not pull away from the muscle.

37. Use a glass pipette when removing liquid because the nerves can stick to plastic pipettes.

38. Samples can be stored in sodium cacodylate at 4 °C for no more than a month.

39. Nerves should be a dark brown/black color.

40. Incubation must be at least 20 min but can go longer than 20 min.

41. Nerves can be stored overnight at 4 °C after washing with 75% ethanol.

42. Carefully handle pipettes when working with propylene oxide so that it does not run out of the pipette.

43. For this step, embedding media can have been previously thawed. Do not use embedding media if it is too viscous due to polymerization. Additionally, we recommend covering your work space with aluminum foil while working with embedding media in case it comes into contact with the fume hood counter.

44. Any embedding media waste or materials that come in contact with embedding media should be baked at 65 °C until the embedding media has become a solid polymer. This usually takes about 2 days.

45. Two nerves can be embedded together in one block.

46. The number of myelinated axons can be visualized by toluidine blue solution on 1000 nm sections on a glass slide under a compound microscope. Place several 1000 nm sections in a small droplet of water on a glass slide. Place slide on a hot plate to evaporate the water and then pipette enough toluidine blue solution on the slide to cover the sections. Heat until a green ring forms around the toluidine blue drop (less than 1 min) and wash off with distilled water. Image sections under a compound microscope.

References

1. Avila RL, Tevlin BR, Lees JP et al (2007) Myelin structure and composition in zebrafish. Neurochem Res 32:197–209

2. Brösamle C, Halpern M (2002) Characterization of myelination in the developing zebrafish. Glia 39:47–57

3. Monk KR, Naylor SG, Glenn TD et al (2009) AG protein–coupled receptor is essential for Schwann cells to initiate myelination. Science 325:1402–1405

4. Langworthy MM, Appel B (2012) Schwann cell myelination requires Dynein function. Neural Dev 7:1

5. Czopka T, Lyons DA (2011) Dissecting mechanisms of myelinated axon formation using zebrafish. Methods Cell Biol 105:25–62

6. Moore AC, Mark TE, Hogan AK et al (2012) Peripheral axons of the adult zebrafish maxillary barbel extensively remyelinate during sensory appendage regeneration. J Comp Neurol 520:4184–4203

7. D'Rozario M, Monk KR, Petersen SC (2016) Analysis of myelinated axon formation in zebrafish. Methods Cell Biol 138:383–414

Chapter 27

Live Imaging of Schwann Cell Development in Zebrafish

Rebecca L. Cunningham and Kelly R. Monk

Abstract

The optical transparency of zebrafish larvae enables live imaging. Here we describe the methodology for live imaging and detail how to mount larvae for live imaging of Schwann cell development.

Key words Live imaging, Transgenes, Mounting, Confocal microscopy

1 Introduction

One of the most powerful aspects of zebrafish as a model organism is optical transparency as larvae. This transparency allows for live imaging of development, including development of Schwann cells [1]. Live imaging in zebrafish has provided great insight into Schwann cell biology [2–5] and is an attractive method to understand how Schwann cells move and behave in vivo. A variety of transgenic zebrafish can mark different stages of Schwann cell development, from Schwann cell precursors migrating out of the ganglion to myelinating Schwann cells (Table 1). In addition to using transgenic zebrafish, Schwann cells can be mosaically labeled by injection of a range of different markers to analyze single cells. Due to the close proximity of Schwann cells along the lateral line nerve, it is more difficult to analyze formation of the peripheral myelin sheath compared to the myelin sheath produced by oligodendrocytes in the central nervous system. Schwann cells can also be visualized on motor nerves emanating from the spinal cord. A range of microscopes can be employed for live imaging; confocal microscopy has been most commonly used to image Schwann cells [2–5]. The methodology presented here is not specific for a certain microscope. Below we outline how to mount larvae for live imaging of Schwann cells.

Paula V. Monje and Haesun A. Kim (eds.), *Schwann Cells: Methods and Protocols*, Methods in Molecular Biology, vol. 1739, https://doi.org/10.1007/978-1-4939-7649-2_27, © Springer Science+Business Media, LLC 2018

Table 1
Transgenic zebrafish that mark different stages of Schwann cell development

Transgenes	Schwann Cell (SC) stage labeled
sox10: *Tg(sox10:mRFP)*[vu234Tg] [6] *Tg(sox10:EGFP)*[ba4Tg] [7] *Tg(sox10:nls-Eos)*[w18Tg] [8] *Tg(sox10:Gal4-VP16)*[co19Tg] [9] *Tg(sox10:KalTA4GI)* [10]	Neural crest, SC precursors, immature SCs, promyelinating SCs, mature SCs
foxd3: *Tg(zFoxd3GFP)* [5] *Gt(foxd3-citrine)*[ct110a] [11]	Neural crest, SC precursors, immature SCs, promyelinating SCs, mature SCs
claudink: *Tg(cldnkGal4)*[e101Tg] [12]	Mature SCs
mbp: *Tg(mbp:EGFP)*[ck1Tg] [13] *Tg(mbp:EGFP)*[ue1Tg] [14] *Tg(mbp:EGFP-CAAX)*[ue2Tg] [14] *Tg(mbp:GAL4-VP16)*[co20Tg] [15]	Mature SCs

2 Materials

1. 0.8% low-melt agarose in egg water (*see* **Note 1**).
2. Glass dish.
3. Hot plate set to 55 °C.
4. 35 mm glass bottom dish.
5. 22 × 22 mm coverslip.
6. Dissecting microscope.
7. Zebrafish larvae (any stage).
8. 1× tricaine in egg water.
9. Glass pipette.
10. Dissecting needle.
11. Vacuum grease.
12. 0.2% (w/v) tricaine in egg water.
13. 0.003% 1-phenyl-2-thiourea (PTU) (*see* **Note 2**).

3 Methods

Make sure that the microscope used for imaging has a heated stage at 28.5 °C so that the larvae develop properly.

1. Melt 0.8% agarose.

2. Once the agarose is melted, pour into a glass dish and set on 55 °C hot plate.

3. Place either a 35 mm dish with glass bottom or glass coverslip under a dissecting microscope (*see* **Note 3**).

4. If larvae have not hatched out of chorions, dechorionate. Anesthetize larvae chosen for live imaging with 1× tricaine in egg water.

5. Using a glass pipette, place one larva in agarose and then quickly pipette up the larva with some agarose.

6. Pipette the larva onto the glass bottom dish or coverslip (*see* **Note 4**)

7. Using a dissecting needle, position the larva in the desired position for imaging, usually lying on their side for imaging of Schwann cells (*see* **Note 5**).

8. Repeat **steps 5–7** until desired numbers of larvae are mounted (*see* **Note 6**).

9. Create a square of two-layer vacuum grease (no greater than the diameter of a 22 × 22 mm coverslip), around the larvae on the glass bottom dish or on the glass slide if using a coverslip (*see* **Note 7**).

10. Fill the square with 0.2% tricaine so that a convex meniscus forms (*see* **Notes 8** and **9**).

11. Place the coverslip on top of the vacuum grease and form a seal. There should not be any air bubbles, and water should not leak out (Fig. 1). If opting to use a glass slide, press coverslip with mounted larvae on top of the vacuum grease on the glass slide.

12. Larvae are now ready for live imaging (*see* **Note 10**).

4 Notes

1. Make egg water by adding 60 μg/mL stock salts to distilled water. Store at room temperature.

2. To make a 50× PTU stock, dissolve 0.75 g PTU in 500 mL distilled water and heat to 60 °C for 1–2 h while stirring in a fume hood. Store at RT. Add 1 mL of 50× PTU to 50 mL of egg water to make 0.003%.

3. To be imaged by an inverted microscope, larvae must be positioned either on the bottom of a 35 mm dish with a glass bottom or on a coverslip on a glass slide. If using a coverslip on a glass slide, the slide must be "upside down" so that the coverslip, and hence the larva, is closest to the objective lens.

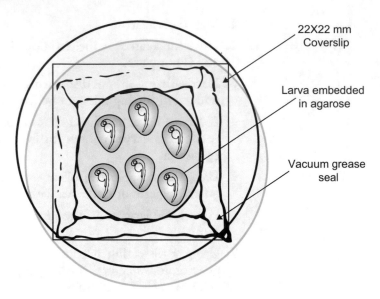

Fig. 1 Larvae are embedded in 0.8% agarose on a 35 mm glass bottom dish. A seal of two layers of vacuum grease is formed around the larvae. The space is then filled with 0.2% tricaine in egg water, and a coverslip is placed on top

4. Make sure the agarose droplet is big enough so that it completely covers the larva.

5. Time is limited to position the larva before the agarose begins to solidify, so work quickly. If larvae are young, the size of the yolk may block the lateral line if the larvae are not positioned correctly on their sides.

6. Six larvae can fit comfortably in the glass bottom dish or on a coverslip.

7. Make sure the vacuum grease forms a seal and will not let any water seep through. This is especially critical if mounting on a coverslip because water could leak onto the objective lens of the microscope.

8. The 0.2% tricaine is used to paralyze the larvae so they do not move during imaging. The larvae should be viable for 12+ hours of imaging.

9. 0.003% PTU can be used to prevent pigmentation if needed.

10. Larvae can be genotyped after imaging or can be dismounted and then imaged later in development.

References

1. Monk KR, Talbot WS (2009) Genetic dissection of myelinated axons in zebrafish. Curr Opin Neurobiol 19:486–490

2. Perlin JR, Lush ME, Stephens WZ et al (2011) Neuronal Neuregulin 1 type III directs Schwann cell migration. Development 138:4639–4648

3. Lyons DA, Pogoda HM, Voas MG et al (2005) erbb3 and erbb2 are essential for Schwann cell migration and myelination in zebrafish. Curr Biol 15:513–524

4. Rosenberg AF, Isaacman-Beck J, Franzini-Armstrong C, Granato M (2014) Schwann cells and deleted in colorectal carcinoma direct regenerating motor axons towards their original path. J Neurosci 34:14668–14681

5. Gilmour DT, Maischein HM, Nüsslein-Volhard C (2002) Migration and function of a glial subtype in the vertebrate peripheral nervous system. Neuron 34:577–588

6. Kirby BB, Takada N, Latimer AJ et al (2006) In vivo time-lapse imaging shows dynamic oligodendrocyte progenitor behavior during zebrafish development. Nat Neurosci 9:1506–1511

7. Dutton KA, Antonellis A, Carney T et al (2008) An evolutionarily conserved intronic region controls the spatiotemproal expression of the transcription factor Sox10. BMC Dev Biol 8:105

8. Prendergast A, Linbo TH, Swarts T et al (2012) The metalloproteinase inhibitor Reck is essential for zebrafish DRG development. Development 139:1141–1152

9. Das A, Crump JG (2012) Bmps and id2a act upstream of Twist1 to restrict ectomesenchyme potential of the cranial neural crest. PLoS Genet 8:e1002710

10. Almeida RG, Lyons DA (2015) Intersectional gene expression in zebrafish using the split KalTA4 system. Zebrafish 12:377–386

11. Hochgreb-Hägele T, Bronner ME (2013) A novel FoxD3 gene trap line reveals neural crest precursor movement and a role for FoxD3 in their specification. Dev Biol 374:1–11

12. Münzel EJ, Schaefer K, Obirei B et al (2012) Claudin k is specifically expressed in cells that form myelin during development of the nervous system and regeneration of the optic nerve in adult zebrafish. Glia 60:253–270

13. Jung SH, Kim S, Chung AY et al (2010) Visualization of myelination in GFP-transgenic zebrafish. Dev Dyn 239:592–597

14. Almeida RG, Czopka T, Lyons DA (2011) Individual axons regulate the myelinating potential of single oligodendrocytes in vivo. Development 138:4443–4444

15. Hines JH, Ravanelli AM, Schwindt R et al (2015) Neuronal activity biases axon selection for myelination in vivo. Nat Neurosci 18:683–689

Part IV

Protocols for Schwann Cell Transplantation

Chapter 28

Transplantation of Adult Rat Schwann Cells into the Injured Spinal Cord

Ying Dai and Caitlin E. Hill

Abstract

Adult Schwann cells (SCs) can provide both a permissive substrate for axonal growth and a source of cells to ensheath and myelinate axons when transplanted into the injured spinal cord. Multiple studies have demonstrated that SC transplants can be used as part of a combinatorial approach to repairing the injured spinal cord. Here, we describe the protocols for collection and transplantation of adult rat primary SCs into the injured spinal cord. Protocols are included for the tissue culture procedures necessary for collection, quantification, and suspension of the cells for transplantation and for the surgical procedures for spinal cord injury at thoracic level nine (T9), reexposure of the injury site for delayed transplantation, and injection of the cells into the spinal cord.

Key words Schwann cells, Rat, Transplantation, Spinal cord injury, Tissue culture, Surgery

1 Introduction

Schwann cells (SCs) are the glial cells of the peripheral nervous system (PNS). Following injury to the PNS, SCs dedifferentiate and help guide axons back to their targets [1, 2]. Unlike injury to PNS axons, injury to CNS axons results in limited axonal regeneration. Transplantation of peripheral nerves into the injured spinal cord creates a permissive environment and allows for CNS axons to regenerate and be myelinated by SCs [3, 4]. Similar to peripheral nerve grafts, the transplantation of SCs isolated and purified from adult rat nerves is sufficient to promote axonal regeneration, and purified SCs both remyelinate and ensheath the axons that grow into the grafts within the injured spinal cord [5–8]. The axons within the grafts are able to conduct action potentials [9, 10] and, in some instances, are able to enhance locomotor recovery [8, 11, 12]. The effects of SC transplants alone on SCI repair are modest [8, 11, 12]. As a result, significant experimental effort has focused on how to combine SC transplants with additional strategies to enhance their efficacy for SCI repair [13, 14]. Several in vivo

Paula V. Monje and Haesun A. Kim (eds.), *Schwann Cells: Methods and Protocols*, Methods in Molecular Biology, vol. 1739, https://doi.org/10.1007/978-1-4939-7649-2_28, © Springer Science+Business Media, LLC 2018

experimental models are available for assessing the ability of SCs to promote different aspects of spinal cord repair. For example, transplantation of SCs mixed with matrix and placed within channels inserted into the completely transected spinal cord enables accurate assessment of axonal regeneration and determination of which axons are able to grow into and potentially out of SC grafts [6, 7, 15]. This can be used to study which additional treatments are required to enhance the growth of specific neuronal populations into and out of SC grafts. Demyelinating lesions have been used extensively to assess the ability of SCs to myelinate CNS axons [16–19]. Although these models enable specific questions regarding the effects of SC transplants to be addressed, they do not model what occurs following most human SCIs. Following human SCI, a cystic lesion forms [20]. Several experimental SCI devices have been developed that reproducibly induce a contusion injury in the rodent spinal cord [21–24]. Although each device is slightly different in terms of the specifics of how the injury is induced, all involve a laminectomy to expose to the spinal cord and subsequent rapid, temporary damage to the spinal cord by a rod. In rats [25], but not mice [26], contusion injuries produce a cystic cavity similar to human SCI and provide a model for testing therapeutic interventions for SCI repair. Injection of SCs directly into the contused rat spinal cord enables assessment of the ability of SCs to promote recovery in a model that mimics the human condition. This model allows for assessment of both the histological and functional changes resulting from the transplantation of SCs [8, 27–29]. It has also provided the basis of the preclinical data for the human trial of SCs for human SCI repair [30]. Moreover, it forms the foundation onto which a variety of combination strategies have been developed to further enhance the utility of SC transplants for SCI repair [13, 14, 31]. Here, we describe the methods for inducing a spinal cord contusion injury at thoracic level nine (T9) using the Infinite Horizon injury device [24], the procedures necessary for collection and transplantation of the cells into the injured spinal cord, and the methods to reexpose the spinal cord (re-laminectomy) to allow for a delay between injury and transplantation, which mimics the clinical situation for patients with spinal cord injury. Figure 1 shows a schematic of the procedures described.

2 Materials

2.1 Tissue Culture

The transplantation procedures described work for a variety of cells. Here, we provide details for growing mitogen-expanded adult rat SCs (adapted from [32]). This method facilitates the generation of a sufficient number of cells for transplantation. All chemicals and solutions should be tissue culture grade and be made up in a tissue culture hood. To ensure sterility, appropriate sterile

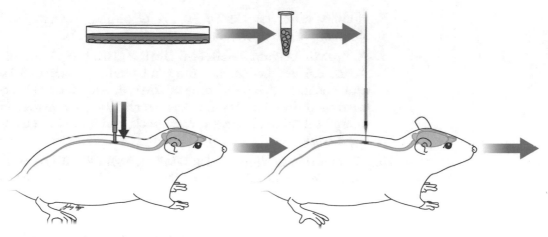

Fig. 1 Steps involved in transplantation of SCs into the injured spinal cord. Animals are injured. Following a delay (typically 7 days), the rats undergo a second surgery, and the cells are transplanted into the epicenter of the injured spinal cord. Rats are then allowed to recover for variable times depending on the purpose of the experiment. SCs are grown in 10 cm dishes until they reach confluence. On the day of transplantation, SCs are collected and placed in an Eppendorf tube. Once each rat is ready for cells, the cells are transferred to a pulled glass capillary tube attached to a Hamilton® syringe and injected into the injury site using a nano-syringe pump injector to regulate the speed of injection

technique should be used. Chemicals weighed outside of the tissue culture hood need to be filtered through a 0.22 μm filter prior to aliquoting.

1. Adult rat SCs. Preferably passage 3 at time of transplantation (*see* **Note 1**).

2. Tissue culture-treated plastic 10 cm dishes coated with poly-L-lysine (PLL). To coat dishes, add 7 ml of 20 mg/ml PLL to dish for 30 min to 1 h (*see* **Note 2**). Remove PLL and wash the dish three times with distilled water before removing the last wash (*see* **Note 3**).

3. DMEM: Dulbecco's Modified Eagle's Medium (DMEM) containing high glucose with L–glutamine, pyruvate, and phenol red.

4. Heat-inactivated fetal bovine serum (FBS): To heat inactivate, heat FBS to 56 °C for 30 min.

5. 40 mg/ml gentamicin stock solution.

6. Forskolin stock: To a 10 mg vial of forskolin (7-deacetyl-7-[O-(*N*-methylpiperazino)-γ-butyryl]-dihydrochloride), add 1.6 ml of 100% ethanol, aliquot, and store at 4 °C.

7. Heregulin stock: To 50 μg of heregulin (recombinant purified heregulin peptide), add 2.836 ml of distilled water, filter, aliquot, and store at −20 °C.

8. Pituitary extract stock: To 50 mg of pituitary extract, add 50 ml of DMEM filter, aliquot, and store at −80 °C.

9. SC growth medium, D-10+3M: DMEM, 10% FBS, 2 μM forskolin, 3.5 μM heregulin, 20 μg/ml pituitary extract, 0.1% gentamicin. To a 500 ml bottle of DMEM, add 55 ml of heat-inactivated FBS, 75 μl of forskolin stock, 500 μl of heregulin stock, 11.2 ml of pituitary extract stock, and 560 μl of gentamicin stock.

10. SC collection medium, D-10: DMEM, 10% FBS, 0.1% gentamicin. To a 500 ml bottle of DMEM, add 55 ml of heat-inactivated FBS and 550 μl gentamicin stock.

11. SC transplantation medium, DMEM-F12: Dulbecco's Modified Eagle's Medium-F12 (DMEM-F12) containing high glucose with L–glutamine, pyruvate, and phenol red.

12. Hank's Balanced Salt Solution (HBSS) without calcium and magnesium.

13. Trypsin-EDTA: 1× in HBSS (*see* **Note 4**).

14. Trypan blue.

15. 50 ml conical tubes (*see* **Note 5**).

16. 14 ml round-bottom polypropylene snap cap tubes (*see* **Note 6**).

17. Vacuum flask.

18. Unplugged glass Pasteur pipettes.

19. Plugged glass Pasteur pipettes: flame-polished (*see* **Notes 7 and 8**).

20. Bunsen burner (*see* **Note 9**).

21. 70% ethanol.

22. 1.5 ml Eppendorf tubes.

23. Pipetman®: p1000, p200, p20.

24. Pipette tips: p1000, p200, p20 (*see* **Note 10**).

25. PipetteAid®.

26. Serological pipettes: 25 ml, 10 ml, 5 ml, 1 ml.

27. 1 plastic beaker (*see* **Note 11**).

28. Hemocytometer (*see* **Note 12**).

29. Refrigerated centrifuge.

30. Water bath or dry bath beads (Lab Armor™).

31. Inverted microscope.

32. CO_2 incubator: water-jacketed, humidified, set to 37 °C and 6% CO_2.

**2.2 Pulled Glass
Injection Pipettes**

1. Pipette puller.

2. Kwik-Fil borosilicate F-glass capillaries (length, 100 mm; outer diameter, 1.5 mm; inner diameter, 0.84 mm).

3. Silicone aquarium sealant.

4. 3 ml or 10 ml luer-lock syringe.

5. 22-gauge needle.

2.3 Surgical: SCI

1. Infinite Horizon (IH) impactor.

2. Surgical microscope.

3. Homeothermic blanket system (*see* **Note 13**).

4. Germinator™ 500 dry sterilizer (*see* **Note 14**).

5. Anesthetic: isoflurane (*see* **Note 15**).

6. Clippers, A5, #50 blade.

7. Handheld vacuum.

8. Bench pads.

9. Sterile surgical gloves.

10. Betadine® and Betadine® Surgical Scrub.

11. 70% ethanol.

12. Paralube® lubricating eye ointment.

13. Lubricating jelly (*see* **Note 16**).

14. Sterile saline.

15. Cotton-tipped applicators.

16. Gauze, 2 × 2″, 8 ply.

17. Gauze bolster (*see* **Note 17**).

18. Absorption triangles.

19. Gelfoam.

20. Surgery tools: scalpel handle #3; scalpel blades (#10, #11, or #15 scalpel blades; we prefer #11); Adson forceps (serrated, straight, 12 cm); Adson forceps (1 × 2 teeth, straight, 12 cm, 2 pairs); Adson-Brown forceps (straight, shark teeth); tissue-separating scissors (straight, blunt-blunt, 11.5 cm); Friedman rongeur (curved, 2.5 mm cup diameter, 13 cm, FST: 16000-14); Friedman-Pearson rongeur (curved, 1 mm cup diameter, 14 cm, FST: 16021-14); Dumont #5 forceps (standard tips, straight, 11 cm); Dumont #5 forceps (standard tips, angled 45, 11 cm); Agricola retractor (3 × 3 sharp, 4 cm, 3.5 cm maximum spread); Olsen-Hegar needle holders with suture cutters (straight, serrated, 12 cm, with lock); Ethicon Vicryl sutures (4-0); Michel suture clip applicator and remover for 11 mm and clips, or autoclip wound clipper 9 mm and wound clips.

2.4 Surgical: Transplantation

1. Surgical SCI materials (listed above).

2. Vertebral stabilizer: Bulldog serrefine clamp (straight, 35 mm, attached to 1 cm diameter rod affixed to a magnetic base with arm (*see* **Note 18**).

3. Stereotaxic nano-syringe pump.

4. Stereotax (*see* **Note 19**).

5. Stereotaxic arm with three-way electrode manipulator and upper bracket clamp (*see* **Note 20**).

6. 10 μl Hamilton® 700 Series Microliter™ syringes: removable needle, 26 gauge, point style 2 (sharp, beveled, curved, non-coring needlepoint).

7. Plunger from Hamilton™ syringe (*see* **Note 21**).

8. p20 and/or p200 Pipetman (*see* **Note 22**).

9. Pulled glass capillaries with silicone plug.

2.5 Postsurgical Animal Care

The specific materials used for animal care will depend on the requirements at your institution and the details outlined and approved in the animal protocol under which the work is being performed. Postsurgical care must conform to the requirements of the International Animal Care and Use Committee (IACUC) at your institution.

1. Record sheet (*see* **Note 23**).

2. Lactated Ringer's (*see* **Note 24**).

3. Antibiotic: gentamicin, 5 mg/kg.

4. Analgesic: buprenorphine, 0.05 mg/kg.

5. Water-jacketed incubator (*see* **Note 25**).

6. Thermoregulated heating pads (*see* **Note 26**).

7. ALPHA-dri® bedding (*see* **Note 27**).

8. Nylabone® extra-small chews (*see* **Note 28**).

9. Mash (*see* **Note 29**).

10. Nutrical® (*see* **Note 30**).

11. Extra-long spout for water bottle (*see* **Note 31**).

3 Methods

3.1 Preparation and Growth of SC Cultures

1. Remove frozen cryovial(s) containing 3×10^6 adult peripheral nerve-derived SCs from liquid nitrogen and place on dry ice.

2. For one vial, put 5 ml of warm D-10 (37 °C) into a 14 ml snap cap tube. For more than one vial, put 5 ml of warm D-10 (37 °C) per vial into a 50 ml conical tube.

3. Take the vials of cells from the dry ice and place them into the water bath/bead bath for a few seconds until almost melted.

4. Transfer cells from the cryovial to the tube containing D-10. Freezing solution contains dimethyl sulfoxide (DMSO). Work quickly. Prolonged contact of the cells with DMSO at the concentrations used can be toxic to the cells.

5. If necessary, add additional warm D-10 to cryovial to remove any remaining cells (*see* **Note 32**).

6. Centrifuge at 4 °C for 5 min at 453 × *g*.

7. While cells are spinning, add 5 ml of D-10+3M to PLL-coated dishes. Make up two dishes for each vial of cells.

8. Remove supernatant and resuspend cell pellet in 2 ? ml of D-10+3M. Add additional D-10+3M to bring final volume to 2 ml/dish to be plated (total volume 4 ml/vial).

9. Determine the number of cells in each vial by counting a sample of the cells using the hemocytometer (*see* **Note 33**). Add 10 μl of cells to each side of the hemocytometer, check that the cells are evenly distributed, and follow the instructions for the hemocytometer. To determine the number of live/dead cells in each vial, trypan blue can be used. When using trypan blue, remove a 100 μl aliquot of cells, and mix with 100 μl of trypan blue in an Eppendorf tube before adding the cell suspension to the hemocytometer. Make sure to adjust the cell number calculation by the dilution factor.

10. Add half of the cells from each cryovial to each 10 cm dish containing 5 ml of D-10+3M (*see* **Note 34**).

11. To obtain even distribution of cells across the dish, rock the dish gently by tilting it front to back followed by side to side immediately upon addition of the cell suspension to the 10 cm dish. Even distribution of the cells onto the dish is important to obtain even growth of the cells across the dish. Swirling the dishes in a circular motion will result in uneven cell distribution. Bubbles will result in empty patches on the dish at early time post-plating and should be avoided. A gradient of cells across the dish can arise if the incubator is not level.

12. Label dishes with the cell information (passage number, number of cells, date plated).

13. Place dishes into the CO_2 incubator.

14. The next day, change the growth medium. Changing the medium helps to remove cellular debris arising from cells that died during thawing. If there are many dead cells, prior to refeeding with D-10+3M, gently rinse the plates twice with HBSS to remove as much cellular debris as possible.

15. Feed the cells every 2–3 days by replacing the medium with 5–7 ml of fresh D-10+3M.

16. When confluent, collect the cells for transplantation or split and plate 1×10^6 to 1.5×10^6 cells per dish depending on desired date of collection (*see* **Note 35**). Figure 2 shows the appearance of SCs at the upper and lower ends of confluency used for experiments.

3.2 Collection of SCs for Transplantation

1. For each dish to be collected, add 1 ml of pre-warmed D-10 to an empty cell collection tube (*see* **Notes 5** and **6**).

2. Rinse each 10 cm dish twice with 5–7 ml of HBSS. Remove the last wash. To remove solutions, tilt the dish away from you. To prevent dislodging cells, touch the Pasteur pipette, attached via tubing to a vacuum flask, to the side of the dish—not the bottom of the dish. Work quickly to ensure that the cells do not dry out.

3. Add 3 ml trypsin-EDTA to each plate.

4. Wait 3–5 min for the cells to detach from the dish. Cells should be exposed to trypsin-EDTA for the least amount of time needed to detach the cells. Fresh enzymes work better than older enzymes as enzymes lose their activity over time when diluted and stored at 4 °C.

5. After 3 min, examine the dish under microscope. If the cells have not detached, rapidly move the dish side to side to help detach the cells. Alternatively, the dishes can be placed at 37 °C to facilitate enzymatic activity.

6. Using a flame-polished pipette, dislodge and dissociate the cells by aspirating the cells and expelling them back onto the dish several times while rotating the dish (*see* **Note 36**).

7. Remove the cells/trypsin-EDTA solution from the dish, and add it to the tube containing 1 ml/dish of D-10 (*see* **step 1**).

8. Add 3 ml of D-10 to the now-empty 10 cm dish.

9. Repeat **step 6** and add the cells/D-10 solution to the tube containing the previously collected cells.

10. Check the dish under the microscope to make sure all the cells have been removed.

11. Centrifuge the tube containing the cells at $453 \times g$ for 10 min at 4 °C.

12. Using a Pasteur pipette attached via tubing to a vacuum flask, carefully remove the supernatant. To prevent accidentally dislodging the cell pellet at the tube bottom, tip the tube 45–90° as the supernatant is removed. Alternatively, a small amount of media can be left at the bottom of the tube and removed with a pipetman.

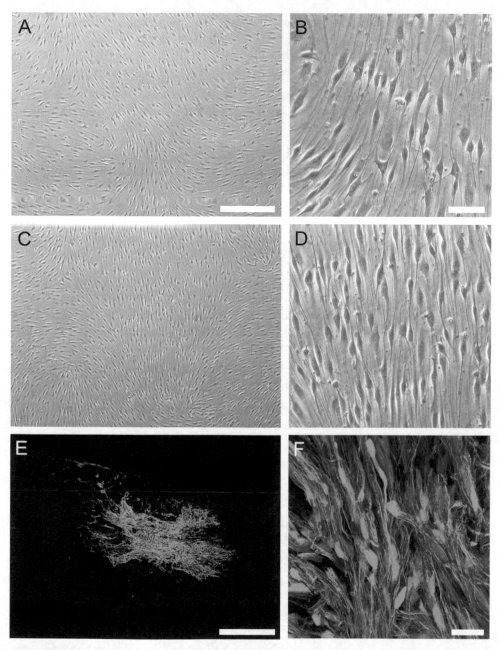

Fig. 2 Appearance of confluent SCs in vitro and transplanted SCs in vivo. (**a–d**) In vitro, SCs have a bipolar morphology with extended processes. When confluent, SCs form whirls on the dish. (**a, c**) Appearance of SCs on the dish at the lower (**a**) and upper (**c**) ends of preferred confluency for use for cell transplantation. (**b, d**) Appearance of individual cells and processes. At the lower end of confluency, SCs have formed whirls (**a**). Individual cells and processes are aligned with each other, but substantial space exists between adjacent cells and processes (**b**). At the upper end of confluency, whirls are more prevalent (**c**), and the individual cells and processes are more tightly packed (**d**). (**e, f**) Appearance of green fluorescent protein (GFP)-positive SCs in vivo 7 days following transplantation. 2×10^6 GFP-SCs were transplanted into a 200 kdyn IH SCI 7 days post-injury. GFP-positive SCs are located within the injury site (**e**) and occupy most of the injury site. Individual SCs within the transplant have elongated, extended processes (**f**). Scale bar: **a, c** = 250 μm; **b, d** = 25 μm; **e** = 1 mm; **f** = 10 μm

13. Add 1 ml of D-10 and resuspend the cells by mixing three to eight times with a large-bore, flame-polished pipette. Mix the cell solution sufficiently to obtain a single-cell suspension but as little as possible to prevent damaging the cells.

14. Add 4 ml of D-10. Make a note of the actual total volume. An accurate volume is needed to calculate the total number of cells collected.

15. Gently mix the cell solution.

16. Remove a sample of the cells to quantify cell number using the hemocytometer. If splitting cells, determine cell number, dilute the cells to 1×10^6 cells/ml, and plate the desired number of cells into PLL-coated 10 cm dishes at this point (*see* **Note 34**). If collecting for transplantation, proceed with the next step.

17. Centrifuge the cells at $453 \times g$ for 5 min at 4 °C.

18. While the cells are spinning, quantify the number of cells using the hemocytometer. First, determine the total number of cells collected, and then calculate the total volume needed to dilute the cells to 2×10^6 cells/6 μl. This is the cell concentration used for transplantation.

19. After centrifugation, aspirate the supernatant using the Pasteur pipette attached to the vacuum.

20. Add 1 ml of D-10 to the cell pellet, and resuspend the cells by mixing three to eight times with a large-bore flame-polished pipette.

21. Transfer the cells to a 1.5 ml Eppendorf tube.

22. Centrifuge the 1.5 ml Eppendorf tube containing the cells to be transplanted at $453 \times g$ for 5 min at 4 °C.

23. After centrifugation, aspirate the supernatant using the vacuum or p1000 pipetman. Leave the pellet containing the cells behind.

24. In separate 1.5 ml tube, add a volume of DMEM-F12 equal to the volume calculated in **step 18** needed to obtain 2×10^6 cells/6 μl. This tube is used as a reference tube for the amount of volume to add when diluting the cells.

25. Slowly add DMEM-F12 to cell pellet. Use the reference tube generated in **step 24** to judge the amount of DMEM-F12 to initially add. The initial volume of DMEM-F12 added to the cells should be equal or less than the final volume needed for cell dilution. Gently mix the cells with the DMEM-F12 by swirling the mixture with a p20 pipette tip. Be cautious with the cells. The concentration of cells is quite high; overmixing will result in damage to the cells, causing them to become viscous and unsuitable for transplantation.

26. Using the Pipetman, check that the volume is accurate. If necessary, add additional media one drop at a time.

27. Place the Eppendorf tube containing the cells on ice.

28. Make up an additional Eppendorf tube containing 500–1000 μl of DMEM-F12. This tube is used for rinsing the injection syringe before and after surgery. It can be kept on ice with the cells.

3.3 Preparation of Glass Injection Pipettes

Instructions are based on setting for the Narishige PE-2 8803 pipette puller. Any pipette puller will work.

1. Set the pipette puller to magnet = 2 and heater = 5.

2. Insert glass capillary tube and lock in place.

3. Hit the red button to initiate the heater coil.

4. Remove the two pulled pipettes that result (*see* **Note 37**).

5. Repeat until sufficient pipettes are made (*see* **Note 38**).

6. To make a silicone plug in the pipette, fill a 3 ml syringe with aquarium sealant (*see* **Note 39**), attach a 22-gauge needle, insert needle into pipette so that it is 2–3 mm away from where the pipette narrows (*see* **Note 40**), push plunger against the hand, or press against a table. As the silicone enters the pipette, turn the pipette to create a 1–2 mm plug. The resulting plug will be approximately 1 mm away from where the pipette narrows (*see* **Notes 41–43**). As the silicone is added, it will move closer to the tip. Withdraw the syringe slightly as the silicone enters to prevent the plug from becoming too close to the tip.

3.4 Preparation of the IH Device

The IH device is capable of producing graded injury magnitudes according to user-defined changes in the amount of force [24]. A laminectomy is performed to expose the underlying spinal cord. A small diameter tip is subsequently used to displace the exposed spinal cord with a preset force. The spinal cord contusion produced results in damage to both the gray and white matter of the spinal cord. Prior to using the IH device, it is advisable to read the IH manual. Understanding how the impactor works and what the output curves produced following the impact mean is important for being able to assess the quality of the injury produced.

1. Prior to starting an experiment, take the IH arms apart, and clean all the joints with alcohol (this can be done in advance). Do not, under any circumstance, grease any of the joints.

2. Make sure the arms are working properly and that there is no give when the screws are tightened.

3. If arms wiggle or give way, it is likely that the brass rings are out of position.

4. Take arms apart and make sure everything is in the correct place/alignment and check again.

5. Turn IH impactor on.

6. Log into the IH program (PSI IH Spinal Cord Impactor v5.0).

7. Set force to desired force. 200 kdyn is good for a moderate injury. *See* [24] for information on injury severities.

8. Do a practice hit on the rubber test block. Check to confirm that the test block produces appropriate force versus time and displacement versus time curves.

3.5 T9 Laminectomy and Induction of SCI

The entire surgical procedure is performed while viewing the surgical site through the surgical microscope to ensure good visibility while performing surgery and to prevent damage to the spinal cord during the laminectomy procedure.

1. Set up the surgical table. Turn on the bead sterilizer at the start of setup; it takes approximately 20 min to reach temperature. Turn on the homeothermic blanket, and place the rectal probe beneath the heating pad to ensure that the heating pad reaches correct temperature. Lay out the tools on a sterile field (*see* **Note 44**). Fill a 10 ml syringe with saline. Wet a 1 cm stack of gauze with saline (*see* **Note 45**). Use a blue bench pad to cover the homeothermic blanket and create an aseptic field.

2. Anesthetize the rat (*see* **Notes 15** and **46**).

3. Using clippers, shave the rat's back from the hip to the shoulders. Work quickly if using isoflurane. Three passes of the clippers, once up the middle and then once on either side, are usually sufficient (*see* **Note 47**).

4. Weigh the rat and note its weight.

5. Disinfect the skin by alternate swabbing of Betadine®, 70% ethanol, Betadine®, 70% ethanol, Betadine® Surgical Scrub.

6. Cover the eyes with ophthalmic ointment.

7. Insert rectal thermometer of homeothermic blanket (*see* **Note 48**).

8. Elevate rat using the bolster.

9. Make an incision through the skin along the midline. It is useful to landmark T8-10 prior to incision to minimize the length of the incision. The incision should be approximately 4–6 cm in length. If the incision is landmarked correctly, a fat pad should be detected beneath the rostral portion of the incision, and the ligaments should appear to form a V-shape beneath the caudal portion of the incision.

10. Use the blunt-tipped, tissue-separating scissors to separate the skin from the underlying muscle using reverse dissection.

11. Find fat pad, located rostrally, and separate fat pad from underlying muscle using reverse dissection. Be careful when inserting the blunt scissors. A large blood vessel penetrates through the fat pad further rostral, which, if damaged, will result in excessive blood loss.

12. The musculature over the vertebral column from T7 (rostrally) to beyond T12 (caudally) should now be visible.

13. Find T10 (*see* **Note 49**).

14. Once T10 is identified, cut the muscle along vertebrae on both sides from around T7 to T11 (*see* **Note 50**) by making an incision parallel to the dorsal vertebral processes. Use sufficient force that only a single incision is needed (*see* **Note 51**). The incision should be as close to the dorsal process as possible and penetrate down to the lateral vertebral processes.

15. Isolate T9 by cutting between T8/T9 and T9/T10 (*see* **Note 52**). Use the Adson forceps with 1 × 2 teeth to gently grasp T8 or T9 rostrally while cutting.

16. Remove the muscle from the lateral vertebral processes from T8 to T10. Using a cotton-tipped applicator, or tip of the scalpel blade, scrape the muscle off of the lateral processes of T8 and T10 (*see* **Note 53**).

17. Remove the top of T9 using large rongeurs. One large nip with the large rongeurs is usually sufficient to remove a large portion of the bone (*see* **Note 54**).

18. Remove the remainder of T9 by making small nips with the smaller rongeurs (*see* **Note 55**). Stabilize the spinal column by holding T8 with forceps (*see* **Note 56**). Work from caudal to rostral to remove the bone. Keep removing more bone until all of the dorsal process of T9 is removed (*see* **Note 57**). Be careful not to damage the underlying spinal cord (*see* **Note 58**). If bleeding occurs, stop the bleeding before proceeding (*see* **Note 59**).

19. Confirm that the laminectomy is wide enough by assessing the size of the laminectomy using the wooden end of a Cotton-tipped applicator which is approximately the size of the impactor probe (*see* **Note 60**). Check that the edges of the laminectomy are smooth to ensure no bone fragments will accidentally damage the spinal cord.

20. Place rat into IH holder (*see* **Notes 61–63**). Start by clamping the rostral side (T8), loosen the joints on the left side, and clamp T8 by the lateral processes (tighten the screw until just tight and then turn one full turn extra) (*see* **Note 64**). Tighten all the joints on the left arm until just tight. Clamp the caudal side (T10) as above (*see* **Note 65**). Once the rat is clamped, tighten all the joints as tightly as possible. Double check to

make sure the arms do not move and that the vertebrae do not wiggle.

21. Move rat to the IH device and use hand clamps to affix the plastic board to the frame (*see* **Note 66**).

22. Center the impactor probe on the laminectomy site (*see* **Notes 67** and **68**).

23. Lower the probe so that it is just above the cord, and then back it off by one full turn.

24. Lock the impactor by flipping the brass switch.

25. Double check that the force is set to the force you want (*see* **Note 69**).

26. Follow the impactor prompts.

27. Watch the probe to make sure it does not hit the bone or tissue prior to impact.

28. Save the impact information (*see* **Note 70**).

29. Check the quality of the impact to ensure it fits the inclusion and exclusion criteria (*see* **Note 71**).

30. Write down the actual force (*see* **Note 72**).

31. Release the rat from the IH clamps holding T8 and T10.

32. Return rat to the thermoregulated heating pad on the surgical table (*see* **Note 73**).

33. Suture the tissue together in two anatomical layers (*see* **Note 74**). For the deep layer, align and suture the muscles previously attached to the vertebral processes (*see* **Note 75**). For the superficial layer, bring the upper layer of tissue directly beneath the skin together (*see* **Note 76**).

34. Bring together the skin on the two sides of the incision. Tent the skin and wipe the edges with gauze pre-wetted with saline to remove any blood (*see* **Note 77**).

35. Close the skin incision with wound clips or Michel clips (*see* **Note 78**).

36. Mark the rat's tail and/or ears with the animal number.

37. Perform acute postsurgical care *see* Subheading 3.7.

3.6 Transplantation

Transplantation can be performed any time following injury. A delay of 7 days is most frequently used [8, 27, 29] and results in better cell survival than acute transplantation [33, 34]. When functional recovery is being assessed following transplantation, testing should be performed on the day of transplantation prior to cell collection. Rats should be randomly assigned to treatment groups. Once assigned, treatment groups should be checked to ensure that the mean of the main functional test being used is equivalent for all groups at the time of intervention. On a given transplant day, it is best practice to transplant all the different treatment combinations. This should then be repeated over multiple transplant sessions. If the number of treatment groups is large, then a limited number of groups should be transplanted each day, and the treatments for any given day should be randomly assigned.

1. Prior to beginning the transplant surgery, set up the pump to inject at a rate of 1 μl/min for a total volume of 6 μl (*see* **Note 79**).

2. Repeat **steps 1–7** from Subheading 3.5, T9 laminectomy and induction of SCI, to prepare the rat for surgery (*see* **Note 80**).

3. Open the skin at the previous incision site. Gentle pulling of the skin on either side of the previous incision is usually sufficient to reopen the skin.

4. Loosen the upper muscle layer by reverse dissection.

5. Find the sutures used to close the upper layer of the muscle, and remove each of the sutures by grasping the sutures with the Adson forceps with 1 × 2 teeth and then cutting the loop of thread below the knot.

6. Separate the top thin layer of the muscle to expose the deep muscle layer.

7. Find and remove the deep muscle layer sutures overlying T8, T9, and T10.

8. Reexpose the laminectomy site. At 7 days post-SCI, the muscles can usually be pulled away from the vertebra along the plane of their previous cut. This will expose the laminectomy site, now covered by scar tissue.

9. Find the ligaments that were brought together over T8–10. Holding the muscle on one side with forceps, grasp the opposite side with the tip of the rongeurs and gently tug laterally. Alternatively, slide the rongeurs in the rostral caudal direction along the vertebrae under the ligaments. If the above approaches do not result in separation of the muscle from the bone, the tissue may be cut with a scalpel.

10. Once the laminectomy site is exposed, carefully remove the scar tissue over the laminectomy site using the rongeurs to expose the spinal cord (*see* **Notes 81** and **82**).

11. Once the spinal cord is exposed, the injury should be visible.

12. Place the rat on the bolster (*see* **Note 83**).

13. Use the retractor to spread apart the tissue adjacent the injury site (*see* **Note 84**).

14. Clamp the dorsal process of T8 with the vertebral stabilizer (*see* **Note 85**).

15. Attach the glass capillary to the Hamilton® syringe (*see* **Notes 86** and **87**).

16. Check that the Hamilton® syringe is patent by inserting the capillary tip into tube containing DMEM-F12 and checking to make sure fluid moves easily into and out of the syringe.

17. Resuspend the cells to be transplanted (*see* **Note 88**).

18. Fill the Hamilton® syringe to the 7 μl mark for a 6 μl injection (*see* **Notes 89** and **90**).

19. Attach Hamilton® syringe to the nano-syringe pump injector.

20. Confirm that the pump is working (*see* **Note 91**). Start the pump and let run until a small drop of the cell mixture is expelled. Remove expelled cells gently with a cotton-tipped applicator before proceeding with injection.

21. Determine the site to inject (*see* **Note 92**).

22. Move the pipette tip to the surface of the dura (*see* **Note 93**).

23. Turn the micromanipulator quickly to puncture the dura. Insert the pipette to a depth of 1 mm (*see* **Notes 94** and **95**).

24. Start the pump. At a rate of 1 μl/min, it will take 6 min for the 6 μl injection.

25. Watch to make sure the cells flow through the pipette and are injected. You should be able to see the cells moving down the pipette (*see* **Note 96**).

26. Once the pump stops, leave the needle in place for an additional 3 min (*see* **Note 97**).

27. Remove the needle slowly (*see* **Note 98** and **99**).

28. Close the rat in anatomical layers *see* Subheading 3.5, **steps 33–35**.

3.7 Acute Postsurgical Animal Care

1. Administer analgesic (*see* **Note 100**).

2. Administer antibiotic (*see* **Note 101**).

3. Administer fluids (*see* **Notes 102** and **103**).

4. Place rat in incubator until awake (*see* **Note 104**).

5. Return to home cage placed on thermoregulated heating pads (*see* **Note 105**).

3.8 Care of SCI Rats

In addition to standard care required postsurgery by most IACUC, SCI rats require specialized additional care. Animals should be carefully examined daily and their health and care documented. Rats are social animals, and they recover better when housed in pairs or triplets. Housing SCI rats singly should be avoided when possible.

1. Manually express bladders twice a day until the animals have regained the ability to express their bladders themselves. Once bladder control is regained, bladders should be manually expressed once per day (*see* **Notes 106–108**).

2. Check for dehydration (*see* **Note 109**).

3. Check for porphyrin around the eyes (*see* **Note 110**).

4. Check for sores or autophagia (*see* **Note 111**).

5. A daily observation sheet/checklist is recommended to facilitate care of SCI rats (*see* **Note 112**).

4 Notes

1. Adult rat nerve-derived Schwann cells can be thawed from frozen early passage cell stocks or alternatively passaged directly to passage 3 from the initial sciatic nerve isolation. Cells should be used at a consistent passage for experiments. Low passage number (passage 3 or passage 4) is preferable. This allows for some amplification of cell numbers while ensuring that the cells retain their phenotype. Primary SCs will transform after 20 passages and become growth factor independent and form foci [35].

2. It is best practice to keep the timing of coating consistent to reduce experimental variability.

3. Dishes are normally coated just prior to plating the cells. If necessary, PLL-coated dishes can be made up in advance, sealed with parafilm, and stored at 4 °C for several days.

4. Must be calcium- and magnesium-free. Bivalent cations will prevent the trypsin-EDTA from working and result in coagulation of the cells. 1× trypsin-EDTA is made from purchased 10× stock of 0.5% trypsin-EDTA.

5. Use a 50 ml conical tube if collecting more than one 10 cm dish.

6. Use a 14 ml snap cap tube if collecting one 10 cm dish. The round bottom is preferable to the conical bottom of 15 ml conical tubes for generation of the cell pellet and results in better retrieval of the cells.

7. Flame polish the pipette to create smooth edges to the glass to prevent damage to the cells during resuspension. Bore should remain as large as possible to prevent damage to the cells.

8. A p1000 tip can be used; however, resuspension is more difficult, and cell yield tends to be lower than when a flame-polished glass pipette is used.

9. For tissue culture hoods without a gas line, a portable butane-powered microburner can be used. To sterilize, wipe with gauze wetted with 70% ethanol. Caution: ensure ethanol has completely evaporated prior to ignition as it is also flammable.

10. Unfiltered pipette tips can be used unless working with cells transduced with virus, in which case filtered tips along with the appropriate BSL2 procedures should be implemented.

11. This is optional. It is frequently easier to dispose of used pipettes and tips into a beaker while you work than into a sharp container.

12. TC 20™ automated cell counter (Bio-Rad) can also be used. For SCs, gating set to 8–13 μm works well and results in counts similar to the hemocytometer.

13. Used to monitor and regulate the rat's temperature. Should be set to 37 °C.

14. Used to sterilize instruments between rats. Be sure to follow the sterilization requirements outlined by IACUC at your institution.

15. We prefer isoflurane because it is short-acting. 5% isoflurane is used for induction; 2.5–3.5% isoflurane is used for maintenance. The level of isoflurane will depend on the size of the rat. The combination of ketamine (80 mg/kg) and xylazine (10 mg/kg) can be used, but rats must be awake and sternally recumbent before the monitoring of vital signs is discontinued.

16. Used to lubricate the probe inserted into the rat rectum for thermoregulation.

17. Bolster is used to elevate the rat and help separate the vertebrae while working. It can be made by rolling a stack of 4 × 4″ gauze and securing the resulting cylindrical-shaped gauze with tape. It should be approximately 2.5 cm in diameter.

18. The vertebral stabilizer can be made by creating a groove in the rod, inserting the end of the clamp into the rod, and securing the clamp to the rod by drilling a hole and affixing it with a screw and bolt. Rod diameter is dependent on the clamp to which it will be attached. We use a magnetic base from Kanetec (MB-B-DG8C).

19. Any stereotax with rail that fits the electrode manipulator arm can be used. A stereotax with a magnetic base plate is beneficial for working with the vertebral stabilizer using the magnetic base from Kanetec.

20. We use the electrode manipulator arm from Kopf (1460-61) and the corresponding upper bracket clamp (1770-C). Manipulator arm and stereotax rail size need to match for the arm to attach securely.

21. An old plunger from a 10 μl Hamilton® syringe is used to make an initial tunnel through the silicone plug to reduce the incidence of clogging of the Hamilton® syringe when inserted through the silicone plug.

22. Used to measure the volume of the cells in the Eppendorf tube to ensure correct cell dilution prior to transplantation. Also used for gently mixing cells between rats.

23. Documentation procedures are dependent on the requirements of the IACUC. Good scientific record keeping practices should be followed.

24. Used to prevent dehydration.

25. Used initially after surgery to maintain body temperature until anesthetic wears off. Line with blue pad and place food pellets in the bottom. A 50 ml conical tube with spout, velcroed to the side, can be used as a water bottle. Set to 35 °C.

26. Used for maintenance of thermoregulation during the acute recovery phase following injury. Cages should be placed half on, half off the heating pad to allow the rats the option to move to a warmer or cooler area.

27. This is more absorbent than corn cob bedding and helps absorb urine. It is beneficial for paralyzed rats.

28. Can be used for required environmental enrichment instead of shredded paper.

29. Food pellets mixed with water. Can be placed in a dish in the cage. Improves amount of food consumed. It helps prevent weight loss and decreases dehydration of the rats following injury.

30. Nutritive supplement to counteract weight loss. Given to rats that lose greater than 10% of body weight following surgery.

31. To facilitate hydration, a longer spout is needed for the water bottle to enable rats to reach their water.

32. This can be done with either a p1000 tip or a 1 ml serological pipette.

33. Typically, each vial of 3×10^6 cells is split into two dishes after thawing resulting in ~1.5×10^6 SCs being plated following freezing. This number of cells helps overcome any death that results from thawing. A more accurate number of cells plated can be achieved following thawing by counting the number of live and dead cells. The number of live cells will determine how fast the cells will grow upon plating. If the number of cells frozen was larger or smaller than expected, or the health of the cells was compromised, the cells may be over-confluent or under-confluent on the anticipated date of collection.

34. Different numbers of cells can be plated onto different dishes if cells are to be collected on different days for transplantation.

35. Although cells can be thawed and used directly for transplantation, it is preferable to thaw cells at an earlier passage, culture

them in SC growth medium, and split them on the day of SCI. This way if the cells grow slower or faster than anticipated upon thawing, confluent dishes with the needed number of cells can still be obtained on the day of transplantation.

36. This will dislodge the cells and create a single-cell suspension. A single-cell suspension when removing cells from the dish is important for obtaining accurate cell counts later.

37. Caution: pipettes will be hot. It is best to use forceps to keep the tips aseptic and prevent burns.

38. At least one good pipette is needed per animal transplanted. It is best to make many more than you think you will need.

39. Sealant comes in colored and clear. Colored sealant makes it easier to visualize the plug. Make sure the sealant is aquarium and not boat sealant. It is important to make sure that the sealant has no additives that might be toxic to the SCs. Silicone sealant can be purchased at most hardware or pet stores.

40. Needle should fit snuggly within capillary tube but still be able to rotate without resistance.

41. Make sure the silicone plug is complete. The bevel in the needle can result in an opening on one side. If not sealed completely, the cells will leak through the opening during injection.

42. Creating the appropriately sized plug takes practice. The plug should not be made too big or it will clog the Hamilton® syringe. Too much space between the plug and the pipette should be avoided as it will result in wastage of cells. Too little space between the plug and the pipette should also be avoided; otherwise the silicone plug is likely to plug the pipette when the Hamilton® syringe is inserted.

43. Tips can be sterilized by UV in a biosafety cabinet. Work cleanly and avoid touching the pipette tip to keep it aseptic.

44. The paper that wraps the sterile surgical gloves provides a useful sterile field. Gloves can be flipped to one side of the paper until surgery commences and they are needed.

45. This is used for removing blood and tissue from tools before placing them back down on the sterile field.

46. Careful monitoring of respiration is required during surgery, especially if anesthetic exposure is prolonged. Careful attention should be paid to the respiration of the rat throughout the surgical procedure. The color of blood, eyes, and hindpaws can also be used as an early indication of decreased respiration or cardiac failure. If problems with respiration are noted, decrease the level of isoflurane and allow the respiration rate to increase.

If necessary, a ventilation bag can be used to increase airflow and restore breathing. If the heart stops, the rat can frequently be resuscitated using chest compressions and the ventilation bag.

47. It is recommended to shave the rat on a blue bench pad and then vacuum the fur from the rat and the blue pad immediately to prevent dispersal of the fur.

48. It is important to keep rats warm during surgery. The temperature of the rat should be maintained at 37 °C. Due to the large surface-to-body-weight ratio, rodents tend to lose heat rapidly and should be kept warm. Heat loss occurs from the tail, ears, and feet. Loss of heat can significantly prolong the duration of anesthetic, which in turn increases the risk of complications. Although we prefer the homeothermic regulated heating pads, which both monitor and adjust to help maintain the rat's temperature, a hot water blanket, warm fluid bag, or glove filled with warm water can also be used.

49. T10 is the smallest vertebrae when moving rostrally from the V-shaped ligaments. It will be little and located between a round-shaped vertebral process rostrally (T9) and flat-shaped vertebral process caudally (T11). T10 is tiny relative to the other vertebrae and is sometimes covered by T11. T9 is round. T11 is flat and protrudes a little over T10. Use the two pairs of Adson forceps with 1 × 2 teeth to gently move T10 and T11 apart if there does not appear to be a small dorsal process. T10 can usually be identified by running the back (non-cutting side) of the scalpel blade gently along the dorsal vertebral process while visualizing the movement through the microscope. An alternative approach is to count caudally from T2 using a blunt probe.

50. Alternatively, smaller incisions (e.g., T8–T10) can be used as long as the incision is long enough for subsequent clamping in the IH device. It is essential that the length and location of the incision be kept consistent across animals in an experiment, particularly if motor function is to be assessed.

51. This will reduce the extent of muscle damage and facilitate surgical recovery.

52. T10/T11 can also be cut. If this is done, it should be done for all animals to maintain consistency, as it will impact the vertebral stability when the injury is induced and therefore has an influence on the resulting injury.

53. All the muscle needs to be removed from the lateral processes of T8 and T10 prior to placement into IH vertebral holders.

54. If successful, a small window in the dorsal vertebral process is created through which the underlying spinal cord is visible.

55. To preserve the small rongeurs, they should only be used for nipping small pieces of the bone. Between nips, any bone on the rongeurs should be removed by wiping them on the saline-soaked gauze.

56. The lateral process of T8 can be held by inserting the Adson forceps with 1 × 2 teeth. Hold gently, as excessive squeezing can result in unwanted damage to the spinal cord rostral to the planned site of injury. Alternatively, the Adson-Brown forceps, straight, shark teeth, can be used to gently clamp the dorsal process of T8. It is essential, however, not to damage the dorsal process of T8, as it is required for stabilization of the spinal cord at the time of transplantation.

57. T8 can project over a portion of the T9 vertebral process. Make sure all of T9 is removed.

58. If the spinal cord is damaged while performing the laminectomy, the rat should be excluded from the study.

59. If bleeding occurs during the laminectomy, use cotton-tipped swabs, absorbent triangles, or Gelfoam to stop the bleeding. If bleeding is extensive, soak up the blood using Gelfoam, and apply gentle pressure to the Gelfoam using the cotton-tipped applicator until the bleeding stops. It is important to have a clear view of the spinal cord while performing the laminectomy to avoid bruising the spinal cord.

60. Be careful not to extend the laminectomy too wide laterally, as damage to the lateral vertebral sinus will result in extensive bleeding.

61. This can be tricky! You must attach the rat by the lateral processes and not by the dorsal processes if you want consistent hits without slippage.

62. To prevent the holder from sliding and moving on the surgical table when clamping the rat, silicone feet can be affixed to the bottom of the plastic IH holder.

63. This step should be performed using the surgical microscope to ensure that no damage occurs to the spinal cord while clamping until one is an expert. Even then, best practice is to always use the surgical microscope.

64. If the muscle on the lateral process has not been completely cleared, you won't get a tight hold and slipping can result.

65. The forceps clamping T10 should be at an angle of 30°–50° to provide sufficient clearance for impactor probe.

66. Hand clamps are not provided with the device. They can be purchased at the hardware store. The hand clamps prevent the board from shifting.

67. Pick a consistent spot for the impact. The widest location at the center of the laminectomy site is frequently the preferred location.

68. It is important that the impactor probe be centered in the medial-lateral direction. If the probe is not centered, the resulting injury will cause more damage on one side than the other. Small differences in the ability to center the impact can have relatively large differences in the extent of tissue sparing and functional recovery. When learning, cut the tissue in cross section. This provides the best feedback on both the consistency and centering of the hits. Performing open field locomotor testing using the BBB can also be useful in determining if the injury is equivalent on both sides of the spinal cord.

69. The default force when the program opens is 100 kdyn. We typically use a moderate SCI of 200 kdyn when doing SC transplants.

70. At the end of the surgery session, print off the impact information, write the animal number on the printout, and file for your records. Information written in the comment section is not included in the printout.

71. The actual force should be within 10% of the desired force, and the change in force over time should have an even slope. Good quality hits are essential when performing studies in which recovery of function is a primary outcome measure.

72. The impactor does not automatically save the impact information. It is advised to write down the impact force.

73. If performing immediate transplantation, proceed to Subheading 3.6, **step 12**.

74. Sutures should be in a straight line and neat. The sutures are knotted by wrapping the sutures clockwise and then counterclockwise around the needle holder. If sutures are not wrapped in opposite directions, the knots will loosen and come undone.

75. Use three sutures for the deep muscle layer that was previously attached to the vertebral column. Place the sutures at T8, T9, and T10. These are used to landmark the laminectomy site at the time of re-laminectomy during the transplant surgery. Deep layers receive three wraps of suture thread in each direction.

76. We like to use a consistent number of sutures (typically 5) to know how many need to be removed at the subsequent transplant surgery. Superficial layers receive two wraps of suture thread in each direction.

77. When blood is present along the wound margin, the rat's cage mate will assist with removal of the blood, which can result in the removal of the wound clips, as well.

78. Both clips work well. Wound clips are faster but much more expensive than Michel clips.

79. The specifics will depend on the type of syringe pump used. For the nano-syringe pump (KD Scientific), set table to 700 (this indicates the type of Hamilton® syringe), rate to 1 μl/min, and volume to 6 μl. Make sure the arrow faces to the right. This indicates that it will advance. If the arrow faces left, it is set to withdraw.

80. If wound clips have not previously been removed, they can be removed either before or after anesthesia.

81. This may result in bleeding. Use Gelfoam or absorbent triangles to stop the bleeding before continuing.

82. Bent #5 forceps can be used to remove scar tissue along the edges of the bone at the laminectomy site.

83. If transplanting immediately following injury, proceed from here.

84. Retractors improve visibility and help create space to work at the injury site. Care should be taken not to spread the retractors too far, as this can cause the rat to stop breathing.

85. The rat should be resting on the bolster. The vertebral stabilizer is used to prevent the rat from shifting. The spinal cord should be parallel to the surgical table.

86. Use the plunger from an old Hamilton® syringe to make a tunnel through the silicone plug in the pulled glass pipette. Once attached, use a pair of #5 forceps to gently push against the tip of the pulled glass capillary to knock off the tip and increase the bore of the pipette. This results in a bore of approximately 50–100 μm in diameter for the cells to pass through. Alternatively, the pipettes can be beveled to a specific diameter in advance, if desired.

87. Use a fresh pipette for each animal.

88. SCs in the DMEM-F12 will settle to the bottom of the Eppendorf tube with time. Before collecting the SCs in the Hamilton® syringe, it is important to thoroughly mix the SCs with DMEM-F12. Swirl the mixture gently with p20 pipetman set to 10 μl, or gently flick the tube. The concentration of cells

is quite high, making the cells more vulnerable to physical damage.

89. Draw cells into the Hamilton® syringe. Then slowly expel the cell solution down to the 7 μl mark. Make sure the cells completely fill the glass capillary below the silicone plug. Any air bubbles will interfere with the flow of cells into the spinal cord.

90. To ensure sufficient cells at the time of transplantation, estimate needing 3×10^6 cells per rat. This will compensate for the cells left in the glass capillary tube at the end of injection. It is good practice to check the viability of an aliquot of cells before and after the transplantation session. Normally, greater than 95% cells are viable at the end of cell collection, and greater than 90% cells are viable at the end of the transplant session [29].

91. Work quickly once the cells are in the Hamilton® syringe. The cells will start to settle once the syringe is attached to the nano-pump injector.

92. If the impact was made at the center of the laminectomy site, the cyst will form primarily in the caudal two thirds of the laminectomy site and extend caudally below T10. In our experience, injection of cells into the spinal cord at the caudal one third of the laminectomy site results in transplants within the injury site.

93. Proceed quickly from this point to ensure that the tip does not become clogged. It is important the spinal cord is visible and free of blood before proceeding to this step.

94. The depth of the transplant will depend on the size of the spinal cord of the rat strain used. It is important that the transplant is centered in the injury site. If the injection is too shallow, the cells will be injected below the dura but not into the injury site. If the injection is too deep, the cells will be injected into the white matter ventrally.

95. When inserting the pipette for injecting cells, go slightly past the desired depth, and then withdraw the pipette to the correct depth. This helps to create a space for the cells to be injected. It also helps lift the spared white matter dorsally, which can collapse if the initial insertion of the pipette does not efficiently puncture the dura.

96. Occasionally the cells will stop moving. This is usually temporary and resolves on its own. If the cells do not appear to move, withdraw the pipette very slightly. If this does not resolve the problem, it may be necessary to completely withdraw the pipette to relieve the clog. A cotton-tipped applicator pre-wetted with saline can be used to gently remove the clog. Go back to **step 22** and proceed.

97. It is important to give the injection pressure time to equilibrate prior to removal of the pipette. Otherwise, some of the cells will leak out when the pipette is removed.

98. Watch the injection site when withdrawing the pipette to make sure cells are not expelled. If cells are present on the surface of the spinal cord, wipe them away with a cotton-tipped applicator.

99. Any problems that arise during the transplantation of the cells should be fully documented. This a priori information is necessary for the justification of exclusion of poor transplants that may result in experimental outliers.

100. Buprenorphine (0.05 mg/kg) is given immediately postsurgery. Additional doses are given twice a day for the first 2 days postsurgery to alleviate pain.

101. Gentamicin (5 mg/kg) is given immediately postsurgery. Additional doses are given once a day for the first 7 days postsurgery to prevent infections.

102. Lactated Ringer's (10 ml) is given immediately postsurgery. 5 ml is injected subcutaneously bilaterally by tenting the skin at the level of the hip. Additional lactated Ringer's is given one to two times per day for the first 2 days postsurgery to prevent dehydration. Lactated Ringer's is also given if there is blood in the urine.

103. Animals can experience extensive fluid loss during surgery. Fluid loss occurs primarily as a result of evaporation from body cavities and due to blood loss. Rodents, because of their small size and smaller total body fluid contents, are particularly vulnerable to intraoperative fluid loss.

104. Rats should be sternally recumbent (i.e., able to lift its head) before returning to their home cage. Rats can stay in incubator longer but should be provided with food and water.

105. Following SCI, the ability of the rats to thermoregulate is compromised. It is important that the rats do not become hypo- or hyperthermic. Cages filled with approximately 2 cm of bedding and placed on top of heating pads set to 33–34 °C help keep the rats warm. Cages should be placed half on and half off the thermoregulated heating pad to give the animals a choice of temperature. The temperature of the thermoregulated heating pads should be carefully monitored. Temperature strips can be used. The temperature of the heating pads should be recorded on a regular basis to ensure that they are working properly.

106. Following thoracic SCI, rats initially lose the ability to void their bladders and must have their bladders emptied for them. The bladder can be found by placing using the forefinger and

thumb between the hipbones and bringing them together. The middle and ring finger can then be placed against the wall of the bladder laterally. The bladder can then be expressed by gently pushing downward with the middle and ring finger while keeping the forefinger and thumb pinched together. A stream of urine should result. It may be necessary to reposition the fingers several times to ensure that the bladder is completely emptied. The bladders will vary in size from large (the size of a grape) to medium (the size of a blueberry) to small (the size of a green pea). Once a rat has a small bladder for several days, it can be switched to once a day bladder expression. If the bladder becomes medium or large again, return it to twice daily expression until a small bladder is present for several consecutive days. A tight bladder is a bladder that is very difficult to express. Caution should be taken not to squeeze too firmly too fast when the bladder is tight as this can cause the bladder to irreparably rupture. A warm wet cloth can be used to help relax a tight bladder to facilitate voiding. It is important that the bladders are completely emptied to prevent bladder infections. Even once rats regain bladder control, bladders should be checked at least once per day throughout the study to ensure complete bladder emptying. This will help prevent bladder infections.

107. Cloudy urine is a sign of an infection. Follow an approved treatment plan provided by your veterinarian.

108. Blood in the urine is sometimes observed. If observed, leave a minimal amount of residual urine in the bladder to prevent a clot from forming. Additional lactated Ringer's should be given until blood is no longer observed when the bladder is expressed.

109. Dehydration can be determined by pinching the skin and pulling up. If upon release the skin remains tented, the animal is dehydrated. Administer up to 10 ml of lactated Ringer's subcutaneously.

110. Porphyrin is a sign of stress. Wipe the rat's face with a damp cloth to remove.

111. Contact your veterinarian for a treatment plan if this occurs.

112. We find it helpful to fill in drug information at the time of surgery and then to have the person who administers the care sign off when the drugs/care is administered.

Acknowledgments

We would like to thank all of the investigators that have trained us over many years on the variety of techniques described. In particular we would like to acknowledge that the SC protocols are based on protocols, guidance, advice, and hands-on training provided by Yelena Pressman, Patrick Wood, and many other Bunge-Wood lab members over the years; the SCI protocols are based on training and guidance provided by Jacqueline Bresnahan, Amy Tovar, Aileen Anderson, Rebecca Nishi, and Chelsea Pagan; and the transplant protocols are based on training and guidance provided by Jacqueline Bresnahan, Gregory Holmes, Martin Oudega, and Andres Hurtado. We would also like to thank Sydney Agger for her artwork (Fig. 1) and Taylor Johns for her comments on the manuscript.

Conflict of Interest: The author has declared that no conflict of interest exists.

Funding: This work was supported by the Burke Foundation, White Plains, NY, and NIH Grant R01-NS075375.

References

1. Scheib J, Hoke A (2013) Advances in peripheral nerve regeneration. Nat Rev Neurol 9(12):668–676. https://doi.org/10.1038/nrneurol.2013.227

2. Painter MW, Brosius Lutz A, Cheng YC, Latremoliere A, Duong K, Miller CM, Posada S, Cobos EJ, Zhang AX, Wagers AJ, Havton LA, Barres B, Omura T, Woolf CJ (2014) Diminished Schwann cell repair responses underlie age-associated impaired axonal regeneration. Neuron 83(2):331–343. https://doi.org/10.1016/j.neuron.2014.06.016

3. Aguayo AJ, David S, Bray GM (1981) Influences of the glial environment on the elongation of axons after injury: transplantation studies in adult rodents. J Exp Biol 95:231–240

4. David S, Aguayo AJ (1981) Axonal elongation into peripheral nervous system "bridges" after central nervous system injury in adult rats. Science 214(4523):931–933

5. Paino CL, Fernandez-Valle C, Bates ML, Bunge MB (1994) Regrowth of axons in lesioned adult rat spinal cord: promotion by implants of cultured Schwann cells. J Neurocytol 23(7):433–452

6. Xu XM, Chen A, Guenard V, Kleitman N, Bunge MB (1997) Bridging Schwann cell transplants promote axonal regeneration from both the rostral and caudal stumps of transected adult rat spinal cord. J Neurocytol 26(1):1–16

7. Xu XM, Guenard V, Kleitman N, Bunge MB (1995) Axonal regeneration into Schwann cell-seeded guidance channels grafted into transected adult rat spinal cord. J Comp Neurol 351(1):145–160. https://doi.org/10.1002/cne.903510113

8. Takami T, Oudega M, Bates ML, Wood PM, Kleitman N, Bunge MB (2002) Schwann cell but not olfactory ensheathing glia transplants improve hindlimb locomotor performance in the moderately contused adult rat thoracic spinal cord. J Neurosci 22(15):6670–6681. 20026636

9. Imaizumi T, Lankford KL, Kocsis JD (2000) Transplantation of olfactory ensheathing cells or Schwann cells restores rapid and secure conduction across the transected spinal cord. Brain Res 854(1-2):70–78

10. Pinzon A, Calancie B, Oudega M, Noga BR (2001) Conduction of impulses by axons regenerated in a Schwann cell graft in the transected adult rat thoracic spinal cord. J Neurosci Res 64(5):533–541. https://doi.org/10.1002/jnr.1105

11. Barakat DJ, Gaglani SM, Neravetla SR, Sanchez AR, Andrade CM, Pressman Y, Puzis R, Garg MS, Bunge MB, Pearse DD (2005) Survival, integration, and axon growth support of glia transplanted into the chronically

contused spinal cord. Cell Transplant 14(4):225–240

12. Schaal SM, Kitay BM, Cho KS, Lo TP Jr, Barakat DJ, Marcillo AE, Sanchez AR, Andrade CM, Pearse DD (2007) Schwann cell transplantation improves reticulospinal axon growth and forelimb strength after severe cervical spinal cord contusion. Cell Transplant 16(3):207–228

13. Fortun J, Hill CE, Bunge MB (2009) Combinatorial strategies with Schwann cell transplantation to improve repair of the injured spinal cord. Neurosci Lett 456(3): 124–132. https://doi.org/10.1016/j.neulet. 2008.08.092

14. Bunge MB (2016) Efficacy of Schwann cell transplantation for spinal cord repair is improved with combinatorial strategies. J Physiol 594(13):3533–3538. https://doi. org/10.1113/JP271531

15. Xu XM, Zhang SX, Li H, Aebischer P, Bunge MB (1999) Regrowth of axons into the distal spinal cord through a Schwann-cell-seeded mini-channel implanted into hemisected adult rat spinal cord. Eur J Neurosci 11(5):1723–1740

16. Blakemore WF (1977) Remyelination of CNS axons by Schwann cells transplanted from the sciatic nerve. Nature 266(5597):68–69

17. Duncan ID, Aguayo AJ, Bunge RP, Wood PM (1981) Transplantation of rat Schwann cells grown in tissue culture into the mouse spinal cord. J Neurol Sci 49(2):241–252

18. Duncan ID, Hammang JP, Jackson KF, Wood PM, Bunge RP, Langford L (1988) Transplantation of oligodendrocytes and Schwann cells into the spinal cord of the myelin-deficient rat. J Neurocytol 17(3):351–360

19. Lankford KL, Imaizumi T, Honmou O, Kocsis JD (2002) A quantitative morphometric analysis of rat spinal cord remyelination following transplantation of allogenic Schwann cells. J Comp Neurol 443(3):259–274

20. Quencer RM, Bunge RP, Egnor M, Green BA, Puckett W, Naidich TP, Post MJ, Norenberg M (1992) Acute traumatic central cord syndrome: MRI-pathological correlations. Neuroradiology 34(2):85–94

21. Noble LJ, Wrathall JR (1987) An inexpensive apparatus for producing graded spinal cord contusive injury in the rat. Exp Neurol 95(2):530–533

22. Somerson SK, Stokes BT (1987) Functional analysis of an electromechanical spinal cord injury device. Exp Neurol 96(1):82–96

23. Basso DM, Beattie MS, Bresnahan JC (1996) Graded histological and locomotor outcomes after spinal cord contusion using the NYU weight-drop device versus transection. Exp Neurol 139(2):244–256. https://doi. org/10.1006/exnr.1996.0098

24. Scheff SW, Rabchevsky AG, Fugaccia I, Main JA, Lumpp JE Jr (2003) Experimental modeling of spinal cord injury: characterization of a force-defined injury device. J Neurotrauma 20(2):179–193. https://doi. org/10.1089/08977150360547099

25. Beattie MS, Bresnahan JC, Komon J, Tovar CA, Van M, Anderson DK, Faden AI, Hsu CY, Noble LJ, Salzman S, Young W (1997) Endogenous repair after spinal cord contusion injuries in the rat. Exp Neurol 148(2):453–463. https://doi.org/10.1006/exnr.1997.6695

26. Jakeman LB, Guan Z, Wei P, Ponnappan R, Dzwonczyk R, Popovich PG, Stokes BT (2000) Traumatic spinal cord injury produced by controlled contusion in mouse. J Neurotrauma 17(4):299–319. https://doi. org/10.1089/neu.2000.17.299

27. Pearse DD, Pereira FC, Marcillo AE, Bates ML, Berrocal YA, Filbin MT, Bunge MB (2004) cAMP and Schwann cells promote axonal growth and functional recovery after spinal cord injury. Nat Med 10(6):610–616. https:// doi.org/10.1038/nm1056

28. Hill CE, Brodak DM, Bartlett Bunge M (2012) Dissociated predegenerated peripheral nerve transplants for spinal cord injury repair: a comprehensive assessment of their effects on regeneration and functional recovery compared to Schwann cell transplants. J Neurotrauma 29(12):2226–2243. https:// doi.org/10.1089/neu.2012.2377

29. Hill CE, Hurtado A, Blits B, Bahr BA, Wood PM, Bartlett Bunge M, Oudega M (2007) Early necrosis and apoptosis of Schwann cells transplanted into the injured rat spinal cord. Eur J Neurosci 26(6):1433–1445. https://doi. org/10.1111/j.1460-9568.2007.05771.x

30. Guest J, Santamaria AJ, Benavides FD (2013) Clinical translation of autologous Schwann cell transplantation for the treatment of spinal cord injury. Curr Opin Organ Transplant 18(6):682–689. https://doi.org/10.1097/ MOT.0000000000000026

31. Tetzlaff W, Okon EB, Karimi-Abdolrezaee S, Hill CE, Sparling JS, Plemel JR, Plunet WT, Tsai EC, Baptiste D, Smithson LJ, Kawaja MD, Fehlings MG, Kwon BK (2010) A sys-

tematic review of cellular transplantation therapies for spinal cord injury. J Neurotrauma 28(8):1611–1682

32. Morrissey TK, Kleitman N, Bunge RP (1991) Isolation and functional characterization of Schwann cells derived from adult peripheral nerve. J Neurosci 11(8):2433–2442

33. Martin D, Robe P, Franzen R, Delree P, Schoenen J, Stevenaert A, Moonen G (1996) Effects of Schwann cell transplantation in a contusion model of rat spinal cord injury. J Neurosci Res 45(5):588–597

34. Hill CE, Moon LD, Wood PM, Bunge MB (2006) Labeled Schwann cell transplantation: cell loss, host Schwann cell replacement, and strategies to enhance survival. Glia 53(3):338–343. https://doi.org/10.1002/glia.20287

35. Haynes LW, Rushton JA, Perrins MF, Dyer JK, Jones R, Howell R (1994) Diploid and hyperdiploid rat Schwann cell strains displaying negative autoregulation of growth in vitro and myelin sheath-formation in vivo. J Neurosci Methods 52(2):119–127

Chapter 29

Schwann Cell Transplantation Methods Using Biomaterials

Christine D. Plant and Giles W. Plant

Abstract

Biomaterials can be utilized to assist in the transplantation of Schwann cells to the central and peripheral nervous system. The biomaterials can be natural or man-made, and can have preformed shapes or injectable formats. Biomaterials can play multiple roles in cellular transplantation; for example, they can assist with cellular integration and protect Schwann cells from cell death initiated by the lack of a substrate, an occurrence known as "anoikis." In addition, biomaterials can be engineered to increase cell proliferation and differentiation by the addition of ligands bound to the substrate. Here, we describe the incorporation of Schwann cells to both man-made and natural matrices for in vitro and in vivo measures relevant to Schwann cell transplantation strategies.

Key words Schwann cells, Biomaterials, Matrigel, Laminin, Collagen, Fibrinogen

1 Introduction

Man-made or naturally occurring biomaterials are being used in both the peripheral nervous system (PNS) and central nervous system in an attempt to repair deficits in tissue loss or induce tissue regeneration. In more recent times, biomaterials have been developed to improve the survival of cellular transplants and improve tissue integration. The design of each biomaterial must allow the cell type of choice to function correctly at molecular and physiological levels. Glial cell populations are one such type that have been successfully transplanted in conjunction with natural and man-made matrices. Adult Schwann cells (SCs) isolated from the PNS can be successfully integrated into a number of biomaterials; these materials can improve SCs survival, induce polarization, and allow long-term cellular integration. The composite of SCs and biomaterial scaffolds also have the ability to induce axonal regeneration, and these axons have the capacity to grow beyond the biomaterial–tissue interface.

The composition of biomaterials must be tailored to the organ or tissue of interest. Two major delivery approaches are injectable liquid matrices that gel in vivo and the placement of preformed

Paula V. Monje and Haesun A. Kim (eds.), *Schwann Cells: Methods and Protocols*, Methods in Molecular Biology, vol. 1739, https://doi.org/10.1007/978-1-4939-7649-2_29, © Springer Science+Business Media, LLC 2018

scaffolds. Fibrin is a gelling matrix that is created when thrombin is added to the glycoprotein fibrinogen; the soluble fibrinogen is converted into insoluble fibrin strands. Fibrin matrices can be delivered in vivo and are advantageous in that active peptide sequences can be inserted within the structure [1]. In addition, fibrin/SC injectable matrices can also be delivered to the damaged spinal cord without the use of thrombin as a gelling agent [2]. A more widely used biomaterial is Matrigel, consisting of a mixture of extracellular matrix molecules such as laminin and nidogen. This has been used to deliver transplants of SCs to the injured spinal cord [3–7] and peripheral nerve [8]. Collagen I is a natural structural matrix protein found within the peripheral nerve that can be used to incorporate cultured SCs [9]. All natural matrices rely on the use of temperature, pH, or the addition of gelling factors to change from a liquid to a gel. Man-made matrices, on the other hand, are preformed by molding polymers into specific shapes to provide a fixed structure prior to transplantation. These man-made structures provide adhesion surfaces for cells and can have either random or aligned channels needed for optimal cellular infiltration and axonal growth. One such substance is Poly(2-hydroxyethyl methacrylate), known as polyHEMA [10]. It is a plastic that can absorb many times its weight in water and is the same hydrogel that is used to make soft contact lenses. The polyHEMA matrix can incorporate SCs by absorption and then subsequently be transplanted into the central nervous system [11, 12]; the structure provides both a mode of delivery and protection for the transplanted SCs (*see* Fig. 1).

In this Chapter, we provide a protocol for the preparation of natural and synthetic matrices and polymers mixed with SCs. We describe the step-by-step process of how to grow SCs in culture, mix them with matrices, or rehydrate polymers with culture medium containing SCs so as to integrate them into the internal porous network. The protocols describe the transplantation of these structures and the post-transplant analysis in the animal. Advantages of the natural compounds are the ease of extraction, purity, degradability, ease of injection, and simplicity for mixing them with SCs. Disadvantages of these matrices can be the animal source of some of the materials (Matrigel) and the need for gelling agents (thrombin for fibrinogen). Synthetic polymers such as poly-HEMA have great advantages, because of the ease of production and the large quantities that can be made. They are biocompatible, highly porous structures that can be handled easily and, when transplanted, allow cellular infiltration. However, there are some drawbacks in using synthetic polymers such as polyHEMA, for example: (1) its nondegradable composition, (2) the presence of random blind-ended channels within the material which can prevent some regenerating axons from exiting, and (3) the polymer also requires additional coating of the surface and interior porous network for SCs to adhere.

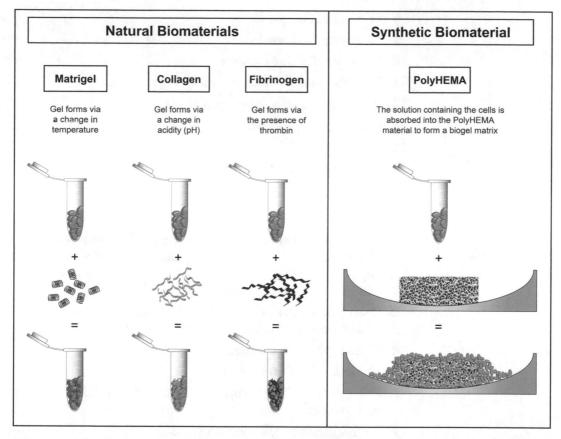

Fig. 1 Schematic representation of natural and synthetic biomaterials combined with purified adult SCs for spinal cord transplantation. The natural biomaterials are mixed with SCs (aqua) under cold conditions (4 °C), and the matrices are gelled via temperature (Matrigel; pink) , pH (collagen I; yellow), and the use of gelling agents such as thrombin (fibrinogen; purple). Mixtures of each biomaterial and SCs are injected into the spinal cord while cold (4 °C). Synthetic biomaterials such as polyHEMA are preformed and cut into required shapes using microscissors. These polymer shapes are placed into culture dishes or flasks and then SCs (in suspension) are added. Over a period of 7–9 days at 37 °C, the SCs will be incorporated within the synthetic structure. The polyHEMA/SC matrices can then be placed into the injury zone within the spinal cord

2 Materials

2.1 Preparation of Schwann Cells from Adult Peripheral Nerves

1. Fetal bovine serum (FBS).

2. 50 mg/ml gentamicin. Prepared in distilled water.

3. Leibowitz's L-15 medium (L-15).

4. Dulbecco's Modified Eagle's Medium (DMEM).

5. L-15/FBS medium. L-15 containing 15% FBS.

6. D-10S medium. DMEM containing 10% FBS and 0.1% gentamicin.

7. Beuthanasia®. Dosage per animal depends on institutional guidelines.

8. 35 mm cell culture dishes.

9. 60 mm cell culture dishes.

10. 100 mm cell culture dishes.

11. Sterilized forceps, scissors, and spring scissors.

12. Sterile graduated serological pipettes.

13. Bent tip glass pipettes.

14. 70% ethanol.

15. Adult female rats. 180–200 g weight or approximately 8 weeks old.

2.2 Dissociation of Adult Peripheral Nerve Explants

1. D-10S medium.

2. Dispase/collagenase digestion solution. DMEM containing 15% FBS, 1.25 U/ml of dispase, and 0.05% collagenase. Add dispase and collagenase to DMEM, and then filter sterilize the solution before adding the FBS.

3. D-10S/mitogens medium. D-10S medium containing 2 μM forskolin and 20 μg/ml pituitary extract (PEX).

4. 35 mm cell culture dishes.

5. 100 mm cell culture dishes.

6. Poly-L-lysine (PLL) solution. Dissolve 20 mg/ml of PLL in sterile water.

7. Hank's balanced salt solution (HBSS).

8. Ca^{2+}/Mg^{2+}-free HBSS (CMF-HBSS).

9. 0.05% trypsin/EDTA solution. Thermo Fisher Scientific, Cat # 25300054.

10. Sterile water.

11. 15 ml centrifuge tubes.

12. Graduated plastic pipettes.

13. Glass pipettes.

2.3 Schwann Cells and Natural Biomaterials

1. DMEM.

2. 18–21 mg/ml Matrigel protein. Store in 10 ml aliquots. It is recommended to use high-concentration Corning Matrigel basement membrane matrix, Cat # 354248 for in vivo studies.

3. Matrigel/DMEM solution. DMEM containing 30% (v/v) Matrigel protein. Prepare immediately before use and keep on ice.

4. 50 mg/ml gentamicin. Prepared in distilled water.

5. 0.04 M acetic acid. Prepared in distilled water (pH 2.9).

6. Collagen I solution. 0.04 M acetic acid containing 3.7 mg/ml of collagen (from the rat tail, Thermo Fisher Scientific, Cat # A1048301). This solution can be kept at 4 °C until used.

7. 15 IU/ml aprotinin. Prepared in distilled water.

8. 100 mg/ml fibrinogen type 1. Mix 100 mg fibrinogen into 1 ml DMEM. It is recommended to use fibrinogen type 1 from human plasma (Sigma-Aldrich, Cat# F3879).

9. Fibrinogen/DMEM. DMEM containing 1% fibrinogen type 1, 15 IU/ml aprotinin, 0.05 mg/ml gentamicin, and 8 mM calcium chloride ($CaCl_2$).

10. 25 units/μl Thrombin. Mix 10,000 units of thrombin into 400 μl of DMEM. The use of thrombin from bovine plasma Sigma-Aldrich, Cat # T6884 is recommended.

11. 8 mM $CaCl_2$ solution. Prepared in distilled water.

12. 10 × HBSS.

13. 1 × HBSS.

14. HBSS–NaOH solution. Mix together 2 ml 10 × HBSS and 1 ml NaOH; filter sterilize and keep on ice until needed.

15. FBS.

16. L-15/FBS medium. Leibovitz's L-15 medium containing 15% FBS.

17. 0.34 M sodium hydroxide (NaOH). Prepared in distilled water.

18. 70% ethanol.

19. Sterilized forceps and scissors.

20. Parafilm.

21. Small and large transfer pipettes.

22. 35 mm cell culture dishes.

23. 100 mm cell culture dishes.

24. Glass Petri dish. Sterile.

25. 2.6 mm gauge metal needle. Sterile.

26. 3 ml plastic syringe. Sterile.

27. Polymer tube molds. 60:40 polyacrylonitrile/polyvinylchloride copolymer (PAN/PVC) tubes or similar.

28. Fibrin sealant. Ethicon, Cat # 3901.

29. Scalpel blade #15.

30. 2.5 ml test tubes.

31. Microcentrifuge tubes.

32. D-10S medium. DMEM containing 10% FBS and 0.1% gentamicin.

33. 0.4% trypan blue.

2.4 Schwann Cells and Man-Made Biomaterials

1. HEMA (2-hydroxyethyl methacrylate; $C_6H_{10}O_3$). Monomer (Sigma-Aldrich, Cat # 128635).

2. Cross-linking agent EDMA (ethylene dimethacrylate; $C_{10}H_{14}O_4$). Acros Organics (Thermo Fisher Scientific, Cat # AC409922500).

3. Polymerization initiators. Ammonium persulfate ($(NH_4)_2S_2O_8$) and sodium metabisulfite ($Na_2S_2O_5$). Both used as 6% (w/v) aqueous solutions in distilled water.

4. 0.04 M acetic acid. Prepared in distilled water (pH 2.9).

5. Collagen I solution.

6. Saline solution. 0.9% sodium chloride (NaCl) prepared in distilled water.

7. Molds. Choose appropriate sizes based on research and/or transplantation requirements.

8. Airtight unit.

9. Measuring beaker.

10. Distilled water.

2.5 Animal Surgery

1. Saline solution or lactated Ringer's solution.

2. Analgesic gas. 40:60 mixture of O_2:N_2O.

3. General anesthetic. 1.5% isoflurane.

4. 2% lidocaine. The 2% lidocaine solution is a liquid that is directly dropped onto the spinal cord using a 1 ml syringe, prior to cutting. Recommended use is 2–4 mg/kg.

5. Penicillin G potassium solution or Pfizerpen®. Recommended use is 0.1 ml or 37.6 units per 200 g animal weight. The penicillin solution is injected intraperitoneally (IP) using a 1 ml syringe with a 26 gauge needle.

6. Ophthalmic lubricant ointment.

7. Antiseptic solution. It is suggested to use 80% ethanol for wound washing.

8. Fine forceps #5 (Fine Science Tools).

9. Durafilm or similar film.

10. Adult female rats. Animals should be 180–200 g weight or approximately 8 weeks old.

3 Methods

3.1 Preparation of Schwann Cells from Adult Peripheral Nerves

3.5.1 Animal Surgery

1. Prepare the laminar flow hood for the procedure. Take out all the instruments needed, and remember to take forceps out of the 70% ethanol.

2. Set up 3 × 100 mm cell culture dishes, and place one 60 mm cell culture dish into one of the 100 mm cell culture dishes.

3. Aliquot 30 ml of L-15/FBS medium into each of the two 100 mm cell culture dishes and 5 ml of L-15/FBS in the 60 mm cell culture dish.

4. Anesthetize the rats with Beuthanasia®; shave the hair from the rear hind limb area and clean with soap, water, and 70% ethanol.

5. Excise the sciatic nerves from the rats. The sciatic nerve branch closest to the spinal cord should be cut first and then lift the nerve carefully and cut the branch near to the knee to give a long piece of peripheral nerve.

6. All excised nerves are placed into one of the 100 mm cell culture dishes.

7. Use a different set of forceps and scissors for each layer tissue that is cut through (skin, muscle, and nerve).

8. Euthanize the rats and safely dispose of them in accordance with institutional guidelines.

3.1.1 Excision of Peripheral Nerves

1. Remove any excess tissue from the nerves.

2. Using fine forceps (such as #5 forceps) dedicated to this procedure, remove any muscle, fat, or blood vessels that were taken with the nerve (*see* **Note 1**).

3. Put the cleaned nerve into a new 100 mm cell culture dish with approximately 7 ml of L-15/FBS medium.

4. Remove the epineurium from the nerve and place the nerve into the 60 mm cell culture dish. Always use fine forceps for this, and start at the end that is the most proximal to the spinal cord.

5. Slightly squeeze the nerve end to expose the nerve and determine where the epineurium is to grasp with forceps.

6. If fraying of the nerve occurs and the end of the nerve cannot easily be distinguished, cut off a small piece and again try to pull the epineurium away.

7. When the epineurium is removed from all the nerves, cut the nerve into approximately 1 mm square pieces using spring scissors.

8. Set up three of the 35 mm cell culture dishes in one of the 100 mm cell culture dishes, and aliquot approximately 0.75 ml of D-10S medium into each 35 mm cell culture dish. Prepare a sufficient number of 35 mm cell culture dishes so that 10–15 pieces of the nerve can be placed in each dish.

9. Using a bent tip glass pipette, place 10–15 pieces of the nerve in each 35 mm cell culture dish and remove any excess media. Leave enough media so that the explants are not floating in the dish, but are still covering the explants completely.

10. These cultures should be fed twice a week with D-10S medium and transferred to new 35 mm cell culture dishes containing fresh medium approximately every week. They should be transferred when a large amount of outgrowth from the explant is observed. The segments will be dissociated when the majority of the migrating cells are SCs.

3.2 Dissociation of Adult Peripheral Nerve Explants

3.2.1 Day 1: Enzymatic Digestion

1. Set up a 35 mm cell culture dish in a 100 mm cell culture dish.

2. Transfer all the explants from one set (i.e., all the explants from one preparation) into a single 35 mm cell culture dish, and then add 1 ml of the dispase/collagenase digestion solution.

3. Place the explants in their digestion solution in the CO_2 incubator overnight.

3.2.2 Day 2: Explant Dissociation

1. Prepare PLL-coated 100 mm cell culture dishes for plating out the cells at the end of the procedure. Coat the 100 mm cell culture dishes with approximately 2–5 ml of PLL solution and incubate in the laminar flow hood for 30 min. Rinse the 100 mm cell culture dishes with sterile water at least three times to remove the excess PLL solution prior to use.

2. Remove digestion enzymes (from Subheading 3.2.3) by transferring the solution containing the explants into a 15 ml centrifuge tube containing 1 ml of D-10S medium.

3. Rinse the 35 mm cell culture dish that contained the explants with 2 ml of D-10S medium to be certain there is no material left on the dish.

4. Centrifuge the cells at $300 \times g$ for 10 min at 4 °C.

5. Remove the supernatant, resuspend the cells in 2 ml of D-10S medium with a small-bore pipette, and then add more D-10S medium to make a total of 5 ml.

6. Centrifuge the cells at $300 \times g$ for 5 min at 4 °C.

7. Repeat **steps 4** and **5** (approximately three times) until the supernatant looks clear. The cells are then ready to be plated out in the PLL-coated dishes prepared in **step 1**.

8. Resuspend the cells for plating in D-10S/mitogens medium within the 15 ml centrifuge tube. Use approximately 2 ml to mechanically triturate the pellet, and then add 2 ml of D-10S/mitogens medium for every 100 mm cell culture dish that is required (e.g., add 16 ml of medium for a total of eight dishes).

9. Plate 2 ml of the resuspended cells into the PLL-coated dishes which contain 5 ml of the D-10S/mitogens medium.

10. Place cells into the CO_2 incubator.

11. Feed the cells two times per week with D-10S/mitogens medium using approximately 7 ml per dish. When feeding the cultures, make sure to tilt the dish and remove the media to the side of the dish; this will avoid disruption of the cells.

12. Split the cells by trypsinization once the SCs have reached confluence and the density reaches to $4–5 \times 10^6$ cells per dish. They can then be passaged 1:2 if they are required for immediate use.

13. Wash the SCs cultures three times using CMF-HBSS to remove any remnants of the feeding medium.

14. Add approximately 2 ml of 0.05% trypsin/EDTA solution per dish and incubate for 5 min at 37 °C.

15. Add 2 ml of D-10S medium to stop the enzymatic reaction.

16. Dislodge the cells using mechanical trituration through a sterile glass pipette, until all cells are dislodged.

17. Transfer the cells to a 15 ml tube and add approximately 8 ml of D-10S medium.

18. Pellet the SCs by low-speed centrifugation at $300 \times g$ for 5 min.

19. Count the cells using a hemocytometer to ensure there are enough cells to use in subsequent experiments.

20. The SCs should be split when the cells have reached confluence and frozen when there is an overabundance of them.

3.3 Schwann Cells and Natural Biomaterials

Natural biomaterials can be surgically placed as a gel or injected, following spinal cord transection or contusion injuries. Natural biomaterials can be used as a liquid injectable and provide a less invasive option for spinal cord contusion injuries where the wound is surrounded by intact white matter. A mixture of natural biomaterials and SCs can be combined to produce an active cellular construct.

3.3.1 Schwann Cell Suspensions

1. Take the 100 mm cell culture dishes containing SCs out of the incubator.

2. Wash the SCs cultures three times using CMF-HBSS to remove any remnants of the feeding medium.

3. Add approximately 2 ml of 0.05% trypsin/EDTA solution per dish and incubate for 5 min at 37 °C.

4. Add 2 ml of D-10S medium to stop the enzymatic reaction.

5. Dislodge the cells using mechanical trituration through a sterile glass pipette, until all cells are dislodged.

6. Transfer the cells to a 15 ml tube and add approximately 8 ml of D-10S medium.

7. Pellet the SCs by low-speed centrifugation at $300 \times g$ for 5 min.

8. Count the cells using a hemocytometer to ensure there are enough cells to mix with the matrix being used (collagen I, fibrinogen, or Matrigel).

3.3.2 Schwann Cells and Matrigel

1. Keep the freshly prepared Matrigel/DMEM solution and the SC suspension on ice at all times prior to use.

2. Using a micropipette, gently mix the SCs with the Matrigel/ DMEM solution at a final density of 120×10^6 cells per ml (*see* **Note 2**).

3. Spray a piece of Parafilm with 70% ethanol to rapidly cool it. Then, place the Parafilm into the center of a sterile glass Petri dish on ice.

4. Pipette the Matrigel/SC matrix onto the cold Parafilm. The matrix mixture will immediately bead into a dome shape.

5. Using a metal needle attached to a plastic syringe, draw up the Matrigel/SC matrix solution into a polymer tube mold.

6. Plug the polymer tubes with fibrin sealant and store overnight at 37 °C (*see* **Note 3**).

7. The Matrigel/SCs matrix can then be removed from the polymer tube by cutting each end of the polymer tube with microscissors or using a scalpel blade (#15).

8. The Matrigel/SCs matrix implant can be stored in a 60 mm cell culture dish placed into a 37 °C incubator (surrounded by DMEM).

9. If used immediately for transplantation surgery, the Matrigel/ SCs matrix can be placed on ice inside a 60 mm cell culture dish containing L-15/FBS medium. The 60 mm cell culture dish containing the implant is placed within a 100 mm culture dish to enable easy handling and transported on ice.

10. Proceed as described in Subheading 3.5.

3.3.3 Schwann Cells and Collagen I

1. Pipette 1–2 ml collagen I into a 2.5 ml test tube, and then add 150–200 μl HBSS–NaOH solution drop by drop. Vortex after each addition, and keep the mixture as cold as possible throughout the process.

2. Stop adding the HBSS–NaOH solution when it becomes neutralized (as evidenced by the solution changing to a light pink color) after vortexing. Keep the neutralized solution on ice.

3. Resuspend the SCs in cold L-15/FBS medium to give a final cell density of 4×10^4 SCs per μl.

4. Pipette 70 μl of cold $1 \times$ HBSS into another labeled microcentrifuge tube; prepare as many tubes as needed.

5. Add 30 μl of the ice-cold neutralized collagen I solution to each microcentrifuge tube containing $1 \times$ HBSS; ensure that each tube is well mixed without incorporating air bubbles.

6. Add 4×10^5 SC (i.e., approximately 10 μl) per microcentrifuge tube.

7. Before using the collagen I/SCs mixture and to check SCs viability after collagen I cross-linking, remove a 10 μl sample from each tube and place each sample onto a designated area into a 35 mm cell culture dish. Because each aliquot should immediately start to gel in place, several samples can be placed into one dish.

8. Place the 35 mm cell culture dish into a larger 100 mm cell culture dish and incubate for 30 min at 37 °C.

9. Then add approximately 2 ml of D-10S medium to the 35 mm cell culture dish and incubate for 30 min at 37 °C.

10. Check each aliquot for cell viability using trypan blue.

11. Once cell viability is confirmed, collagen/SCs matrix can be transported on ice for transplantation.

12. Proceed as described in Subheading 3.5.

3.3.4 Schwann Cells and Fibrinogen

1. Remove the SCs from a confluent plate using approximately 2 ml of 0.05% trypsin/EDTA solution.

2. Pellet the SCs by centrifugation at $300 \times g$ for 5 min, and then dilute in a defined volume of L-15/FBS medium or $1 \times$ HBSS and perform a cell count.

3. Mix 4×10^5 SCs in 5 μl of ice-cold fibrinogen/DMEM.

4. Keep the suspension cold on ice and inject using a Hamilton needle that is also kept as cold as possible (*see* **Notes 4** and **5**).

5. Proceed as described in Subheading 3.5.

3.4 Schwann Cells and Man-Made Biomaterials

The man-made biomaterials can be produced in molds that can be shaped to fit lesion sites in the injured spinal cord. These materials can be combined with cultured SCs to form a biohybrid construct.

3.4.1 Schwann Cells and PolyHEMA

1. To prepare the polyHEMA sponges, the method described by Chirila and colleagues in 1993 is used as following [10]:

2. Mix HEMA (20% wt) and EDMA (0.5% wt of monomer) together in a measuring beaker.

3. Add distilled water (80% wt) to the beaker.

4. Add 6% wt aqueous solutions of ammonium persulfate (0.12% wt of monomer) and sodium metabisulfite (0.12% wt of monomer) to the beaker, and mix all components together well.

5. Place this mixture into the appropriate molds and place the molds into an airtight unit.

6. Replace the air in the unit with nitrogen. Then, place the sealed unit into a water bath. Perform three 10-h incubation cycles at the following temperatures: 30 °C, 40 °C, and 50 °C.

7. After the three temperature cycles are completed, remove the sponges and place them into deionized water for storage before sterilization. Change the water twice daily (*see* **Note 6**).

8. Autoclave the polyHEMA sponges at 130 °C for 20 min before immersing them in a collagen I solution (*see* **Note 7**).

9. Incubate the polyHEMA sponge pieces for 24 h on a shaker at 4 °C.

10. Following this, incubate the polyHEMA sponge pieces in DMEM at 37 °C for 1 h to allow cross-linking of collagen, and then transfer them to saline solution.

11. Place the polyHEMA sponge pieces into sterile culture dishes or flasks (*see* **Notes 8** and **9**).

12. Add SCs in suspension at the desired concentration to enable cells to be absorbed by the material or attach to the surface.

13. Maintain the sponges in vitro for 6–9 days in 10% FBS/DMEM medium prior to transplantation (*see* **Note 10**).

14. Proceed as described in Subheading 3.5.

3.5 Transplantation of Natural or Man-Made Biomaterials

A complete transection spinal cord injury model is ideally suited for the transplantation of preformed matrices such as collagen I, fibrinogen, Matrigel, or polyHEMA [5, 11–14]. Natural biomaterials can also be utilized in a contusion model of spinal cord injury due to their injectable properties. These biomatrices, however, must be injected prior to gelling.

1. Anesthetize the rats with a mixture of analgesic gas and general anesthetic.

2. To prevent wound and bladder infections, Pfizerpen® should be administered subcutaneously.

3. An ophthalmic ointment should also be applied to the eyes to prevent drying and/or infection.

4. The rats should be placed onto a heating pad to maintain body temperature.

5. Perform a multilevel laminectomy to expose the T8–T10 thoracic spinal cord segments. Cut the dura longitudinally and laterally at proximal and distal borders of the laminectomy.

6. Lidocaine should be administered onto the exposed spinal cord, followed by a transection at T9 and removal of a 2 mm segment.

7. Carefully place or inject the transplant material of choice prepared in Subheadings 3.3 and 3.4 to the proximal and distal spinal cord stumps, ensuring that the stumps are either covered by the material or are touching the material if using preformed transplant material.

8. Cover the exposed spinal cord with Durafilm (or similar). Suture the muscles and skin carefully, and then wash the wound with an antiseptic solution.

9. Administer the appropriate preoperative care, including the subcutaneous injection of 5 ml lactated Ringer's solution or saline solution.

10. Administer the antibiotic Pfizerpen®, and perform bladder expression on the rats for 7–10 days.

11. Analyze the transplants after periods of 6–8 weeks using paraformaldehyde fixation and immunohistochemistry [14, 15].

4   Notes

1. It is best to use a dedicated pair of fine forceps solely for this procedure, as common utility forceps easily lose their sharpness.

2. If using Matrigel/SC matrix for contusive injuries where a preformed cable is not required, ensure that all cells and mixes are kept on ice until injection. Needles must also be kept as cold as possible.

3. The SCs will naturally become aligned within the Matrigel matrix. This matrix will contract overnight in the incubator and separate from the tube side walls, enabling it to be easily removed the next day.

4. If using fibrinogen and SCs and injecting without preforming a gel, then take precautions to keep all items as cold as possible, including the Hamilton needle.

5. Thrombin can be used to quickly gel fibrinogen if needed (25 units/µl). If using thrombin, be sure to use a separate needle that does not come into contact with the fibrinogen and the SC mixture. Continual flushing of the needles reduces syringe clogging at future injections.

6. In their fully hydrated state, the polyHEMA sponges have an equilibrium water content of 73% weight, a permeability coefficient of 3.22×10^{-11}, and pore diameters between 10 and 30 µm [10, 12].

7. For optimized SC binding and integration into the polyHEMA structure, the sponges can be pretreated with a collagen I (or collagen IV) solution. This coating allows the matrix to adhere to the polyHEMA surface and provide a cellular matrix anchor for the cell to attach and extend its cytoplasm. Interestingly collagen IV will provide a surface that improves SC differentiation into a myelinating phenotype, whereas collagen I keeps SCs in a nonmyelinating phenotype. Depending on the

researchers needs for the SC biomaterial implant either for myelination or axonal regeneration, the use of different substrate coating will play a major part in the successful outcome.

8. PolyHEMA sponges can be carefully picked up by using a pair of fine # 4 forceps. Care should be taken to carry the pieces very gently and to not squeeze the structure. The same principle is important when placing constructs of polyHEMA and SCs in vivo.

9. The culture dishes or flasks that the polyHEMA sponges are to be placed in can be pre-coated with a PLL solution in order to enhance adherence of the sponge and SCs.

10. The SCs grow well within the polyHEMA structure [12] and are evenly spaced. The sponge is not degradable and therefore provides a continual surface for attachment. Proliferation of the SCs can be monitored by the use of proliferation antibodies or analogs to label proliferation such as bromodeoxyuridine. In vivo, the SCs continue to survive.

References

1. Schense JC, Bloch J, Aebischer P, Hubbell JA (2000) Enzymatic incorporation of bioactive peptides into fibrin matrices enhances neurite extension. Nat Biotechnol 18(4):415–419. https://doi.org/10.1038/74473

2. Oudega M, Xu XM (2006) Schwann cell transplantation for repair of the adult spinal cord. J Neurotrauma 23(3-4):453–467. https://doi.org/10.1089/neu.2006.23.453

3. Chen A, Xu XM, Kleitman N, Bunge MB (1996) Methylprednisolone administration improves axonal regeneration into Schwann cell grafts in transected adult rat thoracic spinal cord. Exp Neurol 138(2):261–276. https://doi.org/10.1006/exnr.1996.0065

4. Oudega M, Xu XM, Guenard V, Kleitman N, Bunge MB (1997) A combination of insulin-like growth factor-I and platelet-derived growth factor enhances myelination but diminishes axonal regeneration into Schwann cell grafts in the adult rat spinal cord. Glia 19(3):247–258

5. Plant GW, Bates ML, Bunge MB (2001) Inhibitory proteoglycan immunoreactivity is higher at the caudal than the rostral Schwann cell graft-transected spinal cord interface. Mol Cell Neurosci 17(3):471–487. https://doi.org/10.1006/mcne.2000.0948

6. Xu XM, Chen A, Guenard V, Kleitman N, Bunge MB (1997) Bridging Schwann cell transplants promote axonal regeneration from both the rostral and caudal stumps of transected adult rat spinal cord. J Neurocytol 26(1):1–16

7. Xu XM, Guenard V, Kleitman N, Bunge MB (1995) Axonal regeneration into Schwann cell-seeded guidance channels grafted into transected adult rat spinal cord. J Comp Neurol 351(1):145–160. https://doi.org/10.1002/cne.903510113

8. Guenard V, Kleitman N, Morrissey TK, Bunge RP, Aebischer P (1992) Syngeneic Schwann cells derived from adult nerves seeded in semipermeable guidance channels enhance peripheral nerve regeneration. J Neurosci 12(9):3310–3320

9. Kromer LF, Cornbrooks CJ (1985) Transplants of Schwann cell cultures promote axonal regeneration in the adult mammalian brain. Proc Natl Acad Sci U S A 82(18):6330–6334

10. Chirila TV, Constable IJ, Crawford GJ, Vijayasekaran S, Thompson DE, Chen YC, Fletcher WA, Griffin BJ (1993) Poly(2-hydroxyethyl methacrylate) sponges as implant materials: in vivo and in vitro evaluation of cellular invasion. Biomaterials 14(1):26–38

11. Plant GW, Chirila TV, Harvey AR (1998) Implantation of collagen IV/poly(2-hydroxyethyl methacrylate) hydrogels containing Schwann cells into the lesioned rat optic tract. Cell Transplant 7(4):381–391

12. Plant GW, Harvey AR, Chirila TV (1995) Axonal growth within poly (2-hydroxyethyl

methacrylate) sponges infiltrated with Schwann cells and implanted into the lesioned rat optic tract. Brain Res 671(1):119–130

13. Joosten EA, Bar PR, Gispen WH (1995) Collagen implants and cortico-spinal axonal growth after mid-thoracic spinal cord lesion in the adult rat. J Neurosci Res 41(4):481–490. https://doi.org/10.1002/jnr.490410407

14. Plant GW, Christensen CL, Oudega M, Bunge MB (2003) Delayed transplantation of olfactory ensheathing glia promotes sparing/regeneration of supraspinal axons in the contused adult rat spinal cord. J Neurotrauma 20(1):1–16. https://doi.org/10.1089/08977150360517146

15. Barbour HR, Plant CD, Harvey AR, Plant GW (2013) Tissue sparing, behavioral recovery, supraspinal axonal sparing/regeneration following sub-acute glial transplantation in a model of spinal cord contusion. BMC Neurosci 14:106. https://doi.org/10.1186/1471-2202-14-106

Chapter 30

Viral Transduction of Schwann Cells for Peripheral Nerve Repair

Christine D. Plant and Giles W. Plant

Abstract

Schwann cells are the primary inducers of regeneration of the peripheral nervous system. Schwann cells can be isolated from adult peripheral nerves, expanded in large numbers, and genetically transduced by viral vectors in vitro prior to their use in vivo. Here we describe how to use lentiviral vectors to transduce primary Schwann cells in vitro. We also describe how cultured Schwann cells can be used in conjunction with decellularized peripheral nerve sheaths prepared by multiple freeze thawing of peripheral nerve tissue. This process depletes all native cells from the nerve sheath but maintains basal lamina integrity and flexibility. A major advantage of using these decellularized nerve sheaths in repair strategies is that they can be obtained from cadaveric tissue and therefore do not require patient matching because the immune response is generated from the intrinsic cells and not the sheath itself. The patient's own cells can then be used to repopulate the decellularized peripheral nerve sheath. Our technique described in this chapter uses decellularized nerve sheaths which are repopulated with extrinsic Schwann cells previously grown in vitro. The Schwann cells can also be engineered in multiple ways, for example, to secrete bioactive proteins beneficial to axonal regeneration.

Key words Lentiviral vectors, Schwann cells, Regeneration, Peripheral nerve, Neurotrophins, Transduction

1 Introduction

Peripheral nerve repair by surgical coaptation is the standard approach after nerve injury. If a larger gap is repaired, then surgical placement of an autograft from the patient's sural nerve is a standard practice. However this has limited effectiveness if the gap is too large, if there is significant degeneration of the distal stump or if there is a failure to recruit Schwann cells (SCs). Harvesting these sural nerves can have additional associated problems, such as neuroma. If allografts are used instead of autografts, the patient will require continual administration of immunorejection drugs, and this frequently does not prevent long-term rejection. A relatively new approach used in peripheral nerve repair is the use of chimeric peripheral nerve grafts which can be composed of essential

Paula V. Monje and Haesun A. Kim (eds.), *Schwann Cells: Methods and Protocols*, Methods in Molecular Biology, vol. 1739, https://doi.org/10.1007/978-1-4939-7649-2_30, © Springer Science+Business Media, LLC 2018

growth-inducing SCs and intrinsic matrix support. Endogenous cells within the peripheral nerve are the primary immunogenic component, and these can be successfully removed by a process of freeze thawing the nerve multiple times to produce an acellular sheath [1]. Previously research has shown that repopulation of these nerve sheaths can be achieved from exogenous grown SCs and will induce long-distance axonal growth [1, 2].

Neurotrophic factors are a key component of successful axonal regeneration and have historically been co-delivered by the use of single injections or osmotic pumps for the repair of peripheral nerves. This delivery of neurotrophic factors is not viable for translational application. One new way to circumvent the problem of neurotrophic delivery is viral vector transduction of SCs [2, 3]. This technique is a critical advancement of peripheral nerve repair by increasing the growth potential of the SCs to stimulate large numbers of axons to regenerate but also enhances the capacity to improve distal nerve innervation over long distances [2].

In this chapter we describe the process by which decellularized peripheral nerve sheaths can be produced using a freeze thawing technique [1, 2, 4]. This technique depletes intrinsic SCs and other support cells from the peripheral nerve, while leaving the bands of Büngner, connective tissue, and matrix largely intact. Following this decellularization process, the flexible nerve sheaths can be stored at −80 °C or used immediately by injecting cultured SCs expanded from the recipient animal.

SCs can be transduced by a number of viral vectors including adenovirus and lentivirus [2–7]. Primary cultures of adult SCs are easily isolated from adult peripheral nerves, [8, 9] expanded by defined culture methods, and transduced by lentiviral vectors encoding both reporter and neurotrophin genes [2]. Here we describe a method using lentiviral vectors to transduce 99% of primary SCs in vitro, which are capable of strongly expressing reporter (GFP, DSRED, and mCherry) and neurotrophin genes (GDNF, BDNF, NGF, CNTF, and NT-3) after 3 days in vitro.

2   Materials

2.1   Lentiviral Vector Production

1. Hank's balanced salt solution (HBSS).

2. 0.05% trypsin/EDTA solution. ThermoFisher Scientific Cat # 25300054.

3. 10 μg/ml Hoechst 33342 (bis-benzimide). Prepared in distilled water.

4. Dulbecco's Modified Eagle's Medium (DMEM).

5. D-10S medium. DMEM containing 10% FBS and 0.1% gentamicin.

6. Leibowitz's L-15 medium (L-15).

7. 0.4% trypan blue.

8. 2 µg/ml pituitary extract (PEX).

9. 60 mm cell culture dishes.

10. 100 mm cell culture dishes.

11. Forskolin solution. Dimethyl sulfoxide (DMSO) containing 0.2 µM forskolin (Sigma Aldrich, Cat # F6886).

12. 70% ethanol.

13. Iscove's Modified Dulbecco's Medium (IMDM).

14. FBS

15. 10% FBS/IMDM. IMDM containing 10% FBS, 100 IU/ml penicillin, 100 µg/ml streptomycin, and 2 mM GlutaMAX™ supplement.

16. HEK 293T cells. Cryopreserved stock or live adherent cultures of 293T cells (American Type Culture Collection, CRL-3216™).

17. Packaging plasmid (pCMVdeltaR8.2, Addgene).

18. Green fluorescent protein (GFP)-expressing gene transfer plasmid (pRRLsin-PPThCMV-GFP-wpre, Addgene).

19. Envelope protein vector plasmid (VSV-G; pMD.G.2, Addgene).

20. 0.22 µm cellulose acetate filter.

2.2 Lentiviral Vector Transduction of Schwann Cells

1. Confluent culture of adult nerve-derived rat SCs. Purified, mitogen-expanded, obtained according to the method of Plant et al. [9]. *See* Chapter 29 for a list of materials and a detailed protocol to prepare cultures of SCs from adult rat sciatic nerve.

2. D-10S medium.

3. D-10S/mitogens medium. D-10S medium containing 2 µM forskolin and 20 µg/ml PEX.

4. D-10S/mitogens (10%) medium. D-10S medium containing 2 µg/ml PEX and 0.2 µM forskolin (i.e., D-10S medium with a tenth of the mitogen concentration as compared to standard D-10S/mitogens medium).

5. 60 mm cell culture dishes.

6. Lentiviral vectors encoding GFP (LV-GFP). Stock of viral particles consisting of 10^7–10^9 transforming units (TU) per ml obtained from a commercial source (or made in the laboratory).

7. L-15 medium.

8. HBSS.

9. 0.05% trypsin/EDTA solution.

10. DMEM.

11. 0.4% trypan blue solution.

12. Poly-L-lysine (PLL) solution. Dissolve 20 mg/ml of PLL in sterile water.

13. 24 well plates.

14. Sterile water.

15. Anti-GFP antibody. Rabbit polyclonal antibody (Millipore, CAT#AB3080) diluted at 1:50 in phosphate buffered saline (PBS), 0.2% triton, and 10% goat serum solution.

16. Goat anti-rabbit secondary fluorescent antibody, such as one with a 488 nm excitation emission wavelength. Prepare the fluorescent secondary antibody at 1:400 in PBS, 0.2% triton, and 10% goat serum solution.

2.3 Preparation of Acellular Nerve Sheaths

1. Beuthanasia®. Dosage per animal depends on institutional guidelines.

2. 100 mm cell culture dishes.

3. Sterilized forceps and scissors.

4. 70% ethanol.

5. Liquid nitrogen.

2.4 Cellular Reconstitution of Peripheral Nerve Conduits

1. 60 mm cell culture dishes.

2. 100 mm cell culture dishes.

3. Glass micropipettes (200 μm diameter tip).

4. Hamilton syringe.

5. DMEM.

6. D-10S medium.

7. D-10S/mitogens medium.

2.5 Peripheral Nerve Surgery

1. Ketamine/xylazine anesthesia. Mix equal volumes of ketamine (100 mg/ml) and xylazine (20 mg/ml). Used for intraperitoneal injection at a dose of 1 ml/kg body weight.

2. Sterilized forceps and scissors.

3. 70% ethanol.

4. 10/0 nylon sutures.

5. 6/0 sutures.

6. Benacillin. Used at a dose of 200 μl/100 g of body weight for intramuscular delivery.

7. Temgesic (buprenorphine). Used at a dose of 20 mg/kg of body weight for subcutaneous delivery.

2.6 Tissue Analysis and Immunohistochemistry

1. 4% paraformaldehyde. Prepared in 0.1 M Sorenson's buffer pH 7.4. Filtered and stored at 4 °C.

2. 30% sucrose solution. Prepared in light buffer. One liter of light buffer contains sodium hydrogen phosphate 16.59 g and 4.28 g of sodium hydroxide.

3. Gelatin-coated slides. Prepare a solution consisting of 1.5 g of gelatin (type A, 220 or 275 Bloom) and 0.25 g of chromium potassium sulfate in 500 ml of distilled water. Heat the water to 60 °C and completely dissolve the gelatin with the aid of a magnetic stirrer. Stir in the chromium potassium sulfate up until the solution turns a pale blue. Add a few crystals of thymol as a preservative. Dip racks of clean slides in the warm gelatin solution (40–50 °C), drain the slides onto paper tissue, and then stand the slides (covered with foil to keep off dust) on end to air-dry or 37 °C incubator overnight. Store in dust-free container at room temperature. Throw out the gelatin mixture after use.

4. PBS. Two PBS tablets (Gibco, Cat # 18912-014) are dissolved into 1 l of deionized water.

5. Blocking solution. PBS containing 10% normal goat serum and 0.2% Triton X-100.

6. Primary antibodies of choice. Prepared in PBS containing 10% normal goat serum and 0.2% Triton X-100. Use anti-neurofilament (PanAxonal, BioLegend, Cat # 837904) for the detection of axonal neurofilaments, anti-S100 (Dako, Cat # Z0311), anti-laminin (Sigma-Aldrich, Cat # L9393) for the detection of basal lamina, and anti-p75 (Abcam, Cat # 61425) for the detection of SCs.

7. Fluorescent secondary antibodies of choice. Prepared in PBS containing 10% normal goat serum and 0.2% Triton X-100 (e.g., goat anti-mouse Cy3).

8. Citifluor™ or fluorescence mounting medium (DAKO).

3   Methods

3.1   Lentiviral Vector Production in Transfected HEK 293T Cells

1. Twenty four hours before transfection, seed a total of 5×10^6 HEK 293T cells in 100 mm cell culture dishes containing 7 ml of 10% FBS/IMDM.

2. Refresh the 10% FBS/IMDM culture medium 2 h prior to the transfection.

3. Transfect HEK 293T cells with 3.5 μg envelope protein vector plasmid, 6.5 μg packaging plasmid, and 10 μg GFP-expressing gene transfer plasmid per 100 mm cell culture dish. It is recommended to use a calcium phosphate lentiviral production method [10] for the transfection of HEK 293T cells.

4. Remove the transfection media and refresh the 10% FBS/IMDM culture medium 24 h post-transfection.

5. Harvest the virion-containing medium 24 h later (i.e., 48 h post-transfection).

6. Remove the debris by low-speed centrifugation at $300 \times g$ for 5 min and then filter through a 0.22 μm cellulose acetate filter. This media-containing virus is then used to ascertain the concentration of viral particles (see Subheading 3.2).

7. Determine the number of transducing units (TU) by infecting a culture of HEK 293T cells and counting the number of GFP-expressing cells 48 h after infection (*see* **Note 1**).

3.2   Lentiviral Vector Transduction of Schwann Cells

Transferring genes into SCs by the use of ex vivo viral vector technology enables better monitoring of the behavior of the SCs within the peripheral nerve and SC association with regenerating axons. In addition transferring neurotrophic genes increases the growth promoting ability of the SCs, particularly in axonal subsets that do not normally regenerate well with SCs in isolation. Lentiviral vectors are good candidates for SCs transduction due to their integrating capacity, which enables all progeny to be labeled with the inserted gene of interest. These lentiviral vectors also allow for continual expression of the gene, unlike in other viral transduction systems such as adenovirus which integrates episomally and reduces its gene expression after each cell division. Ex vivo transduction of SCs using lentiviral vectors is straightforward and can give consistent expression over a long period of time in vivo.

3.2.1   Preparation and Transduction of Schwann Cells

1. Prepare a confluent SC culture in a 100 mm cell culture dish in D-10S/mitogens medium following the protocol described in Chapter 29.

2. Detach primary cultured SCs by adding approximately 2 ml 0.05% trypsin/EDTA solution and incubating for 5 min at 37 °C.

3. Resuspend the cells in D-10S, collect them in a tube, and centrifuge at $300 \times g$ for 5 min. Wash the cells twice in D-10S medium.

4. Resuspend the pellet of SCs in D-10S medium and perform a cell count.

5. Seed the SCs into 60 mm cell culture dishes at a density of 10^6 cells per PLL dish in D-10S/mitogens medium. Incubate the cells in a cell incubator overnight.

6. The next day, replace the medium with D-10S medium containing approximately 10^7 TU/ml LV-GFP. It is recommended to use a multiplicity of infection (MOI) of 50. Incubate the cells with the viral particles in a cell incubator for 16 h.

7. Wash the cultures three times with L-15 medium. Wait at least 3 days in vitro for expression to be maximal for the GFP. After this the cells can be used.

8. Detach the cells by adding approximately 2 ml 0.05% trypsin/ EDTA solution and incubate for 5 min at 37 °C.

9. Centrifuge at 300 × g for 5 min and wash twice with DMEM.

10. Resuspend the cells in DMEM and perform a cell count.

11. Re-pellet the SCs by low-speed centrifugation at 300 × g for 5 min.

12. Resuspend the pelleted cells in an appropriate volume of DMEM, to yield a cell suspension of 10^5 cells/μl.

13. Confirm the viability of the SCs suspension to be in the range of 95–100%, as determined by trypan blue staining.

14. Keep the cell suspension on ice, until the completion of all surgical procedures.

3.2.2 Determination of Lentiviral Vector Transduction Efficiency in Schwann Cells

1. Prepare PLL-coated 24-well plates by adding approximately 300–500 μl of PLL solution in each well, incubating them in the laminar flow hood for 30 min and rinsing them with sterile water at least three times to remove the excess PLL solution.

2. Seed the SCs onto the PLL-coated 24-well plates, at a density of 10^5 cells per well in D-10S/mitogens medium.

3. Culture the SCs overnight in a cell incubator.

4. The next day, replace the medium with D-10S/mitogens (10%) medium to reduce the rate of SC proliferation.

5. Add recombinant LV vectors encoding reporter genes such as GFP. For LV-GFP, perform a dosage curve adding viral particles at a MOI of 1, 5, 10, 25, and 50 to each well containing the SCs (*see* **Note 2**). The best MOI is the concentration that gives the highest percentage of GFP-expressing cells.

6. Analyze the cultures for transgene expression 3 days after transduction by using immunocytochemistry for GFP followed by a secondary conjugate of choice (*see* Fig. 1 and **Notes 3–8**).

3.3 Preparation of Acellular Nerve Sheaths

1. Prepare the laminar flow hood for the procedure. Take out all the instruments needed, and remember to take forceps out of the 70% ethanol.

2. Anesthetize the rats with Beuthanasia®; shave the hair from the rear hind limb area and clean with soap, water, and 70% ethanol.

3. Excise the sciatic nerve or peroneal branch from the rats and place them into one of the 100 mm cell culture dishes. Use a different set of forceps and scissors for each layer tissue that is cut through (skin, muscle, and nerve).

4. The nerve that is excised is placed into L-15.

5. Euthanize the rats and safely dispose of them in accordance with institutional guidelines.

6. Eliminate the cells from the peripheral nerve sheaths by immersing them into liquid nitrogen for 5 min, then thawing them at room temperature for 5 min. Repeat this procedure five times.

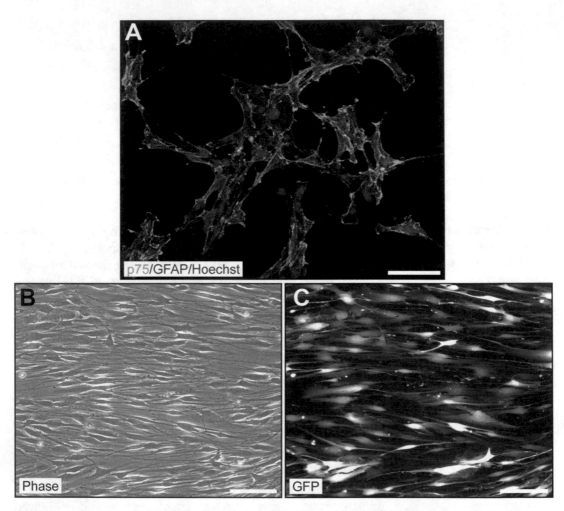

Fig. 1 Phenotype of purified adult Schwann cells and fluorescent reporter gene expression following lentiviral transduction in vitro. (**a**) Immunocytochemical staining for p75 (green), glial fibrillary acidic protein (GFAP; red), and nuclear stain Hoechst 33342 (blue). (**b**) Phase contrast photo of purified Schwann cells from adult peripheral nerves and transduced with lentiviral vectors 1×10^9 encoding GFP (MOI of 50). Scale bar 10 μm

7. Store the acellular nerve sheaths approximately 1.5 cm without any media at −80 °C until needed (*see* **Note 9**).

3.4 Cellular Reconstitution of Peripheral Nerve Conduits

1. Place acellular nerve sheath segments cut into the appropriate size for the lesion gap onto a 60 mm cell culture dish containing a confluent culture of adult SCs (*see* **Note 10**).

2. Using a glass micropipette with a tip diameter of approximately 200 μm attached to a 50 μl Hamilton syringe, slowly inject SCs (using 5×10^4 cells in 1 μl of D-10S medium) into each of the cut ends of each nerve sheath (i.e., use 5×10^4 cells for each nerve sheath end).

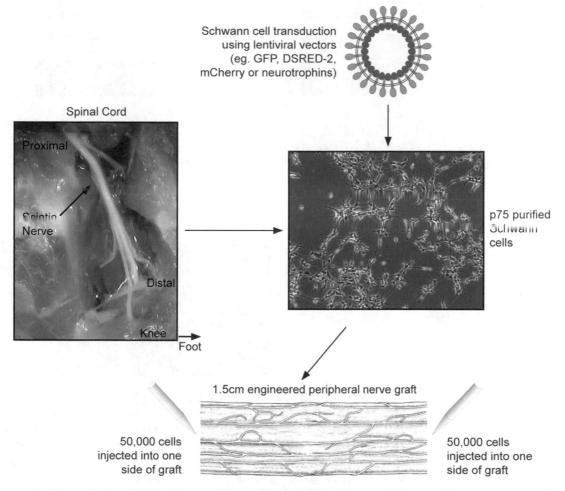

Fig. 2 Schematic describing the transduction of Schwann cells and injection into decellularized peripheral nerves. Lentiviral vectors can encode reporter genes, neurotrophins, or other genes of interest. Purified Schwann cells from adult peripheral nerves are accessible from the main sciatic nerve trunk. Engineered Schwann cells can then be injected into peripheral nerve sheaths devoid of endogenous cells by a freeze thawing procedure and placed into Schwann cell tissue culture plates prior to transplantation

3. Maintain the peripheral nerve pieces in culture in D10-S/ mitogens medium for 2 days prior to grafting (*see* Fig. 2).

3.5 Peripheral Nerve Surgery

1. Anesthetize each host rat with an intraperitoneal injection of ketamine/xylazine anesthesia.

2. Expose either the sciatic nerve or the peroneal nerve branch in the left hind limb and remove an approximately 1 cm segment. The choice of nerve or branch is important to match nerve diameters for repair, and both of these nerve segments are innervated normally by motor and sensory axons.

3. Repair the nerve gap with the graft of choice (as prepared in Subheading 3.4); attach to host nerve stumps using 10/0 nylon sutures, and then close the injury site with 6/0 sutures.

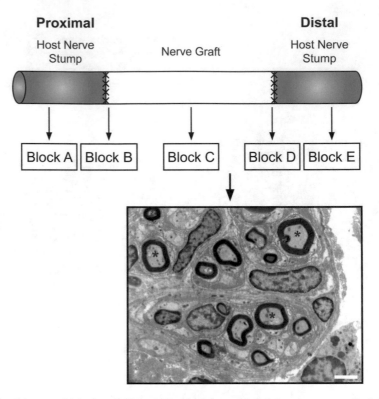

Fig. 3 Immunohistochemical analysis of engineered peripheral nerve grafts following in vivo transplantation. Excised peripheral nerves are straightened using paraformaldehyde fixation on wooden spatulas. Each nerve is then cut into five separate blocks (blocks A–E) incorporating proximal nerve, graft, and distal stump. Electron microscopy (EM) can also be used to ascertain ultrastructural information at the center of the Schwann cell-engineered graft. The image displays a representative electron micrograph showing cross section of a decellularized nerve sheath that has been repopulated with Schwann cells. *, indicates myelinated axons. Scale bar is 2.5 μm

4. Administer appropriate postoperative care, such as Benacillin intramuscularly and Temgesic subcutaneously, or according to institutional guidelines.

3.6 Tissue Analysis and Immunohistochemistry

1. Gather the collected grafted nerves, divide into blocks, and freeze at −80 °C (*see* Fig. 3).

2. To fix fresh grafted nerves, gently straighten them and attach to a wooden spatula.

3. Fix them in 4% paraformaldehyde for 3 h at 4 °C.

4. Cryoprotect the nerves in 30% sucrose solution for 24 h at 4 °C.

5. Using a cryostat, section the frozen blocks. Collect 10 μm sections onto 12 gelatin-coated slides, allow to air-dry, and store at −20 °C until further processing is required.

6. When needed, place slides in a humidified dark chamber at room temperature with gentle agitation.

7. Rinse the sections with PBS for 5 min; repeat twice more.

8. Block the sections for 1 h in blocking solution.

9. Incubate the sections overnight in primary antibodies of choice. This incubation is performed at 4 °C for 2 days for antibodies to S100, p75, and laminin.

10. Perform 3 × 5 min washes with PBS.

11. Add the appropriate secondary antibody dilutions.

12. Wash the sections with PBS.

13. Mount the sections using mounting medium and cover slip.

14. Store the stained sections at 4 °C.

4 Notes

1. Following this method, recombinant stock titers in the order of 10^7–10^9 TU/ml should be achieved.

2. Remove the aliquot of frozen lentiviral vector (generally 10 μl aliquots of 10^8–10^9 concentration at −80 °C), and thaw very rapidly between gloved index finger and thumb. The virus is prone to sticking to the tube walls, so to ensure all virus is mixed at the correct concentration, use a 2–10 μl pipette and appropriate autoclaved plastic tip to resuspend the mixture thoroughly.

3. Ensure that the media is refreshed after 16 h of lentiviral transduction, otherwise cellular integrity can be compromised.

4. Check the viral transduction efficiency of the fluorescent reporter gene daily using a fluorescence microscope; however do not expose the cells to light for lengthy periods of time, otherwise cell fluorescence could be compromised.

5. The choice of promoter used in lentiviral vectors is important to transduction success. CMV is a very effective promoter in lentiviral vectors when used to transduce SCs. These vectors produce strong gene expression in primary SCs.

6. It is worthwhile to remember that when using bicistronic lentiviral vectors, the second gene will always have lower expression levels than the first. To alleviate this issue, one can use two separate lentiviral vectors each coding for a separate gene of interest (such as GFP or neurotrophin, for example).

7. Remember that lentiviral vector aliquots should not be refrozen once thawed, as the viral concentration will decrease significantly. Given their expense, the vector should ideally be frozen in appropriate aliquot sizes for transduction of the required

SCs numbers only. A suggestion is to transduce smaller numbers of SCs and then expand them in vitro, rather than use larger aliquots of lentiviral vectors.

8. Careful consideration must be given when using reporter genes, as some reporter genes can induce unwanted inflammatory responses.

9. The freeze-thaw cycle kills the cells but will maintain basal lamina integrity, providing flexible nerve sheaths that can be effectively repopulated with cultured cells [1, 11].

10. Placement of the cell-injected peripheral nerve sheaths onto beds of the SCs allows for the opportunity for further cellular infiltration and also secreted factor support for the SCs in the nerve to proliferate and survive [11].

References

1. Cui Q, Pollett MA, Symons NA, Plant GW, Harvey AR (2003) A new approach to CNS repair using chimeric peripheral nerve grafts. J Neurotrauma 20(1):17–31. https://doi.org/10.1089/08977150360517155

2. Godinho MJ, Teh L, Pollett MA, Goodman D, Hodgetts SI, Sweetman I, Walters M, Verhaagen J, Plant GW, Harvey AR (2013) Immunohistochemical, ultrastructural and functional analysis of axonal regeneration through peripheral nerve grafts containing Schwann cells expressing BDNF, CNTF or NT3. PLoS One 8(8):e69987. https://doi.org/10.1371/journal.pone.0069987

3. Tannemaat MR, Eggers R, Hendriks WT, de Ruiter GC, van Heerikhuize JJ, Pool CW, Malessy MJ, Boer GJ, Verhaagen J (2008) Differential effects of lentiviral vector-mediated overexpression of nerve growth factor and glial cell line-derived neurotrophic factor on regenerating sensory and motor axons in the transected peripheral nerve. Eur J Neurosci 28(8):1467–1479. https://doi.org/10.1111/j.1460-9568.2008.06452.x

4. Hu Y, Arulpragasam A, Plant GW, Hendriks WT, Cui Q, Harvey AR (2007) The importance of transgene and cell type on the regeneration of adult retinal ganglion cell axons within reconstituted bridging grafts. Exp Neurol 207(2):314–328. https://doi.org/10.1016/j.expneurol.2007.07.001

5. Hu Y, Leaver SG, Plant GW, Hendriks WT, Niclou SP, Verhaagen J, Harvey AR, Cui Q (2005) Lentiviral-mediated transfer of CNTF to schwann cells within reconstructed peripheral nerve grafts enhances adult retinal ganglion cell survival and axonal regeneration. Mol Ther 11(6):906–915. https://doi.org/10.1016/j.ymthe.2005.01.016

6. Li Q, Ping P, Jiang H, Liu K (2005) Nerve conduit filled with GDNF gene-modified Schwann cells enhances regeneration of the peripheral nerve. Microsurgery 26(2):116–121

7. Tannemaat MR, Boer GJ, Verhaagen J, Malessy MJ (2007) Genetic modification of human sural nerve segments by a lentiviral vector encoding nerve growth factor. Neurosurgery 61(6):1286–1294.; discussion 1294-1286. https://doi.org/10.1227/01.neu.0000306108.78044.a2

8. Morrissey TK, Kleitman N, Bunge RP (1991) Isolation and functional characterization of Schwann cells derived from adult peripheral nerve. J Neurosci 11(8):2433–2442

9. Plant GW, Currier PF, Cuervo EP, Bates ML, Pressman Y, Bunge MB, Wood PM (2002) Purified adult ensheathing glia fail to myelinate axons under culture conditions that enable Schwann cells to form myelin. J Neurosci 22(14):6083–6091. doi:20026627

10. Ruitenberg MJ, Plant GW, Christensen CL, Blits B, Niclou SP, Harvey AR, Boer GJ, Verhaagen J (2002) Viral vector-mediated gene expression in olfactory ensheathing glia implants in the lesioned rat spinal cord. Gene Ther 9(2):135–146. https://doi.org/10.1038/sj.gt.3301626

11. Gulati AK, Rai DR, Ali AM (1995) The influence of cultured Schwann cells on regeneration through acellular basal lamina grafts. Brain Res 705(1–2):118–124

Chapter 31

Intraspinal Delivery of Schwann Cells for Spinal Cord Injury

Andrea J. Santamaría, Juan P. Solano, Francisco D. Benavides, and James D. Guest

Abstract

Cell transplant-mediated tissue repair of the damaged spinal cord is being tested in several clinical trials. The current candidates are neural stem cells, stromal cells, and autologous Schwann cells (aSC). Due to their peripheral origin and limited penetration of astrocytic regions, aSC are transplanted intralesionally as compared to neural stem cells that are transplanted into intact spinal cord. Injections into either location can cause iatrogenic injury, and thus technical precision is important in the therapeutic risk-benefit equation. In this chapter, we discuss how we bridged from transplant studies in large animals to human application for two Phase 1 aSC transplant studies, one subacute and one chronic. Preclinical SC transplant studies conducted at the University of Miami in 2009–2012 in rodents, minipigs, and primates supported a successful Investigational New Drug (IND) submission for a Phase 1 trial in subacute complete spinal cord injury (SCI). Our studies optimized the safety and efficiency of intralesional cell delivery for subacute human SCI and led to the development of new simpler techniques for cell delivery into subjects with chronic SCI. Key parameters of delivery methodology include precision localization of the injury site, stereotaxic devices to control needle trajectory, method of entry into the spinal cord, spinal cord motion reduction, the volume and density of the cell suspension, rate of delivery, and control of shear stresses on cells.

Key words Schwann cell, Transplant, Spinal cord injury, Syringe positioning device, Needle-free injection system, Ultrasound, Apnea, Piotomy, Extrusion

1 Introduction

Cell transplantation is a strategy for spinal cord tissue repair, and cell suspensions can be administered into an injury region through needles or small tubes as opposed to the more extensive surgical manipulation required to implant reparative tissue fragments [1, 2]. It is critical to minimize the iatrogenic injury that accompanies cell injections. The lack of evident neurological deficit after transplantation does not mean that tissue injury did not occur [3, 4]. Transplanted autologous human Schwann cells (ahSC) can survive injection and contribute to tissue repair by production

Paula V. Monje and Haesun A. Kim (eds.), *Schwann Cells: Methods and Protocols*, Methods in Molecular Biology, vol. 1739, https://doi.org/10.1007/978-1-4939-7649-2_31, © Springer Science+Business Media, LLC 2018

of extracellular matrix, recruitment of fibroblasts, myelination, and new growth of axonal processes [3]. In some models, tissue formation engendered by the Schwann cell (SC) transplant bridges between the rostral and caudal cord, through the lesion level [5]. The ahSC source tissue is an expendable small peripheral nerve that is dissociated and the SC component purified and expanded within a standardized protocol [6]. The culture purified cells are then concentrated into a high-density cell suspension suitable for spinal cord injection. Autologous cell implants appear not to require administration of immune suppressive medications, a substantial advantage. A large body of data has shown that SC transplantation supports tissue repair in rodent models of SCI [5, 7–17]. Methods for isolating and expanding SC cultures are described in the literature [18, 19]. Notwithstanding this body of supportive literature, SCs are not a normal component of the central nervous system (CNS), and they show little penetration into astrocyte containing parenchyma after implantation. They are better suited for transplantation into the injury epicenter that is devoid of substantial numbers of astrocytes. It is also known that when the glial limiting membrane is breached, endogenous SCs migrate into the injury region [20–23], a phenomenon that has not been sufficiently studied for reparative effects and has not been linked to clinical changes. Injectable SC suspensions are vulnerable to cell death due to their recent trypsinization, multiple media changes, absence of extracellular matrix components, stresses caused by injection, and limited metabolic support within non-vascularized injury tissue. Although transplanted cells may be reparative, their delivery has the potential to disrupt intact tissue, interfere with ongoing tissue healing, and generate inflammation due to the death of implanted cells [15, 24]. To optimize the risk-benefit ratio, these factors must be mitigated as fully as possible. These concerns pertain to all types of cellular transplants delivered into spinal cord tissue (stem cells of any origin, oligodendrocyte precursors [25], neuronal precursors [26], olfactory ensheathing glia [27], activated macrophages [28], etc.).

In our studies of the tolerances of the spinal cord to cell transplants, the most damage occurs when delivering into intact parenchyma with large volumes, fast infusion rates and having poor control over the needle/spinal cord interface [4]. For clinical injections, it is also essential to consider that unpredictable events may occur such as anesthetic emergencies, cardiopulmonary instability, and uncontrolled bleeding. These problems can precipitate an urgent need to stabilize the patient and to halt any other procedure, even requiring turning the patient from prone to supine. Therefore, the ability to rapidly terminate an injection/transplant without risk of additional spinal cord injury is necessary. Three approaches to make spinal cord injections that have been used in clinical trials are handheld syringe injection [28], fixed stereotaxic platform, and floating cannula stereotaxic injections [29, 30]. Each method has advantages and limitations. Based on published and unpublished clinical data, no method has established superiority over another. Safety, convenience, reproducibility, and cost are important concerns, especially if the technique will be used in a multicenter clinical trial. This chapter will summarize our experience in SC transplantation methods, discussing the major lessons from large animal studies and how this influenced human cell delivery.

2   Localizing and Interpreting the Injury Site

Safe, potentially effective, injections require precise evaluation of the injury region. After traumatic SCI, the local region of the primary injury undergoes a series of inflammatory and immunomodulatory cascades known as secondary injury [31–33]. Over time, necrotic debris within the injury epicenter is removed by macrophage phagocytosis [34, 35]. The injury site is transformed into a fluid of inflammatory cells contained by a margin of reactive astroglia [36, 37] with some residual tissue strands whose importance is poorly understood and eventually into a cavity that contains cerebrospinal fluid (CSF) (Fig. 1). The structure of cavities is important for cell delivery. For example, some cavities are unilocular, whereas others are septated into multiple loculations. It is unknown whether tissue strands and the occasional axons found within cavities contribute to neurological function [38, 39]. In the cavity boundary, demyelinated axons may persist chronically [40] and are considered a therapeutic target for ahSC. The reason that ahSC are transplanted into the lesion cavity is because SC-mediated remyelination is improbable in regions of astrogliotic parenchyma but possible where axons persist devoid of astrogliosis. One of the most damaging potential results of spinal cord injections is the dissection of tissue by the suspension volume and injection pressure [4]. These cavities can be hydrodynamically dissected by injection pressure gradients; this tends to occur along the white-gray matter interface leading to further neurological loss. Established gliotic cavities have greater mechanical strength and are less susceptible to this injury. It is important to image the epicenter using real-time ultrasound (US) before initiating injection to identify the injury site and the best tissue pathway to enter it. Observing visible swelling of the spinal cord during injection indicates a high risk for pressure injury. We favor creating a means for cell suspension extrusion if the pressure becomes excessive. Other visual clues to the injury epicenter after dural opening include fine surface neoangiogenesis, hemosiderin deposition, visible gliosis (a change in color or stiffness), and a change in spinal cord size due to malacia. Even though these changes can be visualized using the high magnification and

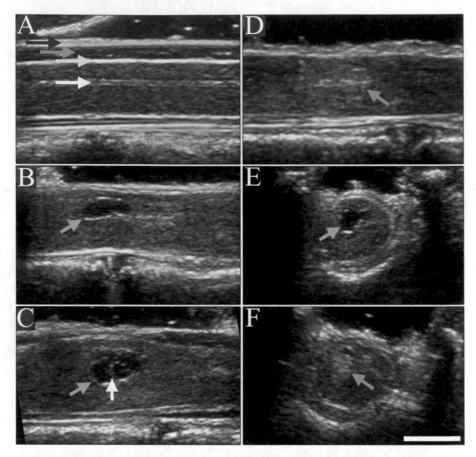

Fig. 1 Ultrasound images from Yucatan minipig spinal cord. (**a**) Naive cord, the red arrow points to the dorsal dura, the blue arrow to the subarachnoid space, the yellow arrow to the pia matter, and the white arrow to the central canal. The differing echogenic properties between structures make their identification easy in the uninjured state. (**b**) The green arrow points to an anechoic cavity with a hyperechoic rim 25 days post-SCI; the same cavity is observed in axial view in (**e**). (**c**) A green arrow points to a cavity 30 days post-SCI; this cavity appears septated. A trabecula is indicated with the white arrow. (**d**) Animal in B was transplanted with 80 μl of aSC, and an immediate sagittal ultrasound displays the change in cavity density with hyperechogenicity inside. Axial view of the transplant filling the cavity is shown in (**f**). Bar 5 mm

illumination afforded by a neurosurgical microscope, the exact position of the epicenter for cell delivery is difficult to define without the use of ultrasound. US is portable, can be used in multiple planes in real time, and is less expensive than other imaging modalities. The visible tissue window for US is limited by the extent of dorsal bone exposure at the margins of the injury site. The probe can be placed very close to the spinal cord making high-frequency imaging in the 10–15 MHz range feasible and increasing signal resolution. To perform US the surgical exposure is filled with warmed sterile saline, and the probe enclosed in a sterile sheath with gel at the tip. Both B-mode and color Doppler can be obtained permitting assessment of blood flow. For subsequent quantitative interpretation of images and post-processing, it is important to maintain standard orientations of the US probe during acquisition of images or video. The transducers have an orientation marker that appears on the screen to indicate their position. This facilitates interpretation of right versus left side and rostral-caudal orientation [41]. It is helpful to always manipulate the probe with the headmark aiming toward the same direction.

During spinal cord US in the sagittal or axial plane, the cord will appear hypoechoic compared to the dura and the pia matter, which are hyperechoic. The subarachnoid space is anechoic under normal conditions but after injury may reveal adhesions. In the midline, the central canal is obvious, and the ependymal walls look hyperechoic with an anechoic lumen. Finding the central canal is important to achieve midline orientation and to interpret lateral ized injuries. The interior of the maturing cavity/injury will look anechoic when compared to the cord. The cavity rim is usually hyperechoic due to gliosis, and trabeculae or septations inside the cavity are also hyperechoic. Examples of these observations are illustrated in Fig. 1.

The size and position of cavities shift because of the surgical exposure and tension on dural sutures. In some instances, the cavity will partially collapse during the lysis of adhesions. Ultrasound is the only feasible real-time tool to observe the injury site after surgical exposure and make suitable modifications before cell injection. It is important to define an external landmark to the injury epicenter as visualized with ultrasound. To do this a safe instrument can be placed next to the cord within the view window and its artifact marked in reference to the lateral tissues. If the dura has been exposed and retracted with dural tack-up sutures, it is also helpful to gently pull on these sutures while scanning to localize the epicenter and positioning relative to the suture. Following injection, US also allows visualization of the transplanted cell suspension as hyperechoic floccular signal.

3   Stereotaxis and Reduction of Spinal Cord Motion for Cell Injection

Transplanting cells usually involves placing a needle into the spinal cord. In rodent experiments, stereotaxic positioning devices with syringes and needles have long been the standard for transplantation. The needle can be well controlled for depth of insertion and tip placement in the three axes, X, Y, and Z. For clinical spinal cord cell transplantation, the rodent devices were not of a suitable scale, and thus human size devices had to be created.

Although cranial stereotaxic devices for brain procedures have been available for decades, there are few that are designed for human spinal cord application [42], and none commercialized. Only a few stereotaxic platforms have been used in clinical trials for the delivery of cells into the spinal cord for SCI or other conditions. These include (1) the delivery of allogenic human fetal spinal cord-derived stem cells for amyotrophic lateral sclerosis (ALS) by Neuralstem Inc. (Germantown, Md.) via a spinal column-mounted platform [43, 44], (2) the delivery of autologous bone marrow mononuclear cells into the spinal cord for ALS via a surgical flexible Yasargil retractor with an attached microtargeting micromanipulator [45, 46], (3) the delivery of olfactory ensheathing cells into the spinal cord for SCI by the Griffith University and the Princess Alexandra Hospital in Australia via a micromanipulator attached to the circular rail of an orthopedic stabilization frame [47], and (4) the delivery of oligodendrocyte precursor cells after subacute SCI by Geron and Asterias Biotherapeutics Inc. (Fremont, CA) via a surgical table-mounted platform coupled to a microdrive injector known as the syringe positioning device or SPD.

In early stages of our large animal studies, we collaborated with Dr. Ed Wirth, M.D., Ph.D., and colleagues to test the SPD for aSC injection into the spinal cord. The SPD (Mizuho OSI, Inc., Union City, CA, USA) followed the small animal archetype of having a stable platform and multiplanar stereotaxis. Geron Corporation provided a supportive letter to the FDA for the use of the SPD in the Miami Project's Phase 1 trial of Safety of Autologous Human Schwann Cells (ahSC) in subacute SCI [6]. The SPD was tested in more than 50 Yucatan pigs and used to implant 6 patients [6]. It is a very robust stereotaxic system that assembles to the rails of conventional surgical tables (Fig. 2) to form a rigid box platform in the

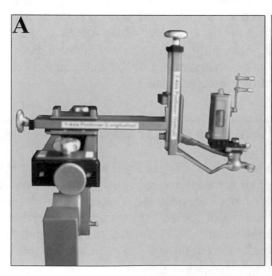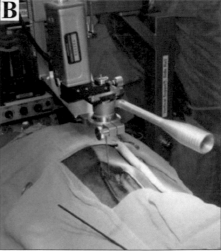

Fig. 2 Syringe positioning device (SPD) assembled. (**a**) The vertical SPD bars are secured to the operating table rails; the horizontal assembly is fitted creating a strong box construct extending with a Y axis support perpendicular to the X axis. The Z axis arm serves to support the ball swivel into which the microdrive is mounted. (**b**) Inset of the microdriver with the SPD in position for a porcine transplantation

X axis. Then Y and Z axis components are sequentially added to bring the microdrive into proximity to the injury site. Each plane has a macro adjustment allowing several centimeters of transition. The system has the capability to be directed through a wide range of trajectories due to a 360-degree ball swivel within the delivery platform. The long needle is stabilized by a reinforcing rigid cannula to reduce bending and vibration. Fine movements are made using a microdrive platform coupled into the head stage, to which is also coupled a syringe holder. For our injections in both porcine experiments and human transplantation, we utilized a Hamilton 26sG 8-inch needle (Part no. 7806-04) attached to a syringe model 710 (100 μl capacity) or model 725 (250 μl capacity) (Hamilton Company, Reno, NV). The syringe plunger was advanced by hand.

Although table fixated rigid platforms bring mechanical reproducibility to the procedures, and our experiments worked well with the platform, several trade-offs became evident through the course of the study related to the use of the SPD.

3.1 Solving the Problem of Spinal Cord Motion

In rodent experiments with sedation and no mechanical ventilation, the spinal cord and body motion generated by breathing is less than in large animals and in people. In porcine experiments at the T9 level, we observed changes in needle-cord distance of a centimeter or more during breathing. This means that a needle placed into the spinal cord during ventilation could travel up and down more than the entire depth of the spinal cord. The constant cyclic motion made stereotaxic injections with fixed needles potentially dangerous. A solution to this problem is to stop breathing, known as apnea, using skeletal muscle paralytics combined with preoxygenation. During apnea, the motion of the exposed thoracic spinal cord is markedly reduced leaving only the pulsatile motion due to the cardiac cycle. Apnea is compatible with needle descent, pial perforation, and injection within a time frame of 5–7 min. Although the practice of apnea sounds risky, we did not experience any significant cardiovascular effects in porcine studies, other than transient carbon dioxide (CO_2) escalation. Additionally, we did not encounter any apnea-related adverse events in the transplanted subjects enrolled in the subacute trial [6]. Nevertheless, methods that obviate a need for apnea are desirable because when CO_2 rises, the drive to breathe increases [48] and it would be disastrous if the subject was aroused from a stable anesthetic depth and moved unexpectedly during the procedure. This did occur in one minipig markedly worsening the SCI. In addition, there are issues that can arise during the injection, prolonging apnea time. For example, the needle deployment and penetration could be unsuccessful on the first try, requiring retraction of the system, ventilation of the subject back to normal end-tidal CO_2, and repetition of the entire process. The effects of apnea may also be more consequential in human subjects with chronically impaired respiratory physiology or disease.

3.2 Resistance of the Cell Suspension to Flow

As the density of a cell suspension increases, the rheological properties are changed due to both shear and colloidal forces affecting the cells. Shear forces arise due to the differences in flow motion between cells and the suspension fluid. Shearing deforms cells and if excessive can disrupt their membranes. Further, shearing is known to activate molecular cascades [49] and trigger apoptosis [50]. The properties of SC suspensions have not been adequately optimized; further characterization might lead to improvements in transplant survival. In addition, SC suspensions often exhibit clumping that may aggravate shear forces [51].

Most of our knowledge regarding flow characteristics of cell suspensions is derived from studies of red blood cells. As viscosity increases, a greater driving pressure is required. We used an 8-in. 26 s–gauge needle with the SPD for both clinical and experimental injections. The narrow diameter and long length of this needle can contribute to the development of significant resistance to flow of concentrated cells escalating shear forces. One factor contributing to shear is that the velocity of cell movement will be increased due to the long narrow path. A second factor is that the cells settle or pack down quite rapidly within a vertically oriented syringe due to gravity, increasing the effective concentration. As the packing fraction increases, the relative viscosity increases exponentially and can make flow impossible. Although the SPD was originally conceived to be driven through a digital interface, we found this to be impractical due to the slow rate of needle depth advance of the microdrive. Thus, we operated the microdrive manually. We also explored using a syringe pump for delivery in association with a coupling to the injection needle. We still encountered vertical packing of the cells with occasional pressure buildup and erratic delivery of potentially harmful suspension boluses. In large animal preclinical studies, we tested aSC suspensions of 200,000 cells/μl in animals. In human subjects the cell density was 100,000 cells/μl. For reference these cell concentrations are 25 and 50 times less than the normal red cell density in the blood. However, SCs are larger than red blood cells.

3.3 Rate of Injection

For clinical injections, the duration of the procedure is an important limitation. While 5 min is a reasonable duration, 30 min is not. This is due to the many risks associated with the duration of surgery [52–54] and the demands for the efficient use of operating room (OR) time. Furthermore, when a needle is deployed within the spinal cord, there is some irreducible needle motion that can damage tissue. Presumably this is duration dependent. While animal surgeries have no fixed time limit, human surgeries need to respect time schedules. The total time for an intraspinal cell transplantation surgery is usually less than 4 h, with transplantation itself taking less than 30 min.

3.4 Preinjection Piotomy

Substantial scar tissue formation after injury was found in all subjects. The scar tissue in the epidural, subdural/arachnoid, and pial layers requires some dissection to create a corridor for injection under direct microscopic visualization. Careful reference to normal tissue planes at the edges of the injury is important for orientation. The scarred arachnoid should only be opened where essential, as there will be new scar formation where it is manipulated. When the injection needle is advanced to penetrate through the pial layer into the spinal cord, there is often resistance that causes spinal cord focal compression to a serious degree. We have found this to be true in both large animal studies and in human subjects. To avoid this, it is important to open the pia enough to allow the injection needle to pass without constriction. We use the tip of a number 11 scalpel blade to create an entry point that is 2 mm long.

3.5   Assembly and Dimensions

Arming and operating the SPD platform requires experience. To optimize operating times, surgeons need to be familiar in advance with the assembly of the frame. The components are brought into the surgical field in cases fully packed and sterile. The vertical components that mount to the operating table can be placed before patient draping or during surgery. It is very important to make an accurate estimate of where the injection site will be in reference to the surgical table. If the vertical posts are not placed within the working range of the SPD, it will be necessary to reposition them during surgery. This incurs some risk of breaching the sterile field. The components are heavy and two people should conduct the assembly together. The microdrive platform sits close to the exposure, at least in the thoracic spine, and this makes observation of the injection process through the neurosurgical microscope more challenging during injection. This is because the microscope must be placed at an angle to the vertical SPD components and some parallax occurs.

In summary, the SPD yielded satisfactory results in our hands. For stereotaxis, it is reliable, very stable, and reproducible. However, the need to use apnea and the fixed needle in the cord represent risk factors that can be improved upon. As we moved forward into our application of ahSC delivery into cavities of persons with chronic SCI, we thought that the SPD was unnecessarily complex and developed a much simpler injection methodology described in the following section.

4   Free-Floating Systems

A free-floating injection system is one that does not require apnea because the injection needle moves with the patient. An excellent example that has been tested in large animals and in clinical trials is the one developed by Boulis and colleagues [29, 55]. Although this system solves the problem of apnea, it has a few limitations. These include the need for fixation to the spine, a fixed depth needle, and a brisk needle pass into the spinal cord to solve the problem of compression-dimpling that was mentioned previously. Because the system uses predominantly tubing and a very short needle, issues of cell shear stress are mitigated, and the system is compatible with an infusion pump delivery that provides a stable rate of injection. Injections into partially preserved spinal cord tissue tend to displace intact axon-containing tissue, and may damage neurons, and cause inflammation. As compared to cell injections that are made into areas of subacute SCI or intact parenchyma, injections into chronically injured spinal cord cavities are less dependent on stereotaxic precision. At the injury epicenter, the injection is made into a space largely devoid of preserved axons. Injection into a space lacking neurologically critical structures permits use of a Needle-Free Injection System (NFIS) that obviates needle-associated risks, apnea and shear stress.

The NFIS was developed by the authors through several iterations in a minipig thoracic spinal cord contusion model for aSC transplantation studies. These included testing several syringes, needles, and tubing configurations for cell delivery in combination with clinical and research grade infusion pumps (Fig. 3). Ultimately, after comparing the performance of these combinations, the NFIS is very simple, consisting only of soft silicone tubing bonded to a small disposable syringe. After studying the chronic injury using ultrasound to visualize the planned injection cavity, the tubing is manually inserted into the SCI cavity, and then the cells are delivered via gentle pressure applied through a syringe plunger. Although initial experiments were performed using a needle attached to the tubing as a delivery interface, we reasoned that making an initial small longitudinal pial incision of 2 mm

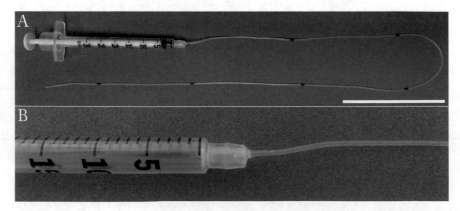

Fig. 3 NFIS delivery system. (**a**) The syringe with bonded silicone tubing, 32 cm in length. Black marks are placed on this tube indicating a distance of 5 cm (3.6 μl), and a final red mark is placed marking the last 2 cm of tubing; this segment is tied once the cells are loaded and then cut once the surgeon is ready to insert the tubing inside the epicenter cavity. (**b**) A close-up of the bonded syringe-tubing interface. Bar 5 cm

would allow extrusion of inflammatory contents from the cavity, if present, increasing available space for the transplant. Having a 2-mm pial incision in place was enough to fit in the tubing into the chronic cavity without any coupled needles. In addition, unlike the tight fit of a needle, the tubing has some space between it and the incised pia mater. This lack of seal allows cells to egress from the spinal cord if the cavity pressure increases during cell delivery. This reduces the risk of hydrodynamic dissection due to intraparenchymal pressure generated by the transplant [4], a serious problem we observed with needle-based injections of larger volumes.

Although often considered undesirable, the extrusion of cell suspension is a good indication that the filling pressure being used to deliver the transplant is too high, or alternatively the cavity space is filled. It is important to have excellent visualization through the microscope at high magnification to observe when extrusion occurs and to carefully clean it. This cavity filling method should allow the injected cell suspension to conform to the boundaries of the injury, placing the cells where they are most likely to be effective, and survive and function if there are accessible axons [54, 55]. The method thus addresses the variability of each patient's lesion characteristics. This simple effective method opens the way for possible admixture of other components, such as biomaterials and extracellular matrix, that could be delivered through tubing but not through a needle. The method requires a relative abundance of cells, as a small quantity are still within the tubing after the procedure. Cavities that are heavily septated, as in four or more loculations, lacking apparent communication are problematic for the NFIS or needle-based transplant methodologies. Further research is needed to determine if the tissue septae that create loculations contain axons contributing to neurological function. If they do not, then consideration could be given to gently penetrating between loculations to allow their contiguous filling with cell suspension.

4.1 NFIS Assembly

The system consisting of the syringe and bonded tubing is assembled under a dissection microscope in a sterile laminar flow hood approximately 5 days before the date of transplantation. The inner diameter of the silicone tubing is 0.30 mm. The length of the tubing is 32 cm. This was determined empirically as an adequate length to manipulate within the surgical sterile field. It is sufficient for loading cells and visualization of the cell movement through the tubing. The tubing could be cut shorter, reducing dead space, to empty residual cells if needed. Black marks are placed every 5 cm (3.6 μl) along the tubing for reference (see Fig. 3a), and a red mark indicates the last 2 cm of tubing. With the last 6 mm, at the distal end of the tubing, there are individual marks every 1 mm. These marks provide insertion length feedback under the operating microscope. The proximal tubing is inserted into the slip tip of the syringe, sealed with cyanoacrylate and cured for 24 h before

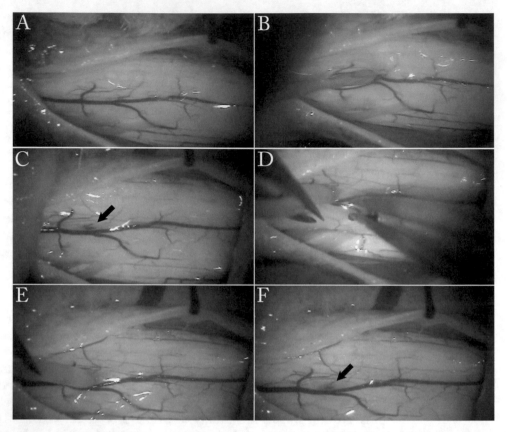

Fig. 4 Piotomy and transplantation using NFIS. Images captured through the surgical microscope show the following: (**a**) The dura is open and the dorsal spinal cord surface is exposed. (**b**) An 11 blade is used to enter the epicenter site resulting in small opening evident in (**c**) (black arrow). (**d**) Small forceps are used to guide the tubing into the injury cavity. (**e**) Once the tubing is in place the infusion of cells begins. (**f**) The tubing is removed after the injury cavity is filled with cells and a pial suture is placed

systematic pressure, leak, and flow testing. Once tested and 2–3 days before the surgery, the systems are double packed, sterilized with ethylene oxide, and delivered to the OR.

4.2 Transplantation Methods Using NFIS

Preoperative magnetic resonance images are used to study the region of tissue cavitation for surgical planning. Epicenter views are used to estimate the injury volume and the quantity of the cell suspension needed for the procedure. The essential steps of the procedure are as follows (*see* Fig. 4).

1. The spinal cord is exposed to the level of the dura. Intraoperative ultrasound is performed to determine the exact site of cavitation and assess for adhesions between the spinal cord and the dura. The dura is then carefully incised over an adequate length. The spinal cord is visually inspected. Ultrasound is repeated and the surface point directly over the maximum cav-

ity epicenter is determined. Generally, this is the thinnest part of the residual spinal cord.

2. Simultaneously, the ahSC cell product arrives to the OR in two sterile microcentrifuge tubes within an iced box. The sterilized NFIS is opened by one surgeon.

3. Before loading, the vial is flicked gently to produce an even cell suspension. The end of the tubing is inserted into the micro-centrifuge tube containing the cells, and the cells are back-loaded into the system slowly. It is very important to prevent air bubbles from entering the tubing and syringe.

4. If substantial air enters the tubing, it may be necessary to extrude the cells and centrifuge them briefly and then reload a new NFIS.

5. The target area around the planned piotomy site is encircled with cottonoid patties to prevent excess ingress of CSF that might float cells into the CSF circulation.

6. A piotomy is performed with a number 11 scalpel blade. Inflammatory fluid, if present, spontaneously escapes from the spinal cord lesion.

7. The loaded system is placed onto the operative field. Under high magnification the tubing is inspected for air gaps or any points of leakage.

8. The tubing is gently inserted into the pial entry point using fine forceps to a depth of 4–5 mm or more if the cavity is very large. If any resistance is encountered, the tubing is not advanced further.

9. The syringe plunger is advanced to deliver the full volume within 5 min. Initially, the injection rate can be relatively fast, such as 50–75 µl per minute. After 2/3 of the estimated injec-tate is delivered, the rate is slowed to deliver 10 µl volumes with 10–20 s of dwell in between. There is constant observa-tion of the injection catheter site by the surgeons. If cell egress is noted, the cell infusion is stopped, and the extruded cells are removed with a cottonoid pattie. A dwell period of 30 s is allowed and the infusion is restarted.

10. After the initial delivery is completed as detected by cell egress, the tubing is slowly extracted from the cavity 1 mm at a time allowing 10 s of dwell in between, with further slow delivery of cells so that all space that was occupied by the tubing should be filled.

11. The pial entry site is closed with one to two small 9–0 Vicryl pial stitches. A watertight dural closure is then performed.

12. The spinal cord injection site is visualized using ultrasound, usually the cellular material is evident as increased echogenicity, and the surgical incision is closed in layers.

4.3 Advantages and Risk Elimination of NFIS

During the testing of the system in large animals and its use in subjects with chronic SCI to deliver ahSC, several advantages of NFIS over other methods have been identified.

1. It is suitable for use in chronic cavitated injuries.

2. A wide range of cellular doses from 50 to 2000 μl, or more, can be delivered.

3. It is sensitive to detect resistance to flow and avoid occlusions and inadvertent boluses of cell suspension.

4. The injection rate may be varied in an efficient manner so that a higher rate of delivery can be used during the early filling of the chronic injury cavity and a slower rate can be used when evidence of cavity filling is observed.

5. It avoids gravitational cell settling and the loss of uniform cell concentration that occurs in vertically mounted syringes and components.

6. Eliminates the need for apnea during cell injection.

7. Eliminates compression of the spinal cord during needle perforation of pia mater.

8. Minimizes potential risk of hemorrhage because there is no needle within the spinal cord that could perforate blood vessels.

9. It is simple and there is less potential for failure of components and reduced setup time.

10. Allows egress of spinal inflammatory debris, which creates additional space for the cell transplant.

11. Eliminates risk of spinal cord damage from the inserted needle due to inadvertent patient movement.

12. Cells are subjected to much less shearing stress and a higher viscosity could be injected. For example, extracellular matrix or fluid biomaterials could be combined with cells. This might increase viability after transplant and would reduce the problem of cell settling leading to increased concentrations.

13. Eliminates the buildup of damaging pressure and hydrodynamic dissection within the spinal cord by allowing egress of the cell suspension through the space between the entry incision and the catheter.

14. NFIS could be prefilled in the cGMP cell culture facility eliminating the need to load cells in the incompletely sterile OR environment.

15. Extremely low component costs.

Some of the potential problems that could arise with the NFIS and that were carefully tested include the following:

1. Infection risk: All parts comprising the tubular injection system are single-use disposable elements, and the systems are preassembled prior to surgery, double wrapped, and ethylene oxide sterilized to be transferred to the OR facilities. The absence of a need to manipulate needles and tubing in preparation for injection reduces infection risk.

2. Unexpected system disengagement or dropping during surgery: Only the surgeons should manipulate the NFIS once the package is opened, and if the system is potentially contaminated in the OR, there should be a switch to a backup system.

3. Tubing failure, interface leakage, or connection occlusion: To permit secure delivery of the cell product, the bonded system should be primed and pressure-tested before cell loading. Several systems are prepared 5 days prior to the planned transplant day. Once bonded, a 24-h period is allotted for the cyanoacrylate to cure for full strength and solvent resistance to develop (as recommended by manufacturer technical specifications). After curing, the systems are tested with sterile water to identify leakage or possible occlusions due to incorrect bonding. Imperfect systems are discarded. Five fully working systems are packed, sterilized, and transferred to the OR. We have observed occasional leakage during this testing at a frequency of approximately 5%.

4. Cell efflux during or after injection: Detection of cell efflux is a component of the NFIS procedure designed to reduce the chance of hydrodynamic cord dissection and detect cavity filling. It is important to adequately remove the effluxed material. This is facilitated by preventing CSF from entering the injection site, enclosing it with cottonoid patties, and suturing the pial incision at the end of the procedure.

5 Conclusions

Needles serve as interfaces between a relatively large delivery syringe and compact tissue. The needle's narrow diameter and cutting tip allow for tissue penetration and advancement with only marginal injury to most non-CNS tissues. For the CNS, needles require precise manipulation that can be afforded through a stereotaxic apparatus. Although direct handheld injection into the spinal cord is feasible and has been used in clinical trials (Proneuron Biotechnologies) [28], the risks for hemorrhage, overpenetration, rapid bolus infusion, and additional injury are higher compared to the use of a controlled system. The move from subacute to chronic SCI provided the impetus to greatly simplify the delivery technology, to obviate the need for apnea, and to reduce shear stress on injected cells. The NFIS is being studied in the current clinical trial

"The Safety of ahSC in Chronic SCI with Rehabilitation." Whether cultured autologous SC evoke an immune response requires further assessment.

Acknowledgments

The experiments were supported by the Miami Project to Cure Paralysis clinical trials initiative and the Buoniconti Fund to Cure Paralysis. Dr. Ed Wirth, M.D., Ph.D., provided access to the SPD, participated in the initial testing in naive animals, and provided helpful advice. Dr. Nicholas Boulis, M.D., Ph.D., demonstrated the floating cannula injection system used in Neuralstem ALS trials. Human aSC transplant procedures were performed together with Dr. Allan Levi, M.D., Ph.D. The clinical trial sponsor is Dr. Dalton Dietrich, Ph.D., and the coordinator is Kim Anderson, Ph.D. The cells were prepared by the aSC team initially supervised by Dr. Pat Wood, Ph.D., and Dr. Gagani Athauda, M.D., and subsequently carried by Aisha Khan, M.B.A., Adriana Brooks-Perez, Maxwell Donaldson, and Risett Silvera-Rodriguez. Yohjans Nunez-Gomez, D.V.M., and Luis Guada Delgado, M.D., assisted with minipig transplant surgeries. The impetus to consider Schwann cell transplantation for human spinal cord injury was an idea of Richard and Mary Bunge in the early 1970s. The achievement of derivation of aSC from a small biopsy and substantial culture expansion required years of scientific exploration and animal testing carried by many members of the Bunge scientific family including Drs: Patrick Wood, James Guest, Allan Levi, Xiao Ming Xu, Giles Plant, Martin Oudega, Damien Pearse, Cristina Fernandez-Valle and Caitlin Hill. Finally, despite her husband's untimely death in 1996, Mary Bartlett Bunge persisted in helping advance this work to clinical trials until her retirement in 2017.

References

1. Wirth ED 3rd et al (1992) In vivo magnetic resonance imaging of fetal cat neural tissue transplants in the adult cat spinal cord. J Neurosurg 76(2):261–274

2. Wirth ED 3rd et al (2001) Feasibility and safety of neural tissue transplantation in patients with syringomyelia. J Neurotrauma 18(9):911–929

3. Guest J, Santamaria AJ, Benavides FD (2013) Clinical translation of autologous Schwann cell transplantation for the treatment of spinal cord injury. Curr Opin Organ Transplant 18(6): 682–689

4. Guest J et al (2011) Technical aspects of spinal cord injections for cell transplantation. Clinical and translational considerations. Brain Res Bull 84(4–5):267–279

5. Golden KL et al (2007) Transduced Schwann cells promote axon growth and myelination after spinal cord injury. Exp Neurol 207(2): 203–217

6. Anderson KD et al (2017) Safety of autologous human Schwann cell transplantation in subacute thoracic spinal cord injury. J Neurotrauma 34(21):2950–2963

7. Pearse DD et al (2004) cAMP and Schwann cells promote axonal growth and functional recovery after spinal cord injury. Nat Med 10(6):610–616

8. Meijs MF et al (2004) Basic fibroblast growth factor promotes neuronal survival but not behavioral recovery in the transected and Schwann cell implanted rat thoracic spinal cord. J Neurotrauma 21(10):1415–1430

9. Barakat DJ et al (2005) Survival, integration, and axon growth support of glia transplanted into the chronically contused spinal cord. Cell Transplant 14(4):225–240

10. Fouad K et al (2005) Combining Schwann cell bridges and olfactory-ensheathing glia grafts with chondroitinase promotes locomotor recovery after complete transection of the spinal cord. J Neurosci 25(5):1169–1178

11. Flora G et al (2013) Combining neurotrophin-transduced schwann cells and rolipram to promote functional recovery from subacute spinal cord injury. Cell Transplant 22(12):2203–2217

12. Kanno H et al (2014) Combination of engineered Schwann cell grafts to secrete neurotrophin and chondroitinase promotes axonal regeneration and locomotion after spinal cord injury. J Neurosci 34(5):1838–1855

13. Williams RR et al (2015) Permissive Schwann cell graft/spinal cord interfaces for axon regeneration. Cell Transplant 24(1):115–131

14. Guest JD et al (1997) The ability of human Schwann cell grafts to promote regeneration in the transected nude rat spinal cord. Exp Neurol 148(2):502–522

15. Hill CE et al (2006) Labeled Schwann cell transplantation: cell loss, host Schwann cell replacement, and strategies to enhance survival. Glia 53(3):338–343

16. Fortun J, Hill CE, Bunge MB (2009) Combinatorial strategies with Schwann cell transplantation to improve repair of the injured spinal cord. Neurosci Lett 456(3):124–132

17. Hill CE et al (2010) A calpain inhibitor enhances the survival of Schwann cells in vitro and after transplantation into the injured spinal cord. J Neurotrauma 27(9):1685–1695

18. Levi AD et al (1995) The influence of heregulins on human Schwann cell proliferation. J Neurosci 15(2):1329–1340

19. Wood PM (1976) Separation of functional Schwann cells and neurons from normal peripheral nerve tissue. Brain Res 115(3):361–375

20. Beattie MS et al (1997) Endogenous repair after spinal cord contusion injuries in the rat. Exp Neurol 148(2):453–463

21. Zhang SX et al (2011) Histological repair of damaged spinal cord tissue from chronic contusion injury of rat: a LM observation. Histol Histopathol 26(1):45–58

22. Blakemore WF, Patterson RC (1978) Suppression of remyelination in the CNS by X-irradiation. Acta Neuropathol 42(2):105–113

23. Gilmore SA, Duncan D (1968) On the presence of peripheral-like nervous and connective tissue within irradiated spinal cord. Anat Rec 160(4):675–690

24. Hill CE et al (2007) Early necrosis and apoptosis of Schwann cells transplanted into the injured rat spinal cord. Eur J Neurosci 26(6):1433–1445

25. Priest CA et al (2015) Preclinical safety of human embryonic stem cell-derived oligodendrocyte progenitors supporting clinical trials in spinal cord injury. Regen Med 10(8):939–958

26. Raore B et al (2011) Cervical multilevel intraspinal stem cell therapy: assessment of surgical risks in Gottingen minipigs. Spine (Phila Pa 1976) 36(3):E164–E171

27. Mackay-Sim A et al (2008) Autologous olfactory ensheathing cell transplantation in human paraplegia: a 3-year clinical trial. Brain 131(Pt 9):2376–2386

28. Lammertse DP et al (2012) Autologous incubated macrophage therapy in acute, complete spinal cord injury: results of the phase 2 randomized controlled multicenter trial. Spinal Cord 50(9):661–671

29. Federici T et al (2012) Surgical technique for spinal cord delivery of therapies: demonstration of procedure in gottingen minipigs. J Vis Exp 70:e4371

30. Riley JP et al (2011) Platform and cannula design improvements for spinal cord therapeutics delivery. Neurosurgery 69(2 Suppl Operative):ons147–ons154. discussion ons155

31. Blesch A, Tuszynski MH (2009) Spinal cord injury: plasticity, regeneration and the challenge of translational drug development. Trends Neurosci 32(1):41–47

32. Blight AR (1994) Effects of silica on the outcome from experimental spinal cord injury: implication of macrophages in secondary tissue damage. Neuroscience 60(1):263–273

33. Dumont RJ et al (2001) Acute spinal cord injury, part I: pathophysiologic mechanisms. Clin Neuropharmacol 24(5):254–264

34. Blight AR (1992) Macrophages and inflammatory damage in spinal cord injury. J Neurotrauma 9(Suppl 1):S83–S91

35. Blight AR (1985) Delayed demyelination and macrophage invasion: a candidate for secondary cell damage in spinal cord injury. Cent Nerv Syst Trauma 2(4):299–315

36. Reier PJ, Houle JD (1988) The glial scar: its bearing on axonal elongation and transplantation approaches to CNS repair. Adv Neurol 47:87–138

37. Silver J, Miller JH (2004) Regeneration beyond the glial scar. Nat Rev Neurosci 5(2):146–156

38. Kakulas BA (1999) The applied neuropathology of human spinal cord injury. Spinal Cord 37(2):79–88

39. Kakulas BA (1999) A review of the neuropathology of human spinal cord injury with emphasis on special features. J Spinal Cord Med 22(2):119–124

40. Guest JD, Hiester ED, Bunge RP (2005) Demyelination and Schwann cell responses adjacent to injury epicenter cavities following chronic human spinal cord injury. Exp Neurol 192(2):384–393

41. Ihnatsenka B, Boezaart AP (2010) Ultrasound: basic understanding and learning the language. Int J Shoulder Surg 4(3):55–62

42. Gabriel EM, Nashold BS Jr (1996) History of spinal cord stereotaxy. J Neurosurg 85(4): 725–731

43. Chen KS, Sakowski SA, Feldman EL (2016) Intraspinal stem cell transplantation for amyotrophic lateral sclerosis. Ann Neurol 79(3): 342–353

44. Boulis NM et al (2011) Translational stem cell therapy for amyotrophic lateral sclerosis. Nat Rev Neurol 8(3):172–176

45. Blanquer M et al (2010) A surgical technique of spinal cord cell transplantation in amyotrophic lateral sclerosis. J Neurosci Methods 191(2):255–257

46. Blanquer M et al (2012) Neurotrophic bone marrow cellular nests prevent spinal motoneuron degeneration in amyotrophic lateral sclerosis patients: a pilot safety study. Stem Cells 30(6):1277–1285

47. Feron F et al (2005) Autologous olfactory ensheathing cell transplantation in human spinal cord injury. Brain 128(Pt 12): 2951–2960

48. Appadu B, Lin T (2017) Respiratory physiology. In: Lin T, Smith T, Pinnock C (eds) Fundamentals of anesthesia. Cambridge University Press, Cambridge, pp 399–400

49. Qazi H, Shi ZD, Tarbell JM (2011) Fluid shear stress regulates the invasive potential of glioma cells via modulation of migratory activity and matrix metalloproteinase expression. PLoS One 6(5):e20348

50. Dimmeler S et al (1996) Shear stress inhibits apoptosis of human endothelial cells. FEBS Lett 399(1–2):71–74

51. Iordan A, Duperray A, Verdier C (2008) Fractal approach to the rheology of concentrated cell suspensions. Phys Rev E Stat Nonlin Soft Matter Phys 77(1 Pt 1):011911

52. Guest JD, Vanni S, Silbert L (2004) Mild hypothermia, blood loss and complications in elective spinal surgery. Spine J 4(2): 130–137

53. Habiba S et al (2017) Risk factors for surgical site infections among 1,772 patients operated on for lumbar disc herniation: a multicentre observational registry-based study. Acta Neurochir 159(6):1113–1118

54. Croft LD et al (2015) Risk factors for surgical site infections after pediatric spine operations. Spine (Phila Pa 1976) 40(2):E112–E119

55. Riley J et al (2014) Intraspinal stem cell transplantation in amyotrophic lateral sclerosis: a phase I trial, cervical microinjection, and final surgical safety outcomes. Neurosurgery 74(1): 77–87

INDEX

Paula V. Monje and Haesun A. Kim (eds.), *Schwann Cells: Methods and Protocols*, Methods in Molecular Biology, vol. 1739,
https://doi.org/10.1007/978-1-4939-7649-2, © Springer Science+Business Media, LLC 2018

Printed in the United States
By Bookmasters